Advances in Geological Science

Series Editors

Junzo Kasahara, Tokyo University of Marine Science and Technology, Tokyo, Japan; Shizuoka University, Shizuoka, Japan

Michael Zhdanov, University of Utah, Utah, USA

Tuncay Taymaz, Istanbul Technical University, Istanbul, Turkey

Studies in the twentieth century uncovered groundbreaking facts in geophysics and produced a radically new picture of the Earth's history. However, in some respects it also created more puzzles for the research community of the twenty-first century to tackle. This book series aims to present the state of the art of contemporary geological studies and offers the opportunity to discuss major open problems in geosciences and their phenomena. The main focus is on physical geological features such as geomorphology, petrology, sedimentology, geotectonics, volcanology, seismology, glaciology, and their environmental impacts. The monographs in the series, including multi-authored volumes, will examine prominent features of past events up to their current status, and possibly forecast some aspects of the foreseeable future. The guiding principle is that understanding the fundamentals and applied methodology of overlapping fields will be key to paving the way for the next generation.

More information about this series at http://www.springer.com/series/11723

Kazuki Koketsu

Ground Motion Seismology

Springer

Kazuki Koketsu
Earthquake Research Institute
The University of Tokyo
Bunkyo-ku, Tokyo, Japan

ISSN 2524-3829 ISSN 2524-3837 (electronic)
Advances in Geological Science
ISBN 978-981-15-8572-2 ISBN 978-981-15-8570-8 (eBook)
https://doi.org/10.1007/978-981-15-8570-8

This Springer imprint is published by the registered company Springer Nature Singapore Pte Ltd.
The registered company address is: 152 Beach Road, #21-01/04 Gateway East, Singapore 189721, Singapore

Preface

This textbook aims to explain the physics behind seismic ground motion and seismic waves for graduate and late undergraduate students or professionals. Both seismic ground motions and seismic waves are terms used for the "shaking" produced by earthquakes. However, in some instances, the shaking in the near-field of an earthquake source may be referred to as seismic ground motion, whereas shaking in the far-field may be referred to as seismic waves. The title "Ground Motion Seismology" indicates that this work explains in detail the equations and methods used for analyses and computations of shaking near an earthquake source. The textbook also details topics related to teleseismic body waves, which are frequently used in the analyses of earthquake sources.

Although there exist several similar books, this textbook is characterized not only by the presentation of the equations and methods, but also by their explanation so that readers can derive them by themselves. In particular, ground motion is often described by tensor equations based on the representation theorem. However, in this textbook, I regard the explicit formulations starting with those from *Treatise on the Mathematical Theory of Elasticity* (2nd edition; Love, 1906) as more important. This approach is based on the idea that these formulations make it easier to capture the physical picture, especially the geometrical aspects of ground motion.

Ground motions are generated at an earthquake source and propagate through the Earth, or through a velocity structure, before reaching us. It can therefore be said that ground motion seismology consists in deriving the effect of the earthquake source and the effect of propagation separately from observed ground motions, and in evaluating them quantitatively. Accordingly, in this textbook, I first explain the physical principles related to these effects in Chap. 1, followed by the effect of the earthquake source in Chap. 2 and the effect of propagation in Chap. 3. Finally, I explain the observation and processing of ground motions in Chap. 4. My aim is to explain the physics behind seismic ground motion; however, the empirical parameters, magnitude, and seismic intensity are also essential for the study of ground motion seismology, and these are explained in the appendix.

The terms and notations used in this textbook follow *Seismology* (3rd edition; Utsu, 2001), as follows:

1. Important terms are shown with American spelling and included in the index. When a more detailed description appears in another section or subsection, the section number is shown in parentheses;
2. The name of an earthquake is followed by the year of its origin time in Coordinated Universal Time (UTC) or by the date if necessary, for example, the Kanto earthquake (1923) or the San Francisco earthquake (1906);
3. References are numbered in alphabetical order by the first author's name at the end of each chapter, and the reference numbers are shown in square brackets where appropriate in the text;
4. To indicate the magnitude of an earthquake, I use a moment magnitude where possible, and otherwise a surface wave magnitude or other magnitude type;
5. Italics are used when Latin or Greek letters represent variables and constants, except for imaginary units. Boldface is used when expressing vectors and matrices;
6. The order of parentheses in the equations is ( ), { }, and [ ], from inside to outside, in principle;
7. When the following characters are used without notes, they have the following meanings.

s	second	gal	cm/s^2
t	time	g	gravitational acceleration
$\dot{}$	time derivative	$\ddot{}$	2nd-order time derivative
$\bar{}$	Fourier transform	$\tilde{}$	Fourier-Hankel transform
f	frequency	ω	angular frequency
log	common logarithm	ln	natural logarithm
e	base of natural logarithm	i	imaginary unit
*	complex conjugate	M_0	seismic moment
M	magnitude	M_w	moment magnitude
M_s	surface wave magnitude	[J]	(in Japanese)

I greatly appreciate the help of Drs. Brian Kennett, Phil Cummins, Teruo Yamashita, Koshun Yamaoka, Hiroaki Yamanaka, Kazuyoshi Kudo, Kazuro Hirahara, and Takuto Maeda. I am also grateful to Editor Yosuke Nishida for giving me the opportunity to publish this work, and Arulmurugan Venkatasalam as Production Administrator.

Tokyo, Japan
September 2020

Kazuki Koketsu

Contents

Chapter 1
Earthquakes and Ground Motion

Abstract In Sect. 1.1, basic concepts and terms in "ground motion seismology" are defined. In Sect. 1.2, the fundamentals of "elastodynamics" are explained since "seismology" is built on a foundation of elastodynamics. The "equation of motion" and "wave equation" are then obtained for ground motion, and their nature as seismic waves with attenuation are discussed. In Sect. 1.3, important principles in ground motion seismology are explained. The final topic described in this chapter, the "representation theorem", demonstrates that ground motion seismology consists in deriving the effect of the earthquake source and the effect of propagation separately from the observed ground motions, and in evaluating them quantitatively.

Keywords Ground motion · Elastodynamics · Equation of motion · Wave equation · Representation theorem

1.1 Definition of Ground Motion

The word **earthquake** has two meanings. An earthquake can refer to the phenomenon of ground shaking, or to the "sudden movement" in the ground that causes such a phenomenon. In most cases, researchers use "earthquake" in the second sense. Especially when we want to express the first concept, we use the term **ground motion**. The shaking within the ground itself is also included in the term "ground motion" [26]. This book describes ground motion at the Earth's surface and in the ground caused by an earthquake, focusing on their physical background. The strong shaking that leads to disasters is termed **strong motion**, or called **strong ground motion**. A representative parameter for earthquakes is the **magnitude** (M, Sect. A.1), and a representative parameter for ground motions is the **seismic intensity** (Sect. A.2). Instruments used to measure ground motion are called seismographs (Sect. 4.1). Among these, an instrument designed so that strong motion can be measured without saturation, is called a strong motion seismograph (Sect. 4.1.2).

Seismology is a science that studies earthquakes and related phenomena, and ground motion is regarded as a vibration phenomenon where the Earth or its parts shake. In this context, in this book the seismic structure of a part of the Earth is

K. Koketsu, *Ground Motion Seismology*, Advances in Geological Science,
https://doi.org/10.1007/978-981-15-8570-8_1

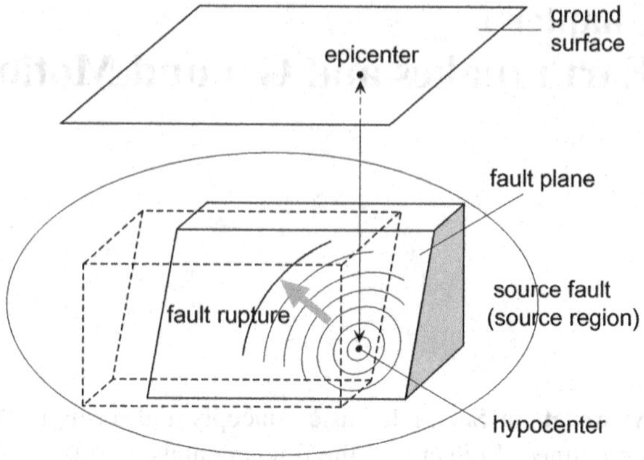

Fig. 1.1 Fault plane, hypocenter, source fault, and source region (modified from Koketsu [9] with permission of Kindai Kagaku)

termed **velocity structure**.[1] The Earth and velocity structures are considered to be **continuums**, and such a continuum is also referred to as a **medium** in the sense that it conveys vibrations. In addition, the "sudden movement" that causes ground motions is a **faulting**, which produces a sharp displacement that occurs along a certain plane of the continuum. This concept was proposed in the early 20th century as **elastic rebound theory** and was established as a sound physical theory in the 1960s (Sect. 2.1.1). The plane, termed the **fault plane**, has a certain extent, in which the point where the fault rupture began is called the **hypocenter**, and its projection on the ground surface is called the **epicenter**. The area of fault rupture is called the **source fault**, or more generally the **source region**, as distinguished from geological faults (Fig. 1.1).

In this book, we will use the term **earthquake source** when we refer to earthquake phenomena in and around the source region. For example, "earthquake source" in the Preface or the title of Chap. 2 is used in this sense. The **hypocentral distance** is the distance between the hypocenter and an observation point, and the **epicentral distance** is the distance between the epicenter and the observation point.

For fast vibrations such as ground motions, the rocks that make up the continuum (medium) basically behave as **elastic bodies** (Sect. 1.2.1) accompanied by **attenuation** (Sect. 1.2.6), therefore theories in seismology are largely based on **elastodynamics** (Sect. 1.2). The ground motion gradually weakens due to the attenuation and geometrical spreading (Sect. 3.2.2) as the motion travels away from the source fault. The weakening by the attenuation is larger when the **period** (the time for one cycle of vibration) is shorter. Short-period components therefore remain in the ground motion close to a source region. In addition, the shorter the period of vibration, the easier it is for the motion to be influenced by the heterogeneity of a velocity structure. For problems relating to three-dimensional velocity structures, **strong motion seismology** has played a more pioneering role than other fields of seismology (Sect. 3.2).

---

[1]"Subsurface structure" is also used, but its frequency of use seems to be lower than "velocity structure". The term "underground structure" is used mostly for engineering structures in the ground.

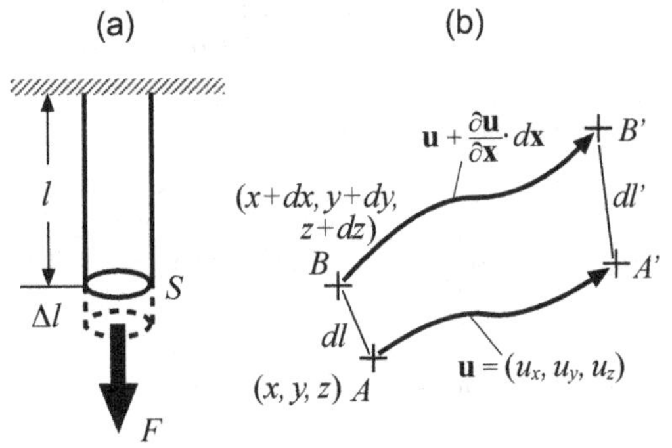

Fig. 1.2 **a** Stretch of the 1-D rubber rod model and **b** deformation of the 3-D continuum (reprinted from Koketsu [9] with permission of Kindai Kagaku)

1.2 Ground Motion and Seismic Waves[2]

1.2.1 Elastic Strain

As mentioned in Sect. 1.1, seismology is built on **elastodynamics**. An **elastic body** is a material that deforms in proportion to an applied force and returns to its original state when the force is removed. This property is called **elasticity** and the proportional relationship is called **Hooke's law**. The simplest case is a one-dimensional (1-D) elastic body such as the rubber rod shown in Fig. 1.2a, which a force F can stretch by Δl. If Hooke's law is $F = k\,\Delta l$, the elastic constant k depends on the length l of the rod, therefore it is better to replace the stretch Δl with a stretch rate of $e = \Delta l / l$. Similarly, the larger the cross section S of the rod, the smaller the stretch at the same F. Thus, if F is replaced by $\tau = F/S$, and Hooke's law is

$$\tau = \gamma\, e\,, \tag{1.1}$$

a common **elastic constant** of γ can be obtained for rods of the single material. **Displacement**, **strain**, and **stress** (Sect. 1.2.1) are extensions of Δl, e and τ for the 1-D elastic body into a three-dimensional (3-D) continuum such as the Earth and velocity structures. The displacement of the Earth or a velocity structure is **ground motion** (Sect. 1.1).

Consider two points A and B close to each other in an elastic continuum subjected to deformation, and their position vectors for any origin O as $\overrightarrow{OA} = (x,\ y,\ z)$ and $\overrightarrow{OB} = (x + dx,\ y + dy,\ z + dz)$. These two points move to A' and B' after the deformation, and the displacement of point A due to the deformation ($\overrightarrow{AA'}$) is assumed to be $\mathbf{u} = (u_x,\ u_y,\ u_z)$, as shown in Fig. 1.2b. Using the distances dl and dl' between the two points before and after the deformation, the stretch rate between A and B is given by

[2]General references for Sect. 1.2 include Fung [5] and Kunio [13].

$$\frac{dl'-dl}{dl} \sim \frac{dl'+dl}{2dl}\frac{dl'-dl}{dl} = \frac{(dl')^2-(dl)^2}{2(dl)^2}. \tag{1.2}$$

From Fig. 1.2b, we obtain

$$(dl)^2 = \left|\overrightarrow{AB}\right|^2 = \sum_{i=1}^{3}(dx_i)^2,\ (dl')^2 = \left|\overrightarrow{A'B'}\right|^2 = \sum_{i=1}^{3}\left(dx_i + \sum_{j=1}^{3}\frac{\partial u_i}{\partial x_j}dx_j\right)^2 \tag{1.3}$$

where $u_1,\ u_2,\ u_3 = u_x,\ u_y,\ u_z$ and $x_1,\ x_2,\ x_3 = x,\ y,\ z$. Therefore,

$$(dl')^2-(dl)^2 = \sum_{i=1}^{3}\left[2dx_i\sum_{j=1}^{3}\frac{\partial u_i}{\partial x_j}dx_j + \left(\sum_{j=1}^{3}\frac{\partial u_i}{\partial x_j}dx_j\right)^2\right]. \tag{1.4}$$

If the derivative of a displacement component u_i is so small that its square can be ignored, then the second term of (1.4) disappears. Here we define **infinitesimal strain** as

$$e_{ij} = \frac{1}{2}\left(\frac{\partial u_j}{\partial x_i} + \frac{\partial u_i}{\partial x_j}\right),\quad e_{ij} = e_{ji}. \tag{1.5}$$

Using this definition and the relation

$$\frac{\partial u_i}{\partial x_j}dx_i dx_j + \frac{\partial u_j}{\partial x_i}dx_j dx_i = (e_{ij}+e_{ji})\,dx_i dx_j, \tag{1.6}$$

Equation (1.2) yields

$$\frac{dl'-dl}{dl} \sim \frac{(dl')^2-(dl)^2}{2(dl)^2} \sim \sum_{i=1}^{3}\sum_{j=1}^{3} e_{ij}\frac{dx_i dx_j}{(dl)^2}. \tag{1.7}$$

Equation (1.7) indicates that the stretch rate of a 3-D continuum can be represented by a linear combination of e_{ij}. For example, if $dy = dz = 0$ so that the points A and B line up along the x-axis, the stretch rate in the x-direction can be calculated. In the case of $dy = dz = 0$, (1.4) yields $dl = dx$, and (1.7) leads to the stretch rate in the x-direction equal to e_{xx}. Since e_{yy} and e_{zz} hold similar relations, e_{ii} $(i = x, y, z)$ represent stretch rates along the x-, y-, or z-axis, termed **normal strains** [13]. Their sum $e_{xx}+e_{yy}+e_{zz}$ corresponds to the **volumetric strain** e_V [5].

Meanwhile, e_{ij} $(i \neq j)$ is termed the **shear strain**,[3] representing the change of an angle due to deformation in the 3-D continuum. For example, on the $x-y$ plane, we assume two line segments, $\overrightarrow{dl_x} = (dx,\ 0,\ 0)$ along the x-axis and $\overrightarrow{dl_y} = (0,\ dy,\ 0)$

[3]In engineering, $\gamma_{ij} = \frac{\partial u_j}{\partial x_i} + \frac{\partial u_i}{\partial x_j}$ is mostly used for shear strain rather than (1.5). Love [16] and Takeuchi and Saito [23] also used this definition.

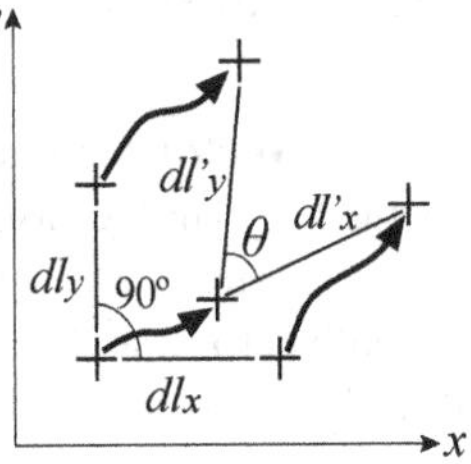

Fig. 1.3 The angles between the two line segments in the 3-D continuum before and after deformation (reprinted from Koketsu [9] with permission of Kindai Kagaku)

along the y-axis, which meet at a right angle (Fig. 1.3). After deformation, these change to $\overrightarrow{dl'_x}$ and $\overrightarrow{dl'_y}$, which do not meet at a right angle, in general. We consider $dy = dz = 0$ or $dx = dz = 0$ in Fig. 1.2b, and obtain

$$\overrightarrow{dl'_x} = \overrightarrow{A'B'}\,|_{dy=dz=0} = \left(\left(1 + \frac{\partial u_x}{\partial x}\right) dx,\ \frac{\partial u_y}{\partial x} dx,\ \frac{\partial u_z}{\partial x} dx\right),$$
$$\overrightarrow{dl'_y} = \overrightarrow{A'B'}\,|_{dx=dz=0} = \left(\frac{\partial u_x}{\partial y} dy,\ \left(1 + \frac{\partial u_y}{\partial y}\right) dy,\ \frac{\partial u_z}{\partial y} dy\right). \tag{1.8}$$

If we define θ as the angle between $\overrightarrow{dl'_x}$ and $\overrightarrow{dl'_y}$ (Fig. 1.3), we find $\overrightarrow{dl'_x} \cdot \overrightarrow{dl'_y} = |\overrightarrow{dl'_x}||\overrightarrow{dl'_y}| \cos\theta \sim 2dxdy\, e_{xy}$. The angle between the line segments changes from 90° to θ, so that the angle decrease is $\alpha = 90° - \theta$. Using $|\overrightarrow{dl'_x}|^2 = (1 + 2e_{xx})(dx)^2$ and $|\overrightarrow{dl'_y}|^2 = (1 + 2e_{yy})(dy)^2$ from (1.3) and (1.5), we find

$$\sin\alpha = \cos\theta = \frac{2e_{xy}}{(1 + 2e_{xx})^{\frac{1}{2}}(1 + 2e_{yy})^{\frac{1}{2}}}. \tag{1.9}$$

Recalling that e_{xy}, e_{xx}, and e_{yy} are infinitesimal, $\alpha \sim \sin\alpha \sim 2e_{xy}$ is obtained. This indicates that the shear strain e_{xy} is half of the angle decrease between the two line segments due to deformation [13].

In addition, strains, which are derived from (1.4) without ignoring the squares of derivatives, are termed **finite strains**. These still hold the symmetry $e_{ij} = e_{ji}$, but the nonlinearity introduced by the squares of derivatives makes the problem particularly complex. However, even in the case of strong motions, the assumption of infinitesimal strain mostly holds, except for large deformations accompanied by plasticity or liquefaction. Consequently, **infinitesimal strain** is what is being referred to in this book, when we simply refer to **strain**.

1.2.2 Balance of Stress

A **stress vector** (or **traction**) is a force acting on a virtual cross section inside the elastic body (in the case of a real cross section, the surface traction should be treated as an external force). When the virtual infinitesimal cross section ΔS has a unit normal vector $\mathbf{n}$ and a force $\Delta\mathbf{F}$ is exerted from the positive side to the negative side of $\mathbf{n}$, the stress vector $\mathbf{T}_n$ is defined as a limit of force per unit area

$$\mathbf{T}_n = \lim_{\Delta S \to 0} \frac{\Delta\mathbf{F}}{\Delta S} = \frac{d\mathbf{F}}{dS}. \tag{1.10}$$

Conversely, assuming the stress vector $\mathbf{T}_{-n}$ exerted from the negative side to the positive side, it is possible to ignore the **body force** (an external force acting on all volume elements of a body, such as gravity) by thinning the peripheral region of ΔS sufficiently. It is necessary that

$$\mathbf{T}_{-n} = -\mathbf{T}_n \tag{1.11}$$

based on the absence of body force and the balance of forces in the peripheral region.

We consider an arbitrary cross section ABC (normal vector $\mathbf{n}$) in an elastic body, and an infinitesimal region surrounded by ABC with sides perpendicular to the x-, y-, and z-axes. Let $\mathbf{T}_n$, $\mathbf{T}_{-x}$, $\mathbf{T}_{-y}$, and $\mathbf{T}_{-z}$ be the stress vectors acting outward from the surrounding surfaces (the last three vectors point in the negative directions of the x-, y-, z-axes), as shown in Fig. 1.4a. We again use the balance of forces and obtain $\mathbf{T}_n \Delta ABC + \mathbf{T}_{-x}\Delta BOC + \mathbf{T}_{-y}\Delta COA + \mathbf{T}_{-z}\Delta AOB = 0$. Since $\mathbf{n} = (n_x, n_y, n_z)$ where $n_x = \Delta BOC/\Delta ABC$ etc., $\mathbf{T}_n$ can be decomposed as $\mathbf{T}_n = n_x\mathbf{T}_x + n_y\mathbf{T}_y + n_z\mathbf{T}_z$. If $\mathbf{T}_x$ etc. are also decomposed as

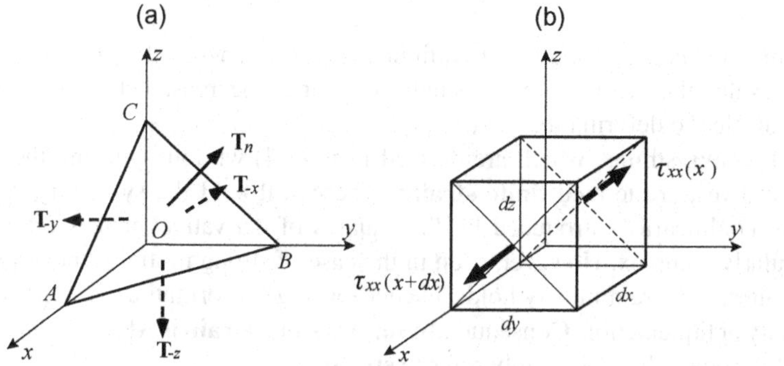

Fig. 1.4 **a** Decomposition of a stress vector and **b** stresses working on an infinitesimal rectangular prism $dxdydz$ (reprinted from Koketsu [9] with permission of Kindai Kagaku)

$$\mathbf{T}_x = \tau_{xx}\mathbf{e}_x + \tau_{xy}\mathbf{e}_y + \tau_{xz}\mathbf{e}_z ,$$
$$\mathbf{T}_y = \tau_{yx}\mathbf{e}_x + \tau_{yy}\mathbf{e}_y + \tau_{yz}\mathbf{e}_z ,$$
$$\mathbf{T}_z = \tau_{zx}\mathbf{e}_x + \tau_{zy}\mathbf{e}_y + \tau_{zz}\mathbf{e}_z \quad (1.12)$$

where τ_{ij} is a **tensor** performing the coordinate transformation between stress vectors $\mathbf{T}_i$ and basis vectors $\mathbf{e}_j$. This tensor is called the **stress tensor**, and its elements are called **stress**. The stress τ_{ij} is the x_j component of the stress vector acting on the cross section perpendicular to the x_i-axis. In the case of $i = j$, τ_{ii} is perpendicular to the cross section and is called the **normal stress**, while τ_{ij} $(i \neq j)$ is parallel to the cross section and is called the **shear stress**. Similarly to (1.11), the stress on the negative side has the opposite sign to the stress on the positive side [13].

If we consider an infinitesimal rectangular prism $dxdydz$ in the elastic body (Fig. 1.4b) and the stresses $\tau_{xx}(x)$ etc. on its side at x, which is perpendicular to the x-axis, the stresses on the other side at $x + dx$ are given as $\tau_{xx}(x + dx) = \tau_{xx} + \dfrac{\partial \tau_{xx}}{\partial x} dx$ etc. We then consider the balance of moments around the line along the z-axis, which extends through the center of the prism, and find that

$$\left(\tau_{xy} + \frac{\partial \tau_{xy}}{\partial x} dx\right) dydz\frac{dx}{2} + \tau_{xy} dydz\frac{dx}{2} = \left(\tau_{yx} + \frac{\partial \tau_{yx}}{\partial y} dy\right) dxdz\frac{dy}{2} + \tau_{yx} dxdz\frac{dy}{2}$$
$$\rightarrow \tau_{xy} + \frac{\partial \tau_{xy}}{\partial x}\frac{dx}{2} = \tau_{yx} + \frac{\partial \tau_{yx}}{\partial y}\frac{dy}{2} . \quad (1.13)$$

At the limit of $dx, dy \rightarrow 0$, (1.13) yields $\tau_{xy} = \tau_{yx}$. The balance of moments around the line along the x- or y-axis also leads to $\tau_{yz} = \tau_{zy}$ or $\tau_{zx} = \tau_{xz}$, so that the stress tensor τ_{ij} is a **symmetric tensor** $\tau_{ij} = \tau_{ji}$.

In addition, similarly to stress vectors, we can define strain vectors using a line segment with length dl and normal vector $\mathbf{n}$. If we assume the segment to be deformed by $\mathbf{u}$, the **strain vector** is defined as $\boldsymbol{\epsilon}_n = d\mathbf{u}/dl$ and can be decomposed as $\boldsymbol{\epsilon}_n = n_x\boldsymbol{\epsilon}_x + n_y\boldsymbol{\epsilon}_y + n_z\boldsymbol{\epsilon}_z$. We further decompose $\boldsymbol{\epsilon}_x$, $\boldsymbol{\epsilon}_y$, and $\boldsymbol{\epsilon}_z$ into components:

$$\boldsymbol{\epsilon}_x = e_{xx}\mathbf{e}_x + e_{xy}\mathbf{e}_y + e_{xz}\mathbf{e}_z ,$$
$$\boldsymbol{\epsilon}_y = e_{yx}\mathbf{e}_x + e_{yy}\mathbf{e}_y + e_{yz}\mathbf{e}_z ,$$
$$\boldsymbol{\epsilon}_z = e_{zx}\mathbf{e}_x + e_{zy}\mathbf{e}_y + e_{zz}\mathbf{e}_z \quad (1.14)$$

where e_{ij} constitutes a symmetric tensor.

1.2.3 Constitutive Law and Equation of Motion

A **constitutive law** describes the relationship between a force and the deformation, based not only on elasticity but also on viscosity and plasticity. The generalized **Hooke's law** (Sect. 1.2.1) is a constitutive law for a **perfectly elastic body**. Similarly to the basic Hooke's law (1.1), a proportional relation holds between the strain tensor e_{kl} and the stress tensor τ_{ij} as

$$\tau_{ij} = \sum_{k=x,y,z} \sum_{l=x,y,z} C_{ijkl}\, e_{kl} \tag{1.15}$$

where C_{ijkl} is a tensor of generalized **elastic constant** (Sect. 1.2.1).

Since τ_{ij} and e_{kl} are both symmetric tensors, C_{ijkl} must have symmetry such that $C_{ijkl} = C_{jikl}$ and $C_{ijkl} = C_{ijlk}$. In addition, the assumption is made that the change of strain due to ground motion is sufficiently fast and there is no room for heat to escape from the medium. Elastic deformation is therefore a reversible adiabatic process, so that there is a **strain energy function**[4]

$$W = \frac{1}{2} \sum_{i=x,y,z} \sum_{j=x,y,z} \tau_{ij} e_{ij} \tag{1.16}$$

corresponding to internal energy, and τ_{ij} and e_{ij} are related by $\tau_{ij} = \dfrac{\partial W}{\partial e_{ij}}$. Thus, there is a third symmetry between τ_{ij} and e_{ij}:

$$C_{ijkl} = \frac{\partial^2 W}{\partial e_{ij} \partial e_{kl}} = \frac{\partial^2 W}{\partial e_{kl} \partial e_{ij}} = C_{klij}\ . \tag{1.17}$$

From the above, 21 out of 81 C_{ijkl} are independent.

If the generalized Hooke's law is symmetric about a certain coordinate axis, there are only 5 independent elastic constants. For example, when the medium is symmetric about the z-axis, according to the notation of Love [16], we have

$$\begin{aligned}
&C_{xxxx} = C_{yyyy} = A\ ,\ C_{xxzz} = C_{zzxx} = C_{yyzz} = C_{zzyy} = F\ ,\\
&C_{zzzz} = C\ ,\ C_{yzyz} = C_{zxzx} = L\ ,\ C_{xyxy} = N\ ,\\
&\text{the other } C_{ijkl} = 0 \ \text{ except for } C_{xxyy} = C_{yyxx} = A - 2N\ .
\end{aligned} \tag{1.18}$$

Such symmetry is termed **transverse isotropy**. Furthermore, if there is symmetry about the x- or y-axis, there are only two independent elastic constants, as shown in

$$A - 2N = F = \lambda\ ,\ L = N = \mu\ ,\ (A = C = \lambda + 2\mu). \tag{1.19}$$

They are called **Lamé's constants** and μ alone is termed the **rigidity**. The generalized Hooke's law yields

$$\tau_{ij} = \lambda\, \delta_{ij}(e_{xx} + e_{yy} + e_{zz}) + 2\mu\, e_{ij}\ . \tag{1.20}$$

[4]See Chap. 12 of Fung [5].

This symmetry is called **isotropy**, and a non-isotropic nature is called **anisotropy**, including transverse isotropy. Using the **volumetric strain** $e_V = e_{xx} + e_{yy} + e_{zz}$ (Sect. 1.2.1) and the **mean stress** $\tau_V = (\tau_{xx} + \tau_{yy} + \tau_{zz})/3$ [5], we define the **deviatoric strain** $e'_{ij} = e_{ij} - \delta_{ij} e_V/3$ and the **deviatoric stress** $\tau'_{ij} = \tau_{ij} - \delta_{ij}\tau_V$. From (1.20), we obtain a proportional relationship similar to (1.1):

$$\tau_V = \kappa \, e_V \,, \quad \tau'_{ij} = 2\mu \, e'_{ij} \tag{1.21}$$

where $\kappa = \lambda + 2\mu/3$ is called the **bulk modulus**. In addition, let us consider the situation in which only forces that pull the elastic body work along a certain line, for example, in the positive and negative directions of the z-axis. In this case, since the stresses other than $\tau_{zz} = \tau$ should be small, from (1.20) all shear strains vanish, and among normal strains there mainly remains the positive $e_{zz} = e$, but negative e_{xx} and e_{yy} also appear. If the elastic body is symmetrical around the z-axis, we can assume $e_{xx} = e_{yy} = e'$. Substituting the above into (1.20), we obtain

$$e = \frac{\lambda + \mu}{\mu(3\lambda + 2\mu)}\tau \,, \quad e' = -\frac{\lambda}{2\mu(3\lambda + 2\mu)}\tau \,. \tag{1.22}$$

The ratio of e and the absolute value of e':

$$\nu = \frac{|e'|}{e} = \frac{\lambda}{2(\lambda + \mu)} \tag{1.23}$$

is called **Poisson's ratio** [13]. Considering the case of $\lambda = \mu$ as a typical value, this becomes $1/4$ from (1.23).

Now, substituting (1.5) into (1.20), we obtain the defining equations for stresses in the Cartesian coordinate system in an isotropic elastic body [22]:

$$\begin{aligned}
\tau_{xx} &= \lambda\left(\frac{\partial u_x}{\partial x} + \frac{\partial u_y}{\partial y} + \frac{\partial u_z}{\partial z}\right) + 2\mu\frac{\partial u_x}{\partial x} \,, \quad \tau_{yy} = \lambda\left(\frac{\partial u_x}{\partial x} + \frac{\partial u_y}{\partial y} + \frac{\partial u_z}{\partial z}\right) + 2\mu\frac{\partial u_y}{\partial y} \,, \\
\tau_{zz} &= \lambda\left(\frac{\partial u_x}{\partial x} + \frac{\partial u_y}{\partial y} + \frac{\partial u_z}{\partial z}\right) + 2\mu\frac{\partial u_z}{\partial z} \,, \\
\tau_{yz} &= \mu\left(\frac{\partial u_z}{\partial y} + \frac{\partial u_y}{\partial z}\right), \quad \tau_{zx} = \mu\left(\frac{\partial u_x}{\partial z} + \frac{\partial u_z}{\partial x}\right), \quad \tau_{xy} = \mu\left(\frac{\partial u_y}{\partial x} + \frac{\partial u_x}{\partial y}\right).
\end{aligned} \tag{1.24}$$

Meanwhile, when the **body force** $\mathbf{f} = (f_x, f_y, f_z)$ works on a unit mass[5] of the infinitesimal rectangular prism in Fig. 1.4b, the balance of forces in the i-direction ($i = x, y, z$) is

[5]Defined by Hudson [6]. Note that $\rho\mathbf{f} \to \mathbf{f}$ in Aki and Richards [1], because their body force works on a unit volume.

$$\rho\, dxdydz \frac{\partial^2 u_i}{\partial t^2} = \left(\tau_{ix} + \frac{\partial \tau_{ix}}{\partial x} dx\right) dydz - \tau_{ix} dydz + \left(\tau_{iy} + \frac{\partial \tau_{iy}}{\partial y} dy\right) dzdx - \tau_{iy} dzdx + \left(\tau_{iz} + \frac{\partial \tau_{iz}}{\partial z} dz\right) dxdy - \tau_{iz} dxdy + \rho\, dxdydz\, f_i \tag{1.25}$$

where ρ is the **density**. Rearranging (1.25) leads to

$$\rho \frac{\partial^2 u_i}{\partial t^2} = \frac{\partial \tau_{ix}}{\partial x} + \frac{\partial \tau_{iy}}{\partial y} + \frac{\partial \tau_{iz}}{\partial z} + \rho f_i\,, \quad i = x, y, z\,. \tag{1.26}$$

Assuming that the elastic body is isotropic, we substitute τ_{xx}, τ_{xy}, and τ_{xz} of (1.24) into (1.26) for $i = x$ and obtain the x-component of the **equation of motion** for displacement (ground motion)

$$\begin{aligned}
\rho \frac{\partial^2 u_x}{\partial t^2} &= \lambda \frac{\partial}{\partial x}\left(\frac{\partial u_x}{\partial x} + \frac{\partial u_y}{\partial y} + \frac{\partial u_z}{\partial z}\right) + \frac{\partial \lambda}{\partial x}\left(\frac{\partial u_x}{\partial x} + \frac{\partial u_y}{\partial y} + \frac{\partial u_z}{\partial z}\right) \\
&+ 2\mu \frac{\partial^2 u_x}{\partial x^2} + 2\frac{\partial \mu}{\partial x}\frac{\partial u_x}{\partial x} + \mu\left(\frac{\partial^2 u_y}{\partial x \partial y} + \frac{\partial^2 u_x}{\partial y^2}\right) + \frac{\partial \mu}{\partial y}\left(\frac{\partial u_y}{\partial x} + \frac{\partial u_x}{\partial y}\right) \\
&+ \mu\left(\frac{\partial^2 u_x}{\partial z^2} + \frac{\partial^2 u_z}{\partial x \partial z}\right) + \frac{\partial \mu}{\partial z}\left(\frac{\partial u_x}{\partial z} + \frac{\partial u_z}{\partial x}\right) + \rho f_x \\
&= (\lambda + \mu)\frac{\partial}{\partial x}\left(\frac{\partial u_x}{\partial x} + \frac{\partial u_y}{\partial y} + \frac{\partial u_z}{\partial z}\right) + \mu\left(\frac{\partial^2}{\partial x^2} + \frac{\partial^2}{\partial y^2} + \frac{\partial^2}{\partial z^2}\right) u_x \\
&+ \frac{\partial \lambda}{\partial x}\left(\frac{\partial u_x}{\partial x} + \frac{\partial u_y}{\partial y} + \frac{\partial u_z}{\partial z}\right) + \frac{\partial \mu}{\partial y}\left(\frac{\partial u_y}{\partial x} - \frac{\partial u_x}{\partial y}\right) \\
&- \frac{\partial \mu}{\partial z}\left(\frac{\partial u_x}{\partial z} - \frac{\partial u_z}{\partial x}\right) + 2\left(\frac{\partial \mu}{\partial x}\frac{\partial}{\partial x} + \frac{\partial \mu}{\partial y}\frac{\partial}{\partial y} + \frac{\partial \mu}{\partial z}\frac{\partial}{\partial z}\right) u_x + \rho f_x\,.
\end{aligned} \tag{1.27}$$

For $i = y$ or z, we cyclically transform the coordinates in (1.27) as $x \to y \to z \to x$, except where x, y, and z appear simultaneously, then obtain

$$\begin{aligned}
\rho \frac{\partial^2 u_y}{\partial t^2} &= (\lambda + \mu)\frac{\partial}{\partial y}\left(\frac{\partial u_x}{\partial x} + \frac{\partial u_y}{\partial y} + \frac{\partial u_z}{\partial z}\right) + \mu\left(\frac{\partial^2}{\partial x^2} + \frac{\partial^2}{\partial y^2} + \frac{\partial^2}{\partial z^2}\right) u_y \\
&+ \frac{\partial \lambda}{\partial y}\left(\frac{\partial u_x}{\partial x} + \frac{\partial u_y}{\partial y} + \frac{\partial u_z}{\partial z}\right) + \frac{\partial \mu}{\partial z}\left(\frac{\partial u_z}{\partial y} - \frac{\partial u_y}{\partial z}\right) \\
&- \frac{\partial \mu}{\partial x}\left(\frac{\partial u_y}{\partial x} - \frac{\partial u_x}{\partial y}\right) + 2\left(\frac{\partial \mu}{\partial x}\frac{\partial}{\partial x} + \frac{\partial \mu}{\partial y}\frac{\partial}{\partial y} + \frac{\partial \mu}{\partial z}\frac{\partial}{\partial z}\right) u_y + \rho f_y\,,
\end{aligned} \tag{1.28}$$

$$\rho\frac{\partial^2 u_z}{\partial t^2} = (\lambda+\mu)\frac{\partial}{\partial z}\left(\frac{\partial u_x}{\partial x}+\frac{\partial u_y}{\partial y}+\frac{\partial u_z}{\partial z}\right)+\mu\left(\frac{\partial^2}{\partial x^2}+\frac{\partial^2}{\partial y^2}+\frac{\partial^2}{\partial z^2}\right)u_z$$
$$+\frac{\partial\lambda}{\partial z}\left(\frac{\partial u_x}{\partial x}+\frac{\partial u_y}{\partial y}+\frac{\partial u_z}{\partial z}\right)+\frac{\partial\mu}{\partial x}\left(\frac{\partial u_x}{\partial z}-\frac{\partial u_z}{\partial x}\right)$$
$$-\frac{\partial\mu}{\partial y}\left(\frac{\partial u_z}{\partial y}-\frac{\partial u_y}{\partial z}\right)+2\left(\frac{\partial\mu}{\partial x}\frac{\partial}{\partial x}+\frac{\partial\mu}{\partial y}\frac{\partial}{\partial y}+\frac{\partial\mu}{\partial z}\frac{\partial}{\partial z}\right)u_z+\rho f_z\,. \quad (1.29)$$

Equations (1.27), (1.28), and (1.29) are equations of motion limited to the Cartesian coordinate system. Using such vector differential operators as the nabla $\nabla = \left(\frac{\partial}{\partial x}, \frac{\partial}{\partial y}, \frac{\partial}{\partial z}\right)$, we can rewrite them as

$$\rho\frac{\partial^2 \mathbf{u}}{\partial t^2} = (\lambda+\mu)\nabla(\nabla\cdot\mathbf{u})+\mu\nabla^2\mathbf{u}+\nabla\lambda(\nabla\cdot\mathbf{u})+\nabla\mu\times(\nabla\times\mathbf{u})+2(\nabla\mu\cdot\nabla)\mathbf{u}+\rho\mathbf{f} \quad (1.30)$$

which are applicable not only to the Cartesian coordinate system but also to orthogonal curvilinear coordinate systems, including the cylindrical and spherical coordinate systems. When the elastic body is homogeneous so that λ, μ, and ρ are constant, we can considerably simplify (1.30) as

$$\rho\frac{\partial^2 \mathbf{u}}{\partial t^2} = (\lambda+\mu)\nabla(\nabla\cdot\mathbf{u})+\mu\nabla^2\mathbf{u}+\rho\mathbf{f} = (\lambda+2\mu)\nabla(\nabla\cdot\mathbf{u})-\mu\nabla\times(\nabla\times\mathbf{u})+\rho\mathbf{f}\,, \quad (1.31)$$

using $\nabla\times\nabla\times = \nabla(\nabla\cdot)-\nabla^2$ [22].

For example, in the **cylindrical coordinate system** (Sect. 2.2)

$$\nabla = \left(\frac{\partial}{\partial r}, \frac{1}{r}\frac{\partial}{\partial\theta}, \frac{\partial}{\partial z}\right), \quad \nabla^2 = \frac{\partial^2}{\partial r^2}+\frac{1}{r}\frac{\partial}{\partial r}+\frac{1}{r^2}\frac{\partial^2}{\partial\theta^2}+\frac{\partial^2}{\partial z^2},$$
$$\nabla\cdot\mathbf{A} = \frac{1}{r}\frac{\partial(rA_r)}{\partial r}+\frac{1}{r}\frac{\partial A_\theta}{\partial\theta}+\frac{\partial A_z}{\partial z},$$
$$\nabla\times\mathbf{A} = \left(\frac{1}{r}\frac{\partial A_z}{\partial\theta}-\frac{\partial A_\theta}{\partial z}, \frac{\partial A_r}{\partial z}-\frac{\partial A_z}{\partial r}, \frac{1}{r}\frac{\partial(rA_\theta)}{\partial r}-\frac{1}{r}\frac{\partial A_r}{\partial\theta}\right) \quad (1.32)$$

are used in (1.30) and (1.31). In the cylindrical coordinate system, the strains [22] are

$$e_{rr} = \frac{\partial u_r}{\partial r}, \quad e_{\theta\theta} = \frac{1}{r}\frac{\partial u_\theta}{\partial\theta}+\frac{u_r}{r}, \quad e_{zz} = \frac{\partial u_z}{\partial z}, \quad (1.33)$$
$$e_{\theta z} = \frac{1}{2}\left(\frac{1}{r}\frac{\partial u_z}{\partial\theta}+\frac{\partial u_\theta}{\partial z}\right), \quad e_{zr} = \frac{1}{2}\left(\frac{\partial u_r}{\partial z}+\frac{\partial u_z}{\partial r}\right), \quad e_{r\theta} = \frac{1}{2}\left(\frac{\partial u_\theta}{\partial r}-\frac{u_\theta}{r}+\frac{1}{r}\frac{\partial u_r}{\partial\theta}\right)$$

and then the stresses [22] are

$$\tau_{rr} = \lambda\nabla\cdot\mathbf{u} + 2\mu\frac{\partial u_r}{\partial r}, \quad \tau_{\theta\theta} = \lambda\nabla\cdot\mathbf{u} + 2\mu\left(\frac{1}{r}\frac{\partial u_\theta}{\partial\theta} + \frac{u_r}{r}\right), \quad \tau_{zz} = \lambda\nabla\cdot\mathbf{u} + 2\mu\frac{\partial u_z}{\partial z},$$
$$\tau_{\theta z} = \mu\left(\frac{1}{r}\frac{\partial u_z}{\partial\theta} + \frac{\partial u_\theta}{\partial z}\right), \quad \tau_{zr} = \mu\left(\frac{\partial u_r}{\partial z} + \frac{\partial u_z}{\partial r}\right), \quad \tau_{r\theta} = \mu\left(\frac{\partial u_\theta}{\partial r} - \frac{u_\theta}{r} + \frac{1}{r}\frac{\partial u_r}{\partial\theta}\right),$$
$$\nabla\cdot\mathbf{u} = \frac{1}{r}\frac{\partial(ru_r)}{\partial r} + \frac{1}{r}\frac{\partial u_\theta}{\partial\theta} + \frac{\partial u_z}{\partial z}. \tag{1.34}$$

1.2.4 Wave Equation and Seismic Wave

According to **Helmholtz's theorem** [17], the displacement vector $\mathbf{u}$ in (1.30) and (1.31) is continuous and finite in the continuum V and is zero outside V, so that $\mathbf{u}$ can be represented by $\mathbf{u} = \nabla\phi + \nabla\times\boldsymbol{\psi}$, where ϕ and $\boldsymbol{\psi}$ are termed the **scalar potential** and the **vector potential**, respectively. We can also set $\nabla\cdot\boldsymbol{\psi} = 0$ for convenience [2]. We then take the **divergence** ($\nabla\cdot$) and **rotation** ($\nabla\times$), swap the left- and right-hand sides, and use $\nabla\cdot\nabla\phi \equiv \nabla^2\phi$, $\nabla\cdot\nabla\times\boldsymbol{\psi} \equiv 0$, and $\nabla\times\nabla\times\boldsymbol{\psi} \equiv \nabla(\nabla\cdot\boldsymbol{\psi}) - \nabla^2\boldsymbol{\psi} = -\nabla^2\boldsymbol{\psi}$, obtaining

$$\nabla^2\phi = \nabla\cdot\mathbf{u}, \quad \nabla^2\boldsymbol{\psi} = -\nabla\times\mathbf{u}. \tag{1.35}$$

Equations (1.35) are the **Poisson equations** for the potentials ϕ and $\boldsymbol{\psi}$. For their general form $\nabla^2 U = -4\pi\sigma$, the solution is [27]

$$U(\mathbf{x}) = \iiint \frac{\sigma(\boldsymbol{\xi})}{r}\,dV(\boldsymbol{\xi}) \tag{1.36}$$

where $\mathbf{x} = (x, y, z)$ and $\boldsymbol{\xi} = (\xi, \eta, \zeta)$ are the position vectors for the evaluation points of the potentials and volume integral, and $r = |\mathbf{x} - \boldsymbol{\xi}|$ is the distance between these points (Fig. 1.5). Since the equations in (1.35) correspond to $\nabla^2 U = -4\pi\sigma$ for $U = \phi$ or each component of $\boldsymbol{\psi}$ and $\sigma = -\nabla\cdot\mathbf{u}/4\pi$, or each component of $\nabla\times\mathbf{u}/4\pi$, their solutions correspond to (1.36) with these substitutions

$$\phi(\mathbf{x}) = -\frac{1}{4\pi}\iiint \frac{\nabla\cdot\mathbf{u}}{r}\,dV(\boldsymbol{\xi}), \quad \boldsymbol{\psi}(\mathbf{x}) = \frac{1}{4\pi}\iiint \frac{\nabla\times\mathbf{u}}{r}\,dV(\boldsymbol{\xi}). \tag{1.37}$$

Similarly, the **body force** (Sect. 1.2.2) $\mathbf{f}$ in (1.30) and (1.31) is also represented with the scalar potential Φ and the vector potential $\boldsymbol{\Psi}$ ($\nabla\cdot\boldsymbol{\Psi} = 0$) as $\mathbf{f} = \nabla\Phi + \nabla\times\boldsymbol{\Psi}$. Φ and $\boldsymbol{\Psi}$ satisfy the Poisson equations

$$\nabla\cdot\mathbf{f} = \nabla^2\Phi, \quad \nabla\times\mathbf{f} = -\nabla^2\boldsymbol{\Psi} \tag{1.38}$$

and their solutions are

Fig. 1.5 Volume integral of the potentials in the region V. S is the perimeter of V and O is the origin (reprinted from Koketsu [9] with permission of Kindai Kagaku)

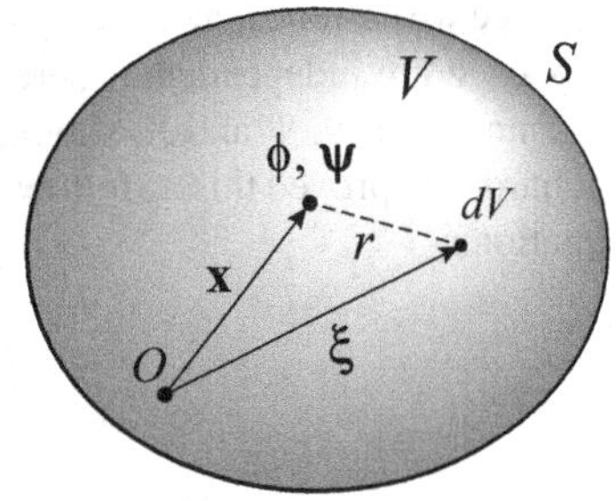

$$\Phi(\mathbf{x}) = -\frac{1}{4\pi}\iiint \frac{\nabla\cdot\mathbf{f}}{r}\,dV(\boldsymbol{\xi})\,, \quad \boldsymbol{\Psi}(\mathbf{x}) = \frac{1}{4\pi}\iiint \frac{\nabla\times\mathbf{f}}{r}\,dV(\boldsymbol{\xi})\,. \tag{1.39}$$

Substituting $\mathbf{u} = \nabla\phi + \nabla\times\boldsymbol{\psi}$ and $\mathbf{f} = \nabla\Phi + \nabla\times\boldsymbol{\Psi}$ into the equation of motion (1.31) for a homogeneous elastic body, we obtain

$$\begin{aligned}&\rho\frac{\partial^2}{\partial t^2}(\nabla\phi + \nabla\times\boldsymbol{\psi}) = \\ &(\lambda+\mu)\nabla(\nabla\cdot(\nabla\phi+\nabla\times\boldsymbol{\psi})) + \mu\nabla^2(\nabla\phi+\nabla\times\boldsymbol{\psi}) + \rho(\nabla\Phi+\nabla\times\boldsymbol{\Psi})\,.\end{aligned} \tag{1.40}$$

We here take the divergence and rotation of (1.40), considering $\nabla^2\nabla h = \nabla\nabla^2 h$, $\nabla^2\nabla\times\mathbf{h} = \nabla\times\nabla^2\mathbf{h}$, $\nabla\times\nabla h \equiv \mathbf{0}$, and $\nabla\cdot\nabla\times\mathbf{h} \equiv 0$, and then obtain

$$\begin{aligned}\nabla\cdot(1.40) &= \rho\nabla^2\frac{\partial^2\phi}{\partial t^2} = (\lambda+\mu)\nabla^2(\nabla^2\phi) + \mu\nabla^2(\nabla^2\phi) + \rho\nabla^2\Phi\,, \\ \nabla\times(1.40) &= \rho\nabla\times\nabla\times\frac{\partial^2\boldsymbol{\psi}}{\partial t^2} = \mu\nabla\times\nabla\times\nabla^2\boldsymbol{\psi} + \rho\nabla\times\nabla\times\boldsymbol{\Psi}\,.\end{aligned} \tag{1.41}$$

The removal of common operators ∇^2 in the first equation and $\nabla\times\nabla\times$ in the second equation of (1.41) leads to **Lamé's theorem** [1]

$$\begin{aligned}\frac{\partial^2\phi}{\partial t^2} &= \alpha^2\nabla^2\phi + \Phi\,, \quad \alpha^2 = \frac{\lambda+2\mu}{\rho}\,, \\ \frac{\partial^2\boldsymbol{\psi}}{\partial t^2} &= \beta^2\nabla^2\boldsymbol{\psi} + \boldsymbol{\Psi}\,, \quad \beta^2 = \frac{\mu}{\rho}\,.\end{aligned} \tag{1.42}$$

Since (1.42) consists of **inhomogeneous wave equations**, the ground motion $\mathbf{u}$ is a wave phenomenon, and can be regarded as a **seismic wave**. Seismic waves by ϕ have no rotation because of $\nabla\times\nabla\phi \equiv 0$, and they are therefore **P waves** propagating with the velocity $\alpha = \sqrt{(\lambda+2\mu)/\rho}$. Other seismic waves by $\boldsymbol{\psi}$ have no divergence because of $\nabla\cdot\nabla\times\boldsymbol{\psi} \equiv 0$, and they are therefore **S waves** propagating with the velocity $\beta = \sqrt{\mu/\rho}$. The P wave velocity is always higher than the S wave velocity, because $\lambda + 2\mu > \mu$ for positive λ and μ.

Since we assume that $\nabla \cdot \boldsymbol{\psi} = 0$ and $\nabla \cdot \boldsymbol{\Psi} = 0$, only two of the three components of $\boldsymbol{\psi}$ or $\boldsymbol{\Psi}$ are independent. Therefore, three scalar potentials including ϕ or Φ, *i.e.*, ϕ, ψ and χ or Φ, Ψ and X, should be able to represent **u** or **f**, respectively. Aki and Richards [1] proved this as follows.[6] Equation (1.42) is equivalent to the three scalar equations:

$$
\begin{aligned}
\frac{\partial^2 \phi}{\partial t^2} &= \alpha^2 \nabla^2 \phi + \Phi \,, \\
\frac{\partial^2}{\partial t^2} (\nabla \times \boldsymbol{\psi})_z &= \beta^2 \nabla^2 (\nabla \times \boldsymbol{\psi})_z + (\nabla \times \boldsymbol{\Psi})_z \,, \\
\frac{\partial^2 \psi_z}{\partial t^2} &= \alpha^2 \nabla^2 \psi_z + \Psi_z
\end{aligned}
\tag{1.43}
$$

where ψ_z and Ψ_z represent the z-components of $\boldsymbol{\psi}$ and $\boldsymbol{\Psi}$. Equation (1.43) without the body force ($\Phi = 0$ and $\boldsymbol{\Psi} = \mathbf{0}$) indicates that arbitrary ground motion **u** can be decomposed into three types, represented by ϕ, $(\nabla \times \boldsymbol{\psi})_z$, and ψ_z.

The first type of ground motion is related to ϕ, which represents a P wave, as described after (1.42). The second type of ground motion related to $(\nabla \times \boldsymbol{\psi})_z$ is derived by setting $\phi = 0$ and $\psi_z = 0$. Due to the assumption that $\nabla \cdot \boldsymbol{\psi} = 0$ and $\psi_z = 0$, $\partial \psi_x / \partial x + \partial \psi_y / \partial y = 0$ (ψ_x and ψ_y are the x- and y-components of $\boldsymbol{\psi}$). This equation indicates that there exists a function ψ which satisfies $\psi_x = \partial \psi / \partial y$ and $\psi_y = -\partial \psi / \partial x$. $\boldsymbol{\psi}$ for the second type of ground motion is therefore $\nabla \times (0, 0, \psi)$. Finally, the third type of ground motion related to ψ_z is derived by setting $\phi = 0$ and $(\nabla \times \boldsymbol{\psi})_z = 0$. As from $\mathbf{u} = \nabla \times \boldsymbol{\psi}$ we know that $\nabla \cdot \mathbf{u} = 0$ and $u_z = (\nabla \times \boldsymbol{\psi})_z = 0$, which are similar to those for the second ground motion type, there exists the function χ which satisfies $\mathbf{u} = \nabla \times (0, 0, \chi)$. These relations also hold true for Φ, $\boldsymbol{\Psi}$, and **f**.

From the above, we know that

$$
\begin{aligned}
\mathbf{u} &= \nabla \phi + \nabla \times \nabla \times (0, 0, \psi) + \nabla \times (0, 0, \chi) \,, \\
\mathbf{f} &= \nabla \Phi + \nabla \times \nabla \times (0, 0, \Psi) + \nabla \times (0, 0, \mathrm{X})
\end{aligned}
\tag{1.44}
$$

are valid, at least in a Cartesian coordinate system or cylindrical coordinate system where the z-component is independent. Substituting these into (1.31), we obtain

$$
\frac{\partial^2 \phi}{\partial t^2} = \alpha^2 \nabla^2 \phi + \Phi \,, \quad \frac{\partial^2 \psi}{\partial t^2} = \beta^2 \nabla^2 \psi + \Psi \,, \quad \frac{\partial^2 \chi}{\partial t^2} = \beta^2 \nabla^2 \chi + X \,. \tag{1.45}
$$

Equations (1.45) indicate that the S wave defined by (1.42) actually consists of two types of seismic waves represented by ψ and χ. The S wave represented by χ is called the **SH wave**, because u_z for this wave type is zero and there are only

[6] Box 6.5 of Aki and Richards [1]. However, although the point is that (1.43) is equivalent to (1.42), they did not prove this equivalence. Kennett's proof using u_V and u_H [7] can be considered to correspond to the point.

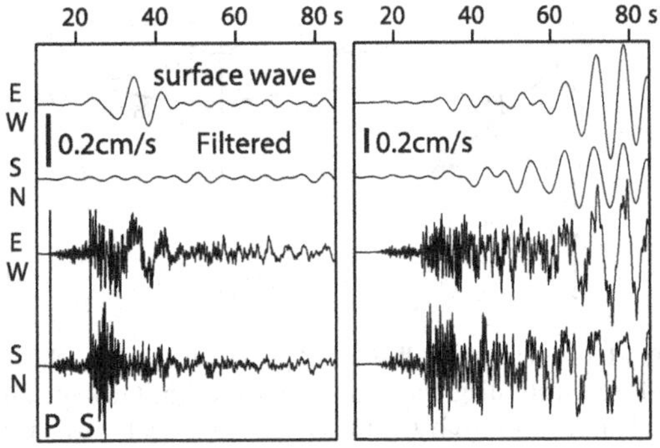

Fig. 1.6 Ground motions recorded at two observation points outside a basin (Left) and inside a basin (Right). The lower seismograms are the East-West and South-North components of the original records, and the upper ones are their long-period components (modified from Koketsu and Kikuchi [10] with permission of AAAS)

horizontal components. Conversely, the S wave represented by ψ is called the **SV wave**, because u_z for this wave type is not zero. Note that ϕ, ψ, and χ are the **displacement potentials** for the P, SV, and SH waves, respectively.

Collectively, these seismic waves are referred to as **body waves**. However, there exist seismic waves of another type, which are termed **boundary waves**. These are generated at **interfaces** (or **boundaries**), where **properties** (Lamé's constants, density, P and S wave velocities, etc.) of the Earth or a velocity structure are discontinuous. In particular, at the **ground surface** (Sect. 3.1.1) where the degree of discontinuity is the strongest, **surface waves** are generated (Sects. 3.1.9 and 3.1.10), and often become prominent. As a typical example, Fig. 1.6 shows the ground motion recorded at two observation points around 40 km away from a shallow earthquake. Surface waves appear after the P and S waves, and most of the long period components are surface waves. These are especially prominent at the observation point located inside the basin in this example.

1.2.5 Wavefronts and Rays

The **homogeneous** wave equations obtained from (1.45) by removing body forces ($\Phi = \Psi = \mathrm{X} = 0$)

$$\frac{\partial^2\phi}{\partial t^2} = \alpha^2\nabla^2\phi\,, \quad \frac{\partial^2\psi}{\partial t^2} = \beta^2\nabla^2\psi\,, \quad \frac{\partial^2\chi}{\partial t^2} = \beta^2\nabla^2\chi \tag{1.46}$$

will be studied in detail in this subsection. Their features are in common so we will study only the first equation for P waves. First, we consider a two-dimensional case, where there is no variation along the y-axis in the Cartesian coordinate system ($\partial/\partial y \equiv 0$). We assume that the medium is uniform and α is constant, and the time dependence is assumed to be **simple harmonic oscillation** $e^{i\omega t}$. From these assumptions, the wave equation and its solution yield

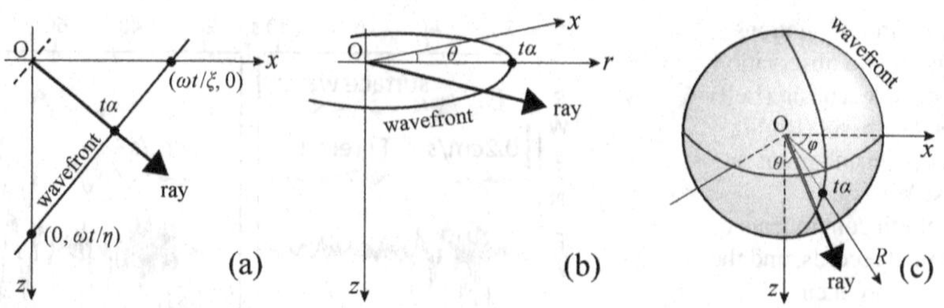

Fig. 1.7 **a** Schematic diagram of plane wave, **b** 3-D schematic diagram of a cylindrical wave, and **c** 3-D schematic diagram of spherical wave. The thick solid lines and arrows represent the wavefronts and rays, respectively (reprinted from Koketsu [9] with permission of Kindai Kagaku)

$$\left(\frac{\partial^2}{\partial x^2}+\frac{\partial^2}{\partial z^2}\right)\phi+\frac{\omega^2}{\alpha^2}\phi=0\,,\quad \phi=He^{\mathrm{i}(\omega t-\xi x-\eta z)}\,,\ \xi^2+\eta^2=\frac{\omega^2}{\alpha^2}\,. \tag{1.47}$$

From (1.44) and $\partial/\partial y \equiv 0$, the ground motion is given as

$$\begin{aligned} u_x &= \frac{\partial\phi}{\partial x} = Ae^{\mathrm{i}(\omega t-\xi x-\eta z)}\,, \quad A=-\mathrm{i}\xi H\,, \\ u_z &= \frac{\partial\phi}{\partial z} = Be^{\mathrm{i}(\omega t-\xi x-\eta z)}\,, \quad B=-\mathrm{i}\eta H\,. \end{aligned} \tag{1.48}$$

Since the phase term of this ground motion is in the form of $\mathrm{i}(\omega t-\xi x-\eta z)$, for the same phase, larger x and z require a longer time t. In other words, the ground motion propagates towards the first quadrant in the $x-z$ plane (Fig. 1.7a). The line connecting points that are in the same phase at the same time is called the **wavefront** [20]. The point of zero phase is $x=\omega t/\xi$ if $z=0$ and $z=\omega t/\eta$ if $x=0$. The line connecting these two points (thick solid line in Fig. 1.7a) is $z=-(\xi/\eta)x+(\omega t)/\eta$, and the phase is zero at any point on this line [21]. This line is therefore a wavefront at $t>0$. In addition, since both points converge to the origin for $t\to 0$, the dashed line with a slope of $-\xi/\eta$ passing through the origin is also a wavefront at $t=0$. Despite it being a line, this wavefront is referred to as a **plane wave**, because there are innumerable lines along the y-axis perpendicular to the page and these form a plane. A line perpendicular to the $t\geq 0$ wavefront is called a **ray** (thick solid arrow). Innumerable rays are possible, but if the wave equation becomes inhomogeneous and the source is determined, the ray passing through the source is realized.

Secondly, we consider the case where there are variations only along the r-axis ($\partial/\partial\theta=\partial/\partial z=0$) in the **cylindrical coordinate system** (r,θ,z) (Sect. 2.2). We again assume a uniform medium with constant α and a simple harmonic oscillation $e^{\mathrm{i}\omega t}$, so that the wave equation yields

$$\left(\frac{d^2}{dr^2}+\frac{1}{r}\frac{d}{dr}\right)\phi+k_\alpha^2\phi=0\,,\quad k_\alpha^2=\frac{\omega^2}{\alpha^2}\,. \tag{1.49}$$

As (1.49) divided by k_α^2 is the zeroth order **Bessel equation** (Sect. 2.2.1) for $r' = k_\alpha r$, we obtain the solution of (1.49):

$$\phi = H \cdot H_0^{(2)}(k_\alpha r)e^{\mathrm{i}\omega t} \sim H\sqrt{\frac{2}{\pi k_\alpha r}}e^{\mathrm{i}(\omega t - k_\alpha r + \pi/4)} \tag{1.50}$$

using the **Hankel function** $H_0^{(2)}$ and its asymptotic expansion (Sect. 3.1.1). From (1.44), $\partial/\partial\theta = \partial/\partial z = 0$, and $dH_0^{(2)}(k_\alpha r)/dr = -k_\alpha H_1^{(2)}(k_\alpha r)$, the ground motion is given as

$$u_r = \frac{d\phi}{dr} \sim A\sqrt{\frac{2}{\pi k_\alpha r}}e^{\mathrm{i}(\omega t - k_\alpha r + \pi/4)}, \quad A = -\mathrm{i}k_\alpha H\,. \tag{1.51}$$

Since the phase term of this ground motion is in the form $\mathrm{i}(\omega t - k_\alpha r + \pi/4) = \mathrm{i}\{\omega(t - r/\alpha) + \pi/4\}$, for the same phase, larger r requires a longer time t. In other words, the ground motion propagates outwards from the z-axis with a velocity of α in the horizontal plane passing through the z-axis (Fig. 1.7b). The wavefront for $t > 0$ is a circle with a radius of $r = t\alpha$ (thick solid line in Fig. 1.7b). There are innumerable circles along the z-axis and they form a **cylindrical wave**. The ray is a line extending from the origin and crossing the wavefront at a right angle (thick solid arrow). The amplitude term of (1.51) includes $1/\sqrt{r}$, and a cylindrical wave decays with **geometrical spreading** in inverse proportion to $\sqrt{r}$. Due to the $1/\sqrt{r}$ term, cylindrical waves diverge at $r = 0$ on the z-axis, but if the wave equation becomes inhomogeneous and the line source is located there, this divergence is canceled [21].

Finally, we consider the case where there are variations only along the R-axis ($\partial/\partial\theta = \partial/\partial\phi = 0$) in the **spherical coordinate system** (R, θ, ϕ) (Sect. 2.3.3). We again assume a uniform medium with constant α and a simple harmonic oscillation $e^{\mathrm{i}\omega t}$, so that the wave equation yields

$$\left(\frac{d^2}{dR^2} + \frac{2}{R}\frac{d}{dR}\right)\phi + k_\alpha^2\phi = 0\,, \quad k_\alpha^2 = \frac{\omega^2}{\alpha^2}\,. \tag{1.52}$$

Substituting $\phi = \phi'/R$ into (1.52), we obtain $d^2\phi'/dR^2 + k_\alpha\phi' = 0$. Its solution is $\phi' = He^{-\mathrm{i}k_\alpha R}$, and so $\phi = \dfrac{H}{R}e^{\mathrm{i}(\omega t - k_\alpha R)}$. According to Satô [22], the displacement potentials in the spherical coordinate system are defined as

$$\mathbf{u} = \nabla\phi + \nabla \times \nabla \times (R\psi, 0, 0) + \nabla \times (R\chi, 0, 0)\,. \tag{1.53}$$

Using this definition and $\partial/\partial\theta = \partial/\partial\phi = 0$, we obtain the ground motion and its approximation for large R:

$$u_R = \frac{d\phi}{dR} = H\left(\frac{-1}{R^2} + \frac{-\mathrm{i}k_\alpha}{R}\right)e^{\mathrm{i}(\omega t - k_\alpha R)} \sim \frac{A}{R}e^{\mathrm{i}(\omega t - k_\alpha R)}, \quad A = -\mathrm{i}k_\alpha H\,. \tag{1.54}$$

Since the phase term of this ground motion is in the form of $i(\omega t - k_\alpha R) = i\omega(t - R/\alpha)$, for the same phase, larger R requires a longer time t. In other words, the ground motion propagates outwards from the origin with a velocity of α (Fig. 1.7c). The wavefront for $t > 0$ is a sphere with a radius of $R = t\alpha$ (thick solid line in Fig. 1.7c), forming a **spherical wave**. The approximated amplitude term of (1.54) includes $1/R$, and a spherical wave decays with geometrical spreading in inverse proportion to R. Due to the $1/R$ term, spherical waves diverge at the origin of $R = 0$, but if the wave equation becomes inhomogeneous and a **point source** (Sect. 2.1.4) is located there, this divergence is canceled [21]. From Chap. 2 onwards, we will mainly deal with the physics of point sources, therefore spherical waves will play a central role (*e.g.*, Sect. 2.1.5). However, to deal with ground motions in **1-D velocity structures** (Sect. 3.1.1), we will decompose a spherical wave into cylindrical waves using the **Sommerfeld integral** (Sect. 2.2.1). We can also decompose the spherical wave into plane waves using the **Weyl integral** [28].

In addition to the general definition in physics mentioned above, Aki and Richards [1] defined a **wavefront** as a propagating discontinuity in the ground motion (displacement) or one of its derivatives. As a result of **causality** (Sect. 1.2.6), the displacement is zero until the seismic wave arrives from the source at time $t = 0$, and the displacement changes along the **source time function** (Sect. 2.1.2) for $t > 0$. As the function can never be constantly zero, the definition of those authors is equivalent to our definition.

1.2.6 Anelasticity

In real media, a perfectly elastic body strictly following Hooke's law (Eqs. (1.15) and (1.20)) is rather rare. Real media have the property of **anelasticity** and deviate more or less from Hooke's law. Crystal defects, dislocation motions, grain boundary processes, etc. in the **crust** (Sect. 2.1.1) and **mantle** (Sect. 2.1.1) where ground motion propagates, result in anelasticity [8]. The anelasticity absorbs a part of the energy of ground motion and exerts **attenuation** on it. In a perfectly elastic medium, no attenuation occurs, except by **geometrical spreading** (Sect. 1.2.5). In addition, small-scale heterogeneities in real media may cause **scattering** and shift a part of the body wave energy to **coda waves** in later portions of the signal [15]. The shift attenuates the short-period component of body waves, and is referred to as **scattering attenuation**. However, this effect is not significant for the medium- and long-period components of ground motion, which we mainly consider. In order to distinguish it from scattering attenuation, that due to by anelasticity is referred to as **intrinsic attenuation**.

Since Hooke's law of **elasticity** (Sect. 1.2.1) does not include time t, this indicates that a change of strain is immediately transmitted to a change of stress, or a change of stress is immediately transmitted to a change of strain. When a certain time elapses without these being transmitted immediately, this property is referred to as **viscosity**. The combined properties of "elasticity" and "viscosity" result in the property of **viscoelasticity**. In the case of a viscoelastic body, attenuation occurs because strain energy is dissipated over time. It is believed that most of the intrinsic attenuation can be expressed by "viscoelasticity" [6]. Considering a 1-D medium for simplicity, instead of (1.1), the stress–strain relationship of a viscoelastic body is described in the form[7]:

$$\tau(t) = \int_0^t \gamma(t-\zeta)\, de(\zeta) = \gamma_0\, e(t) + \int_0^t \dot{R}(t-\zeta) e(\zeta)\, d\zeta\,, \quad \dot{R} = \frac{dR(t)}{dt}. \tag{1.55}$$

In the second equation after the **integration by parts**, γ_0 represents elasticity, and the **relaxation function** $R(t)$ represents viscosity. Taking the **Fourier transform** (Sect. 4.2.2) of (1.55) and using its properties (Table 4.3), we obtain

$$\bar{\tau}(\omega) = \bar{\gamma}(\omega)\, \bar{e}(\omega) = (\gamma_0 + \gamma_1(\omega))\, \bar{e}(\omega)\,, \quad \gamma_1(\omega) = \mathrm{i}\omega\overline{R}(\omega)\,. \tag{1.56}$$

Equation (1.56) implies that, in the frequency domain, viscoelasticity holds a proportional relationship similar to Hooke's law for elasticity.

From (1.5) and (1.26) for a 3-D medium, we can write the equations of strain and balance for a 1-D medium as

$$e = \frac{\partial u}{\partial x}\,, \quad \rho\frac{\partial^2 u}{\partial t^2} = \frac{\partial \tau}{\partial x} \tag{1.57}$$

where x is a spatial coordinate and u is a displacement. Taking the Fourier transforms of both equations in (1.57) and substituting (1.56) and the former equation into the latter equation, we obtain

$$\rho(\mathrm{i}\omega)^2\bar{u} = (\gamma_0 + \gamma_1)\frac{\partial^2 \bar{u}}{\partial x^2}\,. \tag{1.58}$$

Letting $\bar{u} = \overline{U}(\omega)e^{-\mathrm{i}kx}$ be a solution of this equation, using the substitution

$$\gamma_1 = \mathrm{Re}\,\gamma_1 + \mathrm{i}\,\mathrm{Im}\,\gamma_1 = \gamma_1^{\mathrm{R}}(\omega) + \mathrm{i}\omega\gamma_1^{\mathrm{I}}(\omega)\,, \tag{1.59}$$

and taking the **inverse Fourier transform** (Sect. 4.2.2), if γ_1^{I} is almost constant, we obtain

$$\begin{aligned} &m'\frac{d^2U}{dt^2} + c'\frac{dU}{dt} + k'U = 0\,, \\ &m' = \rho\,, \quad c' = k^2\gamma_1^{\mathrm{I}}\,, \quad k' = k^2(\gamma_0 + \gamma_1^{\mathrm{R}})\,. \end{aligned} \tag{1.60}$$

[7]By Hudson [6]. He noted in the footnote on Page 194 of his book that L. Boltzmann made this formulation in 1876 for the first time.

Equation (1.60) is the well-known equation of motion for **damped oscillation** (*e.g.*, Landau and Lifshitz [14], Kreyszig [12]). c' in (1.60) is the **damping coefficient.**[8] Here $U(t)$ is the inverse Fourier transform of $\overline{U}(\omega)$, and its solution for **characteristic oscillation** (Sect. 3.1.9) leads to the real-part solution of ground motion u:

$$u = a\,e^{-\eta t}\cos\left(\sqrt{\omega_0^2 - \eta^2}\,t - kx\right), \tag{1.61}$$

$$\omega_0 = \sqrt{\frac{k'}{m'}} = k\sqrt{\frac{\gamma_0 + \gamma_1^{\mathrm{R}}}{\rho}}, \quad \eta = \frac{c'}{2m'} = \frac{k^2\gamma_1^{\mathrm{I}}}{2\rho}.$$

Q [1, 26] defined by $Q^{-1} = -\Delta E/2\pi E$ (also called the **quality factor** [25]) is most often used as an indicator of attenuation. In a viscoelastic medium, E is elastic energy and $-\Delta E$ is energy lost during one cycle of oscillation due to viscosity. For the damped oscillation of (1.61), E is the energy of simple harmonic oscillation [14] for the elastic part:

$$E = \frac{1}{2}k_0'a^2 \tag{1.62}$$

where ω_0 is the natural angular frequency and $k_0' = k^2\gamma_0$. Between $t = 0$ and $t = T$ (T is the **period**, Sect. 1.1), the amplitude is reduced from a to $ae^{-\eta T}$. If the reduction rate is small and the change of angular frequency is also small,

$$-\Delta E = \frac{1}{2}k'a^2(1 - e^{-2\eta T}) \sim \frac{1}{2}k'a^2(2\eta T), \quad T \sim \frac{2\pi}{\omega_0}. \tag{1.63}$$

Therefore,

$$Q^{-1} = \frac{1}{2\pi}\frac{k'}{k_0'}\frac{4\pi\eta}{\omega_0}. \tag{1.64}$$

If $\gamma_0 \gg \gamma_1^{\mathrm{R}}$, $k' \sim k_0'$ and we obtain

$$Q^{-1} \sim \frac{2\eta}{\omega_0} = \frac{2k^2\gamma_1^{\mathrm{I}}}{2\rho}\frac{1}{\omega_0} \sim \frac{2k^2\mathrm{Im}\,\gamma_1}{2\rho\omega_0}\frac{1}{\omega_0} \sim \frac{\mathrm{Im}\,\gamma_1}{\gamma_0} \tag{1.65}$$

using (1.59) and (1.61).[9] Since we assume that γ_1^{I} is almost constant for (1.60), Q^{-1} in (1.65) is also almost constant. In addition, from (1.65) and (1.61), the **damping constant** c' in (1.60) yields

$$c' = 2m'\eta = \rho\omega_0 Q^{-1} = 2\pi f_0\rho Q^{-1} \tag{1.66}$$

---

[8] Kreyszig [12] called it the **damping constant**, but Aki and Richards [1] and Utsu [26] used this term for a quantity related to η (Sect. 4.1).

[9] In Hudson [6], the numerator of the equation corresponding to (1.65) has a minus sign. The reason for this is that the definition of the Fourier transform is different and $\mathrm{i}\omega$ in the second equation of (1.56) was $-\mathrm{i}\omega$ there.

which is proportional to mass (density), and this is referred to as **mass-proportional damping** [3]. f_0 is the natural frequency.

For example, when a change in strain occurs and it is transmitted to stress, the change in stress cannot occur earlier than the change in strain. This principle is called **causality**. Since in (1.55) only the dR/dt term results in the stress $\tau(t)$ having a different time history compared to the strain $e(t)$, dR/dt should satisfy the causality principle in order to obtain a causal stress. According to "causality" in Table 4.3, the condition for this to hold is that the real and imaginary parts of $\gamma_1(\omega)$, which is the Fourier transform of dR/dt, are related by the Hilbert transform

$$\mathrm{Re}\,\gamma_1(\omega) = \frac{1}{\pi}\int_{-\infty}^{+\infty} \frac{\mathrm{Im}\,\gamma_1(y)}{\omega - y}\,dy\,. \tag{1.67}$$

(Sect. 4.2.2). Furthermore, dR/dt needs to be a real function in order for "viscoelasticity" to be an actual phenomenon. From "real function" in Table 4.3, it is necessary for this requirement that $\gamma_1(-\omega) = \gamma_1^*(\omega)$, that is $\mathrm{Im}\,\gamma_1(-\omega) = -\mathrm{Im}\,\gamma_1(\omega)$. Substituting this into the $\omega < 0$ part of the integral in (1.67) and changing the variable such as that $y' = -y$, we obtain the **Kramers–Kronig relation** [11, 19]

$$\begin{aligned}\mathrm{Re}\,\gamma_1(\omega) &= \frac{1}{\pi}\int_0^{+\infty} \frac{\mathrm{Im}\,\gamma_1(y)}{\omega - y}\,dy + \frac{1}{\pi}\int_{+\infty}^{0} \frac{-\mathrm{Im}\,\gamma_1(y')}{\omega + y'}\,(-dy') \\ &= \frac{2}{\pi}\int_0^{+\infty} \frac{y\,\mathrm{Im}\,\gamma_1(y)}{\omega^2 - y^2}\,dy\,. \end{aligned} \tag{1.68}$$

We then substitute (1.65) into this relation, assume Q^{-1} to be constant for the range $\omega_1 \ll \omega \ll \omega_2$ and zero outside this range, and use the formula $\int 1/(ax+b)\,dx = 1/a \cdot \ln|ax+b|$ [18] and the variable change $x = y^2$, finding

$$\gamma_1^{\mathrm{R}} = \mathrm{Re}\,\gamma_1 = \frac{\gamma_0}{\pi Q}\int_{\omega_1}^{\omega_2} \frac{2y\,dy}{\omega^2 - y^2} = \frac{-\gamma_0}{\pi Q}\ln\frac{\omega_2^2 - \omega^2}{\omega^2 - \omega_1^2} \sim \frac{2\gamma_0}{\pi Q}\ln\frac{\omega}{\omega_2}\,. \tag{1.69}$$

In addition, the ground motion in (1.61) can be rewritten as

$$u = a\,\mathrm{Re}\,e^{\mathrm{i}\omega t - \mathrm{i}kx}, \quad \omega = \sqrt{\omega_0^2 - \eta^2} + \mathrm{i}\eta\,. \tag{1.70}$$

Similarly to (1.63), we again assume the change of natural angular frequency to be small, and then we have $\omega \sim \omega_0 + \mathrm{i}\eta$. Furthermore, we extend the **phase velocity** $c \equiv \omega/k$ for such a complex ω and substitute (1.61) and (1.65) into c, obtaining

$$c^2 = \frac{\omega_0^2}{k^2}\left(1 + \frac{\mathrm{i}\eta}{\omega_0}\right)^2 = \frac{\gamma_0 + \gamma_1^{\mathrm{R}}}{\rho}\left(1 + \frac{\mathrm{i}}{2Q}\right)^2\,. \tag{1.71}$$

The further substitution of (1.69) into the equation above results in

$$c^2 = \frac{\gamma_0}{\rho}\left(1 + \frac{2}{\pi Q}\ln\frac{\omega}{\omega_2}\right)\left(1 + \frac{\mathrm{i}}{2Q}\right)^2 . \tag{1.72}$$

If a real phase velocity c_r is measured for a reference phase velocity ω_r,

$$c_r^2 = \frac{\gamma_0}{\rho}\left(1 + \frac{2}{\pi Q}\ln\frac{\omega_r}{\omega_2}\right) . \tag{1.73}$$

We then substitute this into $\ln\omega_2$ of (1.72) and again assume $\gamma_0 \gg \gamma_1^{\mathrm{R}}$. Since $c_r^2 \sim \gamma_0/\rho$, we obtain

$$c^2 \sim c_r^2\left(1 + \frac{2}{\pi Q}\ln\frac{\omega}{\omega_r}\right)\left(1 + \frac{\mathrm{i}}{2Q}\right)^2 . \tag{1.74}$$

Taking the square roots of both sides of (1.74) finally leads to

$$c = c_r\left(1 + \frac{2}{\pi Q}\ln\frac{\omega}{\omega_r}\right)^{\frac{1}{2}}\left(1 + \frac{\mathrm{i}}{2Q}\right) \sim c_r\left(1 + \frac{1}{\pi Q}\ln\frac{\omega}{\omega_r} + \frac{\mathrm{i}}{2Q}\right) . \tag{1.75}$$

We apply the above results to the proportional relationships (1.21) for a 3-D medium, and define $Q_\alpha^{-1} = \mathrm{Im}\,(\kappa_1 + 4\mu_1/3)/(\kappa_0 + 4\mu_0/3)$ and $Q_\beta^{-1} = \mathrm{Im}\,\mu_1/\mu_0$ based on $\alpha^2 = (\kappa + 4\mu/3)/\rho$, obtaining

$$\alpha(\omega) = \alpha_r\left(1 + \frac{1}{\pi Q_\alpha}\ln\frac{\omega}{\omega_r} + \frac{i}{2Q_\alpha}\right) ,$$
$$\beta(\omega) = \beta_r\left(1 + \frac{1}{\pi Q_\beta}\ln\frac{\omega}{\omega_r} + \frac{i}{2Q_\beta}\right) . \tag{1.76}$$

Therefore, when we cannot ignore attenuation due to viscoelasticity but we can assume constant Qs, damped ground motions can be computed using the complex P and S wave velocities in (1.76) instead of the real α_r and β_r. In this computation, if we forget the term $\ln(\omega/\omega_r)$, the waveform exudes before the arrival of the ground motion, as shown in the right panel of Fig. 1.8.

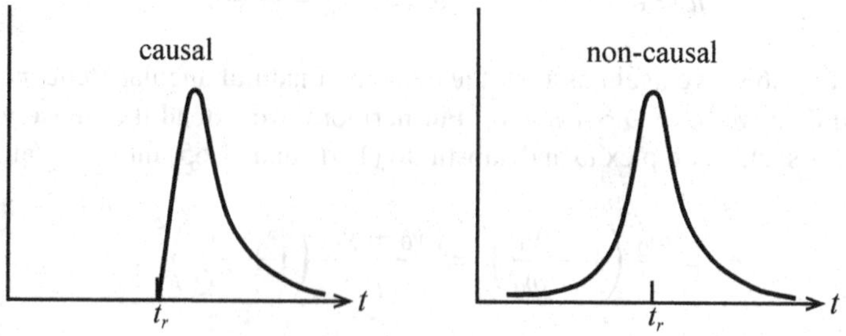

Fig. 1.8 Causal waveform (left) and non-causal waveform (right). t_r indicates the arrival time of the ground motion (modified from Kennett [7] with permission of the author)

In usual media, the attenuation related to **volumetric strain** (Sect. 1.2.3) is small, so that $\kappa_1 \sim 0$. In this case,

$$Q_\alpha^{-1} \sim \frac{\mathrm{Im}\, 4\mu_1/3}{\kappa_0 + 4\mu_0/3} = \frac{4}{3}\frac{\mathrm{Im}\, \mu_1}{\rho\alpha_0^2}, \quad \alpha_0^2 = \frac{\kappa_0 + 4\mu_0/3}{\rho},$$
$$Q_\beta^{-1} = \frac{\mathrm{Im}\, \mu_1}{\mu_0} = \frac{\mathrm{Im}\, \mu_1}{\rho\beta_0^2}, \quad \beta_0^2 = \frac{\mu_0}{\rho} \tag{1.77}$$

and

$$Q_\alpha^{-1} \sim \frac{4}{3}\frac{\beta_0^2}{\alpha_0^2} Q_\beta^{-1} \tag{1.78}$$

can be assumed [7].

1.3 Principles of Ground Motion

1.3.1 Principle of Superposition

All the operators ∇, $\nabla\cdot$, $\nabla\times$, $\nabla\times\nabla\times$, and ∇^2 in the equation of motion (1.30) or (1.31) display **linearity** as

$$\nabla(a\mathbf{A} + b\mathbf{B}) = a\nabla\mathbf{A} + b\nabla\mathbf{B}\,, \tag{1.79}$$

so that the equation of motion itself is also linear. Therefore, the ground motion $\mathbf{u}$ in the case where there are two body forces, $\mathbf{f}_A$ and $\mathbf{f}_B$, in the region V (Fig. 1.5) is given by

$$\mathbf{u} = \mathbf{u}_A + \mathbf{u}_B \tag{1.80}$$

where $\mathbf{u}_A$ is the ground motion in the case where only $\mathbf{f}_A$ exists in V, and $\mathbf{u}_B$ is the ground motion in the case where only $\mathbf{f}_B$ exists in V. This theory is an example of **principle of superposition** in physics [20].

1.3.2 Reciprocity Theorem

In the Preface, I stated that 'ground motion seismology is to derive the effect of the earthquake source and the effect of propagation separately from observed ground motions, and to evaluate them quantitatively'. I will now show that this is possible in principle, as follows.

The most general form of the equation of motion is the one based on

$$\tau_{ij} = C_{ijkl}e_{kl} = C_{ijkl}\frac{\partial u_k}{\partial x_l} \tag{1.81}$$

from the strain equation (1.5) and the generalized Hooke's law (1.15), and

$$\rho\frac{\partial^2 u_i}{\partial t^2} = \frac{\partial \tau_{ij}}{\partial x_j} + \rho f_i \tag{1.82}$$

from the balance equation (1.26).[10] The subscripts i, j, k, and l are either x, y, or z, and x_j or x_l represent the x-, y-, or z-coordinates. We use the **summation convention** [4] (a sum is taken when the same subscript appears repeatedly in one term) to simplify the equations. In addition to this case, we also consider another case where a body force g_i different from f_i exists in the region V (Fig. 1.5). g_i generates ground motions υ_i, strains ϵ_{ij}, and stresses σ_{ij}, which are related by

$$\sigma_{ij} = C_{ijkl}\epsilon_{kl} = C_{ijkl}\frac{\partial \upsilon_k}{\partial x_l} \tag{1.83}$$

and

$$\rho\frac{\partial^2 \upsilon_i}{\partial t^2} = \frac{\partial \sigma_{ij}}{\partial x_j} + \rho g_i\,. \tag{1.84}$$

Here, we introduce the operator "$*$" termed **convolution**

$$f_1(t) * f_2(t) = \int_{-\infty}^{+\infty} f_1(\tau)f_2(t-\tau)d\tau = \int_{-\infty}^{+\infty} f_1(t-\tau)f_2(\tau)d\tau \tag{1.85}$$

which combines two functions into a single function.[11] Taking the convolution of $\rho\frac{\partial^2 u_i}{\partial t^2}$ and υ_i, and substituting (1.82) into it, we have

$$\begin{aligned} &\int_{-\infty}^{+\infty} d\tau \iiint \rho\frac{\partial^2}{\partial \tau^2}u_i(\mathbf{x},\tau)\,\upsilon_i(\mathbf{x},t-\tau)\,dV \\ &= \int_{-\infty}^{+\infty} d\tau \iiint \frac{\partial}{\partial x_j}\tau_{ij}(\mathbf{x},\tau)\,\upsilon_i(\mathbf{x},t-\tau)\,dV + \int_{-\infty}^{+\infty} d\tau \iiint \rho f_i(\mathbf{x},\tau)\,\upsilon_i(\mathbf{x},t-\tau)\,dV\,. \end{aligned} \tag{1.86}$$

We assume that the body forces f_i and g_i begin to operate at $t = 0$, and therefore their ground motions for $t < 0$ vanish due to the causality ($u_i(\mathbf{x}, t) = \upsilon_i(\mathbf{x}, t) = 0,\ t < 0$). Therefore, the integration range of the τ-integral included in (1.86) can be $(-\infty, t]$, and $\partial u_i(\mathbf{x}, t)/\partial t = \partial \upsilon_i(\mathbf{x}, t)/\partial t = 0,\ t < 0$. Furthermore. we perform

---

[10] This book deals mainly with isotropic media, and (1.30) may be used as the equation of motion. However, since the most general expression can be written formally in a shorter form, we use this expression here.

[11] See also Sect. 4.2.2. In this book, the integral variable τ is used consistently for the convolution in the time domain, and should not be confused with the stresses τ_{ij}.

Fig. 1.9 C_{ijkl} and ρ are the properties of the region V in Fig. 1.5. V includes the infinitesimal volume $dV = dxdydz$ in Fig. 1.4b, and infinitesimal areas dS with normal vectors n_i are distributed on S (modified from Udias [25] with permission of Cambridge University Press)

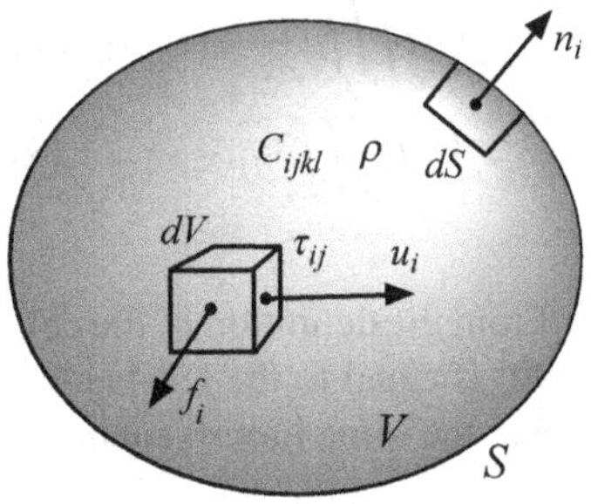

the **integration by parts** for the τ-integral on the left-hand side of (1.86) and obtain

$$\int_{-\infty}^{t} d\tau\, \rho \frac{\partial^2}{\partial \tau^2} u_i(\mathbf{x},\tau)\, \upsilon_i(\mathbf{x}, t-\tau) \tag{1.87}$$
$$= \left[\rho \frac{\partial}{\partial \tau} u_i(\mathbf{x},\tau)\, \upsilon_i(\mathbf{x}, t-\tau)\right]_{-\infty}^{t} - \int_{-\infty}^{t} d\tau\, \rho \frac{\partial}{\partial \tau} u_i(\mathbf{x},\tau) \frac{\partial}{\partial \tau} \upsilon_i(\mathbf{x}, t-\tau)\,.$$

Since $\partial u_i(\mathbf{x}, -\infty)/\partial\tau = 0$, the first term on the right-hand side of (1.87) vanishes if we can set $\upsilon_i(\mathbf{x}, 0) = 0$. Next, we perform the integration by parts for the volume integral of the first term on the right-hand side of (1.86), and apply the **divergence theorem of Gauss** [12] (See Fig. 1.9)

$$\iiint \nabla \cdot \mathbf{F}\, dV = \iint \mathbf{F} \cdot \mathbf{n}\, dS \Rightarrow \iiint \frac{\partial F_j}{\partial x_j} dV = \iint F_j n_j\, dS \tag{1.88}$$

so that

$$\iiint \frac{\partial}{\partial x_j} \tau_{ij}(\mathbf{x},\tau)\, \upsilon_i(\mathbf{x}, t-\tau)\, dV$$
$$= \iiint \frac{\partial}{\partial x_j} \left\{\tau_{ij}(\mathbf{x},\tau)\, \upsilon_i(\mathbf{x}, t-\tau)\right\} dV - \iiint \tau_{ij}(\mathbf{x},\tau) \frac{\partial}{\partial x_j} \upsilon_i(\mathbf{x}, t-\tau)\, dV$$
$$= \iint \tau_{ij}(\mathbf{x},\tau)\, \upsilon_i(\mathbf{x}, t-\tau)\, n_j\, dS - \iiint \tau_{ij}(\mathbf{x},\tau) \frac{\partial}{\partial x_j} \upsilon_i(\mathbf{x}, t-\tau)\, dV\,. \tag{1.89}$$

We summarize the above as

$$-\int_{-\infty}^{t} d\tau \iiint \rho \frac{\partial}{\partial \tau} u_i(\mathbf{x},\tau) \frac{\partial}{\partial \tau} \upsilon_i(\mathbf{x}, t-\tau)\, dV = \int_{-\infty}^{t} d\tau \iint \tau_{ij}(\mathbf{x},\tau)\, \upsilon_i(\mathbf{x}, t-\tau)\, n_j\, dS \tag{1.90}$$
$$- \int_{-\infty}^{t} d\tau \iiint \tau_{ij}(\mathbf{x},\tau) \frac{\partial}{\partial x_j} \upsilon_i(\mathbf{x}, t-\tau)\, dV + \int_{-\infty}^{t} d\tau \iiint \rho f_i(\mathbf{x},\tau)\, \upsilon_i(\mathbf{x}, t-\tau)\, dV\,.$$

Applying the formulation from (1.86) to (1.90) to the convolution of (1.84) and u_i, we obtain

$$-\int_{-\infty}^{t} d\tau \iiint \rho \frac{\partial}{\partial \tau} v_i(\mathbf{x}, \tau) \frac{\partial}{\partial \tau} u_i(\mathbf{x}, t-\tau)\, dV = \int_{-\infty}^{t} d\tau \iint \sigma_{ij}(\mathbf{x}, \tau)\, u_i(\mathbf{x}, t-\tau)\, n_j\, dS \quad (1.91)$$
$$- \int_{-\infty}^{t} d\tau \iiint \sigma_{ij}(\mathbf{x}, \tau) \frac{\partial}{\partial x_j} u_i(\mathbf{x}, t-\tau)\, dV + \int_{-\infty}^{t} d\tau \iiint \rho g_i(\mathbf{x}, \tau)\, u_i(\mathbf{x}, t-\tau)\, dV \,.$$

From the definition of the convolution in (1.85), we find that the left-hand side of (1.90) is equal to the left-hand side of (1.91). In addition, if u_i, $\tau_{ij}n_j$ and v_i, $\sigma_{ij}n_j$ satisfy the same **homogeneous boundary condition** on S, and if

$$l\, u_i + m\, \tau_{ij} n_j = 0\,, \quad l\, v_i + m\, \sigma_{ij} n_j = 0 \quad (1.92)$$

where l and m are constant, the first term on the right-hand side of (1.90) agrees with that of (1.91) [6]. We then substitute (1.81) and (1.83) into the second terms on the right-hand sides of (1.90) and (1.91), obtaining

$$\int_{-\infty}^{t} d\tau \iiint C_{ijkl} \frac{\partial}{\partial x_l} u_k(\mathbf{x}, \tau) \frac{\partial}{\partial x_j} v_i(\mathbf{x}, t-\tau)\, dV \,, \quad (1.93)$$

$$\int_{-\infty}^{t} d\tau \iiint C_{ijkl} \frac{\partial}{\partial x_l} v_k(\mathbf{x}, \tau) \frac{\partial}{\partial x_j} u_i(\mathbf{x}, t-\tau)\, dV \,. \quad (1.94)$$

Using the definition of the convolution in (1.85) again, we find that (1.93) yields

$$\int_{-\infty}^{t} d\tau \iiint C_{ijkl} \frac{\partial}{\partial x_l} v_k(\mathbf{x}, t-\tau) \frac{\partial}{\partial x_j} u_i(\mathbf{x}, \tau)\, dV \,. \quad (1.95)$$

We swap i and k, swap j and l, and use the third symmetry of the generalized elastic constants (1.17), finding that (1.95) agrees with (1.93). In (1.90) and (1.91), the left-hand sides agree, the first terms on the right-hand sides agree, and the second terms on the right-hand sides agree, so that the third terms on the right-hand sides must agree. Rewriting the third terms in the form of a convolution, we obtain the **reciprocity theorem**[12]

$$\iiint \rho f_i * v_i\, dV = \iiint \rho g_i * u_i\, dV \,. \quad (1.96)$$

Note that this simple form is established only in the case that the boundary condition is homogeneous and the initial condition $u_i(\mathbf{x}, 0) = v_i(\mathbf{x}, 0) = 0$ is satisfied. Although the reciprocity theorem is very well known, (1.96) is not often used directly, except in the next section and Sect. 3.3.3.

[12] This is equal to (5.13) of Hudson [6]. The reciprocity theorem is known as Betti's theorem, but Hudson noted that the equation in the form of a convolution as in (1.96) was published by D. Graffi in 1947.

1.3.3 Representation Theorem

$L(\mathbf{u}) = h$ is an ordinary differential equation, or a **partial differential equation** of **hyperbolic type** or of elliptic type, where $\mathbf{u}$ is an n-dimensional vector of independent variables. If it is solved in the region D and under a **homogeneous boundary condition** (Sect. 1.3.2) $B(\mathbf{u}) = 0$, there exists a **Green's function** $G(\mathbf{z}; \boldsymbol{\zeta})$ which satisfies $L(G(\mathbf{z}; \boldsymbol{\zeta})) = \delta(\mathbf{z} - \boldsymbol{\zeta})$ and $B(G(\mathbf{z}; \boldsymbol{\zeta})) = 0$, and

$$\mathbf{u} = \int_D G(\mathbf{z}; \boldsymbol{\zeta})\, h(\boldsymbol{\zeta})\, d\boldsymbol{\zeta} \tag{1.97}$$

is the unique solution of $L(\mathbf{u}) = h$ under $B(\mathbf{u}) = 0$.[13] In the definition of the Green's function, $\delta(\mathbf{z}) = \prod_{i=1}^{n} \delta(z_i)$, where $\delta(z)$ is a **delta function** defined as

$$\int_{z \in D} f(z)\, \delta(z)\, dz = f(0) \tag{1.98}$$

(See also Sect. 4.2.2). We consider a 1-D medium for simplicity as in Sect. 1.2.6, so that $\mathbf{u} \to u$ and the region D consists of the 1-D spatial domain x and time domain t. From (1.1) and (1.57) with a 1-D body force f, the equation of motion yields

$$\rho \frac{\partial^2 u}{\partial t^2} = \frac{\partial}{\partial x}\left(\gamma \frac{\partial u}{\partial x}\right) + \rho f \, . \tag{1.99}$$

Rewriting this in the form of $L(\mathbf{u}) = h$, we obtain

$$\begin{aligned} L(u) &= a\frac{\partial^2 u}{\partial t^2} + 2b\frac{\partial^2 u}{\partial t \partial x} + c\frac{\partial^2 u}{\partial x^2} - \frac{\partial \gamma}{\partial x}\frac{\partial u}{\partial x} \, , \\ h &= \rho f \, , \quad a = \rho \, , \quad b = 0 \, , \quad c = -\gamma \, . \end{aligned} \tag{1.100}$$

Since $b^2 - ac = \gamma\rho > 0$, the equation of motion for ground motions is a partial differential equation of hyperbolic type [24]. In addition, as the **continuity condition** (Sect. 3.1.1) and the **stress-free condition** (Sect. 3.1.2) are homogeneous boundary conditions, there exist Green's functions for ground motions.

In a 3-D case, the region D consists of the spatial region V (Fig. 1.9) and the time domain t, and the equation of motion

$$\rho \frac{\partial^2 u_i}{\partial t^2} = \frac{\partial}{\partial x_j}\left(C_{ijkl} \frac{\partial u_k}{\partial x_l}\right) + \rho f_i \tag{1.101}$$

[13] From *Dictionary of Mathematics, 2nd Edition* [17]. The *Physics Dictionary, Revised Edition* [20] mentions only partial differential equations of elliptic type.

from (1.81) and (1.82) is a set of simultaneous equations. If we apply the stress-free condition $B(u_i(\mathbf{x}, t)) = 0$ on the perimeter S of V, for

$$L_i(u_i(\mathbf{x}, t)) = h_i \,, \quad L_i(u_i(\mathbf{x}, t)) = \rho \frac{\partial^2 u_i}{\partial t^2} - \frac{\partial}{\partial x_j}\left(C_{ijkl} \frac{\partial u_k}{\partial x_l} \right), \quad h_i = \rho f_i \,, \tag{1.102}$$

there exist **tensor Green's functions** $\mathbf{G} = (G_{in}(\mathbf{x}, t; \boldsymbol{\xi}, \tau))$, which satisfy

$$L_i(G_{in}(\mathbf{x}, t; \boldsymbol{\xi}, \tau)) = \delta_{in}\delta(\mathbf{x} - \boldsymbol{\xi})\delta(t - \tau) \tag{1.103}$$

and $B(G_{in}(\mathbf{x}, t; \boldsymbol{\xi}, \tau)) = 0$. In (1.103), i indicates the direction of ground motion, while n indicates the direction of $\delta(\mathbf{x} - \boldsymbol{\xi})\delta(t - \tau)$, which will hereafter be termed **impulse**. $\delta(t - \tau)$ is the delta function defined by (1.98), but $\delta(\mathbf{x} - \boldsymbol{\xi})$ is a 3-D delta function like the n-dimensional $\delta(\mathbf{z})$ above, which can be defined as

$$\iiint A(\boldsymbol{\xi})\delta(\boldsymbol{\xi})dV = A(\mathbf{0}) \,, \quad \mathbf{0} = (0, 0, 0) \,. \tag{1.104}$$

We assume that the stress-free condition does not change in time t. Since (1.103) includes t only in the form of $t - \tau$, we obtain the **reciprocity relation** in the time domain [1]:

$$G_{in}(\mathbf{x}, t; \boldsymbol{\xi}, \tau) = G_{in}(\mathbf{x}, t - \tau; \boldsymbol{\xi}, 0) = G_{in}(\mathbf{x}, -\tau; \boldsymbol{\xi}, -t) \,. \tag{1.105}$$

Next, we compare (1.103) with (1.102) and know that the Green's function is the ground motion for the body force of the impulse. If Green's functions can be assumed to be zero at $t = 0$ (an initial condition), we can use them in (1.96). Two Green's functions $G_{im}(\mathbf{x}, t; \boldsymbol{\xi}_1, 0)$ and $G_{il}(\mathbf{x}, t; \boldsymbol{\xi}_2, 0)$ are substituted into u_i and υ_i in (1.96), and their impulses are also substituted into ρf_i and ρg_i in (1.96). The results of these substitutions, the definition of convolution in (1.85), and those of delta functions in (1.98) and (1.104), yield the reciprocity relation in the space domain

$$G_{lm}(\boldsymbol{\xi}_2, t; \boldsymbol{\xi}_1, 0) = G_{ml}(\boldsymbol{\xi}_1, t; \boldsymbol{\xi}_2, 0) \,. \tag{1.106}$$

When $\boldsymbol{\xi}_1$ and $\boldsymbol{\xi}_2$ are considered to be the locations of the earthquake source and the observation point, (1.106) indicates that the same ground motion is obtained even if these locations are swapped and the direction of impulse is swapped for the direction of displacement. In most cases the reciprocity relation in the space domain is referred to as the **reciprocity theorem**.

Finally, consider replacing only υ_i and ρg_i with the Green function and its impulse and leaving u_i and ρf_i in (1.96). If the initial condition $G_{in}(\mathbf{x}, 0; \boldsymbol{\xi}, \tau) = 0$ can be assumed, we can substitute $\upsilon_i = G_{in}(\mathbf{x}, t; \boldsymbol{\xi}, \tau)$ and $\rho g_i = \delta_{in}\delta(\mathbf{x} - \boldsymbol{\xi})\delta(t - \tau)$ into (1.96). The results of these substitutions, the definition of convolution in (1.85), and those of delta functions in (1.98) and (1.104), again yield

$$u_n(\boldsymbol{\xi}, t) = \int_{-\infty}^{+\infty} d\tau \iiint \rho f_i(\mathbf{x}, \tau) G_{in}(\mathbf{x}, t - \tau; \boldsymbol{\xi}, 0)\, dV(\mathbf{x}) . \qquad (1.107)$$

We then swap $\boldsymbol{\xi}$ and $\mathbf{x}$ and apply the reciprocity relation in the space domain (1.106), finding

$$\begin{aligned} u_n(\mathbf{x}, t) &= \int_{-\infty}^{+\infty} \tau \iiint \rho f_i(\boldsymbol{\xi}, \tau) G_{ni}(\mathbf{x}, t - \tau; \boldsymbol{\xi}, 0)\, dV(\boldsymbol{\xi}) \\ &= \iiint \rho f_i(\boldsymbol{\xi}, t) * G_{ni}(\mathbf{x}, t; \boldsymbol{\xi}, 0)\, dV(\boldsymbol{\xi}) . \end{aligned} \qquad (1.108)$$

This is called the **representation theorem**.[14] Although (1.97) mentioned as a typical solution almost agrees with (1.108), the difference between the directions of the body force and ground motion is not taken into consideration, so $\delta^n(\mathbf{z} - \boldsymbol{\zeta})$ does not include δ_{in}, which swaps the subscripts i and n.

An earthquake is a sudden movement in the ground that causes ground motions (Sect. 1.1), and is represented by the body force $\mathbf{f} = (f_i)$ in the equation of motion (1.30) or (1.101). Therefore, the term ρf_i in the representation theorem stands for the effect of the earthquake source (Sect. 1.1), in a rough sense. Meanwhile, the tensor Green function $\mathbf{G} = (G_{ni})$ is the impulse response of the medium (the Earth or a velocity structure), and represents the effect of ground motion propagation from the earthquake source to the observation point. In other words, the representation theorem indicates that the effect of the earthquake source and the effect of propagation can be separated, and ground motions can be reproduced by combining the effects evaluated individually. However, this meaning is rather symbolic, and it is rare to use (1.108) itself.

In this book, we will deal with the effect of the earthquake source mainly in Chap. 2 and the effect of propagation mainly in Chap. 3. However, since there are cases where the both cannot be distinguished, they are mutually referred to as necessary. The observation and processing required for the study of ground motions are described in Chap. 4.

Problems

1.1 'The medium is symmetric about the z-axis' (Sect. 1.2.3) means that the elastic properties are invariant even if the Cartesian coordinate system is rotated about the z-axis [13]. In other words, the generalized elastic constants $C_{i'j'k'l'}$ in the $x'y'z'$ coordinate system, obtained by rotating the original xyz coordinate system around the z-axis by an arbitrary angle θ, are equal to C_{ijkl} in the xyz coordinate system. Show that Eqs. (1.18) of Love [16] are obtained from this equality.

[14] Equation (1.108) agrees with Equation (3.1) of Aki and Richards [1] in the case of no internal surface.

1.2 In addition to the z-axis symmetry in Problem 1.1, obtain the equations from the x-axis symmetry. Show that these equations and Eqs. (1.18) result in Eqs. (1.19).

References

1. Aki, K., & Richards, P. G. (2002). *Quantitative seismology* (2nd ed., p. 700). Sausalito: University Science Books.
2. Arfken, G. B., & Weber, H. J. (1995). *Mathematical methods for physicists* (4th ed., p. 1029). San Diego: Academic Press.
3. Bathe, K. J. (1996). *Finite element procedures* (p. 1037). Englewood Cliffs: Prentice-Hall.
4. Einstein, A. (1916). Die Grundlage der allgemeinen Relativitätstheorie. *Annalen der Physik, 354*, 769–822.
5. Fung, Y. C. (1965). *Foundation of solid mechanics* (p. 525). Englewood Cliffs: Prentice-Hall.
6. Hudson, J. A. (1980). *The excitation and propagation of elastic waves* (p. 224). Cambridge: Cambridge University Press.
7. Kennett, B. L. N. (1983). *Seismic wave propagation in stratified* (p. 339). Cambridge: Cambridge University Press.
8. Kennett, B. L. N. (2001). *The seismic wavefield* (Vol. 1, p. 370). Cambridge: Cambridge University Press.
9. Koketsu, K. (2018). *Physics of seismic ground motion* (p. 353). Tokyo: Kindai Kagaku. [J]
10. Koketsu, K., & Kikuchi, M. (2000). Propagation of seismic ground motion in the Kanto basin, Japan. *Science, 288*, 1237–1239.
11. Kramers, H. A. (1927). La diffusion de la lumière par les atomes. *Atti del Congresso Internazionale dei Fisici (Como), 2*, 545–557.
12. Kreyszig, E. (1999). *Advanced engineering mathematics* (8th ed., p. 1156). New York: Wiley & Sons.
13. Kunio, T. (1977). *Fundamentals of solid mechanics* (p. 310). Tokyo: Baifukan. [J]
14. Landau, L. D., & Lifshitz, E. M. (1973). *Mechanics* (3rd ed., p. 224). Oxford: Butterworth-Heinemann.
15. Lay, T., & Wallace, T. C. (1995). *Modern global seismology* (p. 517). San Diego: Academic Press.
16. Love, A. E. H. (1906). *Treatise on the mathematical theory of elasticity* (2nd ed., p. 551). Cambridge: Cambridge University Press.
17. Mathematical Society of Japan (Ed.). (1968). *Dictionary of mathematics* (2nd ed., p. 1140). Tokyo: Iwanami Shoten. [J]
18. Moriguchi, S., Udagawa, K., & Hitotsumatsu, S. (1956). *Mathematical formulae I* (p. 318). Tokyo: Iwanami Shoten. [J]
19. Papoulis, A. (1962). *The Fourier integral and its applications* (p. 318). New York: McGraw-Hill.
20. Physics Dictionary Editorial Committee (ed.). (1992). *Physics dictionary* (rev ed., p. 2465). Tokyo: Baifukan. [J]
21. Saito, M. (2016). *The theory of seismic wave propagation* (p. 473). Tokyo: TERRAPUB.
22. Satô, Y. (1978). *Elastic wave theory* (p. 454). Tokyo: Iwanami Shoten. [J]
23. Takeuchi, H., & Saito, M. (1972). Seismic surface waves. *Seismology: Surface waves and earth oscillations* (pp. 217–295). San Diego: Academic Press.
24. Terasawa, K. (1954). *An introduction to mathematics for natural scientists* (rev ed., p. 722). Tokyo: Iwanami Shoten. [J]
25. Udias, A. (1999). *Principles of seismology* (p. 475). Cambridge: Cambridge University Press.
26. Utsu, T. (2001). *Seismology* (3rd ed., p. 376). Tokyo: Kyoritsu Shuppan. [J]
27. Webster, A. G. (1927). *Partial differential equations of mathematical physics* (p. 440). Leipzig: B.G. Teubner.
28. Weyl, H. (1919). Ausbreitung elektromagnetischer Wellen über einem ebenen Leiter. *Annalen der Physik, 365*, 481–500.

Chapter 2
The Effect of Earthquake Source

Abstract This chapter explains the "effect of an earthquake source" on ground motion. Section 2.1 first briefly describes the history of seismology, where an earthquake is interpreted as a sudden movement of rocks along a source fault ("elastic rebound theory") due to strain accumulated as a result of the interaction of the Earth's plates ("plate tectonics"). The representations of source faults, which comprise the "fault geometry", the "body force equivalent" and the concept of a "double couple", are explained first. The theoretical development for ground motion from a point force and a point source are then reviewed. In Sect. 2.2, the results are extended to a "cylindrical wave expansion", which plays a major role in this book. The extension begins with a vertical strike-slip fault and is then applied to an inclined strike-slip fault, a vertical dip-slip fault, and an inclined dip-slip fault. The full set of extensions are combined for an arbitrary fault slip. Section 2.3 explains various methods and models for the analysis of a source fault, such as "hypocenter determination", "fault plane solution", "CMT inversion", "Haskell model", "source inversion", "characterized source model", and "directivity effect".

Keywords Earthquake source · Elastic rebound theory · Source fault · Double couple · Moment tensor · Source inversion

2.1 Representation of Earthquake Source

2.1.1 Discovery of Earthquake Source

As described in Sect. 1.1, an earthquake is a "sudden movement" in the ground that causes ground motions. **Elastic rebound theory** was proposed at the beginning of the 20th century to explain this sudden movement. During that period, a large earthquake, the **San Francisco earthquake** (1906, M 8.3), occurred on the West Coast of the United States. This event caused right-lateral displacements of up to 6.4 m to appear on the ground surface over a distance of 300 km in the northern part of the **San Andreas fault**. From triangular measurements performed before and after the earthquake, Reid [61] analyzed the **crustal deformation** (Sect. 3.1.12)

K. Koketsu, *Ground Motion Seismology*, Advances in Geological Science,
https://doi.org/10.1007/978-981-15-8570-8_2

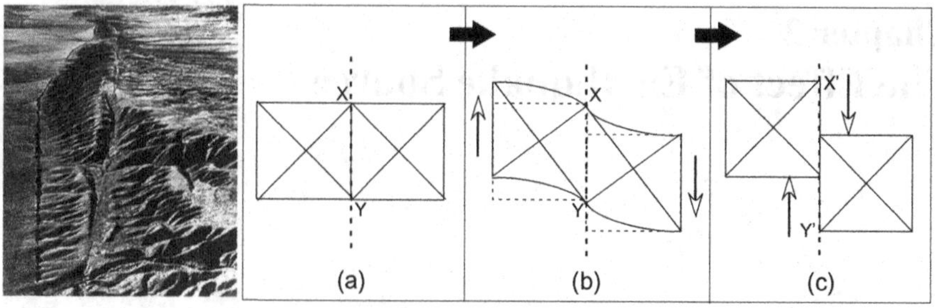

Fig. 2.1 Left: Aerial photograph of the San Andreas fault (reprinted from USGS [72]); Right: Schematic diagrams illustrating the elastic rebound theory (modified from Yamashita [78] with permission of the author)

associated with the earthquake, and proposed the elastic rebound theory as the cause of the earthquake, as follows.

In the area where a fault, that is a weak plane, exists in the ground (Fig. 2.1a; XY is a fault), it is assumed that forces are applied to the rocks on both sides of the fault in opposite directions (Fig. 2.1b). Even if the forces are not very large, they are applied over many years and they will greatly distort the rocks (Fig. 2.1b). When the **strain** due to this distortion reaches a limit, the rocks on both sides move sharply along the fault in the direction which releases the strain (Fig. 2.1c, X'−X or Y−Y' is the amount of displacement). This phenomenon is known as an earthquake and causes ground motions [73, 78]. The sudden movement is termed a **fault rupture** or **faulting**, and this is now an established interpretation of the mechanism operating in the source region.

When the elastic rebound theory was originally proposed, it was straightforwardly accepted in the United States and Europe, but 'in Japan, the dominant view was that earthquakes were not so simple' [73]. However, G. Miyatake, a journalist who published the six-volume book "*Earthquake Disaster Pictorial*" immediately after the **Kanto earthquake** (1923, *M* 7.9), is said to have stated the following in its first volume [69]: 'There are three types of earthquakes: (1) volcanic earthquakes, (2) collapse earthquakes, and (3) fault earthquakes, and Japan's earthquakes generally fall into category (3).[1]' He also wrote that the Kanto earthquake 'is also a fault earthquake. According to expert scholars' opinions, it is a large "landslide" that occurred on the seafloor of the Pacific Ocean, 80 to 120 km south of Tokyo.' In modern terminology, a **landslide** refers to a slow movement of surface materials, but as G. Miyatake is considered to be expressing a rapid slip of rocks, these sentences almost reproduce the faulting that forms an integral part of the elastic rebound theory.

Meanwhile, among Japanese researchers in the 1920s and 30s, a dispute was ongoing as to whether the cause of the earthquake was a fault or magma intrusion. However, through a series of studies by Hirokichi Honda, the elastic rebound theory

[1] According to Ohnaka [59], Professor S. Nakamura of the Tohoku Imperial University advocated a theory of fault earthquakes for the first time.

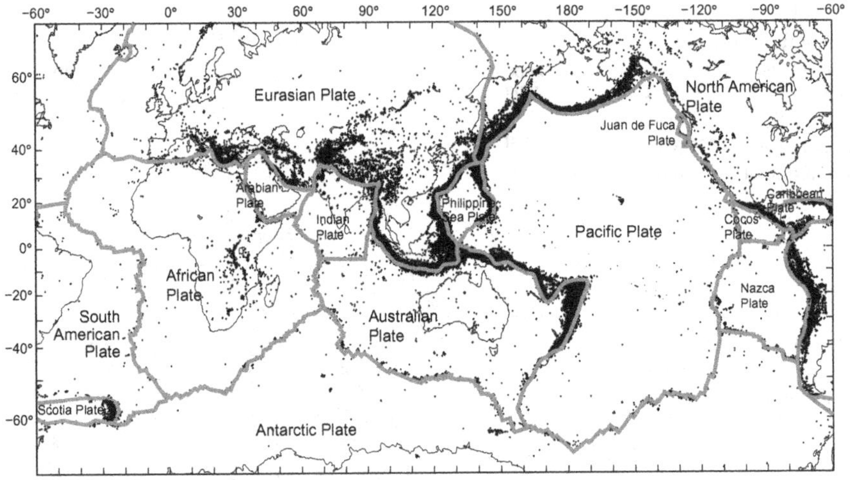

Fig. 2.2 Earth's main tectonic plates (drawn by Toshikatsu Yoshii). The thick gray curves indicate the plate boundaries, and the black dots indicate the epicenters of shallow, medium-sized or larger earthquakes (depth ≤100 km, from 1991 to 2010), as determined by the International Seismological Centre (modified from Koketsu [40])

was gradually accepted in Japan. Honda [26] showed that the mechanical system equivalent to the faulting process consists of two **couples of forces** (Sect. 2.1.2), which later became an established theory. Most studies during this period computed the **radiation patterns** (Sect. 2.3.2) of ground motions for various mechanical systems, and searched for the radiation pattern that best matched the observations of an actual earthquake. Theoretical studies of radiation patterns have been conducted for quite some time. The earliest study by Nakano [57] was published in 1923. This paper was destroyed by fire in the **Kanto earthquake disaster** shortly after printing, but was later rediscovered thanks to the remaining handwritten notes [27].

Following this, researchers in the United States and Europe disagreed with the two couples of forces proposed by Honda [26], insisting on a single couple of forces arising simply from the movement of rocks along a fault (Fig. 2.1c). This second dispute came to an end in the 1960s [73] (See Sect. 2.1.2). As mentioned above, although the elastic rebound theory explaining the occurrence of earthquakes was accepted, there still remained the problem of the nature of the forces in the theory (arrows in Fig. 2.1b, c). The nature of the forces was elucidated through the concept of **plate tectonics**, which was developed rapidly from the 1950s, and was established by the 1960s, similarly to the elastic rebound theory. As shown in Fig. 2.2, the surface of the Earth is made up of dozens of large rocky **plates** with a thickness of several tens of kilometers, up to around 200 km. These plates move in independent directions at a very slow speed of several centimeters per year (**plate motion**), and therefore collide with each other, in the process of which one may be forced (subduct) below the other. It is thought that such a collision or **subduction** is producing the forces responsible for generating earthquakes. In Fig. 2.2, the epicenters of shallow, medium-sized

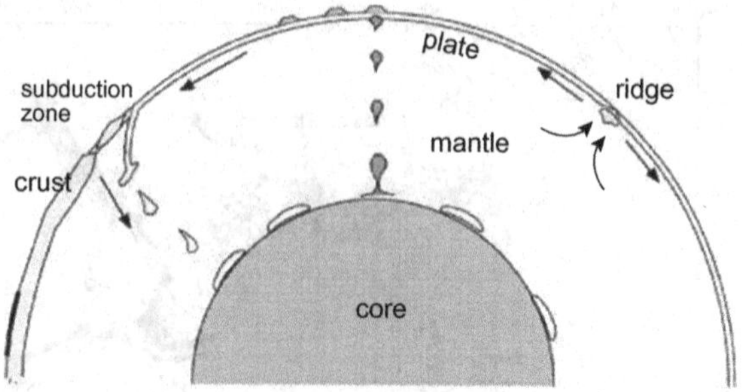

Fig. 2.3 Crust, mantle, and core of the Earth. Arrows indicate circulation within the mantle, which causes plate motions (modified from Yamaoka [77] with permission of the author)

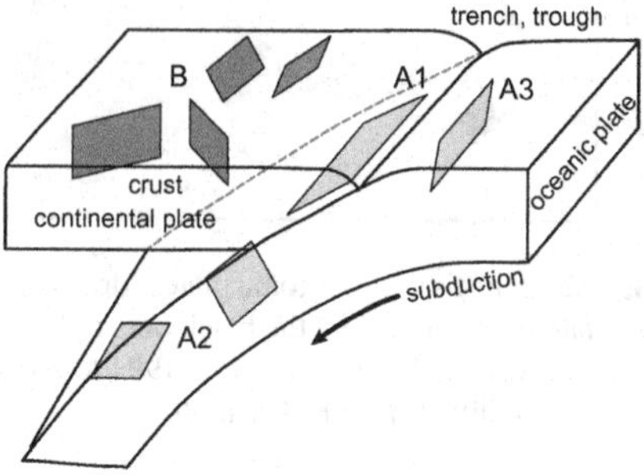

Fig. 2.4 Illustration of earthquake occurrences in a subduction zone (modified from the Earthquake Research Committee [16])

or larger earthquakes (depth $\leq$100 km, from 1991 to 2010), as determined by the **International Seismological Centre**, are also plotted. From their distribution, it can be understood that earthquakes occur predominantly along **plate boundaries**, confirming that earthquakes are caused by the collision or subduction of plates. Plate motions are thought to be caused by circulation within the Earth's **mantle** (Fig. 2.3).

As an example, the occurrence of earthquakes in a **subduction zone** (the area where subduction occurs) illustrates this mechanism (Fig. 2.4). As subduction occurs due to the difference in the densities of two plates, a subduction zone can be formed at the boundary between the ocean and the continent, such as the east coast of Japan. The place where the oceanic plate begins to subduct is called a trench or trough. The subduction directly affects a large fault that constitutes the "plate boundary", producing **plate boundary earthquakes** (A1 in Fig. 2.4). In addition, subduction also directly affects the inside of the oceanic plate near the plate boundary, producing in-slab earthquakes (A2 in Fig. 2.4), where the "slab" is the subducted part of the oceanic plate, and outer rise earthquakes (A3 in Fig. 2.4), where "outer rise" is the rising topography outside the trench or trough. These earthquakes are large in scale because they result from the direct effects of the subduction, and have short **return periods**.

Subduction also indirectly affects the interior of the continental plate away from the plate boundary, generating **crustal earthquakes** (B in Fig. 2.4). These are confined to the **crust** on the continental side, which is the outermost part of the Earth (Fig. 2.3), because the mantle below the crust is very viscous as a result of heat from the decay of radioactive elements and heat transfer from the core, therefore no earthquakes occur there. Since crustal earthquakes are the result of indirect effects of the subduction process, they tend to be smaller and their return periods are shorter than plate boundary earthquakes. Among faults related to crustal earthquakes, those whose existence is recognized from the ground surface are called **active faults**.

2.1.2 Representation of Source Fault

As defined in Sect. 1.1, we replace the "fault" in the elastic rebound theory with the term **source fault**. The simplest representation of a source fault is to treat the fault plane as a flat plane and to geometrically represent the plane and the direction of slip along the plane. The fault plane is first assumed to be a rectangle with horizontal upper and lower sides. Its direction is specified by the **strike**, which is the azimuth ϕ_s measured clockwise from north as viewed from above, on the horizontal plane including the upper side of the rectangle (Fig. 2.5a). The inclination is specified by the angle δ (**dip angle**) with respect to this horizontal plane. We set the right-handed **Cartesian coordinate system** with the z-axis position taken to be downwards and the x-axis matched to the strike.[2] δ is measured from the positive direction of the y-axis in the horizontal plane, so that it is 90° or less. That is, in Fig. 2.5a, the fault plane is not specified as having a strike of 210° and a dip angle of 135°, but as having a strike of 30° and a dip angle of 45°.

It is rare for the dip angle δ to be perfectly 90°, and the fault plane is generally inclined to some extent. Therefore, we can define a **hanging wall** as the rocks lying above the fault plane and a **foot wall**, as those lying below the fault plane. The rocks on the positive side of the y-axis are always the hanging wall, because the dip angle from this side must be 90° or less. The displacement due to the faulting (D in Fig. 2.5a) is called **slip**, **dislocation**, or **fault displacement**. The slip D is measured as a displacement of the hanging wall relative to the foot wall (Fig. 2.5b). The angle λ between the direction of D and the strike (x-axis) is called the **slip angle** or the **rake angle**. The slip angle, the strike, the dip angle, and the depth h of the fault plane (Fig. 2.12) may be collectively referred to as the **fault parameters**.

Furthermore, earthquake faults are classified as shown in Table 2.1 according to the slip angle, and a schematic diagram of each fault type is shown in Fig. 2.6. Left-lateral strike slip faults and right-lateral strike slip faults are collectively referred to as **strike slip faults**, while reverse faults and normal faults are collectively referred to as **dip slip fault**. In Table 2.1, if the slip angle is exactly equal to each value, the source fault will be purely of a single kind, however such cases are rare. For example, if the slip angle is exactly 0°, we have a pure left-lateral strike fault, but components of reverse faulting are involved if there is a slip angle of several degrees, as shown

---

[2]Most other textbooks take the z-axis (x_3-axis) to be upwards. In addition, in observations, the north (N) direction is often set to be the x-axis, but in this book this coordinate system is adopted in order to make the large variations of depth in the Earth easier to be expressed, and not to make solutions complicated. However, for moment tensors, expressions with the north direction as the x-axis are commonly used, which will be described in Sect. 2.3.3.

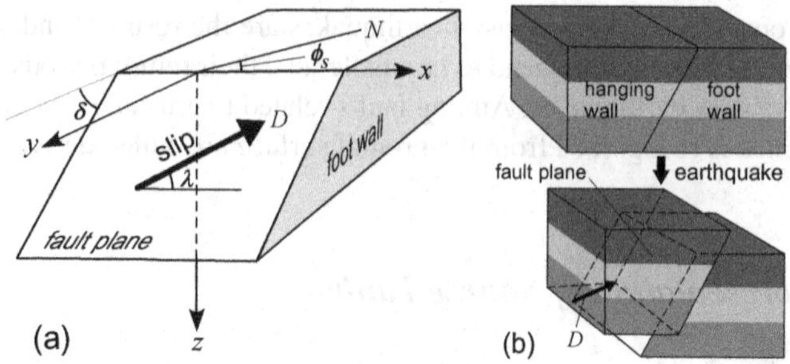

Fig. 2.5 **a** Representation of a source fault, **b** Hanging and foot walls, and the definition of slip (modified from Koketsu [41] with permission of Kindai Kagaku)

Table 2.1 Slip angles and types of source faults

$\lambda \sim 0°$	**left-lateral strike slip fault** foot wall moves to the left as seen from the hanging wall (same from the foot wall)
$\lambda \sim 180°$	**right-lateral strike slip fault** foot wall moves to the right as seen from the hanging wall (same from the foot wall)
$\lambda \sim 90°$	**reverse fault** hanging wall moves upwards against gravity
$\lambda \sim 270°$	**normal fault** hanging wall moves downwards according to gravity

in Fig. 2.5a. In addition, when the slip angle is intermediate and is indistinguishable from a strike slip fault or a dip slip fault, it is sometimes called an **oblique slip fault**.

Among the faults represented in Fig. 2.4, the source faults producing plate boundary earthquakes (A1) are typically reverse faults. The source faults producing in-slab earthquakes (A2) and outer rise earthquakes (A3) include normal faults as well as other types. Those producing crustal earthquakes also include various types, but regional characteristics, such as a predominance of strike slip faults in southwest Japan and reverse faults in northeast Japan, are important and have been observed globally.

If an earthquake is not large, or is large but far away from the observation point, its source fault can be assumed to be a point. Such source faults are termed **point sources**. Even if the source fault cannot be regarded as a point source, it can be divided into **subfaults**, and these can be represented by point sources. Point sources can therefore be said to be fundamental for the modeling of an earthquake source.

For example, we assume a point source for a vertical **left-lateral strike slip fault** (source fault in Fig. 2.5, with $\delta = 90°$ and $\lambda = 0°$) as shown in the Fig. 2.7. As for researchers who opposed the results of Honda [26] (2.1.1), it can be intuitively imagined that the mechanical expression of this **fault rupture** (Sect. 1.1) is a single **couple of forces** corresponding to the fault displacement directions, as indicated by the large and small arrows in Fig. 2.7. However, although a moment is generated by

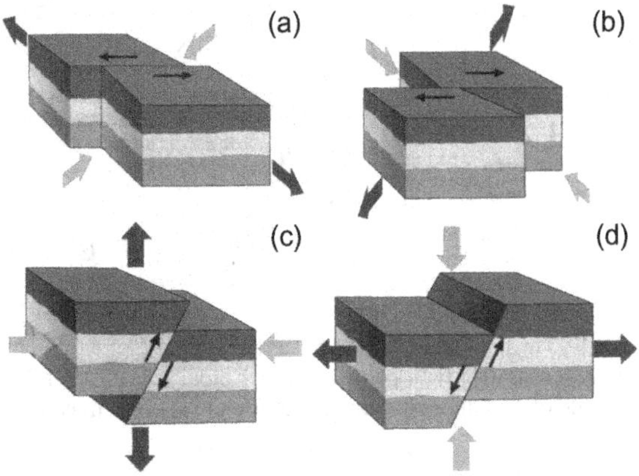

Fig. 2.6 Schematic diagrams of the types of source faults (reprinted from Yamashita [78] with permission of the author). **a** Left-lateral strike slip fault, **b** right-lateral strike slip fault, **c** reverse fault, and **d** normal fault. The small arrows along the fault plane represent the slips, and the large arrows around them represent the equivalent forces (Sect. 2.3.2)

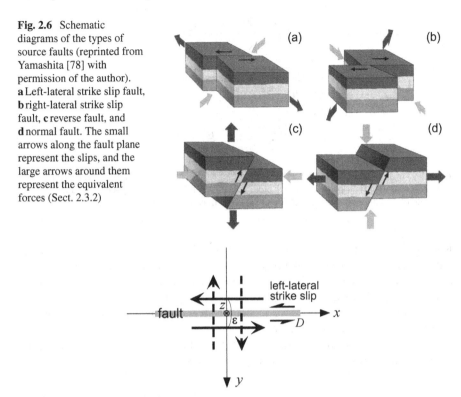

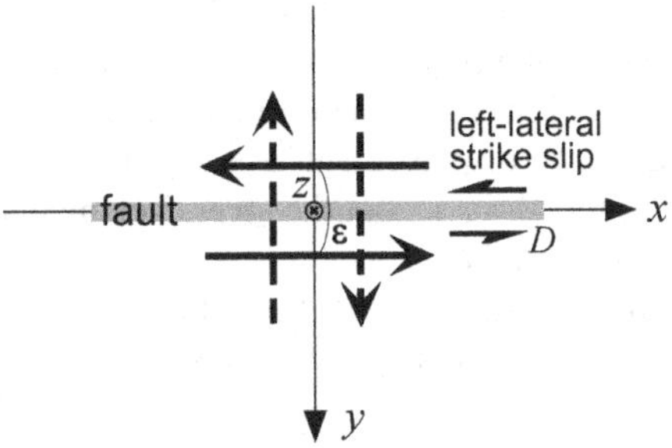

Fig. 2.7 Point source of a vertical left-lateral strike slip fault and its equivalent force system source (bird's eye view, modified from Koketsu [41] with permission of Kindai Kagaku). The force system is a double couple, the arms of which have a length of ε. The z-axis is perpendicular to the page and extends beyond it

this couple of forces, the rocks do not actually rotate. This implies that another couple of forces (dashed arrows in Fig. 2.7) exists, canceling the moment. The second couple has the same magnitude but the opposite direction of rotation to the first couple. Two couples of forces should therefore exist as a force system equivalent to the fault rupture, and this is called a **double couple**. The magnitude of the fault rupture is represented by the magnitude of one of these couples, which is called a **seismic moment**.[3]

Maruyama [52] mathematically proved this physical interpretation, by applying the **representation theorem** (Sect. 1.3.3) to the problem where the Earth includes a source fault as an internal discontinuity. However, his explicit formulation of the Green's function at an early stage made the equations complicated, and makes it difficult to see the results. Here, we will therefore show the proof by Burridge and

[3]It is said that Aki [1] proposed the expression "seismic moment", however in this paper he used only the expression "earthquake moment".

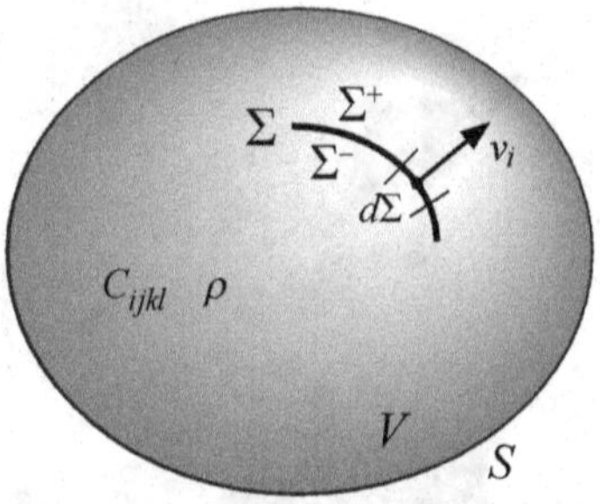

Fig. 2.8 The fault plane Σ is added as an internal discontinuity to the region V of Fig. 1.9 (reprinted from Koketsu [41] with permission of Kindai Kagaku). There are two surfaces, Σ^+ and Σ^-, on each side of Σ, and their normal vector $\boldsymbol{\nu} = (\nu_i)$ is defined to point from Σ^- to Σ^+

Knopoff [8], which was published later[4]. Since the fault plane of the source fault represented by Σ, is a discontinuity in the continuum of the Earth, the surface is Σ^+ and Σ^- on each side of Σ (Fig. 2.8). The normal vector $\boldsymbol{\nu} = (\nu_i)$ of Σ is defined to point from Σ^- to Σ^+. The displacement $\mathbf{u} = (u_i)$ is discontinuous and differs between the Σ^+ side and the Σ^- side, but the **stress vector** in the normal direction $\mathbf{T} = (T_i)$ is assumed to be continuous.

Since, compared with the setting in Sect. 1.3.3 (Fig. 1.9), not only $\iint dS$ but also $\iint d\Sigma$ is added as an area component, (1.89) yields

$$\iiint \frac{\partial}{\partial x_j}\tau_{ij}(\mathbf{x},\tau)\,\upsilon_i(\mathbf{x},t-\tau)\,dV = -\iiint \tau_{ij}(\mathbf{x},\tau)\frac{\partial}{\partial x_j}\upsilon_i(\mathbf{x},t-\tau)\,dV$$
$$+\iint \tau_{ij}(\mathbf{x},\tau)\,\upsilon_i(\mathbf{x},t-\tau)\,n_j\,dS + \iint \tau_{ij}(\mathbf{x},\tau)\,\upsilon_i(\mathbf{x},t-\tau)\,\nu_j\,d\Sigma\,. \quad (2.1)$$

As in (1.89), if S is a free surface, the stress-free condition $\tau_{ij}\,n_j = 0$ is applied[5], the second term on the right-hand side of (2.1) disappears. Therefore, (1.90) yields

$$-\int_{-\infty}^{t} d\tau \iiint \rho\frac{\partial}{\partial \tau}u_i(\mathbf{x},\tau)\frac{\partial}{\partial \tau}\upsilon_i(\mathbf{x},t-\tau)\,dV = -\int_{-\infty}^{t} d\tau \iiint \tau_{ij}(\mathbf{x},\tau)\frac{\partial}{\partial x_j}\upsilon_i(\mathbf{x},t-\tau)\,dV$$
$$+\int_{-\infty}^{t} d\tau \iiint \rho f_i(\mathbf{x},\tau)\,\upsilon_i(\mathbf{x},t-\tau)\,dV + \int_{-\infty}^{t} d\tau \iint \tau_{ij}(\mathbf{x},\tau)\,\upsilon_i(\mathbf{x},t-\tau)\,\nu_j\,d\Sigma \quad (2.2)$$

[4]As mentioned in the Preface, this book aims to emphasize explicit formulations, but this section, where abstract concepts are explained, emphasizes the brevity of expressions, as in Sect. 1.3.3.

[5]This assumption is fine if V is the real Earth or a computational region of the finite element method (Sect. 3.2.5), but in a 1-D velocity structure (Sect. 3.1) we assume that S is far enough away.

Similarly, (1.91) yields

$$-\int_{-\infty}^{t} d\tau \iiint \rho \frac{\partial}{\partial \tau} v_i(\mathbf{x}, \tau) \frac{\partial}{\partial \sigma} u_i(\mathbf{x}, t-\tau)\, dV = -\int_{-\infty}^{t} d\tau \iiint \sigma_{ij}(\mathbf{x}, \tau) \frac{\partial}{\partial x_j} u_i(\mathbf{x}, t-\tau)\, dV$$
$$+ \int_{-\infty}^{t} d\tau \iiint \rho g_i(\mathbf{x}, \tau)\, u_i(\mathbf{x}, t-\tau)\, dV + \int_{-\infty}^{t} d\tau \iint \sigma_{ij}(\mathbf{x}, \tau)\, u_i(\mathbf{x}, t-\tau)\, \nu_j\, d\Sigma \quad (2.3)$$

We then compare (2.2) and (2.3) in the same way as for (1.90) and (1.91), and substitute (1.81) and (1.83) into the results of the comparison, obtaining the **reciprocity theorem** (Sect. 1.3.2) in the case where the internal discontinuity Σ exists in V:

$$\iiint \rho f_i * v_i\, dV + \iint \tau_{ij} \nu_j * v_i\, d\Sigma$$
$$= \iiint \rho g_i * u_i\, dV + \iint C_{ijkl} \frac{\partial v_k}{\partial x_l} \nu_j * u_i\, d\Sigma\,. \quad (2.4)$$

As in Sect. 1.3.3, we substitute $v_i = G_{in}(\mathbf{x}, t; \boldsymbol{\xi}, \tau)$ and $\rho g_i = \delta_{in}\delta(\mathbf{x} - \boldsymbol{\xi})\delta(t - \tau)$ into (2.4), swap $\boldsymbol{\xi}$ and $\mathbf{x}$, and apply the reciprocity relation in the spatial domain (1.106), obtaining the equation[6]

$$u_n(\mathbf{x}, t) = \int_{-\infty}^{+\infty} d\tau \iiint \rho f_i(\boldsymbol{\xi}, \tau) G_{ni}(\mathbf{x}, t-\tau; \boldsymbol{\xi}, 0)\, dV(\boldsymbol{\xi})$$
$$+ \int_{-\infty}^{+\infty} d\tau \iint \nu_j \tau_{ij}(\boldsymbol{\xi}, \tau) G_{ni}(\mathbf{x}, t-\tau; \boldsymbol{\xi}, 0)\, d\Sigma(\boldsymbol{\xi})$$
$$- \int_{-\infty}^{+\infty} d\tau \iint \nu_j u_i(\boldsymbol{\xi}, \tau) C_{ijkl} \frac{\partial}{\partial \xi_l} G_{nk}(\mathbf{x}, t-\tau; \boldsymbol{\xi}, 0)\, d\Sigma(\boldsymbol{\xi}) \quad (2.5)$$

where $\boldsymbol{\xi} = (\xi_x, \xi_y, \xi_z)$. As mentioned earlier, the surface integral $\iint d\Sigma$ is to be replaced with $\iint d\Sigma\big|_{\Sigma^+}$ and $\iint d\Sigma\big|_{\Sigma^-}$. Since the normal vector $\boldsymbol{\nu} = (\nu_i)$ is defined to point from Σ^- to Σ^+, ν_j of (2.5) is fine for $\iint d\Sigma^-$, but $\nu_j \to -\nu_j$ must be applied for $\iint d\Sigma^+$. $\nu_j \tau_{ij}$ is the component T_i of the stress vector in the normal direction.

From the above, the second term on the right-hand side of (2.5) yields

$$+ \int_{-\infty}^{+\infty} d\tau \iint (-\nu_j)\tau_{ij}(\boldsymbol{\xi}, \tau)\big|_{\Sigma^+} G_{ni}(\mathbf{x}, t-\tau; \boldsymbol{\xi}, 0)\, d\Sigma(\boldsymbol{\xi})$$
$$+ \int_{-\infty}^{+\infty} d\tau \iint \nu_j \tau_{ij}(\boldsymbol{\xi}, \tau)\big|_{\Sigma^-} G_{ni}(\mathbf{x}, t-\tau; \boldsymbol{\xi}, 0)\, d\Sigma(\boldsymbol{\xi})$$
$$= -\int_{-\infty}^{+\infty} d\tau \iint [T_i(\boldsymbol{\xi}, \tau)] G_{ni}(\mathbf{x}, t-\tau; \boldsymbol{\xi}, 0)\, d\Sigma(\boldsymbol{\xi})\,. \quad (2.6)$$

[6]This agrees with Eq. (8) of Burridge and Knopoff [8] if $f_i \to \rho f_i$, $p \to k$, $q \to l$, $\mathbf{x} \to \boldsymbol{\xi}$, $\mathbf{y} \to \mathbf{x}$, $s \to t$, $t \to \tau$, and the reciprocity relation in the time domain (1.105) are applied.

Similarly, the third term yields

$$
\begin{aligned}
&-\int_{-\infty}^{+\infty} d\tau \iint (-\nu_j) u_i(\boldsymbol{\xi}, \tau)\big|_{\Sigma^+} C_{ijkl} \frac{\partial}{\partial \xi_l} G_{nk}(\mathbf{x}, t-\tau; \boldsymbol{\xi}, 0)\, d\Sigma(\boldsymbol{\xi}) \\
&-\int_{-\infty}^{+\infty} d\tau \iint \nu_j u_i(\boldsymbol{\xi}, \tau)\big|_{\Sigma^-} C_{ijkl} \frac{\partial}{\partial \xi_l} G_{nk}(\mathbf{x}, t-\tau; \boldsymbol{\xi}, 0)\, d\Sigma(\boldsymbol{\xi}) \\
&= +\int_{-\infty}^{+\infty} d\tau \iint [u_i(\boldsymbol{\xi}, \tau)] \nu_j C_{ijkl} \frac{\partial}{\partial \xi_l} G_{nk}(\mathbf{x}, t-\tau; \boldsymbol{\xi}, 0)\, d\Sigma(\boldsymbol{\xi}) \,. \qquad (2.7)
\end{aligned}
$$

Since $[u_i] = u_i|_{\Sigma^+} - u_i|_{\Sigma^-}$ is non-zero because of the discontinuity of displacement at Σ, and $[T_i] = T_i|_{\Sigma^+} - T_i|_{\Sigma^-} = 0$ because of the continuity of the stress vector at Σ, we obtain the equation[7]

$$
\begin{aligned}
u_n(\mathbf{x}, t) = &\int_{-\infty}^{+\infty} d\tau \iiint \rho f_i(\boldsymbol{\xi}, \tau) G_{ni}(\mathbf{x}, t-\tau; \boldsymbol{\xi}, 0)\, dV(\boldsymbol{\xi}) \\
&+ \int_{-\infty}^{+\infty} d\tau \iint [u_i(\boldsymbol{\xi}, \tau)] \nu_j C_{ijkl} \frac{\partial}{\partial \xi_l} G_{nk}(\mathbf{x}, t-\tau; \boldsymbol{\xi}, 0)\, d\Sigma(\boldsymbol{\xi}) \,. \qquad (2.8)
\end{aligned}
$$

From the definition of the **delta function** (1.104),

$$
G_{ni}(\mathbf{x}, t-\tau; \boldsymbol{\xi}, 0) = \iiint \delta(\boldsymbol{\eta} - \boldsymbol{\xi}) G_{ni}(\mathbf{x}, t-\tau; \boldsymbol{\eta}, 0) dV(\boldsymbol{\eta}) \,. \qquad (2.9)
$$

We partially differentiate both sides of (2.9) by ξ_l and use

$$
\frac{\partial \delta(\boldsymbol{\eta} - \boldsymbol{\xi})}{\partial \xi_l} = -\frac{\partial \delta(\boldsymbol{\eta} - \boldsymbol{\xi})}{\partial \eta_l} \,. \qquad (2.10)
$$

We then have

$$
\frac{\partial}{\partial \xi_l} G_{ni}(\mathbf{x}, t-\tau; \boldsymbol{\xi}, 0) = -\iiint \frac{\partial \delta(\boldsymbol{\eta} - \boldsymbol{\xi})}{\partial \eta_l} G_{ni}(\mathbf{x}, t-\tau; \boldsymbol{\eta}, 0) dV(\boldsymbol{\eta}) \,. \qquad (2.11)
$$

We substitute (2.11) into the second term on the right-hand side of (2.8), swap the order of the integration, change the integral variable from $\boldsymbol{\xi}$ to $\boldsymbol{\eta}$ in the first term on the right-hand side, and change the subscript from i to k, obtaining

$$
\begin{aligned}
u_n(\mathbf{x}, t) = &\int_{-\infty}^{+\infty} d\tau \iiint \rho f_k(\boldsymbol{\eta}, \tau) G_{nk}(\mathbf{x}, t-\tau; \boldsymbol{\eta}, 0)\, dV(\boldsymbol{\eta}) \qquad (2.12) \\
+ \int_{-\infty}^{+\infty} d\tau &\iiint \left\{ -\iint [u_i(\boldsymbol{\xi}, \tau)] C_{ijkl} \nu_j \frac{\partial}{\partial \eta_l} \delta(\boldsymbol{\eta} - \boldsymbol{\xi})\, d\Sigma(\boldsymbol{\xi}) \right\} G_{nk}(\mathbf{x}, t-\tau; \boldsymbol{\eta}, 0)\, dV(\boldsymbol{\eta}).
\end{aligned}
$$

---

[7]This agrees with Eq. (3.3) of Aki and Richards [3] if $p \to i, k, q \to l$, $f_p \to \rho f_i$, and $[T_p] = 0$ are applied.

Bearing in mind that the first term on the right-hand side of (2.12) represents displacements by an arbitrary body force $\mathbf{f} = (f_k)$, we compare this term and the second term, replace τ and η with t and $\mathbf{x} = (x_l) = (x, y, z)$, and then obtain the **body force equivalent**[8]

$$\rho f_k(\mathbf{x}, t) = -\iint [u_i(\boldsymbol{\xi}, t)] C_{ijkl} \nu_j \frac{\partial}{\partial x_l} \delta(\mathbf{x} - \boldsymbol{\xi})\, d\Sigma(\boldsymbol{\xi}) \tag{2.13}$$

which can represent the displacements due to the discontinuity Σ.

If we take the point source in Fig. 2.7 as a source fault Σ, we have $\iint d\Sigma(\boldsymbol{\xi}) = \iint_\Sigma d\xi_x d\xi_z\big|_{\xi_y=0}$. As Σ is sufficiently small because of a point source, we can assume μ to be constant in Σ. The slip in Fig. 2.7 is in the x-direction. This indicates $i \equiv x$ and $j \equiv y$, so that $[u_x] = D(\mathbf{x}, t)$, $[u_y] = 0$, $[u_z] = 0$, and $\boldsymbol{\nu} = (0, 1, 0)$. In addition, if the continuum is isotropic, C_{xykl} becomes zero except for $C_{xyxy} = C_{xyyx} = \mu$. Therefore, in the case of $k = x$, only the terms of $l = y$ remain as

$$\begin{aligned} \rho f_x(\mathbf{x}, t) &= -\iint_\Sigma \mu D(\xi_x, 0, \xi_z, t) \delta(x - \xi_x) \frac{\partial \delta(y)}{\partial y} \delta(z - \xi_z)\, d\xi_x d\xi_z \\ &= -\mu D(\mathbf{x}_0, t) \frac{\partial \delta(y)}{\partial y}, \quad \mathbf{x}_0 = (x, 0, z)\,. \end{aligned} \tag{2.14}$$

Similarly, in the case of $k = y$, only the terms of $l = x$ remain as

$$\begin{aligned} \rho f_y(\mathbf{x}, t) &= -\iint_\Sigma \mu D(\xi_x, 0, \xi_z, t) \frac{\partial \delta(x - \xi_x)}{\partial x} \delta(y) \delta(z - \xi_z)\, d\xi_x d\xi_z \\ &= -\frac{\partial}{\partial x} \iint_\Sigma \mu D(\xi_x, 0, \xi_z, t) \delta(x - \xi_x) \delta(y) \delta(z - \xi_z)\, d\xi_x d\xi_z \\ &= -\mu \frac{\partial D(\mathbf{x}_0, t)}{\partial x} \delta(y)\,. \end{aligned} \tag{2.15}$$

$\rho f_z(\mathbf{x}, t)$ is identical to zero.

We substitute (2.14), (2.15), and $\rho f_z(\mathbf{x}, t) = 0$ into the representation theorem with the summation convention of the subscript i written down, and use the property of the nth derivative of the delta function $\displaystyle\int f(s)\delta^{(n)}(s - \sigma) ds = (-1)^n f^{(n)}(\sigma)$ (Maruyama [53]), to obtain

[8]This agrees with Eq. (3.5) of Aki and Richards [3] if $p \to k$, $q \to l$, $\tau \to t$ and $\eta \to \mathbf{x}$, $f_p \to \rho f_k$.

$$
\begin{aligned}
u_n(\mathbf{x}, t) &= \iiint \left\{\rho f_x(\boldsymbol{\xi}, t) * G_{nx}(\mathbf{x}, t; \boldsymbol{\xi}, 0) + \rho f_y(\boldsymbol{\xi}, t) * G_{ny}(\mathbf{x}, t; \boldsymbol{\xi}, 0)\right\} dV(\boldsymbol{\xi}) \\
&= \iiint \left\{-\mu D(\boldsymbol{\xi}_0, t)\frac{\partial \delta(\xi_y)}{\partial \xi_y} * G_{nx}(\mathbf{x}, t; \boldsymbol{\xi}, 0)\right. \\
&\qquad \left. -\mu \frac{\partial D(\boldsymbol{\xi}_0, t)}{\partial \xi_x}\delta(\xi_y) * G_{ny}(\mathbf{x}, t; \boldsymbol{\xi}, 0)\right\} d\xi_x d\xi_y d\xi_z \\
&= \iint \left\{+\mu D(\boldsymbol{\xi}_0, t)\delta(\xi_y) * \frac{\partial}{\partial \xi_y} G_{nx}(\mathbf{x}, t; \boldsymbol{\xi}_0, 0)\right. \\
&\qquad \left. -\mu \frac{\partial D(\boldsymbol{\xi}_0, t)}{\partial \xi_x} * G_{ny}(\mathbf{x}, t; \boldsymbol{\xi}_0, 0)\right\} d\xi_x d\xi_z
\end{aligned} \tag{2.16}
$$

where $\boldsymbol{\xi}_0 = (\xi_x, 0, \xi_z)$. We apply **integration by parts** on ξ_x to the second term of the integrand, and assume $D(\xi_0, t) = 0$ at $\xi_x^{\min}$ and $\xi_x^{\max}$ on the perimeter of the large region V. From these, $\left[\mu D(\boldsymbol{\xi}_0, t) * G_{ny}(\mathbf{x}, t; \boldsymbol{\xi}_0, 0)\right]_{\xi_x^{\min}}^{\xi_x^{\max}} = 0$. We then obtain

$$
\int \mu \frac{\partial D(\boldsymbol{\xi}_0, t)}{\partial \xi_x} * G_{ny}(\mathbf{x}, t; \boldsymbol{\xi}_0, 0) d\xi_x = -\int \mu D(\boldsymbol{\xi}_0, t) * \frac{\partial}{\partial \xi_x} G_{ny}(\mathbf{x}, t; \boldsymbol{\xi}_0, 0) d\xi_x \,. \tag{2.17}
$$

Using this, (2.16) yields the equation[9]

$$
u_n(\mathbf{x}, t) = \iint \mu D(\boldsymbol{\xi}_0, t) * \left\{\frac{\partial}{\partial \xi_y} G_{nx}(\mathbf{x}, t; \boldsymbol{\xi}_0, 0) + \frac{\partial}{\partial \xi_x} G_{ny}(\mathbf{x}, t; \boldsymbol{\xi}_0, 0)\right\} d\xi_x d\xi_z \,. \tag{2.18}
$$

Furthermore, since it is a point source, the Green's function and its derivatives can assume constant values at $\boldsymbol{\xi} = \mathbf{0}$ in Σ, so that

$$
u_n(\mathbf{x}, t) = \iint \mu D(\boldsymbol{\xi}_0, t) d\Sigma * \left\{\frac{\partial}{\partial \xi_y} G_{nx}(\mathbf{x}, t; \mathbf{0}, 0) + \frac{\partial}{\partial \xi_x} G_{ny}(\mathbf{x}, t; \mathbf{0}, 0)\right\} . \tag{2.19}
$$

In addition, when a force in the x-axis positive direction, the strength of which is $\varepsilon^{-1}\iint \mu D d\Sigma$, acts at the origin shown in Fig. 2.7, its body force equivalent is given by (2.28), described later as

$$
\rho \mathbf{f}(\mathbf{x}) = \delta(\mathbf{x}) \left(\varepsilon^{-1}\iint \mu D d\Sigma, 0, 0\right) . \tag{2.20}
$$

Displacements due to this are given by the representation theorem (1.108) as

[9] This agrees with Eq. (3.13) of Aki and Richards [3] if $[u_1] \to D$, $1 \to x$, $3 \to y$, $d\Sigma \to d\xi_x d\xi_z$, and $\int d\tau \to *$.

$$
\begin{aligned}
u_n^f(\mathbf{x}, t) &= \iiint \delta(\boldsymbol{\xi})\varepsilon^{-1}\iint \mu D d\Sigma * G_{nx}(\mathbf{x}, t; \boldsymbol{\xi}, 0)\, dV(\boldsymbol{\xi}) \\
&= \varepsilon^{-1}\iint \mu D d\Sigma * G_{nx}(\mathbf{x}, t; \mathbf{0}, 0)\,.
\end{aligned} \tag{2.21}
$$

If the point source is not located at the origin but at the point $(0, +\varepsilon/2, 0)$ in Fig. 2.7, $\delta(\boldsymbol{\xi})$ in (2.21) is replaced by $\delta(\xi_x)\delta(\xi_y - \varepsilon/2)\delta(\xi_z)$ and $\varepsilon^{-1}\iint \mu D d\Sigma * G_{nx}(\mathbf{x}, t; 0, +\varepsilon/2, 0, 0)$ is obtained. In the case of the equal and opposite force applied at $(0, -\varepsilon/2, 0)$, the displacement yields $-\varepsilon^{-1}\iint \mu D d\Sigma * G_{nx}(\mathbf{x}, t; 0, -\varepsilon/2, 0, 0)$. If we make a couple of forces from these two forces, the displacement is the sum of the displacements of the two forces, according to the **principle of superposition** (1.3.1). Because we consider a point source, we then take the limit $\varepsilon \to 0$ and use the definition of the partial derivative

$$
\lim_{\varepsilon \to 0} \frac{f(\xi_x, \xi_y + \varepsilon/2, \xi_z) - f(\xi_x, \xi_y - \varepsilon/2, \xi_z)}{\varepsilon} = \frac{\partial f(\xi_x, \xi_y, \xi_z)}{\partial \xi_y}\,, \tag{2.22}
$$

obtaining

$$
\begin{aligned}
&\lim_{\varepsilon \to 0} \iint \mu D d\Sigma * \frac{G_{nx}(\mathbf{x}, t; 0, 0 + \varepsilon/2, 0) - G_{nx}(\mathbf{x}, t; 0, 0 - \varepsilon/2, 0)}{\varepsilon} \\
&= \iint \mu D d\Sigma * \frac{\partial G_{nx}(\mathbf{x}, t; \mathbf{0}, 0)}{\partial \xi_y}
\end{aligned} \tag{2.23}
$$

which agrees the first term on the right-hand side of (2.19). Since the strength of the force is $\varepsilon^{-1}\iint \mu D d\Sigma$ and the length of the arm is ε, the moment of the couple of forces is given as

$$
M_0 = \iint \mu D d\Sigma\,. \tag{2.24}
$$

Similarly, two forces in the y-direction, the strengths of which are $\varepsilon^{-1}\iint \mu D d\Sigma$, are applied at $(\pm\varepsilon/2, 0, 0)$, and we take the limit $\varepsilon \to 0$. The displacement of this couple of forces yields

$$
\begin{aligned}
&\lim_{\varepsilon \to 0} \iint \mu D d\Sigma * \frac{G_{ny}(\mathbf{x}, t; 0 + \varepsilon/2, 0, 0) - G_{nx}(\mathbf{x}, t; 0 - \varepsilon/2, 0, 0)}{\varepsilon} \\
&= \iint \mu D d\Sigma * \frac{\partial G_{ny}(\mathbf{x}, t; \mathbf{0}, 0)}{\partial \xi_x}
\end{aligned} \tag{2.25}
$$

which agrees with the second term on the right-hand side of (2.19). The strength of this couple of forces is equal to (2.24). Summarizing the above, a force system equivalent to the **slip** D in the source fault Σ of a point source is proved to be a **double couple** with seismic moment $M_0 = \iint \mu D d\Sigma$.

Since fault rupture is a nonstationary phenomenon that occurs in a short time, the moment of the couple of forces is a function that varies according to time t. This function $M_0(t)$ or its derivative $\dot{M}_0(t)$ is called the **source time function**, but the latter is more popular [35, 73]. To distinguish between them, the former is called the **moment time function**, and the latter is called the **moment rate function** [35]. In general, the moment starts from 0 and then increases until it finally reaches M_0. This final moment M_0 is often called the seismic moment. In addition, from (2.24), we can define a **slip time function** $D(t)$ or a **slip rate function** $\dot{D}(t)$, which is similar to the moment time function or the moment rate function. As concrete function forms for the moment time function, the moment rate function, the slip time function, and the slip rate function, the **ramp function** and the **triangular function** are often used (Sects. 2.3.4, 2.3.6).

Because earthquakes are fault fractures (Sect. 2.1.1), the scale of earthquakes is most accurately expressed by the seismic moment M_0. However, before the concept of M_0 was established, there was a history of using various types of **magnitude** M (Sect. A.1) which were empirically obtained from the amplitude of ground motion, etc. These have been used for a long time and have been accepted by society, therefore Kanamori [32] proposed to calculate a value close to M from M_0 using the formula[10]

$$\log M_0 = 1.5\ M_w + 16.1\ . \tag{2.26}$$

M calculated using this formula is called the **moment magnitude** M_w (Sect. A.1). One problem is that M_0 cannot be obtained without long-term analysis, but this is being solved through the development of methods such as **CMT inversion** (Sect. 2.3.4), and M_w is now predominantly used instead of the older magnitude scales.

2.1.3 Ground Motion by Point Force

Before considering a point source of double couple, we will consider a **point force** consisting of only one force from a couple of forces. If such a **single force** works with the time function $f(t)$ in the x direction at the Cartesian origin, then its **body force equivalent** (Sect. 2.1.2) **f** must satisfy[11]

$$\rho \iiint \mathbf{f}(\boldsymbol{\xi}) dV(\boldsymbol{\xi}) = f(t)(1, 0, 0)\ . \tag{2.27}$$

Here, V represents a region where **Helmholtz's theorem** is valid (Sect. 1.2.4), $\boldsymbol{\xi} = (\xi, \eta, \zeta)$ is the position vector for the volume integral dV (Fig. 1.5), and $(1, 0, 0)$ is a unit vector in the x-direction. Therefore, from the definition of the **delta function** (1.104) we obtain

[10] The formula was originally proposed by Kanamori [32]. Since the CGS unit system was used at that time, M_0 must be given in dyne·cm.

[11] Refer to p. 19 of Honda [27].

$$\mathbf{f}(\mathbf{x}) = \frac{f(t)}{\rho}\delta(\mathbf{x})(1,0,0) = (f_0, 0, 0)\,. \tag{2.28}$$

Furthermore, in order to obtain the **scalar potential** Φ of $\mathbf{f}$, we substitute (2.28) into the first equation of (1.39)

$$\Phi(\mathbf{x}) = -\frac{1}{4\pi}\iiint \frac{\nabla\cdot\mathbf{f}}{r}\,dV(\boldsymbol{\xi})\,, \quad r = |\mathbf{x}-\boldsymbol{\xi}|\,. \tag{2.29}$$

We then substitute $\mathbf{F} = \mathbf{f}/r$ into the **divergence theorem of Gauss** (1.88) and obtain

$$\iiint \left\{\frac{\nabla\cdot\mathbf{f}}{r} + \nabla\left(\frac{1}{r}\right)\cdot\mathbf{f}\right\} dV = \iint \left(\frac{1}{r}\right)\mathbf{f}\cdot\mathbf{n}\,dS\,. \tag{2.30}$$

If the perimeter S of the region V (Fig. 1.5) is far away from the origin, the surface integral on the right-hand side of (2.30) yields zero. We substitute the left-hand side of (2.30) $= 0$ into (1.39) and obtain

$$\Phi(\mathbf{x}) = \frac{-1}{4\pi}\iiint \left\{-\nabla\left(\frac{1}{r}\right)\cdot\mathbf{f}\right\} dV = \frac{+1}{4\pi}\iiint \frac{\partial}{\partial\xi}\left(\frac{1}{r}\right) f_0\,dV(\boldsymbol{\xi})\,. \tag{2.31}$$

Using the relation[12]

$$\frac{\partial r}{\partial x} = -\frac{\partial r}{\partial \xi} \tag{2.32}$$

we rewrite the partial derivatives, substitute f_0 from (2.28), and perform the volume integral using the definition of the delta function (1.104). (2.31) then yields

$$\Phi(\mathbf{x}) = \frac{-1}{4\pi}\iiint \frac{\partial}{\partial x}\left(\frac{1}{r}\right)\frac{f(t)}{\rho}\delta(\boldsymbol{\xi})\,dV(\boldsymbol{\xi}) = \frac{-1}{4\pi}\frac{\partial}{\partial x}\left(\frac{1}{|\mathbf{x}-\boldsymbol{\xi}|}\right)\frac{f(t)}{\rho}\bigg|_{\boldsymbol{\xi}=\mathbf{0}}\,. \tag{2.33}$$

If we define $|\mathbf{x}-\boldsymbol{\xi}|_{\boldsymbol{\xi}=\mathbf{0}} = |\mathbf{x}| = R(\mathbf{x})$, where $R(\mathbf{x})$ is the distance between the point force and an evaluation point of the potential, (2.33) further yields

$$\Phi(\mathbf{x}) = \frac{-1}{4\pi\rho}f(t)\frac{\partial R^{-1}(\mathbf{x})}{\partial x}\,. \tag{2.34}$$

Beltrami's theorem [76] indicates that the **retarded potential**

$$U = \iiint \frac{\sigma(\boldsymbol{\xi}, t - r/\upsilon)}{r}dV \tag{2.35}$$

[12] Refer to p. 217 of Webster [76].

satisfies the **inhomogeneous wave equation**

$$\frac{1}{\upsilon^2}\frac{\partial^2 U}{\partial t^2} = \nabla^2 U + 4\pi\sigma\ . \tag{2.36}$$

Since the first equation of the wave equations for ground motion (1.42) corresponds to (2.36) with the substitutions of $U = \phi$, $\upsilon = \alpha$, and $4\pi\sigma = \Phi/\alpha^2$, (2.35) with these substitutions

$$\phi(\mathbf{x}, t) = \iiint \frac{\Phi(\boldsymbol{\xi}, t - r/\alpha)}{4\pi\alpha^2 r}\, dV(\boldsymbol{\xi}) \tag{2.37}$$

satisfies the first equation of (1.42).

Substituting (2.34) obtained earlier into this equation and replacing the volume integral with spherical integrals having the center at **x** and radii r (Fig. 2.9), we obtain

$$\phi(\mathbf{x}, t) = \frac{-1}{(4\pi)^2\alpha^2\rho}\int \frac{f(t - r/\alpha)}{r} dr \iint \frac{\partial R^{-1}(\boldsymbol{\xi})}{\partial\xi} dS(\boldsymbol{\xi})\ . \tag{2.38}$$

Here, since the partial derivative rewriting performed on (2.31) is also performed on (2.38), we take the point represented by the position vector $\mathbf{x}'$ close to the origin, and find the solution for this point (Fig. 2.9), and then obtain the solution for the origin by bringing $\mathbf{x}'$ closer to $\mathbf{0}$.[13] Since, for $R'(\mathbf{x}) = |\mathbf{x} - \mathbf{x}'|$ newly defined, $\lim_{\mathbf{x}'\to\mathbf{0}} R'(\mathbf{x}) = R(\mathbf{x})$, and $\partial R'(\boldsymbol{\xi})/\partial\xi = -\partial R'(\boldsymbol{\xi})/\partial x'$ as in (2.32), we can rewrite the spherical integral in (2.38) as

$$\begin{aligned}\iint \frac{\partial R^{-1}(\boldsymbol{\xi})}{\partial\xi} dS(\boldsymbol{\xi}) &= \lim_{\mathbf{x}'\to\mathbf{0}} \iint \frac{\partial R'^{-1}(\boldsymbol{\xi})}{\partial\xi} dS(\boldsymbol{\xi}) \\ &= \lim_{\mathbf{x}'\to\mathbf{0}} \left\{ -\frac{\partial}{\partial x'} \iint \frac{1}{R'(\boldsymbol{\xi})} dS(\boldsymbol{\xi}) \right\}\ .\end{aligned} \tag{2.39}$$

In Fig. 2.9, the spherical element dS is a circular ribbon perpendicular to the page, and the orientation of this ribbon is taken freely if the whole sphere is covered after integration. Therefore, we locate the center of dS on the extension of **x**, as shown in the left diagram of Fig. 2.10. Using the angle θ measured in the direction perpendicular to the page, the radius of the ribbon is $r \sin\theta$ (Fig. 2.10, right panel). The width of the ribbon is $r\, d\theta$, and then $dS(\boldsymbol{\xi}) = 2\pi r^2 \sin\theta\, d\theta$. In the triangle consisting of a point $\boldsymbol{\xi}$ on the ribbon in the direction θ, $\mathbf{x}'$, and **x**, the vectors along its three sides, which are $\boldsymbol{\xi} - \mathbf{x}'$, $\mathbf{x} - \mathbf{x}'$, and $\mathbf{x} - \boldsymbol{\xi}$, are connected as $(\boldsymbol{\xi} - \mathbf{x}') = (\mathbf{x} - \mathbf{x}') - (\mathbf{x} - \boldsymbol{\xi})$. Therefore, since $R'(\boldsymbol{\xi})^2 = R'(\mathbf{x})^2 + r^2 - 2R'(\mathbf{x})r\cos(\pi - \theta)$ $(R'(\boldsymbol{\xi}) = |\boldsymbol{\xi} - \mathbf{x}'|)$, $2R'(\boldsymbol{\xi})dR'(\boldsymbol{\xi}) = -2R'(\mathbf{x})r\sin\theta d\theta$.

[13] Refer to pp. 61–62 of Fukao [19] and pp. 71–72 of Aki and Richards [3].

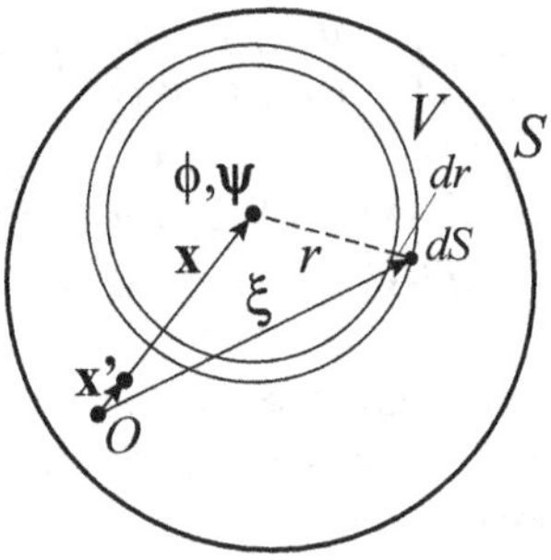

Fig. 2.9 A schematic view of the cross section of region V (Fig. 1.5). The volume integral is replaced with spherical integrals having the center at $\mathbf{x}$ and radii r (reprinted from Koketsu [41] with permission of Kindai Kagaku)

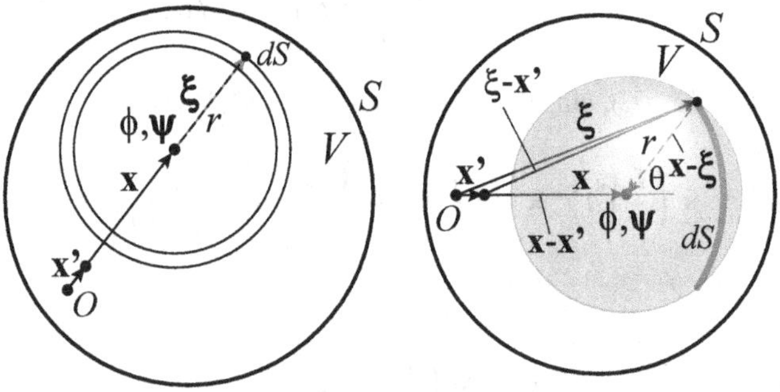

Fig. 2.10 Left: Sectional view of region V (Fig. 1.5) after moving $\boldsymbol{\xi}$; Right: Three-dimensional view from the direction perpendicular to the left diagram (reprinted from Koketsu [41] with permission of Kindai Kagaku). The dark gray ribbon on the right is dS

From the above,

$$\iint \frac{1}{R'(\boldsymbol{\xi})} dS(\boldsymbol{\xi}) = \int \frac{2\pi r^2}{R'(\boldsymbol{\xi})} \sin\theta \, d\theta = -\frac{2\pi r}{R'(\mathbf{x})} \int dR'(\boldsymbol{\xi}) \,. \tag{2.40}$$

The integration range is $[R'(\theta = 0), R'(\theta = \pi)] = [R'(\mathbf{x}) + r, R'(\mathbf{x}) - r]$, but under the condition of $R'(\boldsymbol{\xi}) \geq 0$, this yields $[R'(\mathbf{x}) + r, r - R'(\mathbf{x})]$ if $R'(\mathbf{x}) < r$. Therefore,

$$\iint \frac{1}{R'(\boldsymbol{\xi})} dS(\boldsymbol{\xi}) = \begin{cases} \dfrac{4\pi r^2}{R'(\mathbf{x})} & (R'(x) \geq r) \\ 4\pi r & (R'(\mathbf{x}) < r) \end{cases} . \tag{2.41}$$

Using this and the relation similar to (2.32), that is $\partial R'(\mathbf{x})/\partial x' = -\partial R'(\mathbf{x})/\partial x$, and noting that $r = |\mathbf{x} - \boldsymbol{\xi}|$ does not include $\mathbf{x}'$, (2.39) finally yields

$$\iint \frac{\partial R^{-1}(\boldsymbol{\xi})}{\partial \xi} dS(\boldsymbol{\xi}) = \lim_{\mathbf{x}' \to \mathbf{0}} \left\{ -\frac{\partial}{\partial x'} \iint \frac{1}{R'(\boldsymbol{\xi})} dS(\boldsymbol{\xi}) \right\} \tag{2.42}$$
$$= \lim_{\mathbf{x}' \to \mathbf{0}} \begin{cases} 4\pi r^2 \dfrac{\partial R'^{-1}(\mathbf{x})}{\partial x} & (R'(\mathbf{x}) \geq r) \\ 0 & (R'(\mathbf{x}) < r) \end{cases} = \begin{cases} 4\pi r^2 \dfrac{\partial R^{-1}(\mathbf{x})}{\partial x} & (R(\mathbf{x}) \geq r) \\ 0 & (R(\mathbf{x}) < r) \end{cases} .$$

Substituting this spherical integral into (2.38), setting the integral range of radius r to $[0, R]$ based on $R \geq r$, and applying the change of variables $r = \alpha\tau$ (τ is a variable corresponding to time), we obtain

$$\phi(\mathbf{x}, t) = \frac{-1}{4\pi\alpha^2\rho} \int_0^R rf\left(t - \frac{r}{\alpha}\right) dr \frac{\partial R^{-1}}{\partial x}$$
$$= \frac{-1}{4\pi\rho} \frac{\partial R^{-1}}{\partial x} \int_0^{R/\alpha} \tau f(t - \tau)\, d\tau . \tag{2.43}$$

For $\boldsymbol{\psi} = (\psi_x, \psi_y, \psi_z)$, since the left-hand side of (2.30) is

$$\iiint \nabla \times \left(\frac{\mathbf{f}}{r}\right) dV = \iiint \left\{\frac{\nabla \times \mathbf{f}}{r} - \mathbf{f} \times \nabla \left(\frac{1}{r}\right)\right\} dV , \tag{2.44}$$

then

$$\boldsymbol{\Psi}(\mathbf{x}) = \frac{+1}{4\pi} \iiint \left\{\mathbf{f} \times \nabla \left(\frac{1}{r}\right)\right\} dV$$
$$= \frac{+1}{4\pi} \iiint \left(0, -f_0 \frac{\partial}{\partial \zeta}\left(\frac{1}{r}\right), +f_0 \frac{\partial}{\partial \eta}\left(\frac{1}{r}\right)\right) dV(\boldsymbol{\xi}) . \tag{2.45}$$

From this point on, we perform almost the same derivation as ϕ and obtain

$$\psi_x(\mathbf{x}, t) = 0 ,$$
$$\psi_y(\mathbf{x}, t) = \frac{+1}{4\pi\rho} \frac{\partial R^{-1}}{\partial z} \int_0^{R/\beta} \tau f(t - \tau)\, d\tau ,$$
$$\psi_z(\mathbf{x}, t) = \frac{-1}{4\pi\rho} \frac{\partial R^{-1}}{\partial y} \int_0^{R/\beta} \tau f(t - \tau)\, d\tau . \tag{2.46}$$

When the point force exerts $f(t)$ at the origin of the continuum in the positive direction of the x-axis, the displacements of the continuum, which are ground motions u_x, u_y, and u_z, are given by substituting (2.43) and (2.46) into the equation $\mathbf{u} = \nabla\phi + \nabla \times \boldsymbol{\psi}$ of **Helmholtz's theorem** (Sect. 1.2.4), as

$$u_x = \frac{1}{4\pi\rho}\left[\frac{\partial^2 R^{-1}}{\partial x^2}\int_{R/\alpha}^{R/\beta}\tau f(t-\tau)d\tau + \frac{1}{R}\left(\frac{\partial R}{\partial x}\right)^2\left\{\frac{1}{\alpha^2}f\left(t-\frac{R}{\alpha}\right)-\frac{1}{\beta^2}f\left(t-\frac{R}{\beta}\right)\right\}\right.$$
$$\left.+\frac{1}{\beta^2 R}f\left(t-\frac{R}{\beta}\right)\right], \tag{2.47}$$
$$u_y = \frac{1}{4\pi\rho}\left[\frac{\partial^2 R^{-1}}{\partial x\partial y}\int_{R/\alpha}^{R/\beta}\tau f(t-\tau)d\tau + \frac{1}{R}\frac{\partial R}{\partial x}\frac{\partial R}{\partial y}\left\{\frac{1}{\alpha^2}f\left(t-\frac{R}{\alpha}\right)-\frac{1}{\beta^2}f\left(t-\frac{R}{\beta}\right)\right\}\right],$$
$$u_z = \frac{1}{4\pi\rho}\left[\frac{\partial^2 R^{-1}}{\partial x\partial z}\int_{R/\alpha}^{R/\beta}\tau f(t-\tau)d\tau + \frac{1}{R}\frac{\partial R}{\partial x}\frac{\partial R}{\partial z}\left\{\frac{1}{\alpha^2}f\left(t-\frac{R}{\alpha}\right)-\frac{1}{\beta^2}f\left(t-\frac{R}{\beta}\right)\right\}\right]$$

where α, β, and ρ are the P wave velocity, S wave velocity, and density of the continuum at the origin, respectively. Here, (2.47) is still the ground motion of a point force, but as shown in the next section, it is possible to obtain the ground motion of a point source from (2.47) only by partial differentiation. Therefore, (2.47) and its underlying Eqs. (2.43) and (2.46) are the most essential equations for ground motion seismology. These were the equations that Love [50] had obtained at the latest by 1906, and more surprisingly, according to Love himself, Stokes [70] had already derived equivalent equations in 1849. In other words, ground motion seismology is built on physical mathematics from the mid-19th century.

Hereafter, in order to distinguish them from **fault displacements** (Sect. 2.1.2), displacements of the continuum will be referred to as **elastic displacements** as much as possible.

2.1.4 *Ground Motion by Point Source*

As described in Sect. 2.1.2, in the case of the point source in Fig. 2.7, for the one couple in the double couple indicated by the solid arrows, the arm length is ε. A point force in the positive direction of the x-axis acts at the point on the y-axis $+\varepsilon/2$ away from the origin and the other point force in the opposite direction acts at the point on the y-axis $-\varepsilon/2$ away from the origin. This couple of forces generates $\mathbf{u}(x, y-\varepsilon/2, z) - \mathbf{u}(x, y+\varepsilon/2, z)$, where $\mathbf{u} = (u_x, u_y.u_z)$ and u_x, u_y, u_z are given in (2.47), but because it is a part of the point source, it is necessary to take the limit of $\varepsilon \to 0$. From

$$\lim_{\varepsilon\to 0}\frac{\mathbf{u}(x, y-\varepsilon/2, z) - \mathbf{u}(x, y+\varepsilon/2, z)}{\varepsilon} = -\frac{\partial \mathbf{u}}{\partial y} \tag{2.48}$$

we know that ground motions due to this couple of forces are obtained by partially differentiating (2.47) with respect to y and multiplying by $-\varepsilon$. Similarly, ground motions due to the other couple of forces, indicated by the dashed arrows in Fig. 2.7, are obtained by partial differentiation of the ground motions $\mathbf{u}'$ due to the point force in the positive direction of the y-axis with respect to x, and multiplying by $-\varepsilon$.

To do this, it is necessary to obtain $\mathbf{u}' = (u'_x, u'_y, u'_z)$. First, in the derivation of $\mathbf{u}$ in Sect. 2.1.3, we use $\mathbf{f} = (0, f_0, 0)$, $\nabla r^{-1} \cdot \mathbf{f} = \partial r^{-1}/\partial\eta\, f_0$ for (2.33), and $\mathbf{f} \times \nabla r^{-1} = (+f_0 \partial r^{-1}/\partial\zeta, 0, -f_0 \partial r^{-1}/\partial\xi)$ for (2.45). Hereafter, the derivation is continued along Sect. 2.1.3, and we obtain the scalar potential ψ' of $\mathbf{u}'$, and the vector potential $\boldsymbol{\psi}' = (\psi'_x, \psi'_y, \psi'_z)$

$$
\begin{aligned}
\phi'(\mathbf{x}, t) &= \frac{-1}{4\pi\rho}\frac{\partial R^{-1}}{\partial y}\int_0^{R/\alpha} \tau f(t-\tau)\, d\tau\,, \\
\psi'_x(\mathbf{x}, t) &= \frac{-1}{4\pi\rho}\frac{\partial R^{-1}}{\partial z}\int_0^{R/\beta} \tau f(t-\tau)\, d\tau\,, \\
\psi'_y(\mathbf{x}, t) &= 0\,, \\
\psi'_z(\mathbf{x}, t) &= \frac{+1}{4\pi\rho}\frac{\partial R^{-1}}{\partial x}\int_0^{R/\beta} \tau f(t-\tau)\, d\tau\,.
\end{aligned} \tag{2.49}
$$

We again use the equation $\mathbf{u}' = \nabla\phi' + \nabla \times \boldsymbol{\psi}'$ of Helmholtz's theorem and obtain

$$
\begin{aligned}
u'_x &= \frac{1}{4\pi\rho}\left[\frac{\partial^2 R^{-1}}{\partial x \partial y}\int_{R/\alpha}^{R/\beta} \tau f(t-\tau)d\tau + \frac{1}{R}\frac{\partial R}{\partial x}\frac{\partial R}{\partial y}\left\{\frac{1}{\alpha^2} f\left(t-\frac{R}{\alpha}\right) - \frac{1}{\beta^2} f\left(t-\frac{R}{\beta}\right)\right\}\right], \\
u'_y &= \frac{1}{4\pi\rho}\left[\frac{\partial^2 R^{-1}}{\partial y^2}\int_{R/\alpha}^{R/\beta} \tau f(t-\tau)d\tau + \frac{1}{R}\left(\frac{\partial R}{\partial y}\right)^2\left\{\frac{1}{\alpha^2} f\left(t-\frac{R}{\alpha}\right) - \frac{1}{\beta^2} f\left(t-\frac{R}{\beta}\right)\right\}\right. \\
&\quad \left. + \frac{1}{\beta^2 R} f\left(t-\frac{R}{\beta}\right)\right], \\
u'_z &= \frac{1}{4\pi\rho}\left[\frac{\partial^2 R^{-1}}{\partial y \partial z}\int_{R/\alpha}^{R/\beta} \tau f(t-\tau)d\tau + \frac{1}{R}\frac{\partial R}{\partial y}\frac{\partial R}{\partial z}\left\{\frac{1}{\alpha^2} f\left(t-\frac{R}{\alpha}\right) - \frac{1}{\beta^2} f\left(t-\frac{R}{\beta}\right)\right\}\right].
\end{aligned} \tag{2.50}
$$

From the above, when making the strike of the left-lateral fault coincide with the x-axis as in Fig. 2.7, the x-component of the ground motion $\mathbf{U} = (U_x, U_y, U_z)$ by the **point source** of the **double couple** is given as $U_x = -\varepsilon\partial u_x/\partial y - \varepsilon\partial u'_x/\partial x$. Using $\int_{R/\alpha}^{R/\beta} \tau M_0(t-\tau)d\tau = \int_0^{R/\beta} \tau M_0(t-\tau)d\tau - \int_0^{R/\alpha} \tau M_0(t-\tau)d\tau$ etc., we then obtain

$$
U_x = U_x^1 + U_x^{2P} + U_x^{2S} + U_x^{3P} + U_x^{3S} \tag{2.51}
$$

where

$$
\begin{aligned}
U_x^1 &= \frac{1}{4\pi\rho}\left[-2\frac{\partial^3 R^{-1}}{\partial x^2 \partial y}\right]\int_{R/\alpha}^{R/\beta}\tau M_0(t-\tau)d\tau \,, \\
U_x^{2P} &= \frac{1}{4\pi\rho\alpha^2}\left[+R\frac{\partial^2 R^{-1}}{\partial x^2}\frac{\partial R}{\partial y} + R\frac{\partial^2 R^{-1}}{\partial x \partial y}\frac{\partial R}{\partial x}\right. \\
&\quad \left. -\frac{\partial}{\partial y}\left\{\frac{1}{R}\left(\frac{\partial R}{\partial x}\right)^2\right\} - \frac{\partial}{\partial x}\left(\frac{1}{R}\frac{\partial R}{\partial x}\frac{\partial R}{\partial y}\right)\right] M_0\left(t-\frac{R}{\alpha}\right) \,, \\
U_x^{2S} &= \frac{-1}{4\pi\rho\beta^2}\left[+R\frac{\partial^2 R^{-1}}{\partial x^2}\frac{\partial R}{\partial y} + R\frac{\partial^2 R^{-1}}{\partial x \partial y}\frac{\partial R}{\partial x}\right. \\
&\quad \left. -\frac{\partial}{\partial y}\left\{\frac{1}{R}\left(\frac{\partial R}{\partial x}\right)^2\right\} - \frac{\partial}{\partial x}\left(\frac{1}{R}\frac{\partial R}{\partial x}\frac{\partial R}{\partial y}\right) + \frac{1}{R^2}\left(\frac{\partial R}{\partial y}\right)\right] M_0\left(t-\frac{R}{\beta}\right) \,, \\
U_x^{3P} &= \frac{1}{4\pi\rho\alpha^3 R}\left[+2\left(\frac{\partial R}{\partial x}\right)^2\frac{\partial R}{\partial y}\right]\dot{M}_0\left(t-\frac{R}{\alpha}\right) \,, \\
U_x^{3S} &= \frac{-1}{4\pi\rho\beta^3 R}\left[+2\left(\frac{\partial R}{\partial x}\right)^2\frac{\partial R}{\partial y} - \frac{\partial R}{\partial y}\right]\dot{M}_0\left(t-\frac{R}{\beta}\right) \,.
\end{aligned}
\tag{2.52}
$$

$M_0(t) = \varepsilon f(t)$ is the moment of the couple of forces, which stands for the **seismic moment** of the point source, and $\dot{M}_0(t)$ is its derivative.

Here, we introduce the **direction cosine** $\gamma_{x,\,y,\,z} = x/R,\ y/R,\ z/R$ and partially differentiate R. Since $\partial R/\partial x = \gamma_x$, $\partial R/\partial y = \gamma_y$ and $\partial R/\partial z = \gamma_z$, the above equations yield

$$
\begin{aligned}
U_x^1 &= \frac{30\gamma_x^2\gamma_y - 6\gamma_y}{4\pi\rho R^4}\int_{R/\alpha}^{R/\beta}\tau M_0(t-\tau)d\tau \,, \\
U_x^{2P} &= \frac{12\gamma_x^2\gamma_y - 2\gamma_y}{4\pi\rho\alpha^2 R^2}M_0\left(t-\frac{R}{\alpha}\right),\ U_x^{2S} = -\frac{12\gamma_x^2\gamma_y - 3\gamma_y}{4\pi\rho\beta^2 R^2}M_0\left(t-\frac{R}{\beta}\right) \,, \\
U_x^{3P} &= \frac{2\gamma_x^2\gamma_y}{4\pi\rho\alpha^3 R}\dot{M}_0\left(t-\frac{R}{\alpha}\right),\ U_x^{3S} = -\frac{2\gamma_x^2\gamma_y - \gamma_y}{4\pi\rho\beta^3 R}\dot{M}_0\left(t-\frac{R}{\beta}\right) \,.
\end{aligned}
\tag{2.53}
$$

We perform similar operations for $U_y = U_y^1 + U_y^{2P} + U_y^{2S} + U_y^{3P} + U_y^{3S}$ and obtain

$$
\begin{aligned}
U_y^1 &= \frac{30\gamma_x\gamma_y^2 - 6\gamma_x}{4\pi\rho R^4}\int_{R/\alpha}^{R/\beta}\tau M_0(t-\tau)d\tau \,, \\
U_y^{2P} &= \frac{12\gamma_x\gamma_y^2 - 2\gamma_x}{4\pi\rho\alpha^2 R^2}M_0\left(t-\frac{R}{\alpha}\right),\ U_y^{2S} = -\frac{12\gamma_x\gamma_y^2 - 3\gamma_x}{4\pi\rho\beta^2 R^2}M_0\left(t-\frac{R}{\beta}\right) \,, \\
U_y^{3P} &= \frac{2\gamma_x\gamma_y^2}{4\pi\rho\alpha^3 R}\dot{M}_0\left(t-\frac{R}{\alpha}\right),\ U_y^{3S} = -\frac{2\gamma_x\gamma_y^2 - \gamma_x}{4\pi\rho\beta^3 R}\dot{M}_0\left(t-\frac{R}{\beta}\right) \,.
\end{aligned}
\tag{2.54}
$$

Similarly, for $U_z = U_z^1 + U_z^{2P} + U_z^{2S} + U_z^{3P} + U_z^{3S}$

$$
\begin{aligned}
U_z^1 &= \frac{30\gamma_x\gamma_y\gamma_z}{4\pi\rho R^4}\int_{R/\alpha}^{R/\beta}\tau M_0(t-\tau)d\tau\ ,\\
U_z^{2P} &= \frac{12\gamma_x\gamma_y\gamma_z}{4\pi\rho\alpha^2R^2}M_0\left(t-\frac{R}{\alpha}\right),\ U_z^{2S} = -\frac{12\gamma_x\gamma_y\gamma_z}{4\pi\rho\beta^2R^2}M_0\left(t-\frac{R}{\beta}\right),\\
U_z^{3P} &= \frac{2\gamma_x\gamma_y\gamma_z}{4\pi\rho\alpha^3R}\dot{M}_0\left(t-\frac{R}{\alpha}\right),\ U_z^{3S} = -\frac{2\gamma_x\gamma_y\gamma_z}{4\pi\rho\beta^3R}\dot{M}_0\left(t-\frac{R}{\beta}\right).
\end{aligned}
\tag{2.55}
$$

Summarizing the above, for $n = x, y, z$, we obtain the equations[14]

$$
\begin{aligned}
U_n = &\frac{30\gamma_n\gamma_x\gamma_y - 6\delta_{ny}\gamma_x - 6\delta_{nx}\gamma_y}{4\pi\rho R^4}\int_{R/\alpha}^{R/\beta}\tau M_0(t-\tau)d\tau\\
&+\frac{12\gamma_n\gamma_x\gamma_y - 2\delta_{ny}\gamma_x - 2\delta_{nx}\gamma_y}{4\pi\rho\alpha^2R^2}M_0\left(t-\frac{R}{\alpha}\right)\\
&-\frac{12\gamma_n\gamma_x\gamma_y - 3\delta_{ny}\gamma_x - 3\delta_{nx}\gamma_y}{4\pi\rho\beta^2R^2}M_0\left(t-\frac{R}{\beta}\right)\\
&+\frac{2\gamma_n\gamma_x\gamma_y}{4\pi\rho\alpha^3R}\dot{M}_0\left(t-\frac{R}{\alpha}\right) - \frac{2\gamma_n\gamma_x\gamma_y - \delta_{ny}\gamma_x - \delta_{nx}\gamma_y}{4\pi\rho\beta^3R}\dot{M}_0\left(t-\frac{R}{\beta}\right).
\end{aligned}
\tag{2.56}
$$

The first term of (2.56) decreases rapidly in proportion to R^{-3} with distance from the double couple point source (the factor is R^{-4}, but τ in the integrand of the definite integral contributes approximately R^{+1}). Therefore, this term is important in the vicinity of the point source and is called the **near-field term**. On the other hand, as the fourth and fifth terms only decay with R^{-1}, they have far-reaching effects and are called the **far-field terms**. The second and third terms have an intermediate property between the two above, and are referred to as the **intermediate-field terms**. In addition, the second and fourth terms include only α representing ground motions by **P waves**, while the third and fifth terms include only β representing ground motions by **S waves**.

The time function in the intermediate-field terms is the **moment time function** $M_0(t)$, while the time function in the far-field terms is the **moment rate function** $\dot{M}_0(t)$. In the near-field term, it is in the form of an integral of $M_0(t)$, and this integral is a convolution (1.85) of $M_0(t)$ and the function

$$
g(t) = \begin{cases} t, & R/\alpha \le t \le R/\beta \\ 0, & t < R/\alpha,\ t > R/\beta \end{cases}
\tag{2.57}
$$

shown in Fig. 2.11, which is a gentle function continuing from the P wave arrival time R/α to the S wave arrival time R/β. In general, as it can often be approximated

[14] This agrees with Eq. (4.30) of Aki and Richards [3] if $p = x, q = y, \boldsymbol{\nu} = (0, 1, 0), \mu A\bar{u}_p \to M_0$, and $r \to R$.

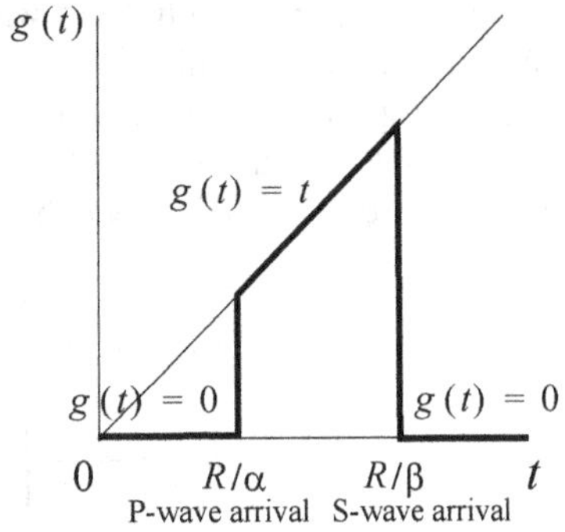

Fig. 2.11 The function convolved with $M_0(t)$ in the near-field term (reprinted from Koketsu [41] with permission of Kindai Kagaku)

by the far-field terms alone, it is said that **body waves** decay due to the **geometrical spreading** inversely proportional to a **hypocentral distance** R, and $\dot{M}_0(t)$ is called a **source time function**. However, in the region near a source, the near- and intermediate-field terms may exceed the far-field terms. In particular, for a very shallow source, very large intermediate-field terms are expected [14]. On the other hand, even in the case of distant seismic waves, the near-field term appears when viewed in the band of very long period. The **W phase** [33] may correspond to this [9].

2.1.5 *Potential Representation*

According to Bessonova et al. [4], for the scalar and vector potentials (hereafter collectively referred to as the **Helmholtz potential**) of ground motion due to a **point force** in the x-direction at the origin, using the second-order integration of the time function of the point force

$$F(t) = \int_0^t ds' \int_0^{s'} f(s)\, ds\,. \tag{2.58}$$

Keilis-Borok [34] proposed the compact representations[15]

$$\begin{aligned}
\phi &= \frac{1}{4\pi\rho}\frac{\partial B_\alpha}{\partial x}\,,\\
\boldsymbol{\psi} &= (\psi_x, \psi_y, \psi_z) = \left(0, \frac{-1}{4\pi\rho}\frac{\partial B_\beta}{\partial z}, \frac{1}{4\pi\rho}\frac{\partial B_\beta}{\partial y}\right) = \nabla \times \left(\frac{-B_\beta}{4\pi\rho}, 0, 0\right),\\
B_\alpha &= \frac{1}{R}F\left(t - \frac{R}{\alpha}\right), \quad B_\beta = \frac{1}{R}F\left(t - \frac{R}{\beta}\right).
\end{aligned} \tag{2.59}$$

[15]Not only Bessonova et al. [4] but also Maruyama [53] and Sato [62] noted that Keilis-Borok [34] derived (2.59). However, having no access to this Russian paper, I work through the proofs of these representations here.

B_α and B_β represent **spherical waves** (Sect. 1.2.5) with the time function (2.58). These are ground motions (elastic displacements) with amplitudes attenuated in inverse proportion to the distance R from the point force, and with phases shifted by R/α or R/β. Substituting these into the equations of Helmholtz's theorem, $\mathbf{u} = \nabla\phi + \nabla \times \boldsymbol{\psi}$ (Sect. 1.2.4) we obtain the equations [62]

$$\begin{aligned} u_x &= \frac{1}{4\pi\rho}\left\{\frac{\partial^2}{\partial x^2}(B_\alpha - B_\beta) + \nabla^2 B_\beta\right\}, \\ u_y &= \frac{1}{4\pi\rho}\left\{\frac{\partial^2}{\partial x \partial y}(B_\alpha - B_\beta)\right\}, \\ u_z &= \frac{1}{4\pi\rho}\left\{\frac{\partial^2}{\partial x \partial z}(B_\alpha - B_\beta)\right\}. \end{aligned} \tag{2.60}$$

We then use the third equation in (2.59) and

$$F'(t) = \frac{dF(t)}{dt} = \int_0^t f(s)\,ds\,,\ F''(t) = \frac{d^2F(t)}{dt^2} = f(t)\,, \tag{2.61}$$

obtaining the equations[16]

$$\begin{aligned} \frac{\partial^2 B_\alpha}{\partial x^2} &= \frac{\partial}{\partial x}\left\{\frac{\partial R^{-1}}{\partial x}F\left(t-\frac{R}{\alpha}\right) + R^{-1}F'\left(t-\frac{R}{\alpha}\right)\left(-\frac{1}{\alpha}\frac{\partial R}{\partial x}\right)\right\} \\ &= \frac{\partial^2 R^{-1}}{\partial x^2}F\left(t-\frac{R}{\alpha}\right) + \left\{2R^{-3}\left(\frac{\partial R}{\partial x}\right)^2 - R^{-2}\frac{\partial^2 R}{\partial x^2}\right\}F'\left(t-\frac{R}{\alpha}\right) + R^{-1}F''\left(t-\frac{R}{\alpha}\right)\left(\frac{1}{\alpha}\frac{\partial R}{\partial x}\right)^2 \\ &= \frac{\partial^2 R^{-1}}{\partial x^2}\left\{F\left(t-\frac{R}{\alpha}\right) + \frac{R}{\alpha}F'\left(t-\frac{R}{\alpha}\right)\right\} + \frac{1}{R}\left(\frac{\partial R}{\partial x}\right)^2\frac{1}{\alpha^2}f\left(t-\frac{R}{\alpha}\right). \end{aligned} \tag{2.62}$$

We now define the function $s'\int_0^{s'} f(s)\,ds$, apply integration by parts to this expression for the s' range $[0, t - R/\alpha]$, and use the change of variables $\tau = t - s$, (2.58), and (2.61) as

---

[16] This agrees with Eq. (3a) on p. 14 of Bessonova et al. [4] if $1/(4\pi\rho) = 1$, $q \rightarrow x$, $a \rightarrow \alpha$, and $K(t) \rightarrow f(t)$.

$$\left[s' \int_0^{s'} f(s)\, ds \right]_0^{t-R/\alpha} = \int_0^{t-R/\alpha} ds' \int_0^{s'} f(s)\, ds + \int_0^{t-R/\alpha} s' f(s')\, ds' ,$$

$$\left(t - \frac{R}{\alpha} \right) \int_0^{t-R/\alpha} f(s)\, ds = \int_0^{t-R/\alpha} ds' \int_0^{s'} f(s)\, ds + \int_0^{t-R/\alpha} s f(s)\, ds ,$$

$$\int_0^{t-R/\alpha} (t-s) f(s)\, ds = \int_0^{t-R/\alpha} ds' \int_0^{s'} f(s)\, ds + \frac{R}{\alpha} \int_0^{t-R/\alpha} f(s)\, ds ,$$

$$-\int_t^{R/\alpha} \tau f(t-\tau)\, d\tau = F\left(t - \frac{R}{\alpha} \right) + \frac{R}{\alpha} F'\left(t - \frac{R}{\alpha} \right) . \tag{2.63}$$

We substitute (2.63) into (2.62) and obtain

$$\frac{\partial^2 B_\alpha}{\partial x^2} = \frac{\partial^2 R^{-1}}{\partial x^2} \left\{ -\int_t^{R/\alpha} \tau f(t-\tau)\, d\tau \right\} + \frac{1}{R} \left(\frac{\partial R}{\partial x} \right)^2 \frac{1}{\alpha^2} f\left(t - \frac{R}{\alpha} \right) . \tag{2.64}$$

Similarly,

$$\frac{\partial^2 B_\beta}{\partial x^2} = \frac{\partial^2 R^{-1}}{\partial x^2} \left\{ -\int_t^{R/\beta} \tau f(t-\tau)\, d\tau \right\} + \frac{1}{R} \left(\frac{\partial R}{\partial x} \right)^2 \frac{1}{\beta^2} f\left(t - \frac{R}{\beta} \right) ,$$

$$\frac{\partial^2 B_\beta}{\partial y^2} = \frac{\partial^2 R^{-1}}{\partial y^2} \left\{ -\int_t^{R/\beta} \tau f(t-\tau)\, d\tau \right\} + \frac{1}{R} \left(\frac{\partial R}{\partial y} \right)^2 \frac{1}{\beta^2} f\left(t - \frac{R}{\beta} \right) ,$$

$$\frac{\partial^2 B_\beta}{\partial z^2} = \frac{\partial^2 R^{-1}}{\partial z^2} \left\{ -\int_t^{R/\beta} \tau f(t-\tau)\, d\tau \right\} + \frac{1}{R} \left(\frac{\partial R}{\partial z} \right)^2 \frac{1}{\beta^2} f\left(t - \frac{R}{\beta} \right) . \tag{2.65}$$

In addition, from $\dfrac{\partial^2 R^{-1}}{\partial x^2} = 3R^{-5}x^2 - R^{-3}$ etc.,

$$\frac{\partial^2 R^{-1}}{\partial x^2} + \frac{\partial^2 R^{-1}}{\partial y^2} + \frac{\partial^2 R^{-1}}{\partial z^2} = 3R^{-5}(x^2 + y^2 + z^2) - 3R^{-3} = 0 . \tag{2.66}$$

When (2.64), (2.65), and (2.66) are substituted, the first equation in (2.60) yields

$$\begin{aligned} u_x &= \frac{1}{4\pi\rho} \left[\left(\frac{\partial^2 B_\alpha}{\partial x^2} - \frac{\partial^2 B_\beta}{\partial x^2} \right) + \left(\frac{\partial^2 B_\beta}{\partial x^2} + \frac{\partial^2 B_\beta}{\partial y^2} + \frac{\partial^2 B_\beta}{\partial z^2} \right) \right] \\ &= \frac{1}{4\pi\rho} \left[\frac{\partial^2 R^{-1}}{\partial x^2} \int_{R/\alpha}^{R/\beta} \tau f(t-\tau) d\tau + \frac{1}{R} \left(\frac{\partial R}{\partial x} \right)^2 \left\{ \frac{1}{\alpha^2} f\left(t - \frac{R}{\alpha} \right) - \frac{1}{\beta^2} f\left(t - \frac{R}{\beta} \right) \right\} \right. \\ &\quad \left. + \frac{1}{\beta^2 R} f\left(t - \frac{R}{\beta} \right) \right] , \end{aligned} \tag{2.67}$$

which agrees with the first equation in (2.47).

As we can show that the second and third equations in (2.60) agree with the second and third equations in (2.47) using a similar procedure to that above, it is proven that (2.59) consists of the Helmholtz potentials for Love's [50] ground motions (2.47) due to a point force. However, the Helmholtz potentials in (2.59) are not completely equivalent to those in (2.43) and (2.46), which were derived by Love [50] himself. We rewrite (2.59) using (2.62), (2.63), etc. as

$$
\begin{aligned}
\phi &= \frac{-1}{4\pi\rho}\frac{\partial R^{-1}}{\partial x}\int_t^{R/\alpha} \tau f(t-\tau)\,d\tau\,, \\
\psi_x &= 0\,, \\
\psi_y &= \frac{+1}{4\pi\rho}\frac{\partial R^{-1}}{\partial z}\int_t^{R/\beta} \tau f(t-\tau)\,d\tau\,, \\
\psi_z &= \frac{-1}{4\pi\rho}\frac{\partial R^{-1}}{\partial y}\int_t^{R/\beta} \tau f(t-\tau)\,d\tau\,.
\end{aligned}
\tag{2.68}
$$

Comparing Keilis-Borok's potentials above with Love's potentials in (2.62) and (2.63), we find that the lower limit of all the definite integrals in Love's potentials is 0 but the lower limit in Keilis-Borok's potentials is t. Therefore, B_α and B_β in Love's potentials (hereafter written as B_α^L and B_β^L) and their second derivatives, such as (2.64) and (2.65), also include the definitive integrals with lower limits of 0. This means that the second derivatives have additional terms as

$$
\begin{aligned}
\frac{\partial^2 B_\alpha^L}{\partial x^2} &= \frac{\partial^2 B_\alpha}{\partial x^2} + \frac{\partial^2 R^{-1}}{\partial x^2}\int_0^t \tau f(t-\tau)\,d\tau\,, \\
\frac{\partial^2 B_\beta^L}{\partial x^2} &= \frac{\partial^2 B_\beta}{\partial x^2} + \frac{\partial^2 R^{-1}}{\partial x^2}\int_0^t \tau f(t-\tau)\,d\tau\,.
\end{aligned}
\tag{2.69}
$$

However, even if B_α^L and B_β^L are used in the first term of (2.67), the additional terms are canceled. They are also canceled in the second term of (2.67) according to (2.66). Both Love's and Keilis-Borok's potentials therefore work as the Helmholtz potentials for ground motion due to a point force in the x-direction at the origin.

We next compare (2.49) with (2.43) and (2.46), and obtain the Helmholtz potentials for ground motion due to a point force in the y-direction at the origin:

$$
\phi' = \frac{1}{4\pi\rho}\frac{\partial B_\alpha}{\partial y}\,, \quad \boldsymbol{\psi}' = \nabla \times \left(0, \frac{-B_\beta}{4\pi\rho}, 0\right)\,.
\tag{2.70}
$$

Using this and $\mathbf{u}' = \nabla\phi' + \nabla \times \boldsymbol{\psi}'$, we obtain

$$
\begin{aligned}
u'_x &= \frac{1}{4\pi\rho}\left\{\frac{\partial^2}{\partial x \partial y}(B_\alpha - B_\beta)\right\}\,, \\
u'_y &= \frac{1}{4\pi\rho}\left\{\frac{\partial^2}{\partial y^2}(B_\alpha - B_\beta) + \nabla^2 B_\beta\right\}\,, \\
u'_z &= \frac{1}{4\pi\rho}\left\{\frac{\partial^2}{\partial y \partial z}(B_\alpha - B_\beta)\right\}\,.
\end{aligned}
\tag{2.71}
$$

Similarly to (2.47) and (2.54), applying operations in Sect. 2.1.4 to (2.60) and (2.71), we obtain the potential representation of the ground motion due to the double couple in Fig. 2.7 [53]

$$
\begin{aligned}
U_x &= -\frac{1}{4\pi\rho}\left\{\frac{2\partial^3}{\partial x^2\partial y}(B_\alpha - B_\beta) + \frac{\partial}{\partial y}\nabla^2 B_\beta\right\}, \\
U_y &= -\frac{1}{4\pi\rho}\left\{\frac{2\partial^3}{\partial x\partial y^2}(B_\alpha - B_\beta) + \frac{\partial}{\partial x}\nabla^2 B_\beta\right\}, \\
U_z &= -\frac{1}{4\pi\rho}\left\{\frac{2\partial^3}{\partial x\partial y\partial z}(B_\alpha - B_\beta)\right\}.
\end{aligned}
\tag{2.72}
$$

However, B_α and B_β in (2.72) are different from those defined by (2.58) and (2.59). They are

$$
B_\alpha = \frac{1}{R}F\left(t - \frac{R}{\alpha}\right), \quad B_\beta = \frac{1}{R}F\left(t - \frac{R}{\beta}\right); \quad F(t) = \int_0^t d\tau \int_0^\tau M_0(s)\,ds\,. \tag{2.73}
$$

Equation (2.73) represents spherical waves with the double integral of the moment time function $M_0(t) = \varepsilon f(t)$, although (2.58) is the double integral of the force time function $f(t)$. In addition, we also apply operations in Sect. 2.1.4 to the potentials in (2.59) and (2.70), obtaining the **Helmholtz potentials** for the ground motion of a point source

$$
\Phi = -\frac{1}{2\pi\rho}\frac{\partial^2 B_\alpha}{\partial x\partial y}, \quad \boldsymbol{\Psi} = \nabla \times \left(\frac{1}{4\pi\rho}\frac{\partial B_\beta}{\partial y}, \frac{1}{4\pi\rho}\frac{\partial B_\beta}{\partial x}, 0\right). \tag{2.74}
$$

In Chap. 1 we used Φ and $\boldsymbol{\Psi}$ for the Helmholtz potentials of a body force, but hereafter we use Φ and $\boldsymbol{\Psi}$ for the Helmholtz potentials of the ground motion due to a point source, similarly to the capital letter **U** for the ground motion due to a point source.

We perform the **Fourier transforms** (Sect. 4.2.2) on B_α and B_β with respect to time t, and using the formulae in Table 4.3 we obtain

$$
\begin{aligned}
\overline{B}_\alpha &= \frac{1}{R(\mathrm{i}\omega)^2}e^{-\mathrm{i}\omega R/\alpha}\overline{M}_0(\omega) = \frac{-1}{\omega^2}A_\alpha \overline{M}_0(\omega), \ A_\alpha = \frac{e^{-\mathrm{i}k_\alpha R}}{R}, \ k_\alpha = \frac{\omega}{\alpha}, \\
\overline{B}_\beta &= \frac{1}{R(\mathrm{i}\omega)^2}e^{-\mathrm{i}\omega R/\beta}\overline{M}_0(\omega) = \frac{-1}{\omega^2}A_\beta \overline{M}_0(\omega), \ A_\beta = \frac{e^{-\mathrm{i}k_\beta R}}{R}, \ k_\beta = \frac{\omega}{\beta}
\end{aligned}
\tag{2.75}
$$

where $\overline{M}_0(\omega)$ is the Fourier transform of $M_0(t)$. A_α and A_β, which are the Fourier transforms of spherical waves, decay in inverse proportion to R and are oscillatory along R (Sect. 1.2.5). k_α and k_β are the **wavenumbers** ($2\pi \times$ number of waves (cycles) in a unit length) of the spherical waves.

2.2 Cylindrical Wave Expansion

2.2.1 Vertical Strike Slip Fault

We here consider a **horizontally layered structure** (Sect. 3.1.1), which is the velocity structure most commonly used. The ground motion in the layer including a point source is a spherical wave. This wave is deformed at the layer boundaries, but should spread concentrically in the horizontal direction around the **epicenter**, which is the point on the ground directly above the point source (Sect. 1.1), because the velocity structure varies only in the depth direction (z-direction). Therefore, the ground motion in this velocity structure can best be handled by using a **cylindrical coordinate system** (r, θ, z) with the origin at the epicenter (Fig. 2.12). r corresponds to an **epicentral distance** (Sect. 1.1). The cylindrical coordinate system also has the merit that the **displacement potentials** can be used instead of the vector potential in the Helmholtz theorem.

In the cylindrical coordinate system, the Fourier transforms of the **spherical waves** A_α and A_β of (2.75) are represented by the equations[17]

$$A_\upsilon = \frac{\exp(-\mathrm{i}k_\upsilon R)}{R} = \int_0^\infty F_\upsilon J_0(kr)\,dk\,, \quad \upsilon = \alpha \text{ or } \beta\,,$$
$$F_\upsilon = \frac{k\,e^{-\mathrm{i}\nu_\upsilon|z-h|}}{\mathrm{i}\nu_\upsilon}\,, \quad \nu_\upsilon = \begin{cases} \sqrt{k_\upsilon^2 - k^2}\,, & k_\upsilon \geq k \\ -\mathrm{i}\sqrt{k^2 - k_\upsilon^2}\,, & k_\upsilon < k \end{cases} \tag{2.76}$$

using the **Sommerfeld integral** [68] and the depth h of a point source (Fig. 2.12). Equation (2.76) indicates that spherical waves can be expressed as the superposition of **cylindrical waves** represented by the zeroth order **Bessel function** $J_0(kr)$, or that **cylindrical wave expansion** can be performed.[18] If we consider the **wavenumber** k of a cylindrical wave as the wavenumber in the horizontal direction of a spherical wave, ν_α and ν_β can be regarded as wavenumbers in the depth direction from their definition in (2.76). When they are imaginary, their sign is taken not to diverge at $z = \pm\infty$.

For the vertical **strike slip fault** in Fig. 2.7, we substitute (2.75) into the Fourier transform of its scalar potential Φ (the first equation in (2.74)), and we use $\partial/\partial x = \cos\theta \cdot \partial/\partial r - \sin\theta/r \cdot \partial/\partial\theta$, $\partial/\partial y = \sin\theta \cdot \partial/\partial r + \cos\theta/r \cdot \partial/\partial\theta$, $\partial A_\alpha/\partial\theta = 0$, and the zeroth order **Bessel equation** $d^2 J_0(q)/dq^2 + 1/q \cdot dJ_0(q)/dq + J_0(q) = 0$, $J_2(q) = -J_0(q) + 2q^{-1}J_1(q) = -J_0(q) - 2q^{-1}dJ_0(q)/dq$, etc.[19] We then have

---

[17]These agree with Eq. (5) of Harkrider [23] if $\nu_\upsilon \to \mathrm{i}\nu_\upsilon$.

[18]More precisely, it should be called a **conical wave** because it also includes a plane wave in the z-direction, $e^{-\mathrm{i}\nu_\upsilon|z-h|}$ [3].

[19]Formulae used in Sects. 2.2.1–2.2.4 are mostly by Moriguchi et al. [54].

Fig. 2.12 Cylindrical coordinate system with the origin at the epicenter of a point source (modified from Koketsu [41] with permission of Kindai Kagaku)

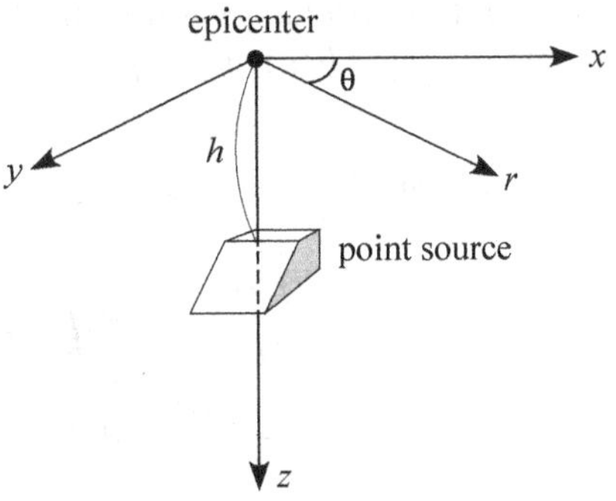

$$
\begin{aligned}
\overline{\Phi} &= \frac{\overline{M}_0(\omega)}{2\pi\rho\omega^2}\left[\frac{\sin 2\theta}{2}\left\{\frac{\partial^2 A_\alpha}{\partial r^2} - \frac{1}{r^2}\left(\frac{\partial^2 A_\alpha}{\partial \theta^2} + r\frac{\partial A_\alpha}{\partial r}\right)\right\} + \frac{\cos 2\theta}{r^2}\left(r\frac{\partial^2 A_\alpha}{\partial r \partial \theta} - \frac{\partial A_\alpha}{\partial \theta}\right)\right] \\
&= \frac{\overline{M}_0(\omega)}{4\pi\rho\omega^2}\sin 2\theta \int_0^\infty F_\alpha \left(\frac{d^2}{dr^2} - \frac{1}{r}\frac{d}{dr}\right) J_0(kr)\, dk \\
&= \frac{\overline{M}_0(\omega)}{4\pi\rho\omega^2}\sin 2\theta \int_0^\infty k^2 F_\alpha J_2(kr)\, dk
\end{aligned} \tag{2.77}
$$

where $J_1(kr)$ and $J_2(kr)$ are the first- and second-order Bessel functions, respectively.

We next substitute (2.75) into the Fourier transform of the vector potential $\boldsymbol{\Psi}$ (the second equation in (2.74)), and we use $\partial/\partial x = \cos\theta \cdot \partial/\partial r - \sin\theta/r \cdot \partial/\partial\theta$, $\partial/\partial y = \sin\theta \cdot \partial/\partial r + \cos\theta/r \cdot \partial/\partial\theta$, $\partial A_\beta/\partial\theta = 0$, $dJ_0(q)/dq = -J_1(q)$, etc. Then, for $\overline{\boldsymbol{\Psi}} = \nabla \times (\overline{\Psi}_r, \overline{\Psi}_\theta, \overline{\Psi}_z)$, we have

$$
\begin{aligned}
\overline{\Psi}_r &= \frac{1}{4\pi\rho}\left(\cos\theta\frac{\partial \overline{B}_\beta}{\partial y} + \sin\theta\frac{\partial \overline{B}_\beta}{\partial x}\right) = \frac{1}{4\pi\rho}\sin 2\theta\frac{\partial \overline{B}_\beta}{\partial r} \\
&= -\frac{\overline{M}_0(\omega)}{4\pi\rho\omega^2}\sin 2\theta\frac{\partial A_\beta}{\partial r} = \frac{\overline{M}_0(\omega)}{4\pi\rho\omega^2}\sin 2\theta\int_0^\infty F_\beta\, k J_1(kr)\, dk\,, \\
\overline{\Psi}_\theta &= \frac{1}{4\pi\rho}\left(-\sin\theta\frac{\partial \overline{B}_\beta}{\partial y} + \cos\theta\frac{\partial \overline{B}_\beta}{\partial x}\right) = \frac{1}{4\pi\rho}\cos 2\theta\frac{\partial \overline{B}_\beta}{\partial r} \\
&= -\frac{\overline{M}_0(\omega)}{4\pi\rho\omega^2}\cos 2\theta\frac{\partial A_\beta}{\partial r} = \frac{\overline{M}_0(\omega)}{4\pi\rho\omega^2}\cos 2\theta\int_0^\infty F_\beta\, k J_1(kr)\, dk\,, \\
\overline{\Psi}_z &= 0\,.
\end{aligned} \tag{2.78}
$$

The Fourier transform of the vector potential is related to those of the displacement potentials of SH and SV waves by

$$
\nabla \times \overline{\boldsymbol{\Psi}} = \nabla \times \nabla \times (0,\, 0,\, \overline{\Psi}) + \nabla \times (0,\, 0,\, \overline{\mathrm{X}})\,. \tag{2.79}
$$

We extract the z-component from the left-hand side of (2.79) and substitute (2.78) into this. Using $dJ_1(q)/dq = q^{-1}J_1(q) - J_2(q)$ etc., we then have

$$
\begin{aligned}
(\nabla \times \overline{\Psi})_z &= (\nabla \times \nabla \times (\overline{\Psi}_r, \overline{\Psi}_\theta, 0))_z = \frac{1}{r}\frac{\partial}{\partial r}\left(r\frac{\partial \overline{\Psi}_r}{\partial z}\right) + \frac{1}{r}\frac{\partial^2 \overline{\Psi}_\theta}{\partial \theta\, \partial z} \\
&= \frac{\overline{M}_0(\omega)}{4\pi\rho\omega^2}\sin 2\theta \int_0^\infty (-\mathrm{i}\epsilon\nu_\beta)F_\beta\, k\left(\frac{dJ_1(kr)}{dr} - \frac{J_1(kr)}{r}\right)dk \\
&= \frac{\overline{M}_0(\omega)}{4\pi\rho\omega^2}\sin 2\theta \int_0^\infty \mathrm{i}\epsilon\nu_\beta F_\beta\, k^2 J_2(kr)\, dk
\end{aligned}
\tag{2.80}
$$

where ϵ stands for the sign of $z - h$, that is $(z-h)/|z-h|$. We then extract the z-component from the right-hand side of (2.79) and obtain

$$
(\nabla \times \nabla \times (0,\, 0,\, \overline{\Psi}))_z + (\nabla \times (0,\, 0,\, \overline{\mathrm{X}}))_z = -\frac{1}{r}\frac{\partial}{\partial r}\left(r\frac{\partial \overline{\Psi}}{\partial r}\right) - \frac{1}{r^2}\frac{\partial^2 \overline{\Psi}}{\partial \theta^2}\,. \tag{2.81}
$$

Since (2.81) includes only $\overline{\Psi}$, the function $\overline{\Psi}$ should take the form

$$
\overline{\Psi} = \frac{\overline{M}_0(\omega)}{4\pi\rho\omega^2}\sin 2\theta \int_0^\infty \mathrm{i}\epsilon\nu_\beta F_\beta \Gamma(kr)\, dk \tag{2.82}
$$

according to (2.80). By substituting (2.82) into (2.81), we have

$$
\begin{aligned}
&-\frac{1}{r}\frac{\partial}{\partial r}\left(r\frac{\partial \overline{\Psi}}{\partial r}\right) - \frac{1}{r^2}\frac{\partial^2 \overline{\Psi}}{\partial \theta^2} = \\
&-\frac{\overline{M}_0(\omega)}{4\pi\rho\omega^2}\sin 2\theta \int_0^\infty \mathrm{i}\epsilon\nu_\beta F_\beta\, k^2\left\{\frac{d^2\Gamma(kr)}{d(kr)^2} + \frac{1}{kr}\frac{d\Gamma(kr)}{d(kr)} - \frac{4}{(kr)^2}\Gamma(kr)\right\}dk\,.
\end{aligned}
\tag{2.83}
$$

If we assume $\Gamma(kr) = J_2(kr)$, the $\{\ \ \}$ part of the integrand in (2.83) yields $d^2J_2(kr)/d(kr)^2 + 1/(kr)\cdot dJ_2(kr)/d(kr) - 4/(kr)^2 J_2(kr)$. This can be further reduced to $-J_2(kr)$, from the second-order Bessel equation $d^2J_2(q)/dq^2 + 1/q\cdot dJ_2(q)/dq + (1 - 4/q^2)J_2(q) = 0$. Therefore, (2.81) is equal to (2.80), so that the assumption $\Gamma(kr) = J_2(kr)$ is correct.

Next, we extract the r-component from the left-hand side of (2.79) and substitute (2.78) into this. Using the first-order Bessel function $d^2J_1(q)/dq^2 + 1/q\cdot dJ_1(q)/dq + (1 - 1/q^2)J_1(q) = 0$, $d^2J_1(q)/dq^2 = -J_1(q) + J_2(q)/q$, $dJ_2(q))/dq = J_1(q) - 2J_2(q)/q$, etc., we have

$$(\nabla \times \overline{\Psi})_r = (\nabla \times \nabla \times (\overline{\Psi}_r, \overline{\Psi}_\theta, 0))_r = \left(\nabla \times \left(-\frac{\partial \overline{\Psi}_\theta}{\partial z}, \frac{\partial \overline{\Psi}_r}{\partial z}, \frac{1}{r}\frac{\partial}{\partial r}(r\overline{\Psi}_\theta) - \frac{1}{r}\frac{\partial \overline{\Psi}_r}{\partial \theta}\right)\right)_r$$

$$= \frac{1}{r}\frac{\partial}{\partial \theta}\left\{\frac{1}{r}\frac{\partial}{\partial r}(r\overline{\Psi}_\theta) - \frac{1}{r}\frac{\partial \overline{\Psi}_r}{\partial \theta}\right\} - \frac{\partial^2 \overline{\Psi}_r}{\partial z^2} = \frac{\partial}{\partial \theta}\left(\frac{1}{r^2}\overline{\Psi}_\theta + \frac{1}{r}\frac{\partial \overline{\Psi}_\theta}{\partial r}\right) - \frac{1}{r^2}\frac{\partial^2 \overline{\Psi}_r}{\partial \theta^2} - \frac{\partial^2 \overline{\Psi}_r}{\partial z^2}$$

$$= \frac{\overline{M}_0(\omega)}{4\pi\rho\omega^2}\sin 2\theta \int_0^\infty F_\beta\, k \left\{-\frac{2}{r^2}J_1(kr) - \frac{2}{r}\frac{dJ_1(kr)}{dr} + \frac{4}{r^2}J_1(kr) + \nu_\beta^2 J_1(kr)\right\} dk$$

$$= \frac{\overline{M}_0(\omega)}{4\pi\rho\omega^2}\sin 2\theta \int_0^\infty F_\beta\, k \left\{\nu_\beta^2 \frac{dJ_2(kr)}{d(kr)} + 2k_\beta^2 \frac{J_2(kr)}{kr}\right\} dk\,. \tag{2.84}$$

We then extract the r-component from the right-hand side of

$$(\nabla \times \nabla \times (0, 0, \overline{\Psi}))_r + (\nabla \times (0, 0, \overline{\mathrm{X}}))_r = \frac{\partial^2 \overline{\Psi}}{\partial r \partial z} + \frac{1}{r}\frac{\partial \overline{\mathrm{X}}}{\partial \theta}\,. \tag{2.85}$$

Equation (2.82) with $\Gamma(kr) = J_2(kr)$ yields

$$\frac{\partial^2 \overline{\Psi}}{\partial r \partial z} = \frac{\overline{M}_0(\omega)}{4\pi\rho\omega^2}\sin 2\theta \int_0^\infty \nu_\beta^2 F_\beta \frac{dJ_2(kr)}{dr}\, dk \tag{2.86}$$

which is equivalent to the first term of (2.84). Therefore, the second term of (2.84) must agree with the second term of (2.85), so that

$$\frac{1}{r}\frac{\partial \overline{\mathrm{X}}}{\partial \theta} = \frac{\overline{M}_0(\omega)}{4\pi\rho\omega^2}\sin 2\theta \int_0^\infty 2k_\beta^2 F_\beta \frac{J_2(kr)}{r}\, dk\,. \tag{2.87}$$

Summarizing (2.77), (2.82) with $\Gamma(kr) = J_2(kr)$, and (2.87), we obtain the displacement potentials[20]

$$\overline{\Phi} = \frac{\overline{M}_0(\omega)}{4\pi\rho\omega^2}\sin 2\theta \int_0^\infty k^2 F_\alpha J_2(kr)\, dk\,,$$
$$\overline{\Psi} = \frac{\overline{M}_0(\omega)}{4\pi\rho\omega^2}\sin 2\theta \int_0^\infty \mathrm{i}\epsilon\nu_\beta F_\beta J_2(kr)\, dk\,,$$
$$\overline{\mathrm{X}} = -\frac{\overline{M}_0(\omega)}{4\pi\rho\omega^2}\cos 2\theta \int_0^\infty k_\beta^2 F_\beta J_2(kr)\, dk\,. \tag{2.88}$$

2.2.2 *Inclined Strike Slip Fault*

Equation (2.88) are valid only for the point source of a vertical strike slip fault perpendicular to a horizontal plane, as shown in Fig. 2.7 and the left panel of Fig. 2.13.

[20] These agree with Eq. (6) of Harkrider [23] if $\phi \to \theta$, $\nu_\gamma \to \mathrm{i}\nu_\gamma$, and $\mu\overline{D}(\omega) \to \overline{M}_0(\omega)$.

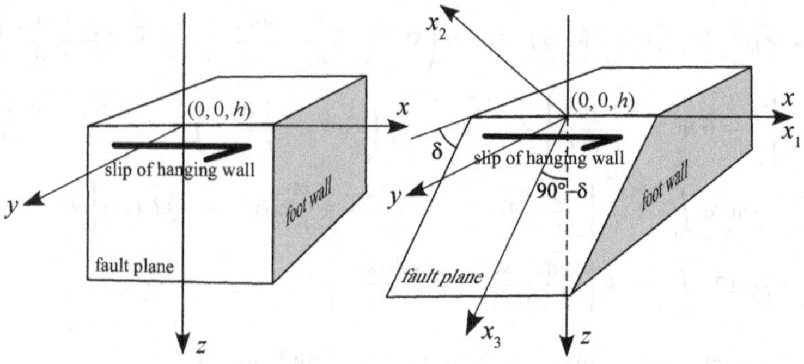

Fig. 2.13 Left: Vertical strike slip fault perpendicular to a horizontal plane (Fig. 2.7); Right: Strike slip fault inclined to an arbitrary dip angle δ (modified from Koketsu [41] with permission of Kindai Kagaku). In the right panel, (x_1, x_2, x_3) is the coordinate system fixed to the fault plane. Here, the intersection point of the x-, y-, and z-axes is not the origin but $(0, 0, h)$ (see Fig. 2.12)

We then extend these equations to the point source of strike slip fault inclined to an arbitrary **dip angle** of δ, as in Sato [62]. To obtain the $x_1x_2x_3$ coordinate system, the x_1-axis is fixed to the x-axis, and the y- and z-axes are rotated by $\delta' = 90° - \delta$ around the x-axis to the x_2- and x_3-axes. In this case, the two coordinate systems are connected by

$$x_1 = x\,, \quad x_2 = y\cos\delta' - z\sin\delta'\,, \quad x_3 = y\sin\delta' + z\cos\delta'\,. \tag{2.89}$$

Since, in the $x_1x_2x_3$ coordinate system, this point source is a left-lateral strike slip fault, perpendicular to the $x_1 - x_2$ plane, we have

$$\Phi = -\frac{1}{2\pi\rho}\frac{\partial^2 B_\alpha}{\partial x_1 \partial x_2}, \quad \boldsymbol{\Psi} = \nabla\times(\Psi_1, \Psi_2, \Psi_3) = \nabla\times\left(\frac{1}{4\pi\rho}\frac{\partial B_\beta}{\partial x_2}, \frac{1}{4\pi\rho}\frac{\partial B_\beta}{\partial x_1} 0\right) \tag{2.90}$$

from (2.74). Using (2.89), we convert the partial derivatives in (2.90) into those in the Cartesian coordinate system (x, y, z) as

$$\begin{aligned}
\Phi &= -\frac{1}{2\pi\rho}\frac{\partial}{\partial x}\left(\cos\delta'\frac{\partial}{\partial y} - \sin\delta'\frac{\partial}{\partial z}\right)B_\alpha = -\frac{1}{2\pi\rho}\frac{\partial}{\partial x}\left(\sin\delta\frac{\partial}{\partial y} - \cos\delta\frac{\partial}{\partial z}\right)B_\alpha\,,\\
\Psi_1 &= \frac{1}{4\pi\rho}\left(\cos\delta'\frac{\partial}{\partial y} - \sin\delta'\frac{\partial}{\partial z}\right)B_\beta = \frac{1}{4\pi\rho}\left(\sin\delta\frac{\partial}{\partial y} - \cos\delta\frac{\partial}{\partial z}\right)B_\beta\,,\\
\Psi_2 &= \frac{1}{4\pi\rho}\frac{\partial B_\beta}{\partial x}, \quad \Psi_3 = 0\,.
\end{aligned} \tag{2.91}$$

We then rewrite (Ψ_1, Ψ_2, Ψ_3) for the Cartesian coordinate system and obtain

$$
\begin{aligned}
\Phi &= \sin\delta\left(-\frac{1}{2\pi\rho}\frac{\partial^2 B_\alpha}{\partial x\partial y}\right) - \cos\delta\left(-\frac{1}{2\pi\rho}\frac{\partial^2 B_\alpha}{\partial x\partial z}\right), \\
\Psi_x &= \sin\delta\left(\frac{1}{4\pi\rho}\frac{\partial B_\beta}{\partial y}\right) - \cos\delta\left(\frac{1}{4\pi\rho}\frac{\partial B_\beta}{\partial z}\right), \\
\Psi_y &= \sin\delta\left(\frac{1}{4\pi\rho}\frac{\partial B_\beta}{\partial x}\right), \quad \Psi_z = -\cos\delta\left(\frac{1}{4\pi\rho}\frac{\partial B_\beta}{\partial x}\right). \qquad (2.92)
\end{aligned}
$$

In (2.92), the terms associated with $\sin\delta$ are identical to (2.74), so that they represent the component of a vertical strike slip fault, and further derivations were given in Sect. 2.2.1. The terms associated with $-\cos\delta$ have the forms

$$
\Phi = -\frac{1}{2\pi\rho}\frac{\partial^2 B_\alpha}{\partial x\partial z}, \quad \boldsymbol{\Psi} = \nabla\times\left(\frac{1}{4\pi\rho}\frac{\partial B_\beta}{\partial z}, 0, \frac{1}{4\pi\rho}\frac{\partial B_\beta}{\partial x}\right). \qquad (2.93)
$$

By comparing these with (2.74), we find them to be the component of a strike slip fault perpendicular to the $x-z$ plane.

Similarly to (2.74), substituting (2.75) into the Fourier transform of the scalar potential Φ, and using $\partial/\partial x = \cos\theta\cdot\partial/\partial r - \sin\theta/r\cdot\partial/\partial\theta, \partial/\partial y = \sin\theta\cdot\partial/\partial r + \cos\theta/r\cdot\partial/\partial\theta$, $\partial A_\alpha/\partial\theta = 0$, $dJ_0(q)/dq = -J_1(q)$, etc., we have

$$
\begin{aligned}
\overline{\Phi} &= \frac{\overline{M}_0(\omega)}{2\pi\rho\omega^2}(-\cos\theta)\frac{\partial^2 A_\alpha}{\partial r\partial z} = \frac{\overline{M}_0(\omega)}{4\pi\rho\omega^2}(-2\cos\theta)\int_0^\infty \frac{dF_\alpha}{dz}\frac{d}{dr}J_0(kr)\,dk \\
&= -\frac{\overline{M}_0(\omega)}{4\pi\rho\omega^2}\cos\theta\int_0^\infty (-2\mathrm{i}\epsilon k\nu_\alpha)F_\alpha J_1(kr)\,dk\,. \qquad (2.94)
\end{aligned}
$$

Next, substituting (2.75) into the Fourier transform of the vector potential $\overline{\boldsymbol{\Psi}} = \nabla\times(\overline{\Psi}_r, \overline{\Psi}_\theta, \overline{\Psi}_z)$ and using $\partial/\partial x = \cos\theta\cdot\partial/\partial r - \sin\theta/r\cdot\partial/\partial\theta$, $\partial/\partial y = \sin\theta\cdot\partial/\partial r + \cos\theta/r\cdot\partial/\partial\theta$, $\partial A_\beta/\partial\theta = 0$, $dJ_0(q)/dq = -J_1(q)$, etc., we have

$$
\begin{aligned}
\overline{\Psi}_r &= \frac{1}{4\pi\rho}\cos\theta\frac{\partial\overline{B}_\beta}{\partial z} = -\frac{\overline{M}_0(\omega)}{4\pi\rho\omega^2}\cos\theta\frac{\partial A_\beta}{\partial z} = \frac{\overline{M}_0(\omega)}{4\pi\rho\omega^2}\cos\theta\int_0^\infty \mathrm{i}\epsilon\nu_\beta F_\beta\, J_0(kr)\,dk\,, \\
\overline{\Psi}_\theta &= \frac{1}{4\pi\rho}(-\sin\theta)\frac{\partial\overline{B}_\beta}{\partial z} = \frac{\overline{M}_0(\omega)}{4\pi\rho\omega^2}\sin\theta\frac{\partial A_\beta}{\partial z} = \frac{\overline{M}_0(\omega)}{4\pi\rho\omega^2}\sin\theta\int_0^\infty (-\mathrm{i}\epsilon\nu_\beta)F_\beta\, J_0(kr)\,dk\,, \\
\overline{\Psi}_z &= \frac{1}{4\pi\rho}\frac{\partial\overline{B}_\beta}{\partial x} = -\frac{\overline{M}_0(\omega)}{4\pi\rho\omega^2}\cos\theta\frac{\partial A_\beta}{\partial r} = \frac{\overline{M}_0(\omega)}{4\pi\rho\omega^2}\cos\theta\int_0^\infty F_\beta\, kJ_1(kr)\,dk\,. \qquad (2.95)
\end{aligned}
$$

As in Sect. 2.2.1, we extract the z-component from (2.79) and have

$$
\begin{aligned}
(\nabla\times\overline{\boldsymbol{\Psi}})_z &= (\nabla\times\nabla\times(\overline{\Psi}_r, \overline{\Psi}_\theta, \overline{\Psi}_z))_z \\
&= \frac{1}{r}\frac{\partial}{\partial r}\left\{r\,(\nabla\times(\overline{\Psi}_r, \overline{\Psi}_\theta, \overline{\Psi}_z))_\theta\right\} - \frac{1}{r}\frac{\partial}{\partial\theta}\left\{(\nabla\times(\overline{\Psi}_r, \overline{\Psi}_\theta, \overline{\Psi}_z))_r\right\} \\
&= \left(\frac{1}{r}\frac{\partial}{\partial z} + \frac{\partial^2}{\partial r\partial z}\right)\overline{\Psi}_r + \frac{1}{r}\frac{\partial^2\overline{\Psi}_\theta}{\partial\theta\partial z} - \left(\frac{\partial^2}{\partial r^2} + \frac{1}{r}\frac{\partial}{\partial r} + \frac{1}{r^2}\frac{\partial^2}{\partial\theta^2}\right)\overline{\Psi}_z\,. \qquad (2.96)
\end{aligned}
$$

We then substitute (2.95), and use $dJ_0(q)/dq = -J_1(q)$ and the first-order Bessel equation, obtaining

$$
\begin{aligned}
(\nabla \times \overline{\boldsymbol{\Psi}})_z &= \frac{\overline{M}_0(\omega)}{4\pi\rho\omega^2}\cos\theta \int_0^\infty \mathrm{i}\epsilon\nu_\beta(-\mathrm{i}\epsilon\nu_\beta)F_\beta \frac{dJ_0(kr)}{dr}\,dk \\
&\quad - \frac{\overline{M}_0(\omega)}{4\pi\rho\omega^2}\cos\theta \int_0^\infty F_\beta\, k\left(\frac{\partial^2}{\partial r^2} + \frac{1}{r}\frac{\partial}{\partial r} - \frac{1}{r^2}\right) J_1(kr)\,dk \\
&= -\frac{\overline{M}_0(\omega)}{4\pi\rho\omega^2}\cos\theta \int_0^\infty \frac{k_\beta^2 - 2k^2}{k} F_\beta\, k^2 J_1(kr)\,dk\,. \qquad (2.97)
\end{aligned}
$$

Conversely, for the z-component (2.81) extracted from (2.79), we assume the form

$$
\overline{\Psi} = -\frac{\overline{M}_0(\omega)}{4\pi\rho\omega^2}\cos\theta \int_0^\infty \frac{k_\beta^2 - 2k^2}{k} F_\beta \Gamma(kr)\,dk \qquad (2.98)
$$

based on (2.97). Using this form, we have

$$
\begin{aligned}
&-\frac{1}{r}\frac{\partial}{\partial r}\left(r\frac{\partial\overline{\Psi}}{\partial r}\right) - \frac{1}{r^2}\frac{\partial^2\overline{\Psi}}{\partial\theta^2} = \qquad (2.99) \\
&-\frac{\overline{M}_0(\omega)}{4\pi\rho\omega^2}\cos\theta \int_0^\infty \frac{k_\beta^2 - 2k^2}{-k} F_\beta\, k^2 \left\{\frac{d^2\Gamma(kr)}{d(kr)^2} + \frac{1}{kr}\frac{d\Gamma(kr)}{d(kr)} - \frac{1}{(kr)^2}\Gamma(kr)\right\} dk\,.
\end{aligned}
$$

The { } part of the integrand yields $-J_1(kr)$ when $\Gamma(kr) = J_1(kr)$, and the first-order Bessel equation $d^2J_1(q)/dq^2 + 1/q \cdot dJ_1(q)/dq + (1 - 1/q^2)J_1(q) = 0$ is used. Therefore, the z-components of both sides of (2.80) agree with each other.

We next examine the r-component of (2.79). The left-hand side is

$$
\begin{aligned}
(\nabla \times \overline{\boldsymbol{\Psi}})_r &= \frac{1}{r}\frac{\partial}{\partial\theta}\left\{(\nabla \times (\overline{\Psi}_r, \overline{\Psi}_\theta, \overline{\Psi}_z))_z\right\} - \frac{\partial}{\partial z}\left\{(\nabla \times (\overline{\Psi}_r, \overline{\Psi}_\theta, \overline{\Psi}_z))_\theta\right\} \\
&= \frac{1}{r}\frac{\partial}{\partial\theta}\left\{\frac{1}{r}\frac{\partial}{\partial r}(r\overline{\Psi}_\theta) - \frac{1}{r}\frac{\partial\overline{\Psi}_r}{\partial\theta}\right\} - \frac{\partial}{\partial z}\left(\frac{\partial\overline{\Psi}_r}{\partial z} - \frac{\partial\overline{\Psi}_z}{\partial r}\right) \\
&= -\left(\frac{1}{r^2}\frac{\partial^2}{\partial\theta^2} + \frac{\partial^2}{\partial z^2}\right)\overline{\Psi}_r + \left(\frac{1}{r^2} + \frac{1}{r}\frac{\partial}{\partial r}\right)\frac{\partial\overline{\Psi}_\theta}{\partial\theta} + \frac{\partial^2\overline{\Psi}_z}{\partial r\partial z}\,. \qquad (2.100)
\end{aligned}
$$

Substituting (2.95) and using $dJ_0(q)/dq = -J_1(q)$, $J_0(q) = dJ_1(q)/dq + J_1(q)/q$, etc., we have

$$(\nabla \times \overline{\Psi})_r = \frac{\overline{M}_0(\omega)}{4\pi\rho\omega^2}\cos\theta \int_0^\infty \mathrm{i}\epsilon\nu_\beta F_\beta \left\{ \nu_\beta^2 J_0(kr) - \frac{1}{r}\frac{dJ_0(kr)}{dr} - k\frac{dJ_1(kr)}{dr} \right\} dk$$

$$= \frac{\overline{M}_0(\omega)}{4\pi\rho\omega^2}\cos\theta \int_0^\infty \mathrm{i}\epsilon\nu_\beta F_\beta \left[\nu_\beta^2 \left\{ \frac{dJ_1(kr)}{d(kr)} + \frac{1}{kr}J_1(kr) \right\} + \frac{k}{r}J_1(kr) - k^2\frac{dJ_1(kr)}{d(kr)} \right] dk$$

$$= \frac{\overline{M}_0(\omega)}{4\pi\rho\omega^2}\cos\theta \int_0^\infty \mathrm{i}\epsilon\nu_\beta F_\beta \left\{ \frac{k_\beta^2 - 2k^2}{k}\frac{dJ_1(kr)}{dr} + \frac{k_\beta^2}{k}\frac{J_1(kr)}{r} \right\} dk \,. \tag{2.101}$$

The right-hand side, which is the first term of (2.85), is obtained from (2.98) and $\Gamma(kr) = J_1(kr)$ as

$$\frac{\partial^2 \overline{\Psi}}{\partial r \partial z} = \frac{\overline{M}_0(\omega)}{4\pi\rho\omega^2}\cos\theta \int_0^\infty \mathrm{i}\epsilon\nu_\beta F_\beta \frac{k_\beta^2 - 2k^2}{k}\frac{dJ_1(kr)}{dr}\, dk \,. \tag{2.102}$$

This is identical to the part of (2.101) related to the first term in { }. Therefore, the other part of (2.101) related to the second term in { } must be identical to the second term of (2.85), namely

$$\frac{1}{r}\frac{\partial \overline{X}}{\partial \theta} = \frac{\overline{M}_0(\omega)}{4\pi\rho\omega^2}\cos\theta \int_0^\infty \frac{\mathrm{i}\epsilon\nu_\beta k_\beta^2}{k} F_\beta \frac{J_1(kr)}{r}\, dk \,. \tag{2.103}$$

Combining (2.95), (2.98), and (2.103), we have

$$\overline{\Phi} = -\frac{\overline{M}_0(\omega)}{4\pi\rho\omega^2}\cos\theta \int_0^\infty (-2\mathrm{i}\epsilon k\nu_\alpha) F_\alpha J_1(kr)\, dk \,,$$

$$\overline{\Psi} = -\frac{\overline{M}_0(\omega)}{4\pi\rho\omega^2}\cos\theta \int_0^\infty \frac{k_\beta^2 - 2k^2}{k} F_\beta J_1(kr)\, dk \,,$$

$$\overline{X} = \frac{\overline{M}_0(\omega)}{4\pi\rho\omega^2}\sin\theta \int_0^\infty \frac{\mathrm{i}\epsilon k_\beta^2 \nu_\beta}{k} F_\beta J_1(kr)\, dk \,. \tag{2.104}$$

From (2.92), we already know that the displacement potentials of an inclined **strike slip fault** are represented by linear combinations of the displacement potentials of strike slip faults perpendicular to the $x-y$ and $x-z$ planes, with coefficients $\sin\delta$ and $-\cos\delta$. Therefore, using (2.88) and (2.104), we have the displacement potentials[21]

[21] These agree with Eqs. (53) and (54) of Sato [62] if $\varphi \to \theta$, $\xi \to k$, $k \to k_\beta$, $\nu_\nu \to \mathrm{i}\nu_\nu$, $|z| \to |z-h|$, and the slip is reversed. Therefore, the equations of Sato [62] give the displacement potentials of a right-lateral strike slip fault.

$$\overline{\Phi} = \frac{\overline{M}_0(\omega)}{4\pi\rho\omega^2}\int_0^{\infty}\left[\sin\delta\sin 2\theta\, k^2 J_2(kr) + \cos\delta\cos\theta\,(-2\mathrm{i}\epsilon k\nu_\alpha)\, J_1(kr)\right] F_\alpha dk, \quad (2.105)$$

$$\overline{\Psi} = \frac{\overline{M}_0(\omega)}{4\pi\rho\omega^2}\int_0^{\infty}\left[\sin\delta\sin 2\theta\,\mathrm{i}\epsilon\nu_\beta J_2(kr) + \cos\delta\cos\theta\frac{k_\beta^2 - 2k^2}{k} J_1(kr)\right] F_\beta\, dk\,,$$

$$\overline{\mathrm{X}} = \frac{\overline{M}_0(\omega)}{4\pi\rho\omega^2}\int_0^{\infty}\left[\sin\delta\frac{d\sin 2\theta}{d\theta}\frac{-k_\beta^2}{2} J_2(kr) + \cos\delta\frac{d\cos\theta}{d\theta}\frac{\mathrm{i}\epsilon k_\beta^2 \nu_\beta}{k} J_1(kr)\right] F_\beta\, dk\,.$$

2.2.3 Vertical Dip Slip Fault

In the case of the point source of a vertical **dip slip fault** perpendicular to the $x - y$ plane, we rotate the Cartesian coordinate system (x, y, z) about the y-axis by 90° to the $x'y'z'$ coordinate system, where the vertical dip slip fault can be seen as a left-lateral strike slip fault perpendicular to the $x' - y'$ plane (Fig. 2.14 Left). Therefore, (2.74) is rewritten for the $x'y'z'$ coordinate system as

$$\Phi = -\frac{1}{2\pi\rho}\frac{\partial^2 B_\alpha}{\partial x'\partial y'}, \quad \boldsymbol{\Psi} = \nabla\times(\Psi_{x'}, \Psi_{y'}, \Psi_{z'}) = \nabla\times\left(\frac{1}{4\pi\rho}\frac{\partial B_\beta}{\partial y'}, \frac{1}{4\pi\rho}\frac{\partial B_\beta}{\partial x'}, 0\right). \quad (2.106)$$

Since the conversion between the two coordinate systems is simply conducted using the replacements $x' \leftrightarrow -z$ and $z' \leftrightarrow x$, (2.106) is further rewritten for the xyz coordinate system as

$$\Phi = \frac{1}{2\pi\rho}\frac{\partial B_\alpha}{\partial y\partial z}, \quad \boldsymbol{\Psi} = \nabla\times(\Psi_x, \Psi_y, \Psi_z)\,,$$

$$\Psi_x = \Psi_{z'} = 0\,,\ \Psi_y = \Psi_{y'} = -\frac{1}{4\pi\rho}\frac{\partial B_\beta}{\partial z}\,,\ \Psi_z = -\Psi_{x'} = -\frac{1}{4\pi\rho}\frac{\partial B_\beta}{\partial y}\,. \quad (2.107)$$

Substituting (2.75) into the Fourier transform of the scalar potential in (2.107) and using $\partial/\partial y = \sin\theta\cdot\partial/\partial r + \cos\theta/r\cdot\partial/\partial\theta$, $\partial A_\alpha/\partial\theta = 0$, $dJ_0(q)/dq = -J_1(q)$, etc., we have

$$\begin{aligned}\overline{\Phi} &= -\frac{\overline{M}_0(\omega)}{2\pi\rho\omega^2}\sin\theta\frac{\partial^2 A_\alpha}{\partial r\partial z} = -\frac{\overline{M}_0(\omega)}{4\pi\rho\omega^2} 2\sin\theta\int_0^{\infty}(-\mathrm{i}\epsilon\nu_\alpha)F_\alpha\frac{dJ_0(kr)}{dr}dk \\ &= \frac{\overline{M}_0(\omega)}{4\pi\rho\omega^2}\sin\theta\int_0^{\infty}(-2\mathrm{i}\epsilon k\nu_\alpha)F_\alpha J_1(kr)\, dk\,. \end{aligned} \quad (2.108)$$

Also substituting (2.75) into the Fourier transform of the vector potential in (2.107) and using $\partial/\partial x = \cos\theta\cdot\partial/\partial r - \sin\theta/r\cdot\partial/\partial\theta$, $\partial/\partial y = \sin\theta\cdot\partial/\partial r + \cos\theta/r\cdot\partial/\partial\theta$, $\partial A_\beta/\partial\theta = 0$, $dJ_0(q)/dq = -J_1(q)$, etc., for $\overline{\boldsymbol{\Psi}} = \nabla\times(\overline{\Psi}_r, \overline{\Psi}_\theta, \overline{\Psi}_z)$, we have

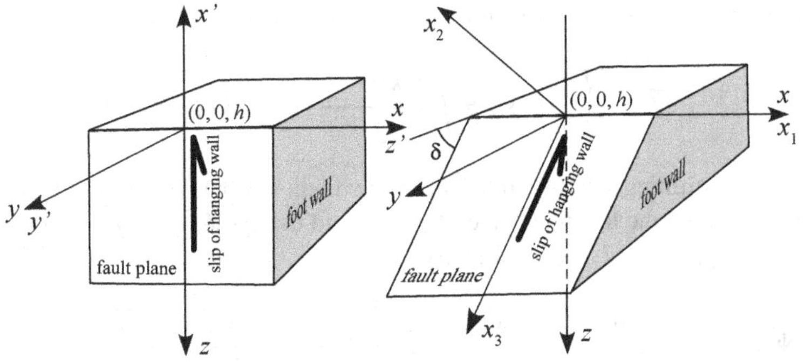

Fig. 2.14 Left: Vertical dip slip fault perpendicular to a horizontal plane; Right: Dip slip fault inclined by an arbitrary dip angle δ (modified from Koketsu [41] with permission of Kindai Kagaku). In the right panel, (x_1, x_2, x_3) is the coordinate system fixed to the fault plane. Here, the intersection point of the x-, y-, and z-axes is not the origin but $(0, 0, h)$ (see Fig. 2.12)

$$\begin{aligned}
\overline{\Psi}_r &= \sin\theta\left(-\frac{1}{4\pi\rho}\frac{\partial\overline{B}_\beta}{\partial z}\right) = \frac{\overline{M}_0(\omega)}{4\pi\rho\omega^2}\sin\theta\int_0^\infty(-\mathrm{i}\epsilon\nu_\beta)F_\beta\, J_0(kr)\,dk\ ,\\
\overline{\Psi}_\theta &= \cos\theta\left(-\frac{1}{4\pi\rho}\frac{\partial\overline{B}_\beta}{\partial z}\right) = \frac{\overline{M}_0(\omega)}{4\pi\rho\omega^2}\cos\theta\int_0^\infty(-\mathrm{i}\epsilon\nu_\beta)F_\beta\, J_0(kr)\,dk\ ,\\
\overline{\Psi}_z &= -\frac{1}{4\pi\rho}\frac{\partial\overline{B}_\beta}{\partial y} = \frac{\overline{M}_0(\omega)}{4\pi\rho\omega^2}\sin\theta\int_0^\infty F_\beta\,\frac{dJ_0(kr)}{dr}\,dk\\
&= -\frac{\overline{M}_0(\omega)}{4\pi\rho\omega^2}\sin\theta\int_0^\infty F_\beta\, kJ_1(kr)\,dk\ .
\end{aligned} \tag{2.109}$$

Substituting (2.109) into (2.96), which is obtained by extracting the z-component from the left-hand side of (2.79) in Sect. 2.2.2, and using $dJ_0(q)/dq = -J_1(q)$, we have

$$\begin{aligned}
(\nabla\times\overline{\boldsymbol{\Psi}})_z &= \left(\frac{1}{r}\frac{\partial}{\partial z}+\frac{\partial^2}{\partial r\partial z}\right)\overline{\Psi}_r + \frac{1}{r}\frac{\partial^2\overline{\Psi}_\theta}{\partial\theta\partial z} - \left(\frac{\partial^2}{\partial r^2}+\frac{1}{r}\frac{\partial}{\partial r}+\frac{1}{r^2}\frac{\partial^2}{\partial\theta^2}\right)\overline{\Psi}_z\\
&= \frac{\overline{M}_0(\omega)}{4\pi\rho\omega^2}\sin\theta\int_0^\infty(-\mathrm{i}\epsilon\nu_\beta)^2F_\beta\,\frac{dJ_0(kr)}{dr}\,dk\\
&+ \frac{\overline{M}_0(\omega)}{4\pi\rho\omega^2}\sin\theta\int_0^\infty F_\beta\, k\left(\frac{\partial^2}{\partial r^2}+\frac{1}{r}\frac{\partial}{\partial r}-\frac{1}{r^2}\right)J_1(kr)\,dk\\
&= \frac{\overline{M}_0(\omega)}{4\pi\rho\omega^2}\sin\theta\int_0^\infty\frac{k_\beta^2-2k^2}{k}F_\beta\, k^2\, J_1(kr)\,dk\ .
\end{aligned} \tag{2.110}$$

We then perform the same derivations as in Sect. 2.2.2 and obtain

$$\overline{\Psi} = \frac{\overline{M}_0(\omega)}{4\pi\rho\omega^2}\sin\theta \int_0^\infty \frac{k_\beta^2 - 2k^2}{k} F_\beta J_1(kr)\,dk\,. \tag{2.111}$$

Similarly, substituting (2.109) into (2.100), which is obtained by extracting the r-component from the left-hand side of (2.79), and using $dJ_0(q)/dq = -J_1(q)$ and $J_0(q) = dJ_1(q)/dq + J_1(q)/q$, we have

$$\begin{aligned}(\nabla \times \overline{\boldsymbol{\Psi}})_r &= -\left(\frac{1}{r^2}\frac{\partial^2}{\partial\theta^2} + \frac{\partial^2}{\partial z^2}\right)\overline{\Psi}_r + \left(\frac{1}{r^2} + \frac{1}{r}\frac{\partial}{\partial r}\right)\frac{\partial\overline{\Psi}_\theta}{\partial\theta} + \frac{\partial^2\overline{\Psi}_z}{\partial r\partial z} \\ &= \frac{\overline{M}_0(\omega)}{4\pi\rho\omega^2}\sin\theta \int_0^\infty \mathrm{i}\epsilon\nu_\beta F_\beta\left\{-\nu_\beta^2 J_0(kr) + \frac{1}{r}\frac{dJ_0(kr)}{dr} + k\frac{dJ_1(kr)}{dr}\right\}dk \\ &= \frac{\overline{M}_0(\omega)}{4\pi\rho\omega^2}\sin\theta \int_0^\infty (-\mathrm{i}\epsilon\nu_\beta)F_\beta\left\{\frac{k_\beta^2 - 2k^2}{k}\frac{dJ_1(kr)}{dr} + \frac{k_\beta^2}{k}\frac{J_1(kr)}{r}\right\}dk\,.\end{aligned} \tag{2.112}$$

Conversely, the first term of (2.85), which is obtained by extracting the r-component from the right-hand side of (2.79), is given as

$$\frac{\partial^2\overline{\Psi}}{\partial r\partial z} = \frac{\overline{M}_0(\omega)}{4\pi\rho\omega^2}\sin\theta \int_0^\infty (-\mathrm{i}\epsilon\nu_\beta)F_\beta \frac{k_\beta^2 - 2k^2}{k}\frac{dJ_1(kr)}{dr}\,dk \tag{2.113}$$

using (2.111). This is identical to the part of (2.112) related to the first term in { }. Therefore, the other part related to the second term in { } must be identical to the second term of (2.85), namely

$$\frac{1}{r}\frac{\partial\overline{\mathrm{X}}}{\partial\theta} = -\frac{\overline{M}_0(\omega)}{4\pi\rho\omega^2}\sin\theta \int_0^\infty \frac{\mathrm{i}\epsilon k_\beta^2\nu_\beta}{k} F_\beta \frac{J_1(kr)}{r}\,dk\,. \tag{2.114}$$

Combining (2.109), (2.111), and (2.114), we have the displacement potentials[22]

$$\begin{aligned}\overline{\Phi} &= \frac{\overline{M}_0(\omega)}{4\pi\rho\omega^2}\sin\theta \int_0^\infty (-2\mathrm{i}\epsilon k\nu_\alpha)F_\alpha J_1(kr)\,dk\,, \\ \overline{\Psi} &= \frac{\overline{M}_0(\omega)}{4\pi\rho\omega^2}\sin\theta \int_0^\infty \frac{k_\beta^2 - 2k^2}{k} F_\beta J_1(kr)\,dk\,, \\ \overline{\mathrm{X}} &= \frac{\overline{M}_0(\omega)}{4\pi\rho\omega^2}\cos\theta \int_0^\infty \frac{\mathrm{i}\epsilon k_\beta^2\nu_\beta}{k} F_\beta J_1(kr)\,dk\,.\end{aligned} \tag{2.115}$$

[22]These agree with Eq. (7) of Harkrider [23] if $\phi \to \theta$, $\nu_\nu \to \mathrm{i}\nu_\nu$, and $\mu\overline{D}(\omega) \to \overline{M}_0(\omega)$.

2.2.4 Inclined Dip Slip Fault

The result of the previous subsection may be extended to situations when the fault plane is inclined from the horizontal plane with an arbitrary **dip angle** δ, as shown in the right diagram of Fig. 2.14. Here, the $x_1x_2x_3$ coordinate system fixed to the fault plane and the xyz coordinate system are connected by (2.89). For the $x_1x_2x_3$ coordinate system, the point source is a **dip slip fault** with reverse faulting perpendicular to the $x_1 - x_2$ plane, therefore (2.107) holds for the $x_1x_2x_3$ coordinate system in the form

$$
\begin{aligned}
\Phi &= \frac{1}{2\pi\rho}\frac{\partial^2 B_\alpha}{\partial x_2 \partial x_3}\,,\\
\boldsymbol{\Psi} &= \nabla \times (\Psi_1, \Psi_2, \Psi_3) = \nabla \times \left(0, -\frac{1}{4\pi\rho}\frac{\partial B_\beta}{\partial x_3}, -\frac{1}{4\pi\rho}\frac{\partial B_\beta}{\partial x_2}\right).
\end{aligned} \tag{2.116}
$$

Using the coordinate conversion (2.89), we rewrite the partial derivatives in (2.116) for the Cartesian coordinate system (x, y, z) and obtain

$$
\begin{aligned}
\Phi &= \frac{1}{2\pi\rho}\left(\cos\delta'\frac{\partial}{\partial y} - \sin\delta'\frac{\partial}{\partial z}\right)\left(\sin\delta'\frac{\partial}{\partial y} + \cos\delta'\frac{\partial}{\partial z}\right)B_\alpha\\
&= \frac{1}{2\pi\rho}\left(\sin\delta\frac{\partial}{\partial y} - \cos\delta\frac{\partial}{\partial z}\right)\left(\cos\delta\frac{\partial}{\partial y} + \sin\delta\frac{\partial}{\partial z}\right)B_\alpha\,,\\
\Psi_1 &= 0\,, \\
\Psi_2 &= -\frac{1}{4\pi\rho}\left(\sin\delta'\frac{\partial}{\partial y} + \cos\delta'\frac{\partial}{\partial z}\right)B_\beta = -\frac{1}{4\pi\rho}\left(\cos\delta\frac{\partial}{\partial y} + \sin\delta\frac{\partial}{\partial z}\right)B_\beta\,,\\
\Psi_3 &= -\frac{1}{4\pi\rho}\left(\cos\delta'\frac{\partial}{\partial y} - \sin\delta'\frac{\partial}{\partial z}\right)B_\beta = -\frac{1}{4\pi\rho}\left(\sin\delta\frac{\partial}{\partial y} - \cos\delta\frac{\partial}{\partial z}\right)B_\beta\,.
\end{aligned} \tag{2.117}
$$

Furthermore, we rewrite (Ψ_1, Ψ_2, Ψ_3) for the xyz coordinate system and obtain

$$
\begin{aligned}
\Phi &= \sin 2\delta\left\{\frac{1}{4\pi\rho}\left(\frac{\partial^2 B_\alpha}{\partial y^2} - \frac{\partial^2 B_\alpha}{\partial z^2}\right\}\right) - \cos 2\delta\left(\frac{1}{2\pi\rho}\frac{\partial^2 B_\alpha}{\partial y \partial z}\right),\\
\Psi_x &= 0\,,\\
\Psi_y &= \Psi_2\sin\delta + \Psi_3\cos\delta = -\sin 2\delta\left(\frac{1}{4\pi\rho}\frac{\partial B_\beta}{\partial y}\right) + \cos 2\delta\left(\frac{1}{4\pi\rho}\frac{\partial B_\beta}{\partial z}\right),\\
\Psi_z &= -\Psi_2\cos\delta + \Psi_3\sin\delta = \cos 2\delta\left(\frac{1}{4\pi\rho}\frac{\partial B_\beta}{\partial y}\right) + \sin 2\delta\left(\frac{1}{4\pi\rho}\frac{\partial B_\beta}{\partial z}\right).
\end{aligned} \tag{2.118}
$$

In (2.118), the terms associated with $-\cos 2\delta$ are identical to (2.107), so that they represent the component of a vertical dip slip fault, and further derivations were given in Sect. 2.2.3.

The terms associated with $\sin 2\delta$ have the forms

$$\Phi = \frac{1}{4\pi\rho}\left(\frac{\partial^2 B_\alpha}{\partial y^2} - \frac{\partial^2 B_\alpha}{\partial z^2}\right), \quad \boldsymbol{\Psi} = \nabla \times \left(0, -\frac{1}{4\pi\rho}\frac{\partial B_\beta}{\partial y}, \frac{1}{4\pi\rho}\frac{\partial B_\beta}{\partial z}\right). \quad (2.119)$$

As $\sin 2\delta = 1$ and $\cos 2\delta = 0$ for $\delta = 45°$, these ought to represent the component of a dip slip fault with $\delta = 45°$.

Similarly to (2.74), substituting (2.75) into the Fourier transform of the scalar potential Φ and using $\partial^2/\partial y^2 = \sin^2\theta \cdot \partial^2/\partial r^2 + \sin 2\theta/r^2(r\partial^2/(\partial r\partial\theta) - \partial/\partial\theta) + \cos^2\theta/r^2(\partial^2/\partial\theta^2 + r\partial/\partial r)$, $\partial A_\alpha/\partial\theta = 0$, $J_2(q) = -J_0(q) + 2q^{-1}J_1(q) = -J_0(q) - 2q^{-1}dJ_0(q)/dq$, $d^2J_0(q)/dq^2 = 1/2 \cdot (J_2(q) - J_0(q))$, etc., we have

$$\begin{aligned}
\overline{\Phi} &= \frac{\overline{M}_0(\omega)}{4\pi\rho\omega^2}\left\{-\sin^2\theta\frac{\partial^2 A_\alpha}{\partial r^2} - \cos^2\theta\frac{1}{r}\frac{\partial A_\alpha}{\partial r} + \frac{\partial^2 A_\alpha}{\partial z^2}\right\} \\
&= \frac{\overline{M}_0(\omega)}{4\pi\rho\omega^2}\int_0^\infty F_\alpha\left\{-\sin^2\theta\frac{d^2J_0(kr)}{dr^2} - \cos^2\theta\frac{1}{r}\frac{dJ_0(kr)}{dr} - \nu_\alpha^2 J_0(kr)\right\}dk \\
&= \frac{\overline{M}_0(\omega)}{4\pi\rho\omega^2}\int_0^\infty F_\alpha\left\{-\sin^2\theta\, k^2\frac{J_2(kr) - J_0(kr)}{2}\right. \\
&\qquad \left. -\cos^2\theta\, k^2\frac{-J_0(kr) - J_2(kr)}{2} - \nu_\alpha^2 J_0(kr)\right\}dk \\
&= \frac{\overline{M}_0(\omega)}{4\pi\rho\omega^2}\frac{1}{2}\int_0^\infty F_\alpha\left\{(k^2 - 2\nu_\alpha^2)J_0(kr) + \cos 2\theta\, k^2 J_2(kr)\right\}dk\,. \qquad (2.120)
\end{aligned}$$

Also substituting (2.75) into the Fourier transform of the vector potential in (2.119) and again using $\partial/\partial x = \cos\theta \cdot \partial/\partial r - \sin\theta/r \cdot \partial/\partial\theta$, $\partial/\partial y = \sin\theta \cdot \partial/\partial r + \cos\theta/r \cdot \partial/\partial\theta$, $\partial A_\beta/\partial\theta = 0$, $dJ_0(q)/dq = -J_1(q)$, etc., for $\overline{\boldsymbol{\Psi}} = \nabla \times (\overline{\Psi}_r, \overline{\Psi}_\theta, \overline{\Psi}_z)$, we have

$$\begin{aligned}
\overline{\Psi}_r &= \sin\theta\left(-\frac{1}{4\pi\rho}\frac{\partial\overline{B}_\beta}{\partial y}\right) = \frac{\overline{M}_0(\omega)}{4\pi\rho\omega^2}\sin^2\theta\int_0^\infty F_\beta\{-kJ_1(kr)\}\,dk\,, \\
\overline{\Psi}_\theta &= \cos\theta\left(-\frac{1}{4\pi\rho}\frac{\partial\overline{B}_\beta}{\partial y}\right) = \frac{\overline{M}_0(\omega)}{4\pi\rho\omega^2}\frac{1}{2}\sin 2\theta\int_0^\infty F_\beta\{-kJ_1(kr)\}\,dk\,, \\
\overline{\Psi}_z &= \frac{1}{4\pi\rho}\frac{\partial\overline{B}_\beta}{\partial z} = \frac{\overline{M}_0(\omega)}{4\pi\rho\omega^2}\int_0^\infty \mathrm{i}\epsilon\nu_\beta F_\beta J_0(kr)\,dk\,. \qquad (2.121)
\end{aligned}$$

Substituting (2.121) into (2.96), which is the z-component extracted from the left-hand side of (2.79) in Sect. 2.2.2, and using the zeroth order Bessel equation $d^2J_0(q)/dq^2 + 1/q \cdot dJ_0(q)/dq + J_0(q) = 0$, $dJ_1(q))/dq = J_0(q) - J_1(q)/q$, $J_2(q) = -J_0(q) + 2q^{-1}J_1(q)$, etc., we have[23]

---

[23]This is a difficult transformation, but can be reached if we work back from the last equation.

$$
\begin{aligned}
(\nabla\times\overline{\Psi})_z &= \left(\frac{1}{r}\frac{\partial}{\partial z}+\frac{\partial^2}{\partial r\partial z}\right)\overline{\Psi}_r+\frac{1}{r}\frac{\partial^2\overline{\Psi}_\theta}{\partial\theta\partial z}-\left(\frac{\partial^2}{\partial r^2}+\frac{1}{r}\frac{\partial}{\partial r}+\frac{1}{r^2}\frac{\partial^2}{\partial\theta^2}\right)\overline{\Psi}_z\\
&=\frac{\overline{M}_0(\omega)}{4\pi\rho\omega^2}\int_0^\infty \mathrm{i}\epsilon\nu_\beta k\,F_\beta\left\{\sin^2\theta\,kJ_0(kr)+\cos 2\theta\frac{J_1(kr)}{r}+kJ_0(kr)\right\}dk\\
&=\frac{\overline{M}_0(\omega)}{4\pi\rho\omega^2}\int_0^\infty \mathrm{i}\epsilon\nu_\beta\,k\,F_\beta\left\{\frac{3}{2}kJ_0(kr)-\frac{1}{2}(1-2\sin^2\theta)\,kJ_0(kr)+\cos 2\theta\frac{J_1(kr)}{r}\right\}dk\\
&=\frac{\overline{M}_0(\omega)}{4\pi\rho\omega^2}\int_0^\infty \mathrm{i}\epsilon\nu_\beta\,k^2F_\beta\left\{\frac{3}{2}kJ_0(kr)+\frac{1}{2}\cos 2\theta\left(-J_0(kr)+2\frac{J_1(kr)}{kr}\right)\right\}dk\\
&=\frac{\overline{M}_0(\omega)}{4\pi\rho\omega^2}\int_0^\infty F_\beta\,k^2\left\{\frac{3}{2}\,\mathrm{i}\epsilon\nu_\beta J_0(kr)+\frac{1}{2}\cos 2\theta\;\mathrm{i}\epsilon\nu_\beta J_2(kr)\right\}dk\,. \qquad (2.122)
\end{aligned}
$$

We then perform the same derivations as in Sect. 2.2.2 and have

$$
\overline{\Psi}=\frac{\overline{M}_0(\omega)}{4\pi\rho\omega^2}\int_0^\infty F_\beta\left\{\frac{3}{2}\,\mathrm{i}\epsilon\nu_\beta J_0(kr)+\frac{1}{2}\cos 2\theta\;\mathrm{i}\epsilon\nu_\beta J_2(kr)\right\}dk\,. \qquad (2.123)
$$

Similarly, substituting (2.121) into (2.100), which is obtained by extracting the r-component from the left-hand side of (2.79), and using $dJ_0(q)/dq=-J_1(q)$, $dJ_1(q)/dq=q^{-1}J_1(q)-J_2(q)$, $dJ_2(q))/dq=J_1(q)-2J_2(q)/q$, etc., we have [24]

$$
\begin{aligned}
(\nabla\times\overline{\Psi})_r &= -\left(\frac{1}{r^2}\frac{\partial^2}{\partial\theta^2}+\frac{\partial^2}{\partial z^2}\right)\overline{\Psi}_r+\left(\frac{1}{r^2}+\frac{1}{r}\frac{\partial}{\partial r}\right)\frac{\partial\overline{\Psi}_\theta}{\partial\theta}+\frac{\partial^2\overline{\Psi}_z}{\partial r\partial z}\\
&=\frac{\overline{M}_0(\omega)}{4\pi\rho\omega^2}\int_0^\infty F_\beta\left[2\cos 2\theta\,k\frac{J_1(kr)}{r^2}-\sin^2\theta\,\nu_\beta^2 kJ_1(kr)\right.\\
&\quad\left.-\cos 2\theta\,k\left\{\frac{J_1(kr)}{r^2}+\frac{1}{r}\frac{dJ_1(kr)}{dr}\right\}+\nu_\beta^2\frac{dJ_0(kr)}{dr}\right]dk \qquad (2.124)\\
&=\frac{\overline{M}_0(\omega)}{4\pi\rho\omega^2}\int_0^\infty F_\beta\left[\nu_\beta^2\frac{dJ_0(kr)}{dr}-\sin^2\theta\,\nu_\beta^2 kJ_1(kr)-\cos 2\theta\frac{\nu_\beta^2-k_\beta^2}{kr}\left\{\frac{J_1(kr)}{r}-\frac{dJ_1(kr)}{dr}\right\}\right]dk\\
&=\frac{\overline{M}_0(\omega)}{4\pi\rho\omega^2}\int_0^\infty F_\beta\left[\frac{3}{2}\nu_\beta^2\frac{dJ_0(kr)}{dr}-\frac{\nu_\beta^2}{2}(1-2\sin^2\theta)\frac{dJ_0(kr)}{dr}-\cos 2\theta\left(\nu_\beta^2-k_\beta^2\right)\frac{J_2(kr)}{r}\right]dk\\
&=\frac{\overline{M}_0(\omega)}{4\pi\rho\omega^2}\int_0^\infty F_\beta\left[\left\{\frac{3}{2}\,\nu_\beta^2\frac{dJ_0(kr)}{dr}+\frac{1}{2}\cos 2\theta\,\nu_\beta^2\frac{dJ_2(kr)}{dr}\right\}+\cos 2\theta\,k_\beta^2\frac{J_2(kr)}{r}\right]dk\,.
\end{aligned}
$$

On the other hand, the first term of (2.85), which is obtained by extracting the r-component from the right-hand side of (2.79) is given as

$$
\frac{\partial^2\overline{\Psi}}{\partial r\partial z}=\frac{\overline{M}_0(\omega)}{4\pi\rho\omega^2}\int_0^\infty F_\beta\left\{\frac{3}{2}\,\nu_\beta^2\frac{dJ_0(kr)}{dr}+\frac{1}{2}\cos 2\theta\,\nu_\beta^2\frac{dJ_2(kr)}{dr}\right\}dk \qquad (2.125)
$$

[24] This is also a difficult transformation like (2.122), but can again be reached working back from the last equation.

using (2.111). This is identical to the part of (2.124) related to the first term in { }. Therefore, the other part related to the second term in { } must be identical to the second term of (2.85), namely

$$\frac{1}{r}\frac{\partial \overline{\mathrm{X}}}{\partial \theta} = \frac{\overline{M}_0(\omega)}{4\pi\rho\omega^2}\cos 2\theta \int_0^\infty k_\beta^2 F_\beta \frac{J_2(kr)}{r}\, dk\,. \tag{2.126}$$

Combining (2.120), (2.123), and (2.126), we have the displacement potentials[25]

$$\begin{aligned}
\overline{\Phi} &= \frac{\overline{M}_0(\omega)}{4\pi\rho\omega^2}\frac{1}{2}\int_0^\infty F_\alpha \left\{(k^2 - 2\nu_\alpha^2)J_0(kr) + \cos 2\theta\, k^2 J_2(kr)\right\} dk\,,\\
\overline{\Psi} &= \frac{\overline{M}_0(\omega)}{4\pi\rho\omega^2}\frac{1}{2}\int_0^\infty F_\beta \left\{3\mathrm{i}\epsilon\nu_\beta J_0(kr) + \cos 2\theta\, \mathrm{i}\epsilon\nu_\beta J_2(kr)\right\} dk\,,\\
\overline{\mathrm{X}} &= \frac{\overline{M}_0(\omega)}{4\pi\rho\omega^2}\frac{1}{2}\int_0^\infty \sin 2\theta\, k_\beta^2 F_\beta J_2(kr)\, dk\,.
\end{aligned} \tag{2.127}$$

We return to (2.118), which indicates that the displacement potentials of an inclined **dip slip fault** are represented by linear combinations of the displacement potentials of dip slip faults with dip angles of 90° and 45°, with coefficients $-\cos 2\delta$ and $\sin 2\delta$. Therefore, from (2.115) and (2.127), we have[26]

$$\begin{aligned}
\overline{\Phi} &= \frac{\overline{M}_0(\omega)}{4\pi\rho\omega^2}\int_0^\infty \left[\frac{1}{2}\sin 2\delta (k^2 - 2\nu_\alpha^2) J_0(kr)\right.\\
&\quad \left. +\frac{1}{2}\sin 2\delta \cos 2\theta\, k^2 J_2(kr) - \cos 2\delta \sin\theta\, (-2\mathrm{i}\epsilon k\nu_\alpha)\, J_1(kr)\right] F_\alpha\, dk\,,\\
\overline{\Psi} &= \frac{\overline{M}_0(\omega)}{4\pi\rho\omega^2}\int_0^\infty \left[\frac{1}{2}\sin 2\delta\, 3\mathrm{i}\epsilon\nu_\beta J_0(kr)\right.\\
&\quad \left. +\frac{1}{2}\sin 2\delta\, \cos 2\theta\, \mathrm{i}\epsilon\nu_\beta J_2(kr) - \cos 2\delta \sin\theta \frac{k_\beta^2 - 2k^2}{k} F_\beta J_1(kr)\right] F_\beta\, dk\,,\\
\overline{\mathrm{X}} &= \frac{\overline{M}_0(\omega)}{4\pi\rho\omega^2}\int_0^\infty \left[\frac{1}{2}\sin 2\delta \frac{d\cos 2\theta}{d\theta}\frac{-k_\beta^2}{2} J_2(kr) - \cos 2\delta \frac{d\sin\theta}{d\theta}\frac{\mathrm{i}\epsilon k_\beta^2 \nu_\beta}{k} J_1(kr)\right] F_\beta\, dk\,.
\end{aligned} \tag{2.128}$$

[25] These agree with Eq. (8) of Harkrider [23] if $\phi \to \theta$, $\nu_\nu \to \mathrm{i}\nu_\nu$, and $\mu\overline{D}(\omega) \to \overline{M}_0(\omega)$. However, in Harkrider's equation, the factor 1/2 of $\overline{\mathrm{X}}$ is missing.

[26] These agree with Eqs. (30) and (32) of Sato [62] if $\varphi \to \theta$, $\xi \to k$, $k \to k_\beta$, $\nu_\nu \to \mathrm{i}\nu_\nu$, $|z| \to |z - h|$, and the slip is reversed. Therefore, the equations of Sato [62] give the displacement potentials of a normal fault.

2.2.5 Extension to Arbitrary Fault Slip

To extend the formulation to an arbitrary fault slip as in Fig. 2.5, based on the **principle of superposition** (Sect. 1.3.1), we use the linear combination of (2.105) for an inclined strike slip fault and (2.128) for an inclined dip slip fault, with coefficients $\cos\lambda$ and $\sin\lambda$, where λ is a **slip angle**. The results of this extension are

$$
\begin{aligned}
\overline{\Phi} &= \frac{\overline{M}_0(\omega)}{4\pi\rho\omega^2}\sum_{l=0}^{2}\Lambda_l\int_0^\infty A_l F_\alpha J_l(kr)dk\ ,\\
\overline{\Psi} &= \frac{\overline{M}_0(\omega)}{4\pi\rho\omega^2}\sum_{l=0}^{2}\Lambda_l\int_0^\infty B_l F_\beta J_l(kr)dk\ ,\\
\overline{\mathrm{X}} &= \frac{\overline{M}_0(\omega)}{4\pi\rho\omega^2}\sum_{l=0}^{2}\frac{\partial\Lambda_l}{\partial\theta}\int_0^\infty C_l F_\beta J_l(kr)dk\ ,\\
\Lambda_0 &= \frac{1}{2}\sin\lambda\sin 2\delta\ ,\\
\Lambda_1 &= \cos\lambda\cos\delta\cos\theta - \sin\lambda\cos 2\delta\sin\theta\ ,\\
\Lambda_2 &= \frac{1}{2}\sin\lambda\sin 2\delta\cos 2\theta + \cos\lambda\sin\delta\sin 2\theta\ ,\\
A_0 &= k^2 - 2\nu_\alpha^2\ ,\quad A_1 = -2\mathrm{i}\epsilon k\nu_\alpha\ ,\quad A_2 = k^2\ ,\\
B_0 &= 3\mathrm{i}\epsilon\nu_\beta\ ,\quad B_1 = (k_\beta^2 - 2k^2)/k\ ,\quad B_2 = \mathrm{i}\epsilon\nu_\beta\ ,\\
C_0 &= 0\ ,\quad C_1 = \mathrm{i}\epsilon k_\beta^2\nu_\beta/k\ ,\quad C_2 = -k_\beta^2/2\ ,
\end{aligned}
\tag{2.129}
$$

which agree with Eqs. (A5) – (A7) of Harkrider [23] if $\phi \to \theta$, $\nu_\nu \to \mathrm{i}\nu_\nu$, and $\mu\overline{D}(\omega) \to \overline{M}_0(\omega)$. Moreover, if $\varphi \to \theta$, $\zeta \to k$, $k \to k_\beta$, $\nu_\nu \to \mathrm{i}\nu_\nu$, $|z| \to |z-h|$, $p \to \mathrm{i}\omega$, and $\rho V_S^2 \bar{f}(p) \to \overline{M}_0(\omega)$, the results almost agree with Eqs. (3) – (5) of Sato [63]. However, his coordinate system is rotated from ours by 180° about the x-axis and the slip is reversed, as mentioned in the footnotes in Sects. 2.2.2 and 2.2.4, so that $\lambda \to 180° - \lambda$ is necessary.

Λ_0, Λ_1, and Λ_2 in (2.129) indicate how the potential amplitudes change according to the coordinate azimuth θ. Therefore, they correspond to the radiation patterns (Sect. 2.3.2) in the displacement potential representations, and are sometimes called the **horizontal radiation patterns**. Among them, Λ_0 does not include θ and so $\partial\Lambda_0/\partial\theta \equiv 0$, therefore the $l = 0$ term of $\overline{\mathrm{X}}$ is always zero. Also, since these displacement potentials include $|z-h|$ and $\epsilon = (z-h)/|z-h|$, the elastic displacements (ground motions) and stresses computed from these have discontinuities at $z = h$. In the ground motion computation for a **1-D velocity structure** (Sect. 3.1), these discontinuities work as source conditions. Thus, the inverse Fourier transform of (2.129), or (2.129) itself is called a **source potential** for short.

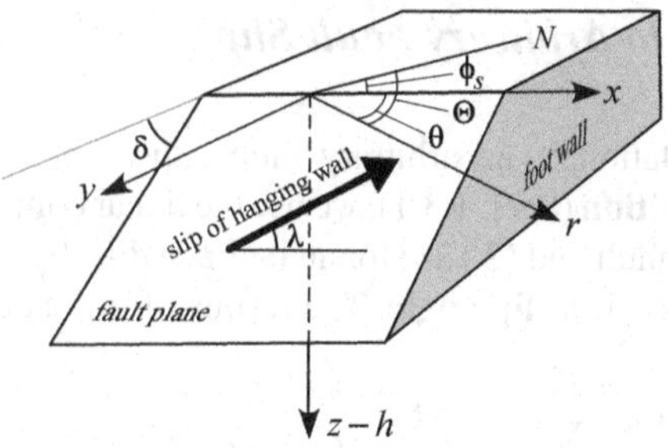

Fig. 2.15 Representation of a source fault in the cylindrical coordinate system (modified from Koketsu [41] with permission of Kindai Kagaku). As this should be a combination of Figs. 2.5 and 2.12, the origin is located at the epicenter h above the intersection point of the coordinate axes. In addition to the coordinate azimuth θ measured from the fault strike, the observer azimuth Θ measured from north is also shown

In addition, in actual problems, the **azimuth** is measured from north (N in Fig. 2.5) to an observation station (observer). Using this observer azimuth Θ and the strike (ϕ_s in Fig. 2.6), the coordinate azimuth ϕ is calculated by

$$\theta = \Theta - \phi_s \tag{2.130}$$

as shown in Fig. 2.15.

2.3 Analysis of the Earthquake Source

2.3.1 Hypocenter Determination

One of the most classical source analyses is the **hypocenter determination**. Beyond simple graphical methods [58], the modern approach was first established by Geiger [21] in 1910. Although he used the spherical coordinate system, here we use a Cartesian coordinate system with the z-axis taken positive downwards, as shown in Fig. 2.12. If the earthquake can be assumed to be a point source, this point source corresponds to a **hypocenter** (Sect. 1.1) located at (x_S, y_S, z_S), while an observation point on the **ground surface** (Sect. 3.1.1) is located at (x_O, y_O, z_O). Considering a **1-D velocity structure** (Sect. 3.1.1), the cross section including the hypocenter and the observation point is as shown in Fig. 3.1.

The **ray** (Sect. 1.2.5) of a **body wave** (P wave or S wave, Sect. 1.2.4) radiates from the hypocenter toward the observation point travels straight within a layer, and is refracted at the boundary between layers according to **Snell's law** (Sect. 3.1.11)

$$\frac{\sin\theta_i}{v_i} = \frac{\sin\theta_{i+1}}{v_{i+1}}, \quad i = 1, 2, \ldots, n-1 \tag{2.131}$$

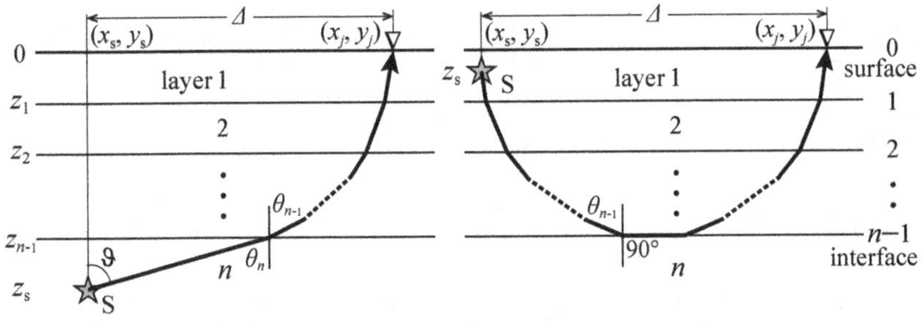

Fig. 2.16 **a** Direct wave propagating from the hypocenter S to the observation point O at the epicentral distance Δ. θ_S represents the angle of incidence at the hypocenter. **b** In the case of a shallow hypocenter, the head wave has a shorter travel time

where θ_i and v_i are the **angle of incidence** and the velocity of a body wave in the ith layer. When the hypocenter is relatively deep, the body wave is emitted upwards as shown in Fig. 2.16a, and reaches the observation point by repeated straight lines and refraction. When such a **direct wave** is emitted from the hypocenter at an angle of incidence θ_S, the **epicentral distance** L and the **travel time** T (time of propagation, Sect. 3.2.2) of the point where the ray reaches the ground surface are given as

$$L = \sum_{i=1}^{n} h_i \tan\theta_i\,, \quad T = \sum_{i=1}^{n} \frac{h_i}{v_i \cos\theta_i}\,, \tag{2.132}$$
$$h_i = z_i - z_{i-1}\,,\ z_0 = 0\,,\ z_n = z_s\,,\ \theta_i = \arcsin\left(\frac{v_i}{v_{i+1}} \sin\theta_{i+1}\right),\ \theta_n = \theta_S$$

from Fig. 2.16a and Snell's law. Generally, it is difficult for the ray to hit the observation point O, that is, $L = \Delta$, with the first given θ_S, but $L = \Delta$ is realized by changing θ_S by trial and error. Since this is a somewhat simple **ray tracing**, detailed implementation is described in Sect. 3.2.3.

In the case where the hypocenter is shallow and has a considerable epicentral distance, as shown in Fig. 2.16b, a head wave (Sect. 3.1.7) propagating horizontally at depth arrives earlier than the direct wave, because deeper parts of the Earth have faster velocities. In this case, θ_n is fixed at 90°, so ray tracing is not necessary and Snell's law (2.131) yields

$$\frac{1}{v_n} = \frac{\sin\theta_i}{v_i}\,, \quad i = 1, 2, \ldots, n-1\,. \tag{2.133}$$

For simplicity, the hypocenter is assumed to be in the first layer. Therefore, if $\Delta > L'$ where

$$L' = \sum_{i=1}^{n-1} h_i \tan\theta_i - z_S \tan\theta_1 + \sum_{i=1}^{n-1} h_{n-i} \tan\theta_{n-i} = 2\sum_{i=1}^{n-1} h_i \tan\theta_i - z_S \tan\theta_1 \,, \tag{2.134}$$

the travel time T can be determined as

$$\begin{aligned} T &= 2\sum_{i=1}^{n-1} \frac{h_i}{\upsilon_i \cos\theta_i} - \frac{z_S}{\upsilon_1 \cos\theta_1} + \frac{\Delta - L'}{\upsilon_n} \\ &= \frac{\Delta}{\upsilon_n} + 2\sum_{i=1}^{n-1} \frac{h_i}{\upsilon_i \cos\theta_i} - 2\sum_{i=1}^{n-1} \frac{h_i \tan\theta_i \sin\theta_i}{\upsilon_i} - \frac{z_S}{\upsilon_1 \cos\theta_1} + \frac{z_S \tan\theta_1 \sin\theta_1}{\upsilon_1} \\ &= \frac{\Delta}{\upsilon_n} + 2\sum_{i=1}^{n-1} \frac{h_i \cos\theta_i}{\upsilon_i} - \frac{z_S \cos\theta_1}{\upsilon_1} \,. \end{aligned} \tag{2.135}$$

There are N observation points, and their coordinates are (x_j, y_j, z_j), $j = 1, 2, \ldots, N$. **hypocenter determination** consists in determining the coordinates of the hypocenter (x_S, y_S, z_S) (Fig. 2.16) and the **origin time** (referred to as t_S), which is the "time when body waves are radiated from the hypocenter" [73], using the arrival times t_j, $j = 1, 2, \ldots, N$ when body waves reach the observation points. In most cases, the arrival times of the initial parts of ground motions (**initial motions**) are used as t_j, for ease of detection. The hypocenter determination is made to ascertain whether the earliest arrival is a direct wave or a head wave, and the theoretical travel time T_j at each observation point is calculated using (2.132) or (2.135) and matched with the observed travel time $t_j - t_S$. However, no perfect match can be expected due to noise and errors involved in t_j and T_j. Thus, the **observation equation** (Sect. 4.4)

$$t_j \simeq T_j(x_S, y_S, z_S) + t_S \tag{2.136}$$

is statistically solved based on the least-squares method (Sect. 4.4).

Since x_S, y_S, and z_S are included in T_j in a nonlinear form, they must be solved with the nonlinear least-squares method (Sect. 4.4.1). Using the estimates $x_S^{(0)}, y_S^{(0)}, z_S^{(0)}, t_S^{(0)}$ the observation equation yields

$$\begin{aligned} &\Delta t_j \simeq \frac{\partial T_j^{(0)}}{\partial x_S}\Delta x_S + \frac{\partial T_j^{(0)}}{\partial y_S}\Delta y_S + \frac{\partial T_j^{(0)}}{\partial z_S}\Delta z_S + \Delta t_S \,, \quad \Delta t_j = t_j - T_j^{(0)} - t_S^{(0)}, \\ &\Delta x_S = x_S - x_S^{(0)}, \ \Delta y_S = y_S - y_S^{(0)}, \ \Delta z_S = z_S - z_S^{(0)}, \ \Delta t_S = t_S - t_S^{(0)} \,. \end{aligned} \tag{2.137}$$

Therefore, the equations for the least-squares method are

$$\begin{aligned} &S = \sum_{j=1}^{N} \frac{1}{\sigma_j^2} \left(\Delta t_j - \frac{\partial T_j^{(0)}}{\partial x_S}\Delta x_S - \frac{\partial T_j^{(0)}}{\partial y_S}\Delta y_S - \frac{\partial T_j^{(0)}}{\partial z_S}\Delta z_S - \Delta t_S \right)^2, \\ &\frac{\partial S}{\partial x_S} = 0, \quad \frac{\partial S}{\partial y_S} = 0, \quad \frac{\partial S}{\partial z_S} = 0, \quad \frac{\partial S}{\partial t_S} = 0 \,. \end{aligned} \tag{2.138}$$

If we define

$$\Delta\mathbf{x} = \begin{pmatrix} \Delta x_S \\ \Delta y_S \\ \Delta z_S \\ \Delta t_S \end{pmatrix}, \quad \Delta\mathbf{y} = \begin{pmatrix} \Delta t_1 \\ \Delta t_2 \\ \vdots \\ \Delta t_N \end{pmatrix}, \quad \mathbf{A} = \begin{pmatrix} \frac{\partial T_1}{\partial x_S} & \frac{\partial T_1}{\partial y_S} & \frac{\partial T_1}{\partial z_S} & 1 \\ \frac{\partial T_2}{\partial x_S} & \frac{\partial T_2}{\partial y_S} & \frac{\partial T_2}{\partial z_S} & 1 \\ \vdots & \vdots & \vdots & \vdots \\ \frac{\partial T_N}{\partial x_S} & \frac{\partial T_N}{\partial y_S} & \frac{\partial T_N}{\partial z_S} & 1 \end{pmatrix}, \tag{2.139}$$

we can rewrite (2.138) to be (4.80), and then determine the hypocenter by solving the simultaneous equations along Sect. 4.4.1. The fourth equation of (2.138), that is $\partial S / \partial \Delta t_S = 0$, leads to

$$\sum_{j=1}^{N} \frac{1}{\sigma_j^2} \left(\Delta t_j - \frac{\partial T_j^{(0)}}{\partial x_S} \Delta x_S - \frac{\partial T_j^{(0)}}{\partial y_S} \Delta y_S - \frac{\partial T_j^{(0)}}{\partial z_S} \Delta z_S - \Delta t_S \right) = 0 \,. \tag{2.140}$$

Assuming that the weights have been normalized as $\Sigma_{j=1}^{N} 1/\sigma_j^2 = 1$, we obtain[27]

$$t_S^{(0)} + \Delta t_S = \langle t_j \rangle - \left\langle T_j^{(0)} \right\rangle - \Delta x_S \left\langle \frac{\partial T_j^{(0)}}{\partial x_S} \right\rangle - \Delta y_S \left\langle \frac{\partial T_j^{(0)}}{\partial y_S} \right\rangle - \Delta z_S \left\langle \frac{\partial T_j^{(0)}}{\partial z_S} \right\rangle, \tag{2.141}$$

where $\langle f_j \rangle = \Sigma_{j=1}^{N} f_j / \sigma_j^2$ is a weighted average. Equation (2.141) indicates that Δt_S can simply be obtained from the weighted averages since only t_S is linearly included in the observation equation (2.136).

We modify the observation equation (2.137) as

$$t_j - T_j^{(0)} \simeq \frac{\partial T_j^{(0)}}{\partial x_S} \Delta x_S + \frac{\partial T_j^{(0)}}{\partial y_S} \Delta y_S + \frac{\partial T_j^{(0)}}{\partial z_S} \Delta z_S + t_S^{(0)} + \Delta t_S \,. \tag{2.142}$$

With the substitution of (2.141), we further rewrite (2.142) as

$$\begin{aligned} &\left(t_j - \langle t_j \rangle\right) - \left(T_j^{(0)} - \left\langle T_j^{(0)} \right\rangle\right) \simeq \\ &\left(\frac{\partial T_j^{(0)}}{\partial x_S} - \left\langle \frac{\partial T_j^{(0)}}{\partial x_S} \right\rangle \right) \Delta x_S + \left(\frac{\partial T_j^{(0)}}{\partial y_S} - \left\langle \frac{\partial T_j^{(0)}}{\partial y_S} \right\rangle \right) \Delta y_S + \left(\frac{\partial T_j^{(0)}}{\partial z_S} - \left\langle \frac{\partial T_j^{(0)}}{\partial z_S} \right\rangle \right) \Delta z_S \,. \end{aligned} \tag{2.143}$$

This rewriting is called **centering** because it is an operation to subtract the weighted average from each term to make its average zero [13]. The centering removes Δt_S from the four variables Δx_S, Δy_S, Δz_S, and Δt_S in the nonlinear least-squares method for hypocenter determination. It is likely that the computation will be stable with only three variables.

---

[27] This agrees with Eqs. (6) and (7) of Linert et al. [49] if $i \to j$ and $w \to 1/\sigma$.

In order to perform the nonlinear least-squares method of the hypocenter determination, it is necessary to formulate and calculate the partial derivatives $\partial T_j^{(0)}/\partial x_{\mathrm{S}}$, $\partial T_j^{(0)}/\partial y_{\mathrm{S}}$, and $\partial T_j^{(0)}/\partial z_{\mathrm{S}}$ in the observation equation. At a point (x, y, z) on the ray, it is assumed that the travel time is T, the body wave velocity is υ, the slowness s is the reciprocal of υ, and the length of the ray is λ. Since time = distance(length of ray)/velocity,

$$T = \int_{\mathrm{S}}^{\mathrm{O}} s\, d\lambda \tag{2.144}$$

in any velocity structure including but not limited to a 1-D velocity structure. As shown in Sect. 3.2.2, if we take λ as an independent variable in the **ray theory**, we have the **Lagrangian** $L = s$ and **Fermat's principle** $\delta T = 0$. Since

$$s = \sqrt{p_x^2 + p_y^2 + p_z^2} \tag{2.145}$$

from (3.237),

$$\delta T = \int_{\mathrm{S}}^{\mathrm{O}} \delta s\, d\lambda = \int_{\mathrm{S}}^{\mathrm{O}} \left(\frac{p_x}{s}\delta p_x + \frac{p_y}{s}\delta p_y + \frac{p_z}{s}\delta p_z \right) d\lambda\,. \tag{2.146}$$

Using the **canonical equations** (3.242)

$$\frac{dx}{d\lambda} = \frac{p_x}{s}, \quad \frac{dy}{d\lambda} = \frac{p_y}{s}, \quad \frac{dz}{d\lambda} = \frac{p_z}{s} \tag{2.147}$$

and performing **integration by parts** for (2.146), we obtain

$$\delta T = \left[p_x\delta x + p_y\delta y + p_z\delta z \right]_{\mathrm{S}}^{\mathrm{O}} - \int_{\mathrm{S}}^{\mathrm{O}} \left(\frac{dp_x}{d\lambda}\delta x + \frac{dp_y}{d\lambda}\delta y + \frac{dp_z}{d\lambda}\delta z \right) d\lambda\,. \tag{2.148}$$

We next apply **Lagrange's equation** [46] for the independent variable λ

$$\frac{d}{d\lambda}\frac{\partial L}{\partial p_i} - \frac{\partial L}{\partial x_i} = 0\,, \quad i = x, y, z \tag{2.149}$$

to $L = s = \sqrt{p_x^2 + p_y^2 + p_z^2}$. The second terms of (2.149) vanish, because L does not include x, y, z. The remaining first terms lead to

$$\frac{1}{s}\frac{dp_x}{d\lambda} = 0\,, \quad \frac{1}{s}\frac{dp_y}{d\lambda} = 0\,, \quad \frac{1}{s}\frac{dp_z}{d\lambda} = 0\,. \tag{2.150}$$

Substituting these into (2.148), we have

$$\delta T = \left[p_x\delta x + p_y\delta y + p_z\delta z \right]_{\mathrm{S}}^{\mathrm{O}}\,. \tag{2.151}$$

In hypocenter determination, observation points O are known and fixed, therefore $\delta x_O = 0$, $\delta y_O = 0$, $\delta z_O = 0$ and we have

$$\delta T = -p_x \delta x_S - p_y \delta y_S - p_z \delta z_S \,. \tag{2.152}$$

Equations (2.152) and (2.147) yield the equations[28]

$$\frac{\partial T}{\partial x_S} = -s\frac{dx_S}{d\lambda}\,, \quad \frac{\partial T}{\partial y_S} = -s\frac{dy_S}{d\lambda}\,, \quad \frac{\partial T}{\partial z_S} = -s\frac{dz_S}{d\lambda}\,. \tag{2.153}$$

We assume that the positional relationship between the ray and the Cartesian coordinate system is represented by azimuth φ and takeoff angle ϑ, as shown in Fig. 3.23. ϑ is a supplementary angle of θ_i in Fig. 2.16, as it is measured from the positive direction of the z-axis. Since the azimuth does not change anywhere along the ray in the 1-D velocity structure,

$$\cos\varphi = \frac{x_O - x_S}{\Delta}\,, \quad \sin\varphi = \frac{y_O - y_S}{\Delta}\,. \tag{2.154}$$

In addition, from the definition of the **direction cosine**, we have

$$\frac{dx}{d\lambda} = \cos\varphi\sin\vartheta\,, \quad \frac{dy}{d\lambda} = \sin\varphi\sin\vartheta\,, \quad \frac{dz}{d\lambda} = \cos\vartheta\,. \tag{2.155}$$

Therefore, we obtain [29]

$$\begin{aligned}
\frac{\partial T}{\partial x_S} &= -\frac{1}{v_S}\cos\varphi\sin\vartheta_S = -\frac{x_O - x_S}{\Delta}\frac{\sin\theta_S}{v_S}\,,\\
\frac{\partial T}{\partial y_S} &= -\frac{1}{v_S}\sin\varphi\sin\vartheta_S = -\frac{y_O - y_S}{\Delta}\frac{\sin\theta_S}{v_S}\,,\\
\frac{\partial T}{\partial z_S} &= -\frac{1}{v_S}\cos\vartheta_S = \frac{\cos\theta_S}{v_S}\,.
\end{aligned} \tag{2.156}$$

Equation (2.156) holds for both the direct waves (Fig. 2.16a, (2.132)) and the head waves (Fig. 2.16b, (2.135)). For the head waves, using Snell's law we can further rewrite (2.156) as[30]

$$\frac{\partial T}{\partial x_S} = -\frac{x_O - x_S}{\Delta}\frac{1}{v_n}\,, \quad \frac{\partial T}{\partial y_S} = -\frac{y_O - y_S}{\Delta}\frac{1}{v_n}\,, \quad \frac{\partial T}{\partial z_S} = -\frac{\sqrt{v_n^2 - v_S^2}}{v_S v_n}\,. \tag{2.157}$$

---

[28]These agree with Eq. (4.66) of Lee and Stuart [48] if $A \to$ S, $u \to s$, $s \to \lambda$.

[29]These agree with Eq. (4.113) of Lee and Stuart [48] if $A \to$ S, $B \to$ O, $j \to$ S, and $\phi \to \theta_S$.

[30]These agree with Eq. (4.110) of Lee and Stuart [48] if $A \to$ S, $B \to$ O, $j \to$ S, and $k \to n$.

No 1-D velocity structure model can completely represent the actual velocity structure, and this incompleteness results in errors in hypocenter determination. Ideally, these errors can be eliminated by constructing a realistic model of the **3-D velocity structure** (Sect. 3.2.1), performing ray tracing in it, and calculating theoretical travel times. However, even in hypocenter determination with a 1-D velocity structure, the errors can be considerably reduced by introducing **station corrections**, which express the effect of the difference between the actual and modeled velocity structures below an observation point, by increasing or decreasing a theoretical travel time [73]. Furthermore, the station corrections for observation points are incorporated into the observation equation for multiple hypocenters as variables, and the hypocenters and station corrections can be simultaneously determined (**joint hypocenter determination**).

When many earthquakes occur in a small area and are observed by similar combinations of observation points, the earthquake for which the hypocenter can be determined with the highest accuracy is termed the master event, and the hypocenters of other earthquakes are determined as relative positions and times compared to those of the master event (**master event method**). This method is known to give good results [73], but has problems such as the limitation of "a small area" and the propagation of the errors of the master event to those of all other events. To overcome these issues, the **double difference method** has been developed [75]. In this method, an observation equation is set up for all combinations of earthquakes, so that the difference in observed travel time is equated with the difference in theoretical travel time, and the equation is solved using the nonlinear least-squares method.

For the kth and lth earthquakes, the coordinates of their hypocenters and their origin times are denoted as x_S, y_S, z_S, t_S with superscripts k and l, respectively. The arrival times, the observed travel times, and the theoretical travel times when the two earthquakes are observed at the jth observation point, are also denoted as t_j, $t_j - t_S = \tau_j, T_j$ with the superscripts k and l. From (2.136), the observation equations of the differences between the observed and theoretical travel times for these two earthquakes are

$$\tau_j^k - T_j^k \simeq 0\,, \quad \tau_j^l - T_j^l \simeq 0 \tag{2.158}$$

and, the observation equation for the difference between them is

$$\left(\tau_j^k - T_j^k\right) - \left(\tau_j^l - T_j^l\right) \simeq 0\,. \tag{2.159}$$

The observation equation (2.159) is referred to as a "double difference" because it is the difference between the differences [75]. The name of the method comes from this. When we rewrite (2.159) in the form of (2.137), we obtain the observation equation[31]

[31] This agrees with Eq. (6) of Waldhouser and Ellsworth [75] if $k \to j, i \to k, j \to l, \Delta\tau \to \Delta t$, $dr_k^{ij} \to \Delta t_j^k - \Delta t_j^l$, and the subscript S is added to Δx, Δy, Δz, and Δt.

$$\Delta t_j^k - \Delta t_j^l \simeq \frac{\partial T_j^{k(0)}}{\partial x_S^k}\Delta x_S^k + \frac{\partial T_j^{k(0)}}{\partial y_S^k}\Delta y_S^k + \frac{\partial T_j^{k(0)}}{\partial z_S^k}\Delta z_S^k + \Delta t_S^k$$
$$- \frac{\partial T_j^{k(0)}}{\partial x_S^l}\Delta x_S^l - \frac{\partial T_j^{k(0)}}{\partial y_S^l}\Delta y_S^l - \frac{\partial T_j^{k(0)}}{\partial z_S^l}\Delta z_S^l - \Delta t_S^l . \quad (2.160)$$

If the total number of earthquakes is M, the number of combinations of (k, l) is ${}_MC_2 = M(M-1)/2$. Therefore, (4.80) becomes $M(M-1)N/2$ simultaneous equations and the number of variables is $4M$. The double difference method minimizes errors due to a 1-D velocity structure model without using station corrections [75].

Finally, it is effective to impose a **non-negative condition** [38] and various **constraints** (Sect. 4.4.2) on hypocenter determination as well as on source inversion (Sect. 2.3.6).

2.3.2 Radiation Pattern and Fault Plane Solution

We extract only the **far-field terms** from (2.53) $\sim$ (2.55), and convert them to the spherical coordinate system whose origin is located at the hypocenter, as shown in the center of Fig. 2.17, using $\gamma_x = \sin\theta\cos\phi$, $\gamma_y = \sin\theta\sin\phi$, and $\gamma_z = \cos\theta$, which are obtained from $x = R\sin\theta\cos\phi$, $y = R\sin\theta\sin\phi$, and $z = R\cos\theta$. A result of the conversion is

$$U_R = (U_x^{3P} + U_x^{3S})\sin\theta\cos\phi + (U_y^{3P} + U_y^{3S})\sin\theta\sin\phi + (U_z^{3P} + U_z^{3S})\cos\theta$$
$$= \frac{1}{4\pi\rho\alpha^3 R}\dot{M}_0\left(t - \frac{R}{\alpha}\right)(2\sin^4\theta\sin\phi\cos^3\phi$$
$$+ 2\sin^4\theta\sin^3\phi\cos\phi + 2\sin^2\theta\cos^2\theta\sin\phi\cos\phi)$$
$$- \frac{1}{4\pi\rho\beta^3 R}\dot{M}_0\left(t - \frac{R}{\beta}\right)(2\sin^4\theta\sin\phi\cos^3\phi - \sin^2\theta\sin\phi\cos\phi$$
$$+ 2\sin^4\theta\sin^3\phi\cos\phi - \sin^2\theta\sin\phi\cos\phi + 2\sin^2\theta\cos^2\theta\sin\phi\cos\phi)$$
$$= \frac{1}{4\pi\rho\alpha^3 R}\dot{M}_0\left(t - \frac{R}{\alpha}\right)2\sin^2\theta\sin\phi\cos\phi$$
$$- \frac{1}{4\pi\rho\beta^3 R}\dot{M}_0\left(t - \frac{R}{\beta}\right)(2\sin^2\theta\sin\phi\cos\phi - 2\sin^2\theta\sin\phi\cos\phi) , \quad (2.161)$$

where the second term related to S waves vanishes because of $2\sin^2\theta\sin\phi\cos\phi - 2\sin^2\theta\sin\phi\cos\phi = 0$. Similarly, in the other results for

$$U_\theta = (U_x^{3P} + U_x^{3S})\cos\theta\cos\phi + (U_y^{3P} + U_y^{3S})\cos\theta\sin\phi - (U_z^{3P} + U_z^{3S})\sin\theta ,$$
$$U_\phi = -(U_x^{3P} + U_x^{3S})\sin\phi + (U_y^{3P} + U_y^{3S})\cos\phi ,$$

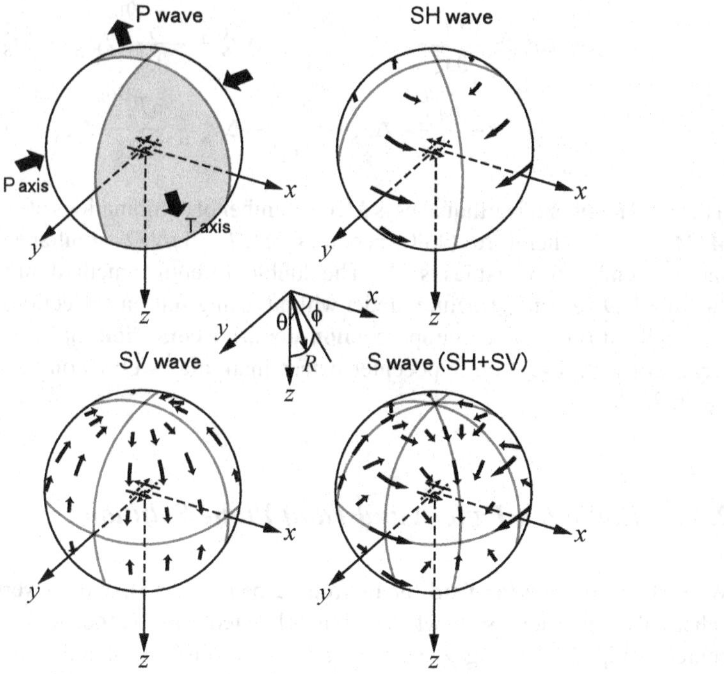

Fig. 2.17 Radiation patterns due to the point source of a left-lateral strike slip at the origin based on Kikuchi [35], and the spherical coordinate system shown in the center (modified from Koketsu [41] with permission of Kindai Kagaku)

the first terms related to P waves vanish. We then have[32]

$$
\begin{aligned}
U_R &= \frac{1}{4\pi\rho\alpha^3 R}\dot{M}_0\left(t - \frac{R}{\alpha}\right)\sin^2\theta\sin 2\phi\,,\\
U_\theta &= \frac{1}{4\pi\rho\beta^3 R}\dot{M}_0\left(t - \frac{R}{\beta}\right)\frac{1}{2}\sin 2\theta\sin 2\phi\,,\\
U_\phi &= \frac{1}{4\pi\rho\beta^3 R}\dot{M}_0\left(t - \frac{R}{\beta}\right)\sin\theta\cos 2\phi\,.
\end{aligned}
\tag{2.162}
$$

In (2.162), P and S waves are so completely separated that U_R includes only P waves whereas U_θ and U_ϕ include only S waves. Since U_ϕ is parallel to the $x - y$ plane and has no vertical component, it corresponds to **SH waves** according to the definition in Sect. 1.2.4. U_θ corresponds to **SV waves** because it lies in a plane perpendicular to the $x - y$ plane. The parts of U_R, U_θ, and U_ϕ related to θ and ϕ are called **radiation patterns**, and they are drawn with arrows on a virtual sphere around the hypocenter (**focal sphere**) in Fig. 2.17.

---

[32]These agree with Eq. (6.55) of Kikuchi [35].

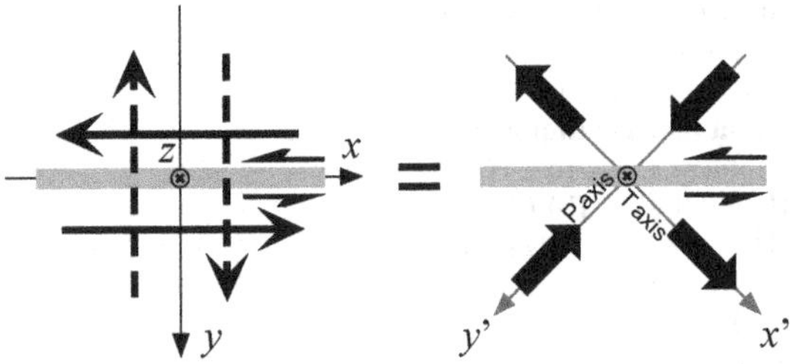

Fig. 2.18 Left: Double couple (two couples of forces) due to the point source of a left-lateral strike slip; Right: The force system equivalent to this double couple (modified from Koketsu [41] with permission of Kindai Kagaku)

U_R of the P wave and U_θ of the SV wave have the same ϕ-dependence, that is $\sin 2\phi$. Since $\sin 2\phi = 0$ in the cases of $\phi = 0°$, $90°$, $180°$, and $270°$, the radiation patterns of P and SV waves are divided by the $0° - 180°$ plane and the $90° - 270°$ plane (gray arcs in Fig. 2.17) into four quadrants. The two planes are called **nodal planes** because the ground motion is zero, and one of them corresponds to the fault plane. The other nodal plane is called the **auxiliary plane**, which is **conjugate** to the fault plane. In Fig. 2.17, the plane along the x-axis is the fault plane as the double couple at the origin indicates, while the plane along the y-axis is the auxiliary plane. U_ϕ of the SH wave also has a radiation pattern of **four-quadrant type** because of the factor $\cos 2\phi$, though its pattern is shifted by 45° compared to those of the P wave and SV wave. In addition, for the SV wave, the plane of $\theta = 90°$ is also a nodal plane because of the factor $\sin 2\theta$. Since the SV and SH waves have different patterns as described above, combining these patterns results in a complex pattern, as shown in the lower right of Fig. 2.17. In particular, the points where the ground motion is zero are distributed not over planes (nodal planes), but along lines from the hypocenter. However, the ground motion near the points is small along the nodal planes of the SH and SV waves, and therefore nodal planes can still be seen in observations.

In the upper left panel of Fig. 2.17 for the P wave, the two quadrants corresponding to positive or negative U_R are colored in gray or white, respectively. A line connecting the centers of the gray quadrants is called the **T axis**, while one connecting the centers of the white quadrants is called the **P axis**. The directions of these two axes are identical to those of the forces (Fig. 2.18, right) obtained by performing vector summations of the single forces that compose the two couples of the double couple (Figs. 2.7; 2.18, left) in their respective quadrants. Since the forces along the T axis act to pull the hypocenter and a pulling force is named "tension", the "T" of the T-axis stands for "tension". The forces along the P axis act to apply pressure to the hypocenter, therefore the "P" of the P-axis stands for "pressure". Figure 2.17 and the right panel of Fig. 2.18 are valid only for point sources of vertical left-lateral strike slip ($\delta = 90°$, $\lambda = 0°$). However, we can easily obtain the equivalent forces obtained for different types of source fault by simply rotating in three dimensions, as shown in Fig. 2.6.

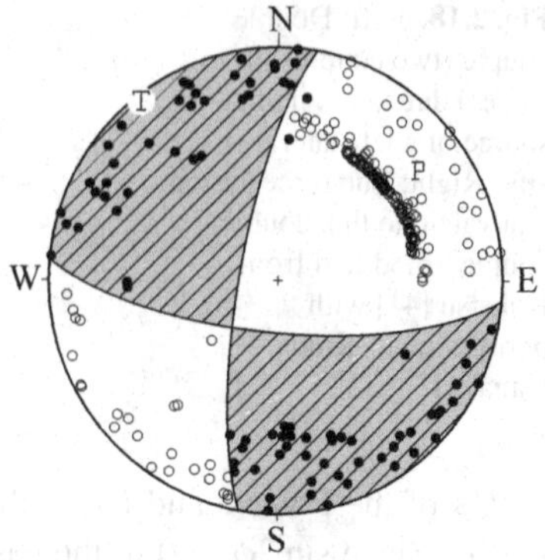

Fig. 2.19 An example of a focal mechanism solution, from the analysis of the **Kumamoto earthquake** (February 15, 2016 UTC, M_w 7.0) by the **JMA** (Sect. A.2.1) [31]. The black and white circles indicate "push" and "pull", respectively. T and P represent the T and P axes (reprinted from JMA [31])

The **fault parameters** of a fault plane and its auxiliary plane determined using the radiation pattern of the P wave are termed a **fault plane solution** or **focal mechanism solution**. When the P wave is radiated from the hypocenter of an earthquake, passes through the gray quadrant of positive U_R, and reaches a point beyond it, observed ground motions are in the direction indicated by the large arrow in Fig. 2.17, i.e., the direction in which the observation point is pushed by the earthquake ("**push**"). Conversely, in the case when the P wave passes through the white quadrant of negative U_R, observed ground motions are in the direction in which the observation point is pulled by the earthquake ("**pull**"). Since the hypocenter is in the ground and the observation point is on the ground surface, the vertical component of a seismograph at the observation point records a "push" as an upward ground motion and a "pull" as a downward ground motion. Therefore, we plot the points where the P wave passes through the focal sphere on the lower projection of the sphere, as shown in Fig. 2.19. In this projection, the boundary between the distribution of "push" points and the distribution of "pull" points is drawn to determine the nodal planes, and from these the P and T axes are then determined. Determining which nodal plane is the fault plane is assessed from the **aftershock distribution**[33] and other information.

Usually, points and quadrants corresponding to "push" are colored as shown in Fig. 2.19, while those corresponding to "pull" are not colored. The diagram representing a focal mechanism solution therefore looks like a beach ball and is sometimes called a **beach ball solution**. The type of source fault (Table 2.1, Fig. 2.6) can be recognized from the appearance of the beach ball pattern, as shown in Fig. 2.20. However, actual earthquakes seldom belong to a single type, but rather comprise a mix to some extent of different types. The focal mechanism solutions shown in Fig. 2.20 take this reality into account.

[33]Immediately after a large earthquake, many aftershocks occur mainly in the area of its source fault. As opposed to aftershocks, the large earthquake itself is called the **mainshock**.

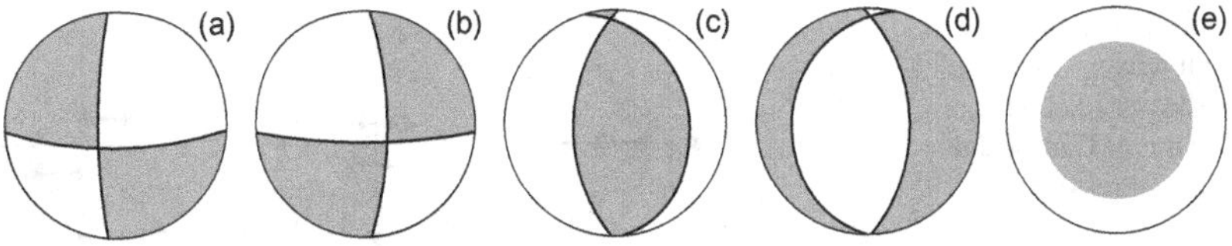

Fig. 2.20 The focal mechanism solutions corresponding to different source fault types: **a** left-lateral strike slip, **b** right-lateral strike slip, **c** reverse fault, **d** normal fault, and **e** CLVD (Sect. 2.3.3) (based on Kikuchi [35, 36], reprinted from Koketsu [41] with permission of Kindai Kagaku). In **a** and **b**, the nodal planes extending horizontally are assumed to be fault planes

When the intersection of two nodal planes is visible near the center of the focal sphere, the source fault is a strike slip fault. However, whether it is left-lateral or right-lateral depends on which of the two nodal planes is a fault plane. Panels (a) and (b) in Fig. 2.20 represent types of source fault where the nodal plane extending horizontally is a fault plane. If instead the nodal plane extending up and down is a fault plane, the left-lateral and right-lateral cases shown in (a) and (b), respectively, would be switched. If one of the colored "push" quadrants appears large and in the center of the focal sphere, the source fault is a reverse fault (Fig. 2.20c). Conversely, if one of the white "pull" quadrants appears large, the source fault is a normal fault (Fig. 2.20d). For (c) and (d), no matter which nodal plane is a fault plane, this will not change whether the source fault is a normal fault or a reverse fault. The **CLVD** pattern is explained in Sect. 2.3.3.

2.3.3 Moment Tensor

If we generalize pairs of forces that are opposite to each other, including **couples of forces**, we can classify them as shown in Fig. 2.21. The classification is based on the direction of the forces and the alignment of the points of action of forces relative to the coordinate axes in a Cartesian coordinate system. The strengths of the pairs of forces compose a matrix referred to as a **moment tensor** and represented by $\mathbf{M}_0$. The couple of forces drawn with solid arrows in Fig. 2.7 has the direction of the forces along the x-axis and the direction of the arm (the vector connecting the points of action of the forces) along the y-axis, so that the strength of the couple is represented by M_{xy}. Similarly, the couple of forces drawn with dashed arrows in Fig. 2.7 is represented as M_{yx}. These strengths correspond to a seismic moment M_0 or its time function $M_0(t)$. In addition, pairs of forces M_{ij} shown in Fig. 2.21 may exist, but M_{xx}, M_{yy}, and M_{zz} are not couples of forces because the direction of the forces and the direction of the arm are along a same coordinate axis. They are instead termed **dipoles** or **vector dipoles**.

The moment tensor is a symmetric tensor, as $M_{xy} = M_{yx}$ in Fig. 2.7. The diagonal components are dipoles, and the average of the diagonal components $I = (M_{xx} + M_{yy} + M_{zz})/3$ or I times the unit tensor is called an **isotropic com-**

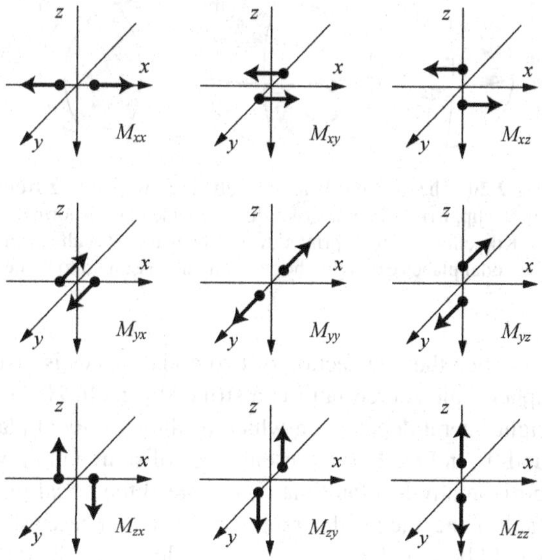

Fig. 2.21 Elements of a moment tensor (modified from Shearer [67] with permission of Cambridge University Press)

ponent, which represents forces that cause the volume around the point source to increase or decrease. The moment tensor minus the isotropic component in a tensor form is called a **deviatoric component**, which is further decomposed into a **double couple** and a **CLVD (compensated linear vector dipole)**. The latter represents forces that cause expansion (or shrinking) in one direction and shrinking (or expansion) in another, perpendicular direction, such that the volume remains unchanged [35].

When the point source is purely a double couple with a seismic moment of M_0 (or its time function $M_0(t)$), the moment tensor can be diagonalized by rotating the coordinate system with the principal axis transformation. After the diagonalization, the three diagonal components are $M_0, -M_0$, and 0, in any order. The new coordinate axis related to M_0 corresponds to the **T axis** of the focal mechanism solution and that related to $-M_0$ corresponds to the **P axis**. For example, when the point source corresponds to a left-lateral strike slip, as shown in Fig. 2.7 or Fig. 2.18 (left), its moment tensor has only the M_{xy} and M_{yx} elements of M_0. The xyz coordinate system is rotated 45° around the z-axis in the direction that points the x-axis toward the y-axis (Fig. 2.18, right). Using the **rotation matrix**

$$\mathbf{R} = \begin{pmatrix} \cos 45^\circ & -\sin 45^\circ & 0 \\ \sin 45^\circ & \cos 45^\circ & 0 \\ 0 & 0 & 1 \end{pmatrix} = \begin{pmatrix} 1/\sqrt{2} & -1/\sqrt{2} & 0 \\ 1/\sqrt{2} & 1/\sqrt{2} & 0 \\ 0 & 0 & 1 \end{pmatrix} \tag{2.163}$$

and the matrix transpose (superscript T), we obtain

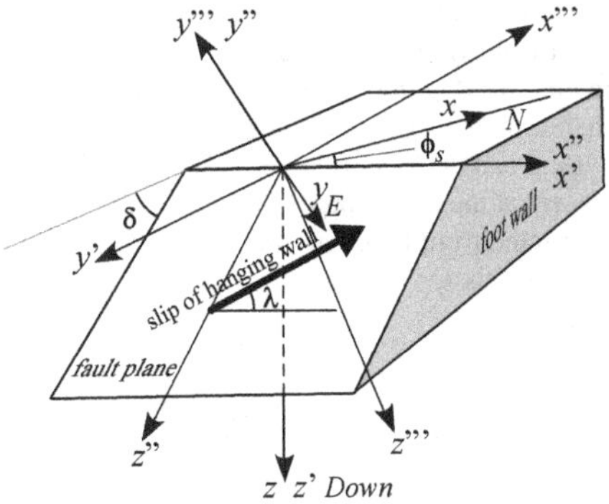

Fig. 2.22 The rotations of the Cartesian coordinate system by the dip angle δ, the slip angle λ, and the strike ϕ_s (modified from Koketsu [41] with permission of Kindai Kagaku)

$$\mathbf{M}_0 = \mathbf{R}^{\mathrm{T}} \begin{pmatrix} 0 & M_0 & 0 \\ M_0 & 0 & 0 \\ 0 & 0 & 0 \end{pmatrix} \mathbf{R} = \begin{pmatrix} M_0 & 0 & 0 \\ 0 & -M_0 & 0 \\ 0 & 0 & 0 \end{pmatrix} \tag{2.164}$$

for the moment tensor in the new $x'y'z'$ coordinate system. This indicates that the x'-axis is the T-axis and the y'-axis is the P-axis, and that the force pairs associated with these axes are outward and inward dipoles. It is mathematically proved that the left and right panels of Fig. 2.18 are equivalent.

Figure 2.22 shows the point source of an arbitrary fault slip (dip angle δ, slip angle λ, strike ϕ_s) in Sect. 2.2.5. This is almost the same as Fig. 2.15, but here the x- and y-axes are rotated by $-\phi_s$ around the z-axis so that the xyz coordinate system is fixed to the Earth, and the x-axis is oriented in the N (north) direction. In this case, the y-axis naturally points toward E (east), and the z-axis continues to point down. In addition, we assume $h = 0$ for simplicity. The xyz coordinate system in Fig. 2.15 is the $x'y'z'$ coordinate system here. In the $x'''y'''z'''$ coordinate system fixed to the fault plane and adjusted to the direction of slip, the point source is a left-lateral strike slip fault perpendicular to the $x''' - y'''$ plane. This coordinate system is rotated by $-\lambda$ around the y'''-axis in the direction that points the z'''-axis, toward the x'''-axis. The obtained $x''y''z''$ coordinate system is then rotated by $90° - \delta$ around the x''-axis. The obtained $x'y'z'$ coordinate system is further rotated by $-\phi_s$ around the z'-axis, and we finally obtain the xyz coordinate system.

Therefore, from the moment tensor of the left-lateral strike slip in Fig. 2.7 and the rotation matrices

$$\mathbf{M}_0''' = \begin{pmatrix} 0 & M_0 & 0 \\ M_0 & 0 & 0 \\ 0 & 0 & 0 \end{pmatrix}, \quad \mathbf{R}_\lambda = \begin{pmatrix} \cos\lambda & 0 & -\sin\lambda \\ 0 & 1 & 0 \\ \sin\lambda & 0 & \cos\lambda \end{pmatrix},$$

$$\mathbf{R}_\delta = \begin{pmatrix} 1 & 0 & 0 \\ 0 & \sin\delta & -\cos\delta \\ 0 & \cos\delta & \sin\delta \end{pmatrix}, \quad \mathbf{R}_\phi = \begin{pmatrix} \cos\phi_s & \sin\phi_s & 0 \\ -\sin\phi_s & \cos\phi_s & 0 \\ 0 & 0 & 1 \end{pmatrix} \tag{2.165}$$

Fig. 2.23 Spherical coordinate system (R, θ, ϕ) in gray fixed at the center of the Earth, the xyz coordinate system and $\theta\phi R$ coordinate system at the point source (reprinted from Koketsu [41] with permission of Kindai Kagaku)

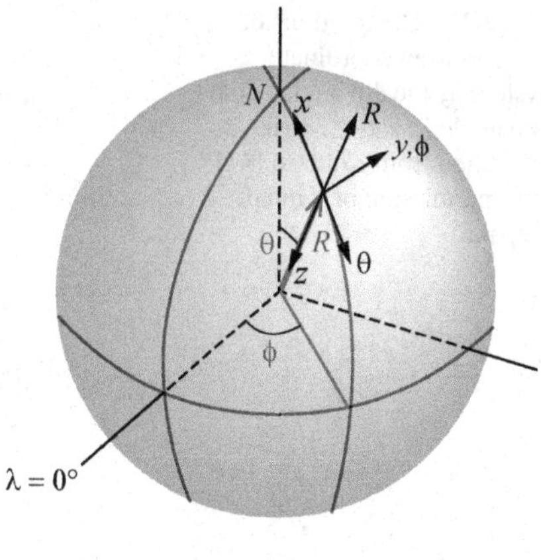

through

$$\mathbf{M}_0'' = \mathbf{R}_\lambda^{\mathrm{T}} \mathbf{M}_0''' \mathbf{R}_\lambda = \begin{pmatrix} 0 & M_0 \cos\lambda & 0 \\ M_0 \cos\lambda & 0 & -M_0 \sin\lambda \\ 0 & -M_0 \sin\lambda & 0 \end{pmatrix}, \tag{2.166}$$

$$\mathbf{M}_0' = \mathbf{R}_\delta^{\mathrm{T}} \mathbf{M}_0'' \mathbf{R}_\delta = \begin{pmatrix} 0 & M_0 \sin\delta \cos\lambda & -M_0 \cos\delta \cos\lambda \\ M_0 \sin\delta \cos\lambda & -M_0 \sin 2\delta \sin\lambda & M_0 \cos 2\delta \sin\lambda \\ -M_0 \cos\delta \cos\lambda & M_0 \cos 2\delta \sin\lambda & M_0 \sin 2\delta \sin\lambda \end{pmatrix}$$

we have the moment tensor [34]

$$\mathbf{M}_0 = \mathbf{R}_\phi^{\mathrm{T}} \mathbf{M}_0' \mathbf{R}_\phi = \begin{pmatrix} M_{xx} & M_{xy} & M_{xz} \\ & M_{yy} & M_{yz} \\ & & M_{zz} \end{pmatrix},$$

$$\begin{cases} M_{xx} = -M_0(\sin\delta \cos\lambda \sin 2\phi_s + \sin 2\delta \sin\lambda \sin^2 \phi_s) \\ M_{xy} = +M_0(\sin\delta \cos\lambda \cos 2\phi_s + \frac{1}{2} \sin 2\delta \sin\lambda \sin 2\phi_s) \\ M_{xz} = -M_0(\cos\delta \cos\lambda \cos\phi_s + \cos 2\delta \sin\lambda \sin\phi_s) \\ M_{yy} = +M_0(\sin\delta \cos\lambda \sin 2\phi_s - \sin 2\delta \sin\lambda \cos^2 \phi_s) \\ M_{yz} = -M_0(\cos\delta \cos\lambda \sin\phi_s - \cos 2\delta \sin\lambda \cos\phi_s) \\ M_{zz} = +M_0 \sin 2\delta \sin\lambda \end{cases} \tag{2.167}$$

where lower left elements are omitted because this is a symmetric tensor.

In the 1980s, Dziewonski et al. [15] developed a procedure that can routinely determine moment tensors, when the propagation characteristics are known for a given Earth model (velocity structure). **Broadband seismographs** (Sect. 4.1.4) began to be deployed globally around the same time, and seismograms observed by these

[34] This agrees with Eq. (1) in Box 4.4 of Aki and Richards [3].

instruments are used for the determination (the details of the method are explained in Sect. 2.3.4). The **Global CMT Project** started by Dziewonski et al. uses the **spherical coordinate system** (R, θ, ϕ) fixed at the center of the Earth (Fig. 2.23) because it is a global project. A Cartesian coordinate system is constructed from the unit vectors in the R, θ, and ϕ directions at the point source. When the constructed coordinate system is compared to the xyz coordinate system in Fig. 2.22, the ϕ-axis is found to coincide with the y-axis, and the R- and θ-axes are opposite to the z- and x-axis, respectively, as shown in Fig. 2.23. This means that rotating the xyz coordinate system around the y-axis by 180° in the direction that points the z-axis toward the x-axis results in a $\theta\phi R$ coordinate system. Therefore, using (2.167) we have

$$\begin{pmatrix} M_{\theta\theta} & M_{\theta\phi} & M_{\theta R} \\ & M_{\phi\phi} & M_{\phi R} \\ & & M_{RR} \end{pmatrix} = \mathbf{R}_R^{\mathrm{T}} \begin{pmatrix} M_{xx} & M_{xy} & M_{xz} \\ & M_{yy} & M_{yz} \\ & & M_{zz} \end{pmatrix} \mathbf{R}_R \,, \ \mathbf{R}_R = \begin{pmatrix} \cos 180^\circ & 0 & \sin 180^\circ \\ 0 & 1 & 0 \\ -\sin 180^\circ & 0 & \cos 180^\circ \end{pmatrix} . \tag{2.168}$$

We rearrange the result of (2.168) in the order of R, θ, ϕ and obtain the moment tensor[35]

$$\mathbf{M}_0 = \begin{pmatrix} M_{RR} & M_{R\theta} & M_{R\phi} \\ & M_{\theta\theta} & M_{\theta\phi} \\ & & M_{\phi\phi} \end{pmatrix} = \begin{pmatrix} M_{zz} & M_{xz} & -M_{yz} \\ & M_{xx} & -M_{xy} \\ & & M_{yy} \end{pmatrix} . \tag{2.169}$$

2.3.4 CMT Inversion

Assuming M_0 and M_{ij} to be time functions, we take the **Fourier transform** (Sect. 4.18) of (2.167). From the results of this, the Fourier transforms $\overline{M}_0$ and $\overline{M}_{ij}$, as well as a_k defined by Nabelek [55] as[36]

$$\begin{aligned} a_1 &= \overline{M}_{xx} \cos^2 \Theta + \overline{M}_{xy} \sin 2\Theta + \overline{M}_{yy} \sin^2 \Theta \,, \quad a_2 = \overline{M}_{zz} \,, \\ a_3 &= \overline{M}_{xz} \cos \Theta + \overline{M}_{yz} \sin \Theta \,, \quad a_4 = \frac{1}{2} \left(\overline{M}_{yy} - \overline{M}_{xx} \right) \sin 2\Theta + \overline{M}_{xy} \cos 2\Theta \,, \\ a_5 &= -\overline{M}_{xz} \sin \Theta + \overline{M}_{yz} \cos \Theta \,, \end{aligned} \tag{2.170}$$

and (2.130), Kuge [45] obtained

---

[35] This agrees with Eq. (4) in Box 4.4 of Aki and Richards [3] if $r \to R$ and $\Delta \to \theta$.

[36] Equation (A11) of Nabelek [55] with the substitutions of $\phi \to \Theta$ and $\hat{M}_{ij} \to \overline{M}_{ij}$.

$$\begin{aligned}
\frac{1}{2}a_2 &= \overline{M}_0\frac{1}{2}\sin\lambda\sin 2\delta = \overline{M}_0\Lambda_0\,,\\
-a_3 &= -\left\{-\overline{M}_0(\cos\delta\cos\lambda\cos\phi_s+\cos 2\delta\sin\lambda\sin\phi_s)\cos\Theta\right\}\\
&\quad -\left\{-\overline{M}_0(\cos\delta\cos\lambda\sin\phi_s-\cos 2\delta\sin\lambda\cos\phi_s)\sin\Theta\right\}\\
&= \overline{M}_0\,(\cos\delta\cos\lambda\cos(\Theta-\phi_s)-\cos 2\delta\sin\lambda\sin(\Theta-\phi_s))\\
&= \overline{M}_0\,(\cos\delta\cos\lambda\cos\theta-\cos 2\delta\sin\lambda\sin\theta) = \overline{M}_0\Lambda_1\,,\\
a_1+\tfrac{1}{2}a_2 &= \left\{-\overline{M}_0(\sin\delta\cos\lambda\sin 2\phi_s+\sin 2\delta\sin\lambda\sin^2\phi_s)\right\}\cos^2\Theta\\
&\quad +\overline{M}_0(\sin\delta\cos\lambda\cos 2\phi_s+\frac{1}{2}\sin 2\delta\sin\lambda\sin 2\phi_s)\sin 2\Theta\\
&\quad +\overline{M}_0(\sin\delta\cos\lambda\sin 2\phi_s-\sin 2\delta\sin\lambda\cos^2\phi_s)\sin^2\Theta+\frac{1}{2}\overline{M}_0\sin\lambda\sin 2\delta\\
&= \overline{M}_0\sin\delta\cos\lambda(-\sin 2\phi_s\cos 2\Theta+\cos 2\phi_s\sin 2\Theta)+\frac{1}{2}\overline{M}_0\sin 2\delta\sin\lambda\\
&\quad \left(-2\sin^2\phi_s\cos^2\Theta+\sin 2\phi_s\sin 2\Theta-2\cos^2\phi_s\sin^2\Theta+\sin^2\Theta+\cos^2\Theta\right)\\
&= \overline{M}_0\sin\delta\cos\lambda\sin 2(\Theta-\phi_s)+\frac{1}{2}\overline{M}_0\sin 2\delta\sin\lambda\cos 2(\Theta-\phi_s)\\
&= \overline{M}_0\left(\sin\delta\cos\lambda\sin 2\theta+\frac{1}{2}\sin 2\delta\sin\lambda\cos 2\theta\right) = \overline{M}_0\Lambda_2\,,\\
-a_5 &= -\left[-\left\{-\overline{M}_0(\cos\delta\cos\lambda\cos\phi_s+\cos 2\delta\sin\lambda\sin\phi_s)\sin\Theta\right\}\right]\\
&\quad -\left[+\left\{-\overline{M}_0(\cos\delta\cos\lambda\sin\phi_s-\cos 2\delta\sin\lambda\cos\phi_s)\cos\Theta\right\}\right]\\
&= \overline{M}_0\,(-\cos\delta\cos\lambda\sin(\Theta-\phi_s)-\cos 2\delta\sin\lambda\cos(\Theta-\phi_s))\\
&= \overline{M}_0\,(-\cos\delta\cos\lambda\sin\theta-\cos 2\delta\sin\lambda\cos\theta) = \overline{M}_0\frac{\partial\Lambda_1}{\partial\theta}\,,\\
2a_4 &= \left[\left\{+M_0(\sin\delta\cos\lambda\sin 2\phi_s-\sin 2\delta\sin\lambda\cos^2\phi_s)\right\}\right.\\
&\quad \left.-\left\{-M_0(\sin\delta\cos\lambda\sin 2\phi_s+\sin 2\delta\sin\lambda\sin^2\phi_s)\right\}\right]\sin 2\Theta\\
&\quad +2M_0(\sin\delta\cos\lambda\cos 2\phi_s+\frac{1}{2}\sin 2\delta\sin\lambda\sin 2\phi_s)\cos 2\Theta\\
&= \overline{M}_0\,\{2\sin\delta\cos\lambda\cos 2(\Theta-\phi_s)-\sin 2\delta\sin\lambda\sin 2(\Theta-\phi_s)\}\\
&= \overline{M}_0\,(2\sin\delta\cos\lambda\cos 2\theta-\sin 2\delta\sin\lambda\sin 2\theta) = \overline{M}_0\frac{\partial\Lambda_2}{\partial\theta}\,,
\end{aligned}\tag{2.171}$$

where Θ is the observer azimuth measured from the north (Eq. (2.130), Fig. 2.15).

Equation (2.171) indicates that the coefficients of the **source potentials** (2.129) are linear combinations of the Fourier transforms of the moment tensor elements $\overline{M}_{ij}$. Therefore, the coefficients in the **discontinuity vectors** (Sects. 3.1.2, 3.1.3) are also linear combinations of $\overline{M}_{ij}$ because the discontinuity vectors are made from the source potentials. Furthermore, the Fourier transforms of ground motions in a **1-D velocity structure** are represented by the sum of the products of the **propagator matrix** elements (Sect. 3.1.4) and the discontinuity vector elements as shown in

(3.25). The coefficients of the Fourier transforms of the ground motions are therefore also linear combinations of $\overline{M}_{ij}$.

If $M_{ij}(t)$ is assumed to be proportional to the origin-shifted ramp function $U_0(t)$ as in the Haskell model (Sect. 2.3.5),

$$\overline{M}_{ij}(\omega) = M_{ij}\overline{U}_0(\omega) \tag{2.172}$$

where M_{ij} is a variable independent of t and ω. In this case, from the nature of "linearity" of the Fourier transform (Table 4.3), the ground motion at the point $\mathbf{x}$ in the 1-D velocity structure is given as $M_{ij} f_{ij}(t, \mathbf{x})$. $f_{ij}(t, \mathbf{x})$ is the ground motion at the point $\mathbf{x}$ due to the unit moment tensor element $M_{ij}(t) = U_0(t)$ $(M_{ij} = 1)$. $f_{ij}(t, \mathbf{x})$ therefore includes the effect of the point source, and differs from the Green's function described in Sect. 1.3.3 in a strict sense. However, it is also called **"Green's function"** because of the similarity between the unit moment tensor element and the delta function.

Here, to simplify the following description, the notation of the subscript of a moment tensor is changed as [15]

$$M_{RR} = M_{zz} \to M_1\,, \quad M_{\theta\theta} = M_{xx} \to M_2\,, \quad M_{\phi\phi} = M_{yy} \to M_3\,,$$
$$M_{R\theta} = M_{xz} \to M_4\,, \; M_{R\phi} = -M_{yz} \to M_5\,, \; M_{\theta\phi} = -M_{xy} \to M_6\,. \tag{2.173}$$

The kth component (usually either the north-south component, the east-west component, or the vertical component) of the ground motion of the jth observation point at $\mathbf{x}_j$ and t_i, obtained by **sampling** (Sect. 4.2.3) over a time t, is the **principle of superposition** (Sect. 1.3.1), which can be synthesized as

$$F_k(t_i, \mathbf{x}_j) = \sum_{m=1}^{6} M_m f_{mk}(t_i, \mathbf{x}_j) \tag{2.174}$$

using the Green's functions calculated by a method in Chap. 3. $f_{mk}(t, \mathbf{x})$ is an extension of the Green's function $f_{ij}(t, \mathbf{x}) \equiv f_m(t, \mathbf{x})$ to the kth component in multi-component observations (Sect. 4.1.1). The synthesized ground motions $F_k(t_i, \mathbf{x}_j)$ are called **synthetic seismograms**. Conversely, if the kth component of the ground motion $F_k^o(t_i, \mathbf{x}_j)$ is observed at the jth observation point (**observed seismogram**), M_m is determined so that the synthetic seismograms are matched to the observed seismograms. However, since both the synthetic and observed seismograms actually contain errors and noise, and cannot be completely matched, to determine M_m,

$$S = \sum_i \sum_j \sum_k \frac{1}{\sigma_{jk}^2} \left\{F_k^o(t_i, \mathbf{x}_j) - F_k(t_i, \mathbf{x}_j)\right\}^2 \tag{2.175}$$

is minimized based on the least-squares method (Sect. 4.4). $1/\sigma_{jk}$ is the weight for each observed seismogram estimated from the **observational error** (Sect. 4.4) or other factors. S is minimized, so that

$$\frac{\partial S}{\partial M_m} = 0, \quad m = 1, 2, \ldots, 6. \tag{2.176}$$

Since M_m are included in $F_k(t_i, \mathbf{x}_j)$ only in the form of coefficients, (2.176) yields simultaneous linear equations (**linear least-squares method**, Sect. 4.4) for M_m, which are solved by various numerical methods.

It is possible to make not only the moment tensor but also the location of the point source $\mathbf{x}_0$ variables. However, since $\mathbf{x}_0$ is embedded in $F_k(t_i, \mathbf{x}_j)$ instead of in the form of coefficients, (2.176) yields simultaneous nonlinear equations for M_m, and they are iteratively solved using the nonlinear least-squares method (Sect. 4.4) [15]. $\mathbf{x}_0$ obtained in this way is the center of ground motion radiation in the source fault, and is called the **centroid**. On the other hand, the location of the **hypocenter** is not determined from seismograms but from the arrival times of **initial motions** (Sect. 2.3.1). Therefore, the **hypocenter determination** does not determine the location of the centroid but the location of the **rupture initiation point** (Sect. 2.3.5). In the case of a large earthquake, the centroid and the rupture initiation point usually do not coincide.

Needless to say, the ground motion is a result and the earthquake is a cause, therefore estimating the ground motion by giving an earthquake model (here, the moment tensor) is a **forward problem**. Conversely, estimating an earthquake model by giving the ground motion is an **inverse problem**. Solving the inverse problem with the least-squares method or another method is called an **inversion**, and the inversion for the moment tensor at the centroid as described here is called the centroid moment tensor inversion, or the **CMT inversion** for short. The analysis that the **Global CMT Project** performs routinely is a CMT inversion using the Green's functions of the **free oscillation** of the Earth (e.g., Chap. 8 of Aki and Richards [3]). The CMT inversion of the **F-net** [20] is performed using the same method as explained here.

2.3.5 Ground Motion from a Finite Fault[37]

In reality, a source fault is not a point and has a physical extent. A **fault rupture** does not occur at the same time throughout the extent of the fault, but begins in a small area and propagates at a certain speed to the surrounding parts of the fault plane. The effects of such a physical extent ("**finite fault**") and **rupture propagation** are particularly prominent near the source fault, and a model of an earthquake that takes these factors into account is called a **fault plane model**.

[37] For more information on Sect. 2.3.5, see Sato [65].

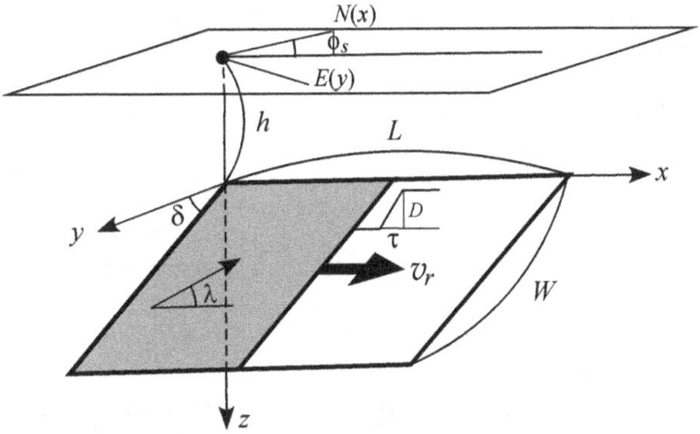

Fig. 2.24 Haskell model (reprinted from Koketsu [41] with permission of Kindai Kagaku)

The linear approximation of a source fault was achieved in the early 1960s, but the first realistic rectangular fault model was published by Haskell [25]. This relatively simple model, the **Haskell model**, has long been used as the basis for subsequent model developments. As shown in Fig. 2.24, the fault plane of the Haskell model is obtained by expanding the point source into a rectangle, the long side of which is horizontal. The geometrical parameters of the rectangle include its **strike** ϕ_s, **dip angle** δ, **slip angle** λ, and depth h, which define the geometry of the point source, as well as its length L and width W, which define the geometry of the rectangle. h is often represented by the depth at the top of the fault plane.

The time history of the slip on the fault plane, that is, the slip time function (Sect. 2.1.2), is represented by the origin-shifted ramp function

$$U_0(t) = \begin{cases} 0 & t < 0 \\ t/\tau & 0 \le t \le \tau \\ 1 & t > \tau \end{cases} \tag{2.177}$$

(Figure 2.25) built from the ramp function $U(t)$ in Sect. 4.2.2. It is assumed that the slip begins at $t = 0$, reaches the final **slip amount** D after a certain time τ (the **rise time**), and stops. D is simply called a **slip** as for a point source (Sect. 2.1.2). τ and D are assumed to be constant throughout the fault plane. The fault rupture occurs simultaneously in the width direction of the rectangle, and propagates in the length direction with a constant **rupture velocity** v_r from the line at the start of the rupture.

The **seismic moment** M_0 is given as $M_0 = \iint \mu D d\Sigma$ for a point source, as proven in Sect. 2.1.2. If, for simplicity, the finite fault in Fig. 2.24 is assumed to be a vertical strike-slip fault like the point source in Fig. 2.7, and $v_r = \infty$, the above integration is

$$\iint d\Sigma(\boldsymbol{\xi}) = \int_0^L \int_0^W d\xi_x d\xi_z \bigg|_{\xi_y = 0},$$

Fig. 2.25 Origin-shifted ramp function $U_0(t)$ (reprinted from Koketsu [41] with permission of Kindai Kagaku)

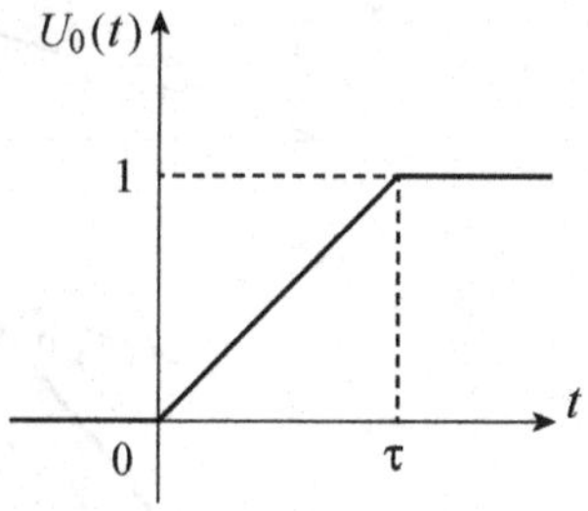

and $[u_x] = D(\mathbf{x}, t)$, $[u_y] = 0$, $[u_z] = 0$, $\rho f_z(\mathbf{x}, t) = 0$ in the **body force equivalent** (2.13). Using these and $\int \delta(\xi) d\xi = H(\xi)$, where $H(x)$ is a **step function** (Sect. 4.2.2), we have from (2.14) that

$$\begin{aligned}\rho f_x(\mathbf{x}, t) &= -\mu D(t) \int_0^L \int_0^W \delta(x - \xi_x) \frac{\partial \delta(y - \xi_y)}{\partial y} \delta(z - \xi_z)\, d\xi_x d\xi_z \bigg|_{\xi_y = 0} \\ &= -\mu D(t) \int_{x-L}^{x} \int_{z-W}^{z} \delta(\eta_x) \frac{\partial \delta(y)}{\partial y} \delta(\eta_z)\, d\eta_x d\eta_z \\ &= -\mu D(t) \frac{\partial \delta(y)}{\partial y} \left[H(x) - H(x - L)\right]\left[H(z) - H(z - W)\right] . \end{aligned} \tag{2.178}$$

Similarly, (2.15) yields

$$\rho f_y(\mathbf{x}, t) = -\mu D(t) \frac{\partial \delta(x)}{\partial x} \left[H(x) - H(x - L)\right]\left[H(z) - H(z - W)\right] . \tag{2.179}$$

Substituting (2.178), (2.179), and $\rho f_z(\mathbf{x}, t) = 0$ into the representation theorem (1.108) with the summation expanded, we obtain

$$\begin{aligned}u_n(\mathbf{x}, t) &= \iiint_{-\infty}^{+\infty} \left\{\rho f_x(\boldsymbol{\xi}, t) * G_{nx}(\mathbf{x}, t; \boldsymbol{\xi}, 0) + \rho f_y(\boldsymbol{\xi}, t) * G_{ny}(\mathbf{x}, t; \boldsymbol{\xi}, 0)\right\} d\xi_x d\xi_y d\xi_z \\ &= \mu D(t) LW * \left\{-\frac{\partial}{\partial \xi_y} G_{nx}(\mathbf{x}, t; \mathbf{0}, 0) - \frac{\partial}{\partial \xi_x} G_{ny}(\mathbf{x}, t; \mathbf{0}, 0)\right\} . \end{aligned} \tag{2.180}$$

Comparing (2.180) with (2.19), we find that (2.180) represents displacements due to the **double couple** with a seismic moment of

$$M_0 = \mu D S\,, \quad S = LW\,. \tag{2.181}$$

$S = LW$ is a **fault area** and μ is the **rigidity** (Sect. 1.2.3) of the rock constituting the fault plane, which has a value of around 30 GPa.

As for the point source in Sect. 2.1.2, the parameters used here are collectively referred to as the fault parameters of the Haskell model. Among these, h, ϕ_s, δ, L, W, M_0, etc. represent the position, geometry, and scale of the entire source fault and are

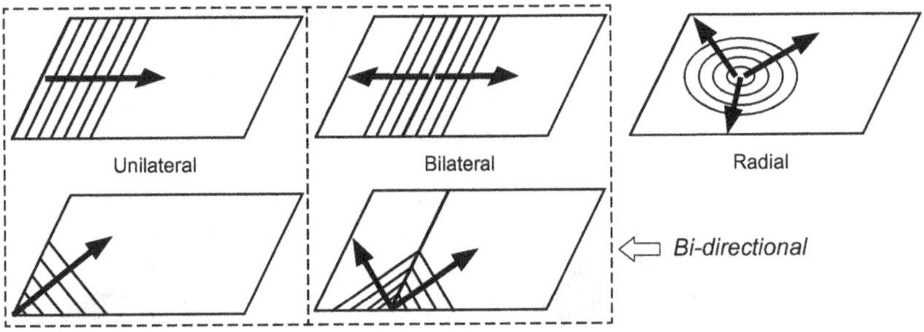

Fig. 2.26 Definitions of rupture propagation modes based on Sato et al. [64] (reprinted from Koketsu [41] with permission of Kindai Kagaku)

termed the **outer fault parameters** [29]. In addition, in order to identify the source fault as a source of ground motions, it is necessary to provide the start location and mode of the **rupture propagation**. As shown in Fig. 2.26, the mode in which the rupture starts from one of the left or right edges of the fault plane and propagates to the other edge is termed **unilateral faulting**. The mode in which the rupture starts near the center of the fault plane and propagates in two directions toward both ends is termed **bilateral faulting**. The faulting mode where the rupture propagation direction does not match the strike or dip direction of the fault plane, regardless of unilateral or bilateral behavior, is sometimes called **bi-directional**. However, from studies on the dynamics of faulting, rupture propagation in concentric circles is considered to be physically valid, with a small area around the hypocenter representing the rupture initiation point. This mode is now the most frequently used and is termed "radial", as shown in Fig. 2.26. The location of the hypocenter is determined by hypocenter determination (Sect. 2.3.1). Of M 6 or larger earthquakes that occurred in and around Japan between 1498 and 1987, the fault parameters for 92 earthquakes were compiled by Sato et al. [64].

Among the fault parameters, the **rise time** τ and the mode of rupture propagation both affect the temporal characteristics of ground motion. If a particular function is not assumed for the **moment time function**, the shape of the function itself also has an effect. Therefore, the **amplitude spectrum** (Sect. 4.2.2) that combines the **spectrum** (Sect. 4.2.2) of the moment-time function and the spectrum of the rupture propagation mode is called the **source spectrum** [22]. For example, a point source has no rupture propagation mode, so the spectrum of the moment time function is the source spectrum.

Hereafter, considering the ground motion that consists mainly of the **far-field term**, the moment time function is replaced by the **moment rate function** (Sect. 2.1.4). Furthermore, if the slip time function is the origin-shifted **ramp function** $U_0(t)$ (2.181), from (2.181) the moment rate function is

$$\frac{dM_0(t)}{dt} = \mu \frac{dD(t)}{dt} S = \mu DS \frac{dU_0(t)}{dt} = \mu DS \frac{dU(t-\tau/2)}{dt}. \tag{2.182}$$

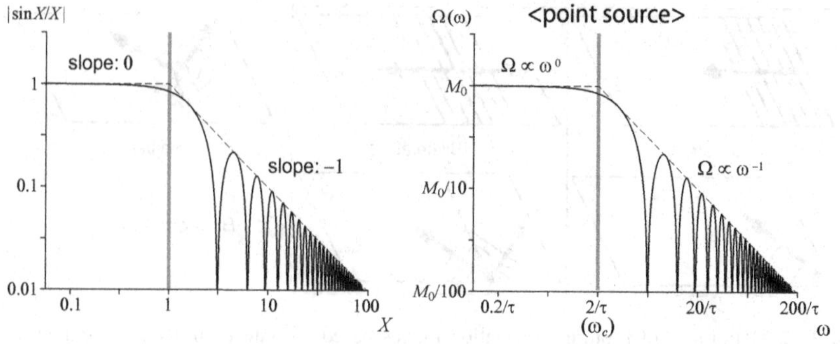

Fig. 2.27 Left: Function $|\sin X/X|$, based on Sato [65], Right: Source spectrum $\Omega(\omega)$ of a point source with M_0 (the parts where the amplitude is less than 0.01 or $M_0/100$ are not shown) (reprinted from Koketsu [41] with permission of Kindai Kagaku)

Using the formulas in Tables 4.3 and 4.4, the source spectrum is given as

$$\Omega(\omega) = \mu DS \left| \mathrm{i}\omega\, e^{-\mathrm{i}\omega\tau/2} \frac{2\sin(\omega\tau/2)}{\mathrm{i}\omega^2\tau} \right| = M_0 \left| \frac{\sin(\omega\tau/2)}{\omega\tau/2} \right| . \tag{2.183}$$

$|\sin X/X|$, $X = \omega\tau/2$, which is the source spectrum excluding M_0, has the shape shown in Fig. 2.27 (left) on a log-log plot. Since the derivative of the ramp function divided by τ is a **rectangular function** $r(t)$ (Sect. 4.2.3), it can also be seen that Fig. 2.27 (left) represents the amplitude spectrum of the rectangular function. Furthermore, based on this, the source spectrum of the point source $\Omega(\omega)$ (2.183) is drawn in Fig. 2.27 (right). The source spectrum is almost flat until $\omega_c = 2/\tau$ corresponding to $X = 1$, and attenuates in proportion to ω^{-1} at higher frequencies, with overlaid vibration of $\sin(\omega\tau/2)$. Therefore,

$$f_c = \frac{\omega_c}{2\pi} = \frac{1}{\pi\tau} \tag{2.184}$$

is called a **corner frequency**.

When the source fault is linear with length L and moment density M_0/L (**line source**), the time lapse of the ground motion at an observation point located at a distance R in the direction φ from the fault is given for **unilateral faulting** as

$$\frac{M_0}{L}\int_0^L r_0\left(t - \frac{\xi}{v_r} - \frac{R - \xi\cos\varphi}{\beta}\right)\frac{d\xi}{\tau} = \frac{M_0}{L}\int_0^L r_0\left(t - \frac{R}{\beta} - \frac{\tau_L}{L}\xi\right)\frac{d\xi}{\tau} . \tag{2.185}$$

Here, $r_0(t)$ is an origin-shifted rectangular function of width τ, and v_r is the rupture velocity. At a distant observation point, the direction does not change even if the rupture moves, but the hypocentral distance is assumed to become $\xi\cos\varphi$ shorter according to the rupture distance ξ (Fig. 2.28). $\tau_L = L(1/v_r - \cos\varphi/\beta)$ is

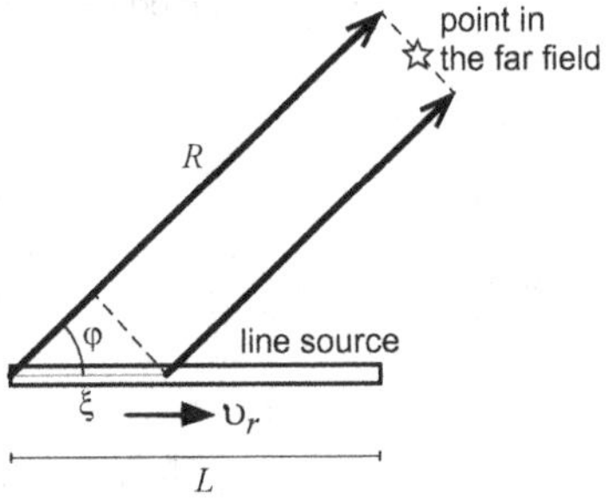

Fig. 2.28 Line source and distant observation point based on Sato [65] (reprinted from Koketsu [41] with permission of Kindai Kagaku)

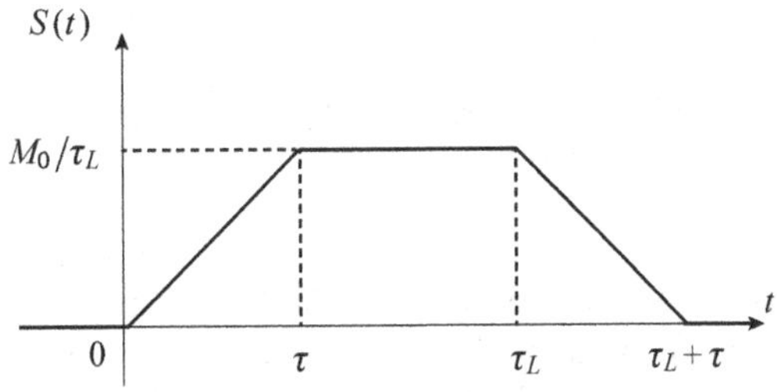

Fig. 2.29 Trapezoidal function, which is the moment rate function of the line source (reprinted from Koketsu [41] with permission of Kindai Kagaku)

the apparent **rupture duration** as viewed from the direction φ. The main part of the ground motion is the S wave and the S wave velocity β is used, but for the P wave β can be replaced by α in (2.185). Noting that $r_0(t)/\tau$ is the derivative of $U_0(t)$, and performing the integration in (2.185), the result is

$$\frac{M_0}{\tau_L}\left\{U_0\left(t-\frac{R}{\beta}\right)-U_0\left(t-\frac{R}{\beta}-\tau_L\right)\right\}. \tag{2.186}$$

This corresponds to the moment rate function of the line source, and is a **trapezoidal function** $S(t)$ shown in Fig. 2.29 [35]. The height of the trapezoid is M_0/τ_L, and its area is M_0 regardless of φ.

The integration in (2.185) can be rewritten as

$$\begin{aligned}&\frac{M_0}{L}\int_{-\infty}^{+\infty} r_0\left(t-\frac{R}{\beta}-\frac{\tau_L}{L}\xi\right) r_0\left(\frac{\tau\xi}{L}\right)\frac{d\xi}{\tau}\\ &= M_0\int_{-\infty}^{+\infty} r_0\left(t-\frac{R}{\beta}-\eta\right) r_0\left(\frac{\tau\eta}{\tau_L}\right)\frac{d\eta}{\tau_L\tau}\end{aligned} \tag{2.187}$$

using the rectangular function $r_0(\tau t/L)$ of width L. Equation (2.187) is just the **convolution** (1.85) of a rectangular function of width τ and a rectangular function of width τ_L [65]. Noting again that the amplitude spectrum of $r_0(t)/\tau$ is given by (2.183), we have

$$\Omega(\omega) = M_0\left|\frac{\sin(\omega\tau/2)}{\omega\tau/2}\right|\left|\frac{\sin(\omega\tau_L/2)}{\omega\tau_L/2}\right| \tag{2.188}$$

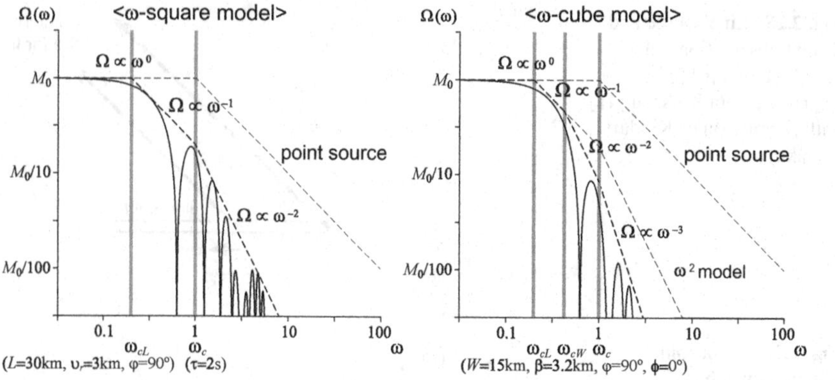

Fig. 2.30 Left: ω-square model, Right: ω-cube model (reprinted from Koketsu [41] with permission of Kindai Kagaku)

for the source spectrum of the line source. Since (2.188) is the multiplication of (2.183) and a similar function, its logarithmic plot has two corner frequencies, as shown in Fig. 2.30 (left). These are f_c of (2.188) and $f_{cL} = 1/(\pi\tau_L)$, the latter being the corner frequency related to the fault length L. In general, if the source fault is sufficiently long, $\tau_L > \tau$, therefore the source spectrum is flat at $[0,\ f_{cL}]$, and is attenuated in proportion to ω^{-1} at $[f_{cL},\ f_c]$ and ω^{-2} at $[f_c,\ \infty]$. When the source spectrum is proportional to ω^{-2} at the higher frequencies, it is called the **ω-square model** [2].

Furthermore, if the source fault is a unilateral **Haskell model**, a new integration related to the fault width W and the accompanying corner frequency $f_{cW} = 1/(\pi\tau_W)$ are added, resulting in the **ω-cube model**, as shown in Fig. 2.30 (right) [22]. When the azimuth of the observation point to look at the source fault is defined as shown in Fig. 2.31, the apparent rupture duration for L coincides with that for the line source, and that for W is given by $\tau_W = W\cos\phi\,\sin\varphi/\beta$. The corner frequencies of the unilateral Haskell model are therefore f_c, f_{cL}, and f_{cW}. In general, $L > W$ and $f_{cL} < f_{cW}$ in many cases, as shown in Fig. 2.30 (right), although this depends on the azimuth.

When a circular fault ruptures concentrically from the center, there is only a single integration of the radius therefore it has the source spectrum of the ω-square model, like the line source. The **Brune model** [7] was established by approximating this integration with the function [6]

$$\left(t - \frac{R}{\beta}\right) e^{-a(t-R/\beta)}, \quad \frac{a}{2\pi} = 4.9 \times 10^6\, \beta \left(\frac{\Delta\sigma}{M_0}\right)^{\frac{1}{3}}. \tag{2.189}$$

In the second equation of (2.189), β, $\Delta\sigma$, and M_0 are given in km/s, bar, and dyne·cm, respectively. The amplitude spectrum of this time function is proportional to $1/(\omega^2 + a^2)$ and is approximately a ω-square model, where $a/2\pi$ is the corner frequency τ_{cL}.

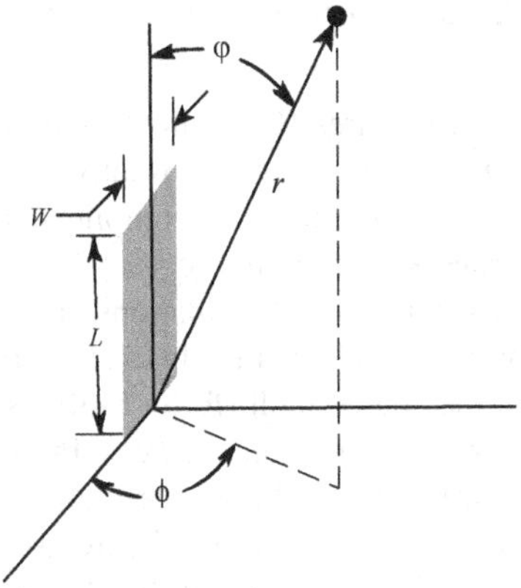

Fig. 2.31 Definition of the direction of the observation point with respect to the source fault (gray rectangle), based on Savage [66] (reprinted from Koketsu [41] with permission of Kindai Kagaku)

In the line source, if **bilateral faulting** occurs (in Fig. 2.28, the rupture propagates not only in the $+\xi$ direction for L but also in the $-\xi$ direction for L'), $|\sin(\omega\tau_L/2)/(\omega\tau_L/2)|$ in (2.188) is replaced with [66]

$$\frac{1}{L+L'}\left\{L^2\frac{\sin^2(\omega\tau_L/2)}{(\omega\tau_L/2)^2}+L'^2\frac{\sin^2(\omega\tau_{L'}/2)}{(\omega\tau_{L'}/2)^2}+\right.$$
$$\left.2LL'\frac{\sin(\omega\tau_L/2)}{\omega\tau_L/2}\frac{\sin(\omega\tau_{L'}/2)}{\omega\tau_{L'}/2}\cos\frac{\omega(\tau_L-\tau_{L'})}{2}\right\}^{\frac{1}{2}}. \tag{2.190}$$

This replacement introduces a new rupture duration $\tau_{L'} = L'(1/v_r + \cos\phi/\beta)$, which results in a new corner frequency $f_{cL'} = 1/(\pi\tau_{L'})$.

In many cases, the ω-square model can be applied to observed source spectra (**scaling law**), though the portion that is proportional to ω^{-1} in the intermediate frequency band is often unclear [2]. The shape of the source fault is considered to be a rectangle rather than a circle, therefore even if the fault rupture spreads concentrically, the finiteness of the integration associated with the rectangular shape appears in both the length and width. The source spectrum is thereby thought to be represented by a^n ω^{-3} model, however the observed spectrum is actually represented by a^n ω^{-2} model, implying that the high frequency component is enhanced. The reason for this high-frequency enhancement is that the source process (Sect. 2.3.6) of a real source fault is not as simple as the Haskell model but includes complexity, which leads to this enhancement.

In addition, the highest frequency part of the observed spectrum deviates from the ω-square model, and is often largely attenuated. The corner frequency at which this large attenuation begins is called $\mathbf{f_{max}}$, but it is not clear whether the large attenuation results from the earthquake source or the propagation.

2.3.6 Source Processes and Source Inversion

Fault ruptures (fault slips) in simple ways as shown in Fig. 2.24 are rare for real earthquakes. Ruptures usually follow more complicated time histories and are spatially distributed in a complicated fashion. This time history is called the **source process** (or rupture process), therefore source models representing such time histories and spatial distributions are referred to as "source process models". A source process model is mainly expressed in two ways. In both expressions, the finite area of the source fault (**finite fault**) is divided into **subfaults** that are small enough to be replaced by point sources. In the first method, subfaults are arranged to cover the entire plane of the source fault (Fig. 2.32a), whereas only a small number of subfaults, called "subevents", are used in the second method (Fig. 2.32b). The second method was adopted in the 1980s, when computer capabilities were limited (e.g., Kikuchi and Fukao [37]). However, since then most source process models have been constructed using the first method, albeit with some number of subfaults. Even with the first method, if a sufficient number of subfaults cannot be employed, the approximation of a subfault using a simple point source may involve errors. A point source that considers rupture propagation in a subfault is therefore also used [74].

When the **outer fault parameters** of the source fault plane are known in advance from the aftershock distribution (Sect. 2.3.2), the **CMT inversion** of teleseismic data (Sect. 2.3.4), etc., and when the positions of the subfaults are fixed as in the first method, determining the source process model consists in determining the **moment time function** $M_0(t)$ or the **slip time function** $D(t) = M_0(t)/(\mu S)$ (from Eq. (2.181)) of the point source of each subfault, where S is the area of a subfault and μ is the rigidity at that point. If the forms of the time functions are fixed to the origin-shifted ramp function $U_0(t)$ (Eq. (2.177), Fig. 2.25) as in Sect. 2.3.5, $M_0(t) = M_0U_0(t)$ or $D(t) = DU_0(t)$ results in the problem of determining the seismic moment M_0 or the slip D of each subfault.

The source potentials (2.129) of the point source contain the Fourier transform of the moment time function, that is $\overline{M}_0(\omega)$, as an independent coefficient. The **discontinuity vector** (Sects. 3.1.2, 3.1.3) at the source made from these potentials therefore also includes $\overline{M}_0(\omega)$ as an independent coefficient. Furthermore, the Fourier transforms of ground motions in a 1-D velocity structure are expressed as the sum of

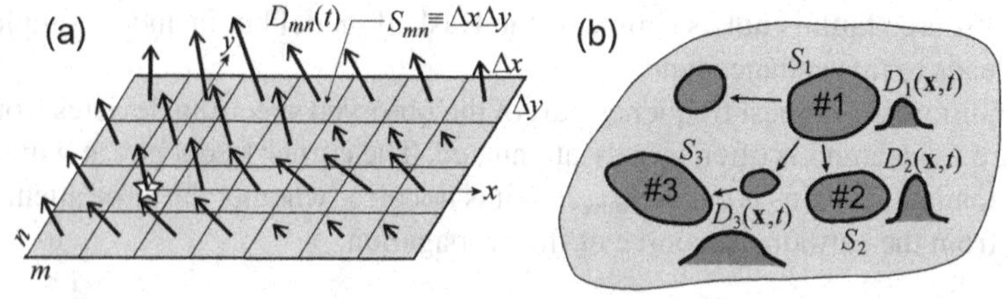

Fig. 2.32 Two source process models based on Kikuchi [35]. In **a**, the asterisk indicates the rupture initiation point, and the xy coordinate system is set on the fault plane with the point as the origin (reprinted from Koketsu [41] with permission of Kindai Kagaku)

the products of the **propagator matrix** elements (Sect. 3.1.4) and the discontinuity vector elements, as shown in (3.25). As mentioned above, assuming that the form of the time function is $U_0(t)$ etc., we have

$$\overline{M}_0(\omega) = M_0\overline{U}_0(\omega) = \mu DS\overline{U}_0(\omega)\,, \tag{2.191}$$

and, from the nature of the "linearity" of the Fourier transform (Table 4.3), the ground motion at the point $\mathbf{x}$ in the 1-D velocity structure is given as $Df(t,\mathbf{x})$. $f(t,\mathbf{x})$ in this is commonly called the **Green's function** as in the CMT inversion (Sect. 2.3.4). Here, it is the ground motion at the point $\mathbf{x}$ due to the point source with a unit slip ($D = 1$), that is, with a moment time function of $\mu SU_0(t)$.

Since the actual slip time function is quite complicated and can rarely be expressed by a single function such as $U_0(t)$, a sequence of functions with varying slip are used, with possible overlap in time. This formulation is called a **multi-time window** [24]. As shown in 2.32a, the fault plane is divided by the horizontal interval Δx and the dip direction interval Δy, and the subfaults are identified by horizontal order m and the dip direction order n. Assuming that the center of a subfault is (x_m, y_n) in the xy coordinate system on the fault plane with the origin at the **rupture initiation point** (the point where the fault rupture starts, that is, the hypocenter in Fig. 1.1), and assuming the **rupture velocity** v_r of the concentric rupture propagation (Sect. 2.3.5) is constant, the rupture start time of the subfault is given by

$$T_{mn} = \frac{\sqrt{x_m^2 + y_n^2}}{v_r}\,. \tag{2.192}$$

It is rare for an actual fault rupture to propagate at a constant velocity, but in the multi-time window formulation, variable rupture propagation can be expressed even if v_r is constant. That is, if v_r is given a velocity higher than the average velocity of rupture propagation, a large slip is obtained for the first time function of a subfault in the case of a high velocity in front of the subfault, otherwise a small slip is obtained.

The formulation of the slip time function described thus far, that is arranging $U_0(t)$ consecutively, is the method of Yoshida et al. [80]. On the other hand, Hartzell and Heaton [24] combine the integrals of the origin-shifted triangular functions (Sect. 4.2.2) so that they partially overlap. This time function corresponds to an approximation of the ramp function with two quadratic functions, convex downward and convex upward (Fig. 2.33).

It is also usual that the actual **slip angle** λ varies for each subfault. Therefore, λ is not included in the outer fault parameters and is obtained for each subfault from analyses. However, λ is not an independent coefficient like D, but embedded in $f(t,\mathbf{x})$, rendering the analyses complicated. To avoid this, the following formulation was elaborated. Since any slip can be decomposed into a slip X in the x-direction and a slip Y in the y-direction in the xy coordinate system on the fault plane, the ground motion for an arbitrary slip can be obtained as $Xf(t,\mathbf{x}) + Yg(t,\mathbf{x})$, using the Green's function $f(t,\mathbf{x})$ for the unit slip in the x-direction and the Green's function $g(t,\mathbf{x})$ for the unit slip in the y-direction [24]. If an approximate slip angle λ_0 is known from

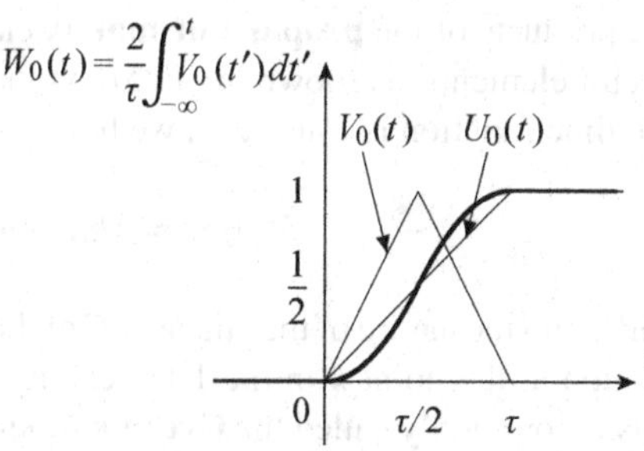

Fig. 2.33 $W_0(t)$ obtained by integrating $2/\tau$ times the origin-shifted triangular function $V_0(t)$. The origin-shifted ramp function $U_0(t)$ and triangular function $V_0(t)$ are also shown (reprinted from Koketsu [41] with permission of Kindai Kagaku)

the CMT inversion, etc., the slip is decomposed into the $\lambda_0 - 45°$ direction and the $\lambda_0 + 45°$ direction, and the introduction of the conditions under which the slips X and Y do not become negative (**non-negative condition**) increases the accuracy of this formulation [80].

Summarizing so far and applying the **principle of superposition** (Sect. 1.3.1), we obtain the kth component (normally, either the north-south component, the east-west component, or the up-down component) of the ground motion at the jth observation point at $\mathbf{x}_j$ and t_i, obtained by the **sampling** (Sect. 4.2.3) of time t

$$F_k(t_i, \mathbf{x}_j) = \sum_m \sum_n \sum_l X_{mnl} f_{mnk}(t_i - (l-1)\tau - T_{mn}, \mathbf{x}_j) + \sum_m \sum_n \sum_l Y_{mnl} g_{mnk}(t_i - (l-1)\tau - T_{mn}, \mathbf{x}_j) . \quad (2.193)$$

Here, the slip time function of the mnth subfault is assumed to be a sequence of multiple $U_0(t)$ from T_{mn} in (2.192), the lth of which has X_{mnl} or Y_{mnl} for the slip in the $\lambda_0 \pm 45°$ direction, and $(l-1)\tau$ (τ is the rise time of $U_0(t)$) for the time delay (Fig. 2.34) [80]. f_{mnl} and g_{mnl} are Green's functions computed by the method in Chap. 3 for the unit slips in the $\lambda_0 \pm 45°$ directions on the mnth subfault.

Similarly to the CMT inversion (Sect. 2.3.4), when the kth component $F_k^o(t, \mathbf{x}_j)$ of the ground motion is observed at the jth observation point, X_{mnl} and Y_{mnl} are determined using the **linear least-squares method** (Sect. 4.4) so that the **observed seismogram** matches the **synthetic seismogram** computed by (2.193). This determination is called a **source inversion**. Assuming that the number of subfaults is M in the x-direction and N in the y-direction in Fig. 2.32a, and the number of slip time functions is L, the total number of variables is as large as $M \times N \times L \times 2$. As a result, numerical calculations are likely to be unstable, therefore **constraints** (Sect. 4.4.2) are introduced. For example, the conditions that X_{mnl} and Y_{mnl} are smooth, are given as

$$\nabla^2 X_{mnl} \to \min , \quad \nabla^2 Y_{mnl} \to \min \quad (2.194)$$

where ∇^2 is a Laplacian in the space-time domain.

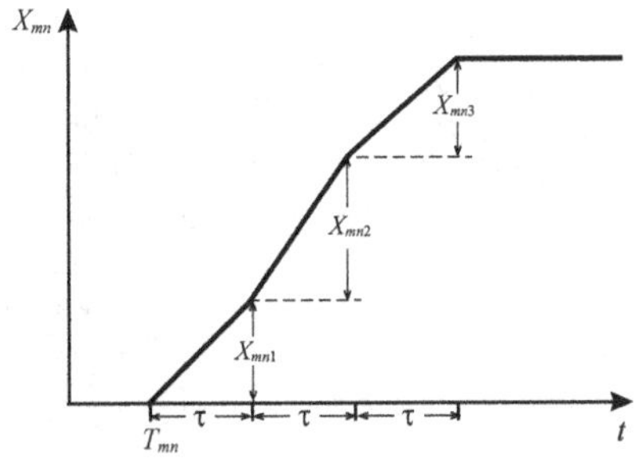

Fig. 2.34 Slip time function $X_{mn}(t)$ in the $\lambda_0 - 45°$ direction of the mnth subfault. In this example, three $U_0(t)$ are multiplied by the slips X_{mn1}, X_{mn2}, and X_{mn3}, respectively, and are summed, shifting the time by the rise time τ (based on Yoshida et al. [80], reprinted from Koketsu [41] with permission of Kindai Kagaku)

In addition, ground motions involve the following problems. Ground motions at the observation point are similar in the case when a distant subfault ruptures at an early time and in the case when a nearby subfault ruptures at a later time. It is therefore inherently difficult to distinguish these scenarios from the source inversion [74]. To solve this **tradeoff**, time-independent data related to **crustal deformation** (Sect. 3.1.12) must be added. The addition of data such as teleseismic body waves (Sect. 3.1.11) with different frequency components also has a certain effect, even though these are ground motions. Performing an inversion using multiple types of data in this way, is called a **joint inversion**. It is desirable for the observation points used for the source inversion to be distributed so as to surround the source fault. Since the station distribution for crustal deformation data is different than for ground motion data, joint inversion has the effect of compensating for the paucity in both station distributions.

The **synthetic deformation**

$$U_k(\mathbf{x}_j) = \sum_m \sum_n \left(\sum_l X_{mnl} \right) u_{mnk}(\mathbf{x}_j) + \sum_m \sum_n \left(\sum_l Y_{mnl} \right) \upsilon_{mnk}(\mathbf{x}_j) \tag{2.195}$$

is obtained using the principle of superposition and the Green's functions u_{mnk} and υ_{mnk}, calculated for a unit amount of **final slip** using the method described in Sect. 3.1.12. Therefore, the constrained joint inversion is formulated so that X_{mnl} and Y_{mnl} are determined with the **least-squares method** (Sect. 4.4) by minimizing

$$S = \frac{1}{\sigma_f^2} \sum_i \sum_j \sum_k \left\{ F_k^o(t_i, \mathbf{x}_j) - F_k(t_i, \mathbf{x}_j) \right\}^2 + \frac{1}{\sigma_u^2} \sum_j \sum_k \left\{ (U_k^o(\mathbf{x}_j) - U_k(\mathbf{x}_j) \right\}^2$$
$$+ \frac{1}{\rho^2} \left\{ \sum_m \sum_n \sum_l \left(\nabla^2 X_{mnl} \right)^2 + \sum_m \sum_n \sum_l \left(\nabla^2 Y_{mnl} \right)^2 \right\}, \tag{2.196}$$

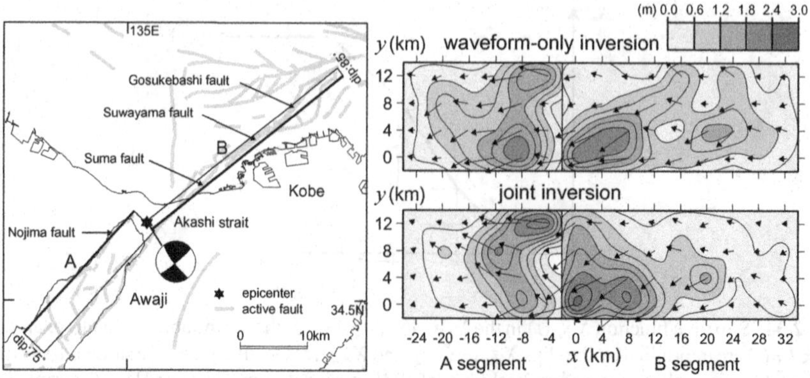

Fig. 2.35 Source inversion results for the Kobe earthquake using the fault model shown on the left. The upper right panel shows the slip vectors and slip amount distribution from the inversion of strong motion data alone, and the lower right panel shows those from the joint inversion of strong motion data, teleseismic body wave data, and crustal deformation data (based on Yoshida et al. [80], reprinted from Koketsu [41] with permission of Kindai Kagaku)

where $U_k^o(t, \mathbf{x}_j)$ is the **observed deformation**. In the case when a spatio-temporal Laplacian such as $\nabla^2 X_{mnl}$ is discretized by the central difference (3.307) as

$$\nabla^2 X_{mnl} = X_{m+1,n,l} + X_{m,n+1,l} + X_{m,n,l+1} - 6X_{mnl} + X_{m-1,n,l} + X_{m,n-1,l} + X_{m,n,l-1}, \tag{2.197}$$

there is no nonlinear term, therefore the joint source inversion can be solved using the **linear least-squares method** (Sect. 4.4). If the non-negative condition is imposed, the constraints for that condition are added to (2.196), or it is iteratively solved by adjusting the amount of correction [47].

σ_f and σ_u in (2.196) formally correspond to the **observational error** of ground motion data and crustal deformation data, as in the case of CMT inversion (Sect. 2.3.4). These also control the relative weight between the first term of the ground motion data and the second term of the crustal deformation data. ρ controls the relative weight of the third term of the constraints with respect to the first and second terms, which are data terms. However, there is still no definitive criterion for what value to give to ρ, although **empirical Bayes estimation** and **ABIC** approaches have been proposed (Sect. 4.4).

Figure 2.35 shows an example of applying the source inversion method described here to the **Kobe earthquake** (Sect. 2.3.8). The two-segment fault model shown on the left was constructed from the epicenter of the **mainshock** and the **aftershock distribution**.

In addition to strong motion data from strong motion seismographs, teleseismic body wave data from global observation networks are available,[38] in addition to crustal deformation data from surveying and GPS, so that the inversion of each

[38] Strong motion data and teleseismic body wave data are both ground motion data, but they are hereafter referred to by those names in order to distinguish them.

dataset could be performed, as well as the joint inversion of the three datasets [80]. Figure 2.35 (right panel) shows the results of the inversion of the strong motion data (top) and the joint inversion (bottom) using the distributions of slip vectors and slip amounts recovered from $\sum_l X_{mnl}$ and $\sum_l Y_{mnl}$. Normally, slip vectors on the **hanging wall** are displayed. However, for this earthquake, as the dip directions are different between segments A and B, the vectors are unified on the southeast side, which is the hanging wall of A.

2.3.7 Stress Drop and Slip Rate Function

The simplest model for a finite source fault is a Haskell model, and the most realistic model is a source process model. However, an intermediate model between the two may be required. In particular, when making ground motion predictions, it is almost impossible to construct a realistic model in advance. However, a small number of parameters that have large effects on ground motions can be extracted from the realistic model, and we will construct an intermediate model using these. The extracted parameters are called the **inner fault parameters**, in contrast to the outer fault parameters of the Haskell model. Extracting inner fault parameters is termed **characterization**, and the model constructed using those parameters is called a **characterized source model** [29].

One of the typical inner fault parameters is the **stress drop**. According to the **elastic rebound theory** (Sect. 2.1.1), an earthquake is a phenomenon in which a source fault slips so as to release the strain accumulated in the rocks over for many years. The slip during the earthquake should therefore be related to the released strain and also to the released stress via the generalized **Hooke's law** (Sect. 1.2.1). The stress released is called the "stress drop". The stress drop and the effective stress estimated from it are important parameters that determine the levels and spectra of ground motions in methods for synthesizing short-period ground motion [10, 28].

If the stress drop $\Delta\sigma$ is constant over the elliptical source fault or an elliptical part of the source fault (Fig. 2.36), the analytical solution of the longitudinal slip D corresponding to $\Delta\sigma$ is given as [39]

$$D = \frac{2\Delta\sigma b}{\eta\,\mu}\sqrt{1 - \frac{x_1^2}{a^2} - \frac{x_2^2}{b^2}}, \quad \eta = \begin{cases} E + \dfrac{\nu}{1-\nu}\dfrac{K-E}{\kappa^2 - 1}, & a > b \\ \dfrac{\pi(2-\nu)}{4(1-\nu)}, & a = b \end{cases}. \tag{2.198}$$

[39]Equation (2.198) is written in the notation of Kikuchi [35], but if $k = (1 - 1/\kappa^2)^{1/2}$ and $k' = 1/\kappa$, it matches Eq. (5.3) of Eshelby [18] with the correction of Eshelby (1963, *Phys. Stat. Sol.*, **3**, 2057–2060).

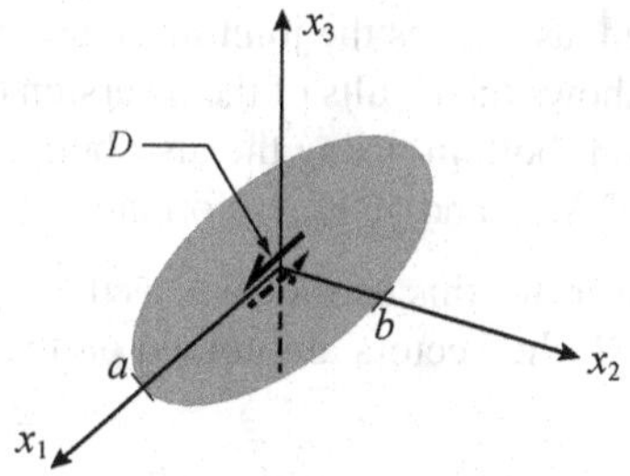

Fig. 2.36 An elliptical source fault, or an elliptical part of a source fault, in the x_1x_2 plane of the Cartesian coordinate system. This source fault has an elliptical shape with the center at the origin and with major and minor axes of length $2a$ and $2b$ (based on Kikuchi [35], reprinted from Koketsu [41] with permission of Kindai Kagaku)

κ represents a/b, and K and E represent complete elliptic integrals of the first and second kinds with $\sqrt{1-1/\kappa^2}$ (corresponding to k in the footnote) as variables. ν is the Poisson's ratio of the medium (Sect. 1.2.3).

In particular, if it is a circle with $a = b$ and ν takes a typical value of 1/4 (in the case of $\lambda = \mu$), $\eta = 7\pi/12$ so that

$$D = \frac{24\Delta\sigma a}{7\pi\,\mu}\sqrt{1 - \frac{x_1^2}{a^2} - \frac{x_2^2}{a^2}} = \frac{24\Delta\sigma}{7\pi\,\mu}\sqrt{a^2 - r^2}\,, \tag{2.199}$$

where r is a distance from the center of the circle. Integrating this over the entire circle and dividing by the area πa^2 gives the average slip

$$\begin{aligned}\overline{D} &= \frac{1}{\pi a^2}\int_0^a 2\pi r D dr = \frac{48\Delta\sigma}{7\pi^2\mu a^2}\int_0^a r\sqrt{a^2 - r^2}dr \\ &= \frac{48\Delta\sigma}{7\pi\mu a^2}\left[-\frac{(a^2 - r^2)^{\frac{3}{2}}}{3}\right]_0^a = \frac{16\Delta\sigma}{7\pi\,\mu}a\,.\end{aligned} \tag{2.200}$$

When the definition of seismic moment $M_0 = \mu DS$ (Eq. (2.181)) is extended to the case where D is a variable, we have

$$M_0 = \mu\iint DdS\,. \tag{2.201}$$

Furthermore, if we let S be a circular region with the radius of a and substitute (2.199) for (2.201), we obtain[40]

$$M_0 = \mu\int_0^a 2\pi r D dr = \mu\pi a^2\frac{16\Delta\sigma}{7\pi\,\mu}a = \frac{16}{7}a^3\Delta\sigma\,. \tag{2.202}$$

[40]This agrees with Eq. (1) of Irikura and Miyake [30] if $R \to a$.

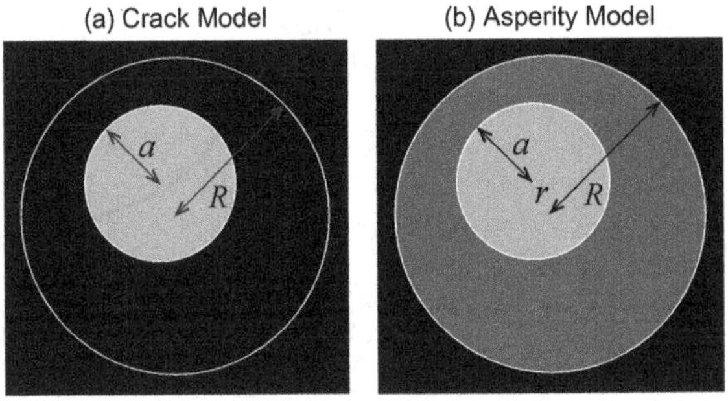

Fig. 2.37 **a** Circular crack model and **b** asperity model (based on Boatwright [5], reprinted from Koketsu [41] with permission of Kindai Kagaku). Following Irikura and Miyake [30], a circle with a large radius R indicates a source fault, a circle with a small radius a indicates an area of constant stress drop, and an area colored in black indicates zero slip. r represents the distance between the centers of the two circles

A situation where the stress drop is considered to be constant in a certain area within the source fault and the slip is zero in other areas, is called a **crack model** (Fig. 2.37a), and the area of constant stress drop is called a **crack**. Conversely, a situation where there is slip even outside the area of constant stress drop, is called an **asperity model** (Fig. 2.37b), and in this model the area of constant stress drop is called an **asperity** [11].

We consider two states in the asperity model. The first state is the standard state of the asperity model, where the asperity has a stress drop $\Delta\sigma$ and a slip D has occurred. In the second state, it is assumed that a stress drop $\Delta\tau = \mu$ and a slip E occur across the source fault. Since this state is equivalent to a crack model where the entire source fault is cracked, from (2.199) we have

$$E(r) = \frac{24}{7\pi}\sqrt{R^2 - r^2}\,, \tag{2.203}$$

where r is the distance from the center of the source fault. We apply the **reciprocity theorem** for the case of an internal discontinuity (Eq. (2.4)) to these two states [51]. There is no body force, and neither stress drop nor slip includes time. Thus, the convolution in (2.4) is simply a product and we have

$$\mu \iint D\,dS = \iint \Delta\sigma E\,dS\,. \tag{2.204}$$

From (2.201), we know that the left-hand side of this equation is M_0 of the asperity model. Also, assuming that the stress drop is $\Delta\sigma$ only within the asperity, and that the stress drop outside the asperity is sufficiently small, S on the right-hand side is the circular region of the asperity. Furthermore, if $a \ll R$, then

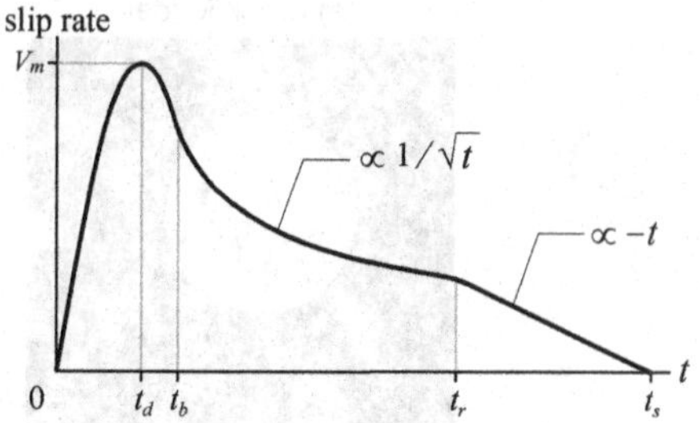

Fig. 2.38 Model of slip rate function (based on Nakamura and Miyatake [56], reprinted from Koketsu [41] with permission of Kindai Kagaku)

$$M_0 = \iint \Delta\sigma E \, dS \sim \Delta\sigma E(r) \, \pi a^2 \, , \tag{2.205}$$

where r is the distance between the center of the source fault and the center of the asperity. Therefore, M_0 of the asperity model changes depending on the position of the asperity, and its average value is[41]

$$\begin{aligned} M_0 &= \Delta\sigma \cdot \overline{E(r)} \, \pi a^2 = \Delta\sigma \frac{1}{\pi R^2} \frac{24}{7\pi} \int_0^R 2\pi R \sqrt{R^2 - r^2} dr \cdot \pi a^2 \\ &= \frac{16}{7} \Delta\sigma a^2 R = \frac{16}{7} \Delta\sigma R^3 \frac{S_a}{S} \, , \quad S_a = \pi a^2 \, , \quad S = \pi R^2 \, . \end{aligned} \tag{2.206}$$

In contrast to M_0 of the asperity model (Eq. (2.205)), M_0 of the crack model (Eq. (2.202)) is constant regardless of the crack position.

The source fault of an actual earthquake can be represented by a **composite model** with multiple cracks or asperities. Boatwright [5] showed that the composite asperity model explained the **scaling law** (Sect. 2.3.5) of observed ground motion spectra better than the composite crack model (also called the **barrier model**). For the composite asperity model, it is assumed that the entire seismic moment M_0 is the sum of the seismic moments M_{0k} of the individual asperities, and that the asperities are circular with radii a_k and constant stress drops $\Delta\sigma_a$. Thus,

$$M_0 = \sum_k M_{0k} = \frac{16}{7} \Delta\sigma_a R^3 \frac{\sum_k S_{ak}}{S} \, , \tag{2.207}$$

so that (2.206) still holds as it is if $S_a = \sum_k S_{ak}$ and $\Delta\sigma = \Delta\sigma_a$.

Another example of an inner fault parameter is the **slip rate function** (Sect. 2.1.2) $\dot{D}(t)$, the physical background of which has been relatively well examined, similar to the stress drop. Nakamura and Miyatake [56] summarized the results of numerical

[41] According to Boatwright [5]. This agrees with Eq. (2) of Irikura and Miyake [30] if $r \to a$. Although (2.206) plays an important role in the characterized source model, this model does not satisfy the condition that the stress drop outside the asperity is sufficiently small. This problem is discussed in Irikura and Miyake [29].

simulations of fault rupture and proposed the slip rate function model shown in Fig. 2.38. It is the friction on the source fault surface that controls the temporal change of slip. Rock experiments have shown that **slip weakening**, in which friction decreases as slip proceeds, is dominant in fault rupture during earthquakes [59]. Two-dimensional simulations incorporating this effect were performed, and the results were approximated by an upwardly convex quadratic function from $t = 0$ (the start of rupture) to $t = t_b$. After that, the part up to $t = t_r$ was approximated by the reciprocal of $\sqrt{t}$ (this is called a Kostrov function [44]).

As the slip dependence of friction is not as effective after slipping has occurred to some extent, three-dimensional simulations without slip weakening were also performed on the crack and asperity models described above. The results show that the part before t_r can be approximated by a Kostrov function similar to that in the two-dimensional simulations. After t_r, the slip rate rapidly decreases, then becomes zero and the slip stops. Summarizing the above, we obtain

$$\dot{D}(t) = \begin{cases} 0, & t \le 0 \text{ or } t \ge t_s \\ \dfrac{2V_m}{t_d} t \left(1 - \dfrac{t}{2t_d}\right), & 0 \le t \le t_b \\ \dfrac{b}{\sqrt{t - \varepsilon}}, & t_b \le t \le t_r \\ c - a_r(t - t_r), & t_r \le t \le t_s \end{cases} . \tag{2.208}$$

If the width W, slip amount D, rupture velocity u_r (Sect. 2.3.5), and stress drop $\Delta\sigma$ (this section) are separately determined in advance, the following four parameters in (2.208) must be determined independently.

1. $t_d = 1/(\pi f_{\max})$. $f_{\max}$ is estimated from the source spectra obtained by removing the propagation effects from observed spectra (**source-controlled $f_{\max}$**).[42] Nakamura and Miyatake [56] obtained the empirical formula $t_d = 2/(2\pi f_{\max})$ from the waveforms and angular frequency spectra of the simulation results.
2. $t_r = W/(2v_r)$. W is the width of the asperity or source fault. Day [12] called t_r the **rise time** and obtained this empirical formula.
3. $t_s = 3/2 \cdot t_r$. Nakamura and Miyatake [56] simply assumed that $t_s - t_r$ was half of t_r, because the effect of the part $[t_r, t_s]$ on the ground motion is small.
4. $V_m = \Delta\sigma/\mu \cdot \sqrt{2 f_c W v_r}$. Day [12] obtained this parameter from theoretical approximations. W is the same as in 2., $\Delta\sigma$ is the stress drop in the asperity or elsewhere, and f_c is assumed to be equal to the source-controlled $f_{\max}$.

Once these four parameters have been determined, the remaining parameters can be determined to be the functions of t_b represented as

[42] As mentioned in Sect. 2.3.6, there are difficulties in $f_{\max}$ observation, therefore a fixed value such as 6 Hz is often used [17].

$$\varepsilon = \frac{5t_b - 6t_d}{4(1 - t_d/t_b)},$$
$$b = \frac{2V_m t_b}{t_d}\sqrt{t_b - \varepsilon}\left(1 - \frac{t_b}{2t_d}\right), \quad c = \frac{b}{\sqrt{t_r - \varepsilon}}, \quad a_r = \frac{c}{t_s - t_r} \tag{2.209}$$

using the condition for ε that the quadratic function and Kostrov function are smoothly continuous at $t = t_b$. Finally, numerically solving the equation

$$F(t_b) = \int_0^{t_s} \dot{D}(t)dt = D \tag{2.210}$$

gives t_b.

As the work of Nakamura and Miyatake [56] was published in Japanese with an English abstract, it is not well known in the international community. Instead, Tinti et al.'s model [71], where the Yoffe solution [79] substitutes for the Kostrov solution [44], has sometimes been used.

2.3.8 Directivity Effect

On January 16 (local time: January 17), 1995, the **Kobe earthquake**[43] (1995, M_w 6.9) occurred, causing the **Hanshin-Awaji earthquake disaster**. This event is remembered, first and foremost, for its devastating damage (6,434 deaths, 104,906 collapsed houses [40]), but also for bringing a major turning point in ground motion seismology. **Strong motion seismographs** (Sect. 4.1.2) must be widely deployed in order to characterize the strong motions that cause major damage, however at the time of the Kobe earthquake there were only a few nearby stations. The strong motion records of the Kobe earthquake were therefore limited, but they include observational facts that were contrary to the common knowledge on ground motion at that time. Subsequently, nationwide networks of strong motion seismographs such as **K-NET** were built in Japan by **NIED** (National Research Institute for Earth Science and Disaster Resilience) to provide much better coverage.

The lower trace in Fig. 2.39 shows a strong motion record at the Kushiro observatory from the Kushiro-oki earthquake (1993, M_w 7.6). In general, ground motions of several Hz were prominent and lasted for more than 20 s, and most strong motion records observed in Japan before that showed a similar tendency. However, in the records of the Kobe earthquake shown in the upper and middle traces, the situation was dramatically different, with a long period of 1 to 2 s and a duration of only several seconds. Interestingly, it is no exaggeration to say that the record at Kobe University shown in the upper trace consists of two **seismic pulses** with a period of about 2 s. In addition, when the trajectories were drawn by combining the two horizontal

[43] The JMA officially named this event the "Hyogo-ken Nanbu jishin" (Southern Hyogo-Prefecture earthquake).

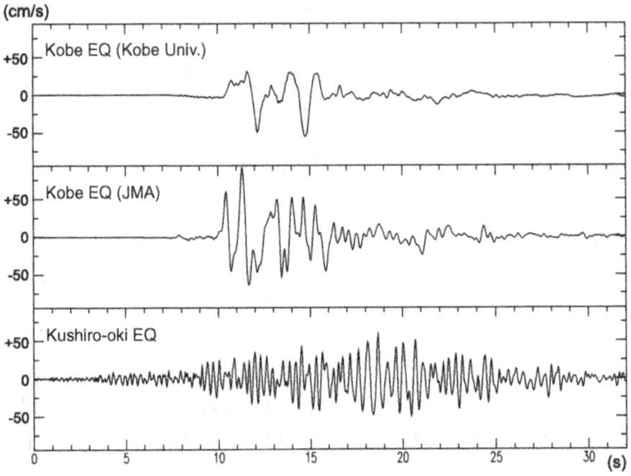

Fig. 2.39 Velocity seismograms of strong motions from the Kobe earthquake (north-south components; upper and middle panels) and Kushiro-oki earthquake (east-west component; lower panel), by the CEORKA (Committee of Earthquake Observation and Research in the Kansai Area) and **JMA** (Sect. A.2.1). Modified from Koketsu [41] with permission of Kindai Kagaku

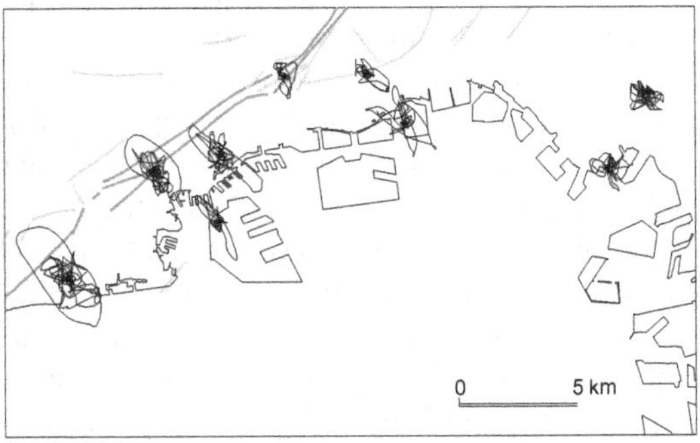

Fig. 2.40 Trajectories of horizontal strong ground motions near the source fault of the Kobe earthquake, as observed by CEORKA, JMA, the City of Kobe, JR Warning System, Osaka Gas, and the Public Works Research Institute. The active faults are indicated by the gray lines, with the darker gray lines corresponding to the source fault (reprinted from Koketsu [41] with permission of Kindai Kagaku)

components of the Kobe earthquake records, it was observed that the strong ground motions close to the source fault were biased in the direction perpendicular to the source fault (Fig. 2.40). Such directivity of strong ground motion had not previously been observed in Japan.

The two major characteristics of the strong motions from the Kobe earthquake, that are the "long period" and "directivity", are collectively referred to as the **directivity effect**. As the directivity effect was significantly different from conventional strong motion records, including the Kushiro-oki earthquake record, the source process of the Kobe earthquake was considered to be very distinctive immediately after the

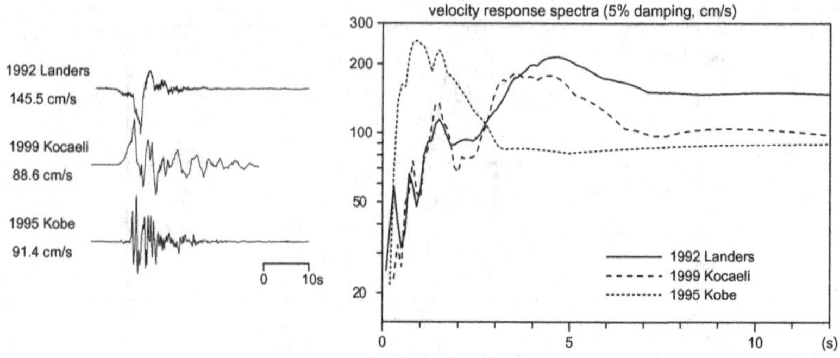

Fig. 2.41 Examples of the directivity effect. Velocity seismograms (left) and velocity response spectra (right). Modified from Koketsu and Miyake [43] with permission of Springer Nature

earthquake. However, subsequent studies have revealed the directivity effects in the **Landers earthquake** (1992, M_w 7.3) and **Imperial Valley earthquake** (1979, M_w 6.5) in California, as well as the **Kocaeli earthquake** (1999, M_w 7.6) in Turkey (Fig. 2.41). This indicates that the directivity effect is a universal phenomenon close to source faults, and has a physical explanation [39].

In the center diagram of Fig. 2.42, the situation around the northern segment of the source fault of the Kobe earthquake (Fig. 2.35 (left); Fig. 2.40) is schematically drawn as viewed from above the ground surface. The northern segment, extending from west-southwest to east-northeast in Fig. 2.40, is represented by a line source, which consists of right-lateral strike-slip point sources and extends from left to right on the diagram. Since the rupture initiation point of the Kobe earthquake was located beneath the Akashi Strait in the west-southwest part of the segment (Fig. 2.40), it corresponds to the ★ symbol at the left end of the line source in the center diagram of Fig. 2.42. The **rupture propagation** begins from that point and progresses to the right end. In this process, the point sources constituting the line source sequentially generate ground motions. Since the difference between the S wave velocity and the **rupture velocity** is small, the ground motions arriving at the observation point ▲ are clustered as shown in the right diagram of Fig. 2.42. As a result, **constructive interference** occurs and the ground motions form a long-period pulse, in the **forward direction** of rupture propagation.

The distribution of ground motions due to each point source forms a pattern of **four-quadrant type**, as shown in the left panel of Fig. 2.42. This can be obtained from the radiation pattern of the S wave in Fig. 2.17 by reversing the direction of ground motion, because Fig. 2.17 is for a left-lateral strike slip but Fig. 2.42 is for a right-lateral strike slip. In the left and center panels of Fig. 2.42, constructive interference due to the rupture propagation is observed in one of the four quadrants, where ground motions are oriented in the direction orthogonal to the source fault (the **fault-normal direction**), so that the directivity pulses are oriented in the fault-normal direction. The directivity effect is therefore created by the combination of the rupture propagation and the radiation pattern, which are among the source effects.

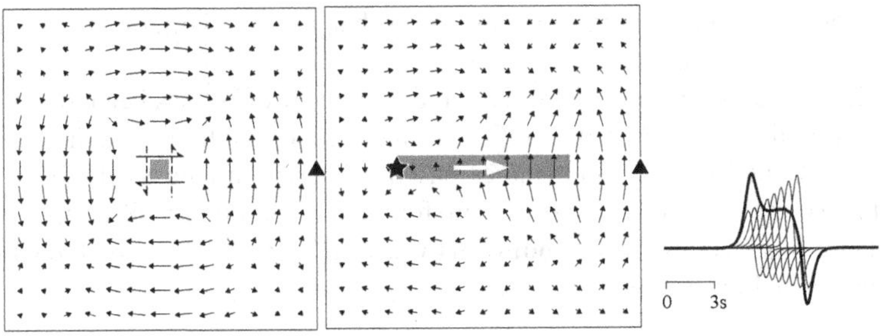

Fig. 2.42 Bird's-eye view of the ground motion patterns for the point and line sources of right-lateral strike slip (left and center), and the formation of a long-period pulse (right). ★ and ▲ indicate the rupture initiation point and observation point, respectively (based on Koketsu [39], reprinted from Koketsu [41] with permission of Kindai Kagaku)

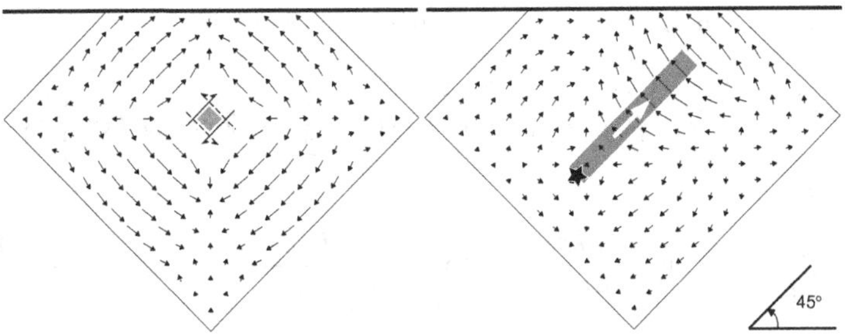

Fig. 2.43 Cross sections of the ground motion patterns for the point (left) and line sources for a reverse fault (right). ★ and the thick solid line indicate the rupture initiation point and the ground surface, respectively (reprinted from Koketsu [41] with permission of Kindai Kagaku)

For the Kobe earthquake, the directivity effect is observed in the components of ground motions with periods longer than 0.5 to 1 s, but is not clearly seen in the components with shorter periods [60]. Although the boundary period changes variably, this feature is not limited to the Kobe earthquake but is rather common, and results from the fact that at high frequencies random characteristics are more prominent than deterministic characteristics, such as rupture propagation with a constant rupture velocity.

So far, the description has focused on the case of a **strike slip fault** earthquake, but the directivity effect was also observed during the **Northridge earthquake** (1994, M_w 6.6), which is a **dip slip fault** earthquake with reverse faulting (a reverse fault earthquake). This mechanism can be explained by rotating the left and center diagrams in Fig. 2.42, 45° counterclockwise (Fig. 2.43). However, Fig. 2.42 shows a bird's-eye view of the strike slip fault, while Fig. 2.43 shows a cross-section view of the reverse fault from the side. The oblique upwards rupture propagation generates a long-period pulse at an observation point on the extension of the ground surface due to constructive interference, and the pulse is oriented in the fault-normal direction. These characteristics are the same as in the case of a strike slip fault earthquake, but the difference is that the fault-normal direction changes from horizontal to obliquely upwards by the 45° rotation.

In addition to the above typical cases, the directivity effect has been identified in various ground motions from theoretical studies related to the Southern California Earthquake Center (SCEC) and studies based on observed records (Koketsu et al. [42]). For example, directivity pulses were found in the fault-parallel components of ground motions from the **Kumamoto earthquake** (2016, M_w 7.0) and the **Chi-Chi earthquake** in Taiwan (1999, M_w 7.6).

Problems

2.1 Show in detail that $\overline{\Psi}_r$, $\overline{\Psi}_\theta$, and $\overline{\Psi}_z$ in (2.121) result in

$$(\nabla \times \overline{\Psi})_z = \frac{\overline{M}_0(\omega)}{4\pi\rho\omega^2}\int_0^\infty F_\beta\, k^2 \left\{ \frac{3}{2}\, \mathrm{i}\epsilon\nu_\beta J_0(kr) + \frac{1}{2}\cos 2\theta\, \mathrm{i}\epsilon\nu_\beta J_2(kr) \right\} dk\,,$$

$$(\nabla \times \overline{\Psi})_r =$$

$$\frac{\overline{M}_0(\omega)}{4\pi\rho\omega^2}\int_0^\infty F_\beta \left[\left\{ \frac{3}{2}\nu_\beta^2 \frac{dJ_0(kr)}{dr} + \frac{1}{2}\cos 2\theta\, \nu_\beta^2 \frac{dJ_2(kr)}{dr} \right\} + \cos 2\theta\, k_\beta^2 \frac{J_2(kr)}{r} \right] dk$$

in (2.122) and (2.124).

2.2 In order to obtain graphically the fault plane solution for an earthquake, we have to map the focal sphere onto the paper as shown in Fig. 2.44a, and the **equal-area projection** is mainly used for this mapping. When projecting the point A on the focal sphere to A' on the paper, the circle of A' is determined so that its area is equal to the surface area of the dark gray part of the focal sphere below A (Fig. 2.44b). Represent the radius r of the circle with the radius R of the focal sphere and the angle of emergence α.

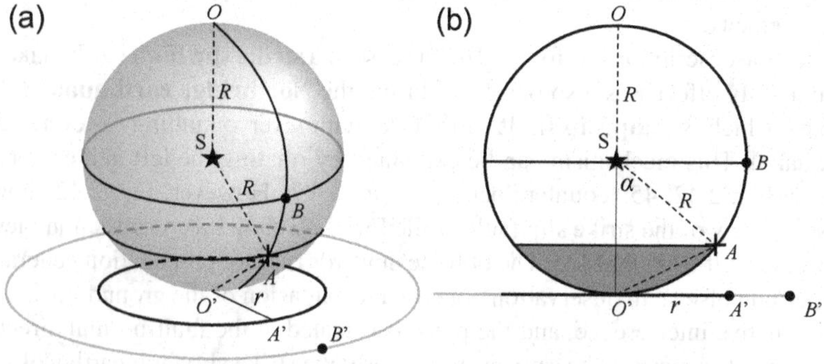

Fig. 2.44 **a** Focal sphere and its projection on the horizontal plane, and **b** their cross section

References

1. Aki, K. (1966). Generation and propagation of G waves from the Niigata earthquake of June 16, 1964. Part 2. Estimation of earthquake moment, released energy, and stress-strain drop from the G wave spectrum. *Bulletin of the Earthquake Research Institute*, *44*, 73–88.
2. Aki, K. (1967). Scaling law of seismic spectrum. *Journal of Geophysical Research*, *71*, 1217–1231.
3. Aki, K., & Richards, P. G. (2002). *Quantitative seismology* (2nd ed., p. 700). Sausalito: University Science Books.
4. Bessonova, É. N., Gotsadz, O. D., Keilis-Borok, V. I., Kililova, I. V., Kogan, S. D., Kikhtikova, T. I., et al. (1960). *Investigation of the mechanism of earthquakes* (p. 201). Washington DC: American Geophysical Union.
5. Boatwright, J. (1988). The seismic radiation from composite models of faulting. *Bulletin of the Seismological Society of America*, *78*, 489–508.
6. Boore, D. M. (1983). Stochastic simulation of high-frequency ground motions based on seismological models of the radiated spectra. *Bulletin of the Seismological Society of America*, *73*, 1865–1894.
7. Brune, J. N. (1970, 1971). Tectonic stress and spectra of seismic shear waves from earthquakes, *Journal of Geophysical Research*, *75*, 4997–5009; Correction, *Journal of Geophysical Research*, *76*, 5002.
8. Burridge, R., & Knopoff, L. (1964). Body force equivalents for seismic dislocations. *Bulletin of the Seismological Society of America*, *54*, 1875–1888.
9. Cummins, P. R. (1997). Earthquake near field and W phase observations at teleseismic distances. *Geophysical Research Letters*, *24*, 2857–2860.
10. Dan, K., & Sato, T. (1998). Strong-motion prediction by semi-empirical method based on variable-slip rupture model of earthquake fault, *J. Struct. Constr. Eng. AIJ*, *509*, 49–60. [J]
11. Das, S., & Kostrov, V. (1986). Fracture of a single asperity on a finite fault: A model for slow earthquake? *Earthquake source mechanics* (pp. 91–96). Washington DC: American Geophysical Union.
12. Day, S. M. (1982). Three-dimensional finite difference simulation of fault dynamics: Rectangular faults with fixed rupture velocity. *Bulletin of the Seismological Society of America*, *72*, 705–727.
13. Draper, N. R. & Smith H. (1966). *Applied regression analysis* (p. 407). New York: Wiley.
14. Dreger, D., Hurtado, G., Chopra, A., & Larsen, S. (2011). Near-field across-fault seismic ground motions. *Bulletin of the Seismological Society of America*, *101*, 202–221.
15. Dziewonski, A. M., Chou, T.-A., & Woodhouse, J. H. (1981). Determination of earthquake source parameters from waveform data for studies of global and regional seismicity. *Journal of Geophysical Research*, *86*, 2825–2852.
16. Earthquake Research Committee. (1999). *Seismic Activity in Japan —Regional perspective on the characteristics of destructive earthquakes—* (rev. ed.). Headquarters for Earthquake Research Promotion, https://www.hp1039.jishin.go.jp/eqchr/eqchrfrm.htm. [J]
17. Earthquake Research Committee. (2017). *Strong ground motion prediction method for earthquakes with specified source faults* ("*Recipe*") (April 2017 version). Headquarters for Earthquake Research Promotion, https://www.jishin.go.jp/main/chousa/17_yosokuchizu/recipe.pdf. [J]
18. Eshelby, J. D. (1957). The determination of the elastic field of an ellipsoidal inclusion, and related problems. *Proceedings of the Royal Society of London A*, *241*, 316–396.
19. Fukao, Y. (1978). Classical seismology, *Physics of earthquake*, (Chap. 2, pp. 19–87). Tokyo: Iwanami Shoten. [J]
20. Fukuyama, E., Ishida, M., Dreger, D. S., & Kawai, H. (1998). Automated seismic moment tensor determination by using on-line broadband seismic waveforms, *Zisin* (*Journal of the Seismological Society of Japan*), *51*, 149–156. [J]
21. Geiger, L. (1910). Herdbestimung bei Erdbeben aus Ankunftszeiten, Kachr. *Köninglichen Gesell. Wiss. Göttingen Mathematical Physics*, *4*, 331–349.

22. Geller, R. J. (1976). Scaling relations for earthquake source parameters and magnitudes. *Bulletin of the Seismological Society of America, 66*, 1501–1523.
23. Harkrider, D. G. (1976). Potentials and displacements for two theoretical seismic sources. *Geophysical Journal of the Royal Astronomical Society, 47*, 97–133.
24. Hartzell, S. H., & Heaton, T. H. (1983). Inversion of strong ground motion and teleseismic waveform data for the fault rupture history of the 1979 Imperial Valley, California, earthquake. *Bulletin of the Seismological Society of America, 73*, 1553–1583.
25. Haskell, N. A. (1969). Elastic displacements in the near-field of a propagating fault. *Bulletin of the Seismological Society of America, 59*, 865–908.
26. Honda, H. (1931, 1932). On the initial motion and the types of the seismograms of the North Idu and the Itô earthquakes, *Geophysical Magazine, 4*, 185–213; On the mechanisms and the types of the seismograms of shallow earthquakes. *Geophysical Magazine, 5*, 69–88.
27. Honda, H. (1954). *Seismic waves* (p. 230). Tokyo: Iwanami Shoten. [J]
28. Irikura, K. (1986). Prediction of strong acceleration motion using empirical Green's function. *Proceedings of 7th Japan Earthquake Engineering Symposium, Tokyo* (pp. 151–156). [J]
29. Irikura, K., & Miyake, H. (2001). Prediction of strong ground motions for scenario earthquakes. *Journal of Geography, 110*, 849–875. [J]
30. Irikura, K., & Miyake, H. (2011). Recipe for predicting strong ground motion from crustal earthquake scenarios. *Pure Applied Geophysics, 168*, 85–104.
31. JMA (Japan Meteorological Agency). (2016). *On the 2016 Kumamoto earthquake (7th Report)* (p. 17). Tokyo: JMA. [J]
32. Kanamori, H. (1977). The energy release in great earthquakes. *Journal of Geophysical Research, 82*, 2981–2987.
33. Kanamori, H. (1993). W-phase. *Geophysical Research Letters, 20*, 1691–1694.
34. Keilis-Borok, V. I. (1950). Determination of the dynamic characteristics of the focus of an earthquake, *Akad. Nauk SSSR, Trudy Instituta geologicheskikh nauk, 9*, 3–19.
35. Kikuchi, M. (2001). Seismic radiation pattern, Fault model: Source process. *Encyclopedia of earthquakes* (2nd ed., pp. 248–283). Tokyo: Asakura Shoten. [J]
36. Kikuchi, M. (2003). *Realtime eismology* (p. 222). Tokyo: University of Tokyo Press. [J]
37. Kikuchi, M., & Fukao, Y. (1985). Iterative deconvolution of complex body waves from great earthquakes Tokachi-oki earthquake of 1968. *Physics of the Earth and Planetary Interiors, 37*, 235–248.
38. Koketsu, K. (1989). Hypocenter determination with non-negative depth, *Zisin* (*Journal of the Seismological Society of Japan*), *42*, 325–331. [J]
39. Koketsu, K. (1996). Damaging earthquakes in California and 1995 Kobe earthquake. *Science Journal, 66*, 93–97. [J]
40. Koketsu, K. (2016). Japanese Damaging Earthquakes List and World Seismicity & Plate Boundaries, *Chronological Scientific Tables 2017* (pp. 728–761, 788–789). Tokyo: Maruzen. [J]
41. Koketsu, K. (2018). *Physics of seismic ground motion* (p. 353). Tokyo: Kindai Kagaku. [J]
42. Koketsu, K., Kobayashi, H., & Miyake, H. (2018). Irregular modes of rupture directivity found in recent and past damaging earthquakes. *11th National Conference on Earthquake Engineering*, Paper No. 645.
43. Koketsu, K., & Miyake, H. (2008). A seismological overview of long-period ground motion. *Journal of Seismological, 12*, 133–143.
44. Kostrov, B. V. (1964). Selfsimilar problems of propagation of shear cracks. *Journal of Applied Mathematics and Mechanics, 28*, 1077–1087.
45. Kuge, K. (2003). Source modeling using strong-motion waveforms: Toward automated determination of earthquake fault planes and moment-release distributions. *Bulletin of the Seismological Society of America, 93*, 639–654.
46. Landau, L. D., & Lifshitz, E. M. (1973). *Mechanics* (3rd ed., p. 224). Butterworth-Heinemann.
47. Lawson, C. L., & Hanson, R. J. (1974). *Solving least squares problems* (p. 337). Englewood Cliffs: Prentice-Hall.
48. Lee, W. H. K. & Stuart, S. W. (1981). *Principles and applications of microearthquake networks* (p. 293). New York: Academic Press.

49. Linert, B. R., Berg, E., & Frazer, L. N. (1986). HYPOCENTER: An earthquake location method using centered, scaled, and adaptively damped least squares. *Bulletin of the Seismological Society of America, 76*, 771–783.
50. Love, A. E. H. (1906). *A treatise on the mathematical theory of elasticity* (2nd ed., p. 551). Cambridge: Cambridge University Press.
51. Madariaga, R. (1979). On the relation between seismic moment and stress drop in the presence of stress and strength heterogeneity. *Journal of Geophysical Research, 84*, 2243–2250.
52. Maruyama, T. (1963). On the force equivalents of dynamical elastic dislocations with reference to the earthquake mechanism. *Bulletin of the Earthquake Research Institute, 41*, 467–486.
53. Maruyama, T. (1968). Seismic waves –Basic theories–. *Earthquakes · Volcano · Rock property* (pp. 1–62). Tokyo: Kyoritsu. [J]
54. Moriguchi, S., Udagawa, K., & Hitotsumatsu, S. (1960). *Mathematical formulae III* (p. 298). Tokyo: Iwanami Shoten. [J]
55. Nabelek, J. L. (1984). Determination of earthquake source parameters from inversion of body waves, *Ph.D.* thesis (p. 361), Massachusetts Institute of Technology.
56. Nakamura, H., & Miyatake, T. (2000). An approximate expression of slip velocity time function for simulation of near-field strong ground motion, *Zisin* (*Journal of the Seismological Society of Japan*), *53*, 1–9. [J]
57. Nakano, H. (1923). Notes on the nature of the forces which give rise to the earthquake motions. *Seismological Bulletin, 1*, 92–120.
58. Nishide, N. (2001). Determination of hypocenters, *Encyclopedia of earthquakes* (2nd ed., pp. 46–53). Tokyo: Asakura Shoten. [J]
59. Ohnaka, M. (2001). History of exploring the causes of earthquakes. *Encyclopedia of earthquakes* (2nd ed., pp. 212–217). Tokyo: Asakura Shoten. [J]
60. Ohno, S., Takemura, M., & Kobayashi, Y. (1998). Directionality of near-source strong ground motion as derived from observation records. *Proceedings of 10th Japan Earthquake Engineering symposium* (vol. 1, pp. 133–138). [J]
61. Reid, H. F. (1910). The Mechanics of the earthquake, The California earthquake of April 18, 1906. *Report of the state earthquake investigation commission* (Vol. II, p. 192). Washington, DC: Carnegie Inst.
62. Sato, R. (1969). Formulations of solutions for earthquake source models and some related problems. *Journal of Physics of the Earth, 17*, 101–110.
63. Sato, R. (1972). Seismic waves in the near field. *Journal of Physics of the Earth, 20*, 357–375.
64. Sato, R., Abe, K., Okada, Y., Shimazaki, K., & Suzuki, Y. (1989). *Handbook of parameters of earthquake source faults in Japan* (p. 390). Tokyo: Kajima Institute Publishing. [J]
65. Sato, T. (1994). Theoretical evaluation of seismic ground motion. *Seismic Ground Motion: Its synthesis and waveform processing* (pp. 21–88). Tokyo: Kajima Institute Publishing. [J]
66. Savage, J. G. (1972). Relation of corner frequency to fault dimensions. *Journal of Geophysical Research, 77*, 3788–3795.
67. Shearer, P. M. (2009). *Introduction to seismology* (2nd ed., p. 396). Cambridge: Cambridge University Press.
68. Sommerfeld, A. (1909). Über die Ausbreitung der Wellen in der drahtlosen Telegraphie. *Annalen der Physik, 28*, 665–736.
69. Sotooka, H. (1997). *Earthquake and society* (Vol. 1, p. 366). Tokyo: Misuzu Shobo. [J]
70. Stokes, G. G. (1851). On the dynamical theory of diffraction. *Transactions of the Cambridge Philosophical Society, 9*, 1–62.
71. Tinti, E., Fukuyama, E., Piatanesi, A., & Cocco, M. (2005). A kinematic source-time function compatible with earthquake dynamics. *Bulletin of the Seismological Society of America, 95*, 1211–1223.
72. USGS (United States Geological Survey). (1997). Photo by R. E. Wallace, *The San Andreas fault* (p. 17), Reston: USGS.
73. Utsu, T. (2001). *Seismology* (3rd ed., p. 376). Tokyo: Kyoritsu Shuppan. [J]
74. Wald, D. J., & Heaton, T. H. (1994). Spatial and temporal distribution of slip for the 1992 Landers, California, earthquake. *Bulletin of the Seismological Society of America, 84*, 668–691.

75. Waldhauser, F., & Ellsworth, W. L. (2000). A double-difference earthquake location algorithm: Method and application to the northern Hayward fault, California. *Bulletin of the Seismological Society of America, 90*, 1353–1368.
76. Webster, A. G. (1927). *Partial differential equations of mathematical physics* (p. 440). Leipzig: B. G. Teubner.
77. Yamaoka, K. (2008). *Encyclopedia of earthquakes, tsunamis and volcanoes* (pp. 1–18). Tokyo: Maruzen. [J]
78. Yamashita, T. (2008). What is an earthquake?. *Encyclopedia of earthquakes, tsunamis and volcanoes* (pp. 19–35). Tokyo: Maruzen. [J]
79. Yoffe, E. (1951). The moving Griffith crack. *Philosophical Magazine, 42*, 739–750.
80. Yoshida, S., Koketsu, K., Shibazaki, B., Sagiya, T., Kato, T., & Yoshida, Y. (1996). Joint Inversion of near- and far-field waveforms and geodetic data for the rupture process of the 1995 Kobe earthquake. *Journal of Physics of the Earth, 44*, 437–454.

Chapter 3
The Effect of Propagation

Abstract This chapter explains the "effect of propagation" on ground motion and the interaction between wave propagation phenomena and the "effect of the earthquake source". One-dimensional (1-D) velocity structures are commonly used for evaluating the "effect of propagation" (Sect. 3.1). The methods used for this evaluation, such as the "propagator matrix", "reflection/transmission matrix", "wavenumber integration", "surface wave", "teleseismic body wave", and "crustal deformation", are then reviewed. The "discontinuity vector" is explained as a representation of the "effect of the earthquake source". In Sect. 3.2, three-dimensional (3-D) velocity structures are introduced as more realistic models for the effect of propagation. The methods of evaluation in a 3-D velocity structure, such as the "ray theory", "ray tracing", "finite difference method", "finite element method", and "Aki-Larner method", are then reviewed. Finally, Sect. 3.3 explains various methods and models for the analysis of propagation, such as "long-period ground motion", "microtremors", "seismic interferometry", and "seismic tomography".

Keywords Effect of propagation · 1-D velocity structure · Propagator matrix · Surface wave · 3-D velocity structure · Numerical computation

3.1 Propagation in 1-D Media

3.1.1 1-D Velocity Structure

As changes in depth are generally greater than horizontal changes within velocity structures, **1-D velocity structure** (where "1-D" is an abbreviation for "one dimensional") with properties varying only in the depth direction, are often used as a first approximation to evaluate the effect of propagation on ground motion. In particular, the most commonly used model is a **horizontally layered structure** that consists of homogeneous layers separated by horizontal **interface** (Sect. 1.2.4). If a veloc-

K. Koketsu, *Ground Motion Seismology*, Advances in Geological Science,
https://doi.org/10.1007/978-981-15-8570-8_3

ity structure is homogeneous in a piecewise manner such as this, the displacement potentials of P, SV, and SH waves exist in each layer (Sect. 1.2.4). In the case where ground motions pass through horizontal interfaces, the existence of the displacement potentials is maintained but **coupling** is introduced between the P and SV waves.

Aki and Richards [3] demonstrate this coupling as follows.[1] When the horizontal discontinuity passing through the origin is the $x-y$ plane of the Cartesian coordinate system, the three components u_x, u_y, and u_z of the ground motion (elastic displacement) $\mathbf{u}$ must be continuous at the boundary. Because these should be differentiable with respect to x and y, these **continuity conditions** can be replaced with those for $\partial u_y/\partial x - \partial u_x/\partial y$, $\partial u_x/\partial x + \partial u_y/\partial y$, and u_z, which are

$$(\nabla \times \mathbf{u})_z\,, \quad \nabla \cdot \mathbf{u} - \partial u_z/\partial z\,, \quad u_z\,. \tag{3.1}$$

For the continuity of force at the interface, from (1.11) we have

$$\mathbf{T}_n = \text{continuous} \tag{3.2}$$

where $\mathbf{T}_n$ is a stress vector normal to the interface. Here, as the interface is horizontal, $\mathbf{n}$ is in the z-axis direction, and the components of $\mathbf{T}_n$ are the stresses τ_{zx}, τ_{zy}, and τ_{zz}. Similar to (3.1), these continuity conditions can be replaced with those for $\partial\tau_{zy}/\partial x - \partial\tau_{zx}/\partial y$, $\partial\tau_{zx}/\partial x - \partial\tau_{zy}/\partial y$, and τ_{zz}, which are rewritten as

$$\mu\partial(\nabla \times \mathbf{u})_z/\partial z\,, \quad \mu(\partial\nabla \cdot \mathbf{u}/\partial z - 2\partial^2 u_z/\partial z^2 + \nabla^2 u_z)\,, \quad \lambda\nabla \cdot \mathbf{u} + 2\mu\partial u_z/\partial z \tag{3.3}$$

using the generalized Hooke's law (1.20). We therefore have the first, second, and third continuity conditions for the three terms in (3.1), and the fourth, fifth, and sixth continuity conditions for those in (3.3).

If the ground motion is an SH wave in the layer in front of the interface, $(\nabla \times \mathbf{u})_z \neq 0$, $\nabla \cdot \mathbf{u} = 0$, and $u_z = 0$ there. From the second and third continuity conditions, $\nabla \cdot \mathbf{u} = 0$ and $u_z = 0$ in the layer on the other side of the interface, so that the ground motion remains an SH wave even after transmission through the interface. The fifth and sixth conditions are trivially satisfied, and the amplitudes of the reflected and transmitted SH waves are determined from the remaining first and fourth conditions. If the ground motion is a P wave or an SV wave in the layer in front of the interface, $(\nabla \times \mathbf{u})_z = 0$ there. From the first continuity condition, $(\nabla \times \mathbf{u})_z = 0$ in the layer on the other side of the interface, so that the ground motion remains a P wave or an SV wave even after transmission through the interface. In addition, though the fourth condition is automatically satisfied, the four conditions still remain. When only the P or SV wave exists, there are only two unknowns, which are the amplitudes of the

[1]Box 6.5 of Aki and Richards [3]. As with the proof of the former in Sect. 1.2.4, no proof has been provided for the replacements of the conditions; however, there does not seem to be problem because the three conditions are simply replaced with three other conditions, and the three variables in the others are independent of each other.

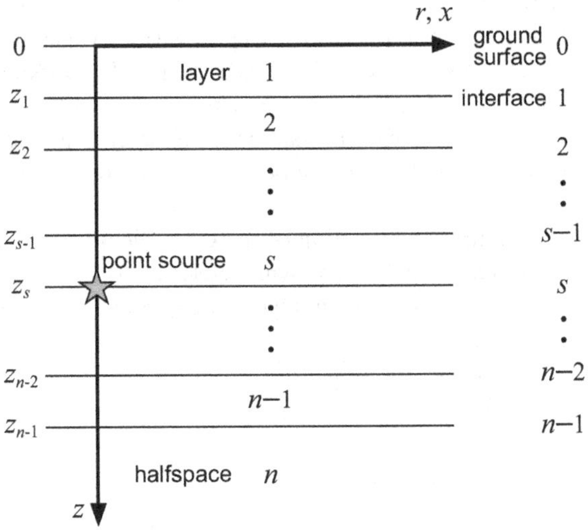

Fig. 3.1 Indexing of a horizontally layered structure. $d_i = z_i - z_{i-1}$, α_i, β_i, and ρ_i are the thickness, P wave velocity, S wave velocity, and density of layer i, which is the ith layer (modified from Koketsu [72] with permission of Kindai Kagaku)

reflected and transmitted P or SV waves. As only two variables cannot satisfy the four conditions, the P and SV ground motions must be coupled.

As shown in Sect. 2.2, ground motions in a horizontally layered structure are easy to manage using a **cylindrical coordinate system** (r, θ, z), the origin of which is the epicenter of a point source. Although the ground motion radiated from a point source is a combination of **spherical waves** (Sect. 1.2.5) as shown in (2.59) and (2.72), these are not simple spherical waves in a horizontally layered structure. However, assuming that its **Fourier transform** (Sect. 4.2.2) can be represented by the **superposition** of **cylindrical harmonics**,

$$J_l(kr) \begin{matrix} \cos l\theta \\ \sin l\theta \end{matrix} e^{\mp i\nu_\upsilon z}, \tag{3.4}$$

the same treatment as spherical waves can be used. J_l is the lth order **Bessel function** (Sect. 2.2.1), and k is the **wavenumber** (Sect. 2.1.5) of a **cylindrical wave** (Sect. 1.2.5). ν_υ (Eq. (2.76)) is termed the wavenumber in the depth direction.

As already shown in (2.129), the ground motion radiated from a point source can be represented by the zeroth-, first-, and second-order cylindrical harmonics, such that this assumption holds in the layer containing a point source. In other layers without any sources, the **wave equation** to be satisfied by the displacement potentials is **homogeneous** as in (1.45) with $\Phi = \Psi = \mathrm{X} = 0$, and its Fourier transform is in the form of the **Helmholtz equation** [82] $\nabla^2\mathcal{H} + \omega^2\mathcal{H} = 0$. The cylindrical harmonic function is one of the solutions of the Helmholtz equation using the method of separating variables [104]; therefore, the same assumption also holds for other layers.

A horizontally layered structure is indexed as shown in Fig. 3.1. The layers are numbered from the shallowest to the deepest, as layer 1, layer 2, and so on. The interfaces are similarly numbered, but the ground surface, which is the shallowest interface, is numbered 0. The depth of interface i is z_i ($z_0 = 0$), and the thickness of layer i is $d_i = z_i - z_{i-1}$. In the real Earth the structure can extend to several thousand kilometers depth. Below the deepest surface we assume a uniform medium. Such a structure lacking a lower interface is termed a **halfspace**. It is also assumed that the point source always lies on interface s. In reality, an earthquake source is usually buried within a layer. In that case, the layer is divided into two virtual layers with the same properties and the point source is considered to be at the interface between them.

Using the superposition of the cylindrical harmonics and the **horizontal radiation pattern** (Sect. 2.2.5) with reference to (2.129), the Fourier transforms of the displacement potentials in layer i are defined as follows[2]:

$$
\begin{aligned}
\bar{\phi}_i &= \sum_{l=0}^{2} \bar{\phi}_{li} = \sum_{l=0}^{2} \Lambda_l(\theta) \int_0^\infty \tilde{\phi}_{li}(z)\, J_l(kr)\, dk\,,\\
\bar{\psi}_i &= \sum_{l=0}^{2} \bar{\psi}_{li} = \sum_{l=0}^{2} \Lambda_l(\theta) \int_0^\infty \tilde{\psi}_{li}(z)\, J_l(kr) \frac{\mathrm{i}}{k} dk\,,\\
\bar{\chi}_i &= \sum_{l=0}^{2} \bar{\chi}_{li} = \sum_{l=0}^{2} \frac{d\Lambda_l(\theta)}{d\theta} \int_0^\infty \tilde{\chi}_{li}(z)\, J_l(kr)\, dk\,,\\
\tilde{\phi}_{li} &= A_{li} e^{+\mathrm{i}\nu_{\alpha i} z} + B_{li} e^{-\mathrm{i}\nu_{\alpha i} z} = \phi_{li}^- + \phi_{li}^+\,,\ \tilde{\psi}_{li} = C_{li} e^{+\mathrm{i}\nu_{\beta i} z} + D_{li} e^{-\mathrm{i}\nu_{\beta i} z} = \psi_{li}^- + \psi_{li}^+\,,\\
\tilde{\chi}_{li} &= E_{li} e^{+\mathrm{i}\nu_{\beta i} z} + F_{li} e^{-\mathrm{i}\nu_{\beta i} z} = \chi_{li}^- + \chi_{li}^+\,,\ \ k_{\alpha i} = \omega/\alpha_i\,,\ \ k_{\beta i} = \omega/\beta_i\,,\\
\nu_{\alpha i} &= \begin{cases} \sqrt{k_{\alpha i}^2 - k^2}\ k_{\alpha i} \ge k \\ -\mathrm{i}\sqrt{k^2 - k_{\alpha i}^2}\ k_{\alpha i} < k \end{cases},\ \ \nu_{\beta i} = \begin{cases} \sqrt{k_{\beta i}^2 - k^2}\ k_{\beta i} \ge k \\ -\mathrm{i}\sqrt{k^2 - k_{\beta i}^2}\ k_{\beta i} < k \end{cases}.
\end{aligned}
\tag{3.5}
$$

Assuming that the horizontal radiation patterns of the point source are not disturbed by the horizontal interfaces, $\Lambda_l(\theta)$ in (2.129) are used as they are. As mentioned above, α_i and β_i are the P and S wave velocities in layer i, respectively. Next, the Fourier transforms of (1.44):

$$
\begin{aligned}
\bar{u}_r &= \frac{\partial \bar{\phi}}{\partial r} + \frac{\partial^2 \bar{\psi}}{\partial r \partial z} + \frac{1}{r}\frac{\partial \bar{\chi}}{\partial \theta}\,,\\
\bar{u}_\theta &= \frac{1}{r}\frac{\partial \bar{\phi}}{\partial \theta} + \frac{1}{r}\frac{\partial^2 \bar{\psi}}{\partial \theta \partial z} - \frac{\partial \bar{\chi}}{\partial r}\,,\\
\bar{u}_z &= \frac{\partial \bar{\phi}}{\partial z} + \frac{\partial^2 \bar{\psi}}{\partial z^2} - \nabla^2 \bar{\psi}
\end{aligned}
\tag{3.6}
$$

link the Fourier transforms of the displacement potentials with the Fourier transforms of the ground motion components.

In (3.5), the displacement potentials with superscripts $+$ and $-$, such as ϕ_{li}^+, are related to ground motions propagating in the $+$ and $-$ directions of the z-axis, which

[2]The coefficient i/k attached to the integrand of $\bar{\psi}_i$ is to absorb the difference from the displacement potential $\bar{\psi}_i'$ of plane wave (Sect. 3.1.4). The $l = 0$ terms of $\bar{\chi}_i$ are always zero (Sect. 2.2.5). As mentioned in the Preface, $\tilde{}$ indicates the **Fourier–Hankel transform**, but does not follow the strict definition of the **Hankel transform** (Sect. 4.2.2, Page 847 of Arfken and Weber [5]).

correspond to the downward and upward directions, respectively. According to the definition of the Fourier transform in this book (Sect. 4.2.2), the inverse Fourier transform

$$\frac{1}{2\pi}\int_{-\infty}^{+\infty}\phi_{li}^{+}e^{\mathrm{i}\omega t}d\omega=\frac{1}{2\pi}\int_{-\infty}^{+\infty}B_{li}e^{-\mathrm{i}(\nu_{\alpha i}z-\omega t)}d\omega \tag{3.7}$$

is performed in order to convert ϕ_{li}^{+} in (3.5) to the time domain. In this inverse transformation of ϕ_{li}^{+}, the phase term has the form $-\mathrm{i}(\nu_{\alpha_i}z-\omega t)$; therefore, in order for the phase to be the same, the time t must be longer if z increases. This indicates that the ground motion is propagating in the + direction of the z-axis. Hence, the displacement potentials that include $e^{-\mathrm{i}\nu_{\alpha i}z}$ or $e^{-\mathrm{i}\nu_{\beta i}z}$, such as ϕ_{li}^{+}, are given a + superscript. Conversely, those containing $e^{+\mathrm{i}\nu_{\alpha i}z}$ or $e^{+\mathrm{i}\nu_{\beta i}z}$ are related to ground motions propagating in the − direction of the z-axis; therefore, a − superscript is added.

As previously mentioned in the footnote on Sect. 1.2.6, in this book, we consistently use the common Fourier transform defined in Sect. 4.2.2. However, Aki and Richards [3] and Kennett [65] use a different definition for just the time domain, and this difference affects the interpretation of the propagation direction of ground motions. When using their definition, the phase term is $-\mathrm{i}(\nu_{\alpha i}z+\omega t)$. In order for this phase to be the same, the time t must be shortened as z increases. This therefore represents propagation in the – direction of the z-axis.[3]

The **Bessel function** $J_l(kr)$ (Sect. 2.2.1), another component of the cylindrical harmonics (Eq. (3.4)), represents the horizontal propagation of ground motion. The Bessel function can be decomposed into the **Hankel function** of the first and second kind as

$$J_l(kr)=\frac{1}{2}\left\{H_l^{(1)}(kr)+H_l^{(2)}(kr)\right\}. \tag{3.8}$$

When kr is large, these functions can be approximated as[4]

$$H_l^{(1)}(kr),\,H_l^{(2)}(kr)\sim\sqrt{\frac{2}{\pi kr}}\exp\left\{\pm\mathrm{i}\left(kr-\frac{l\pi}{2}-\frac{\pi}{4}\right)\right\}, \tag{3.9}$$

where + and – correspond to (1) and (2), respectively. As $H_l^{(2)}(kr)$ in (3.9) contains a complex exponential function similar to $e^{-\mathrm{i}\nu_{\alpha i}z}$, it represents outward propagation from the point source in the direction of increasing r. $H_l^{(1)}(kr)$ is converted to $H_l^{(2)}$ with a negative wave number k using the formula[5]

$$H_l^{(2)}(-kr)=-e^{+l\mathrm{i}\pi}H_l^{(1)}(kr), \tag{3.10}$$

[3]Aki and Richards [3] argue in Box 5.2 that this is convenient because the sign of the exponent matches the sign of the propagation direction.

[4]Equations (D.32) and (D.33) of Ben-Menahem and Singh [10].

[5]Equation (D.6) of Ben-Menahem and Singh [10].

and, for large r, the contribution from negative wavenumbers is small,[6] therefore $H_l^{(1)}(kr)$ is often ignored. To summarize, the Bessel function $J_l(kr)$ mainly represents the propagation of ground motion from the point source.

3.1.2 SH Wave

We first consider only the SH wave, which is not coupled with other waves. We have

$$\bar{u}_r = \frac{1}{r}\frac{\partial \bar{\chi}}{\partial \theta}\,, \quad \bar{u}_\theta = -\frac{\partial \bar{\chi}}{\partial r}\,, \quad \bar{u}_z = 0 \tag{3.11}$$

by substituting $\bar{\phi} = \bar{\psi} = 0$ for (3.6). According to the proof of Aki and Richards [3] (Sect. 3.1.1) for solving the problem of the SH wave propagation, it is sufficient to have one boundary condition of ground motion and one of stress at an interface. Here, we therefore consider the boundary conditions related to u_θ and $\tau_{z\theta}$. By obtaining $\tau_{z\theta}$ from the definitions of the stresses (Eq. (1.34)) and substituting (3.11) into its Fourier transform, we have

$$\bar{\tau}_{z\theta} = \mu\frac{\partial \bar{u}_\theta}{\partial z} = -\mu\frac{\partial^2 \bar{\chi}}{\partial r \partial z}\,. \tag{3.12}$$

$\partial\bar{\chi}/\partial r$ is included in $\bar{\tau}_{z\theta}$ (Eq. (3.12)), and $\bar{u}_\theta$ in (3.11). Thus, using the partial derivative with respect to r of the superposition of the cylindrical harmonics for $\bar{\chi}_i$ in (3.5), we represent $\bar{u}_\theta$ and $\bar{\tau}_{z\theta}$ in layer i as

$$\begin{aligned}
\bar{u}_{\theta i} &= \sum_{l=0}^{2} \bar{\upsilon}_{li} = \sum_{l=0}^{2} \frac{d\Lambda_l(\theta)}{d\theta} \int_0^\infty \tilde{\upsilon}_{li}(z)\frac{dJ_l(kr)}{dr}dk \\
\bar{\tau}_{z\theta i} &= \sum_{l=0}^{2} \bar{p}_{li} = \sum_{l=0}^{2} \frac{d\Lambda_l(\theta)}{d\theta} \int_0^\infty \tilde{p}_{li}(z)\frac{dJ_l(kr)}{dr}dk\,.
\end{aligned} \tag{3.13}$$

$\tilde{\upsilon}_{li}(z)$ can be regarded as the $\omega - k$ spectrum or **Fourier–Hankel transform** of the SH ground motion. By substituting (3.13) into (3.11) and (3.12), we obtain

$$\begin{aligned}
\tilde{\upsilon}_{li} &= -\chi_{li}^- - \chi_{li}^+ \\
\tilde{p}_{li} &= -\mathrm{i}\mu_i\nu_{\beta i}\chi_{li}^- + \mathrm{i}\mu_i\nu_{\beta i}\chi_{li}^+\,,
\end{aligned} \tag{3.14}$$

where $\mu_i = \rho_i\beta_i^2$ is the rigidity of layer i.

We represent the above relationships in vector and matrix notation using the **potential vector** of $\mathbf{\Phi}_{li} = \left(\chi_{li}^-, \chi_{li}^+\right)^{\mathrm{T}}$ and the **motion-stress vector** (Aki and Richards [3])

[6] Refer to Page 177 of Kennett [64].

of $\mathbf{S}_{li} = (\tilde{v}_{li}, \tilde{p}_{li})^{\mathrm{T}}$. We have

$$\mathbf{S}_{li} = \mathbf{T}_i \, \mathbf{\Phi}_{li} \,, \quad \mathbf{T}_i = \begin{pmatrix} -1 & -1 \\ -\mathrm{i}\mu_i \nu_{\beta i} & +\mathrm{i}\mu_i \nu_{\beta i} \end{pmatrix}, \quad \mathbf{T}_i^{-1} = \frac{1}{2} \begin{pmatrix} -1 & \dfrac{-1}{\mathrm{i}\mu_i \nu_{\beta i}} \\ -1 & \dfrac{+1}{\mathrm{i}\mu_i \nu_{\beta i}} \end{pmatrix}. \tag{3.15}$$

In addition, the relation

$$\mathbf{\Phi}_{li}(z_i) = \mathbf{E}_i \, \mathbf{\Phi}_{li}(z_{i-1}) \,, \quad \mathbf{E}_i = \begin{pmatrix} e^{+\mathrm{i}\nu_{\beta i}(z_i - z_{i-1})} & 0 \\ 0 & e^{-\mathrm{i}\nu_{\beta i}(z_i - z_{i-1})} \end{pmatrix} \tag{3.16}$$

holds between $\mathbf{\Phi}(z_{i-1})$ at the upper interface and $\mathbf{\Phi}(z_i)$ at the lower interface of layer i. Using the matrices $\mathbf{T}_i$ and $\mathbf{E}_i$, $\mathbf{S}_{li}$ at the upper interface and $\mathbf{S}_{li}$ at the lower interface are therefore related as[7]

$$\begin{aligned} &\mathbf{S}_{li}(z_i) = \mathbf{G}_i \, \mathbf{S}_{li}(z_{i-1}) \,, \quad \mathbf{G}_i = \mathbf{T}_i \mathbf{E}_i \mathbf{T}_i^{-1} = \begin{pmatrix} (G_i)_{11} & (G_i)_{12} \\ (G_i)_{21} & (G_i)_{22} \end{pmatrix}, \\ &(G_i)_{11} = (G_i)_{22} = \cos Q_i \,, \quad d_i = z_i - z_{i-1} \,, \\ &(G_i)_{12} = (\mu_i k r_{\beta i})^{-1} \sin Q_i \,, \; Q_i = \nu_{\beta i} d_i \,, \\ &(G_i)_{21} = -\mu_i k r_{\beta i} \sin Q_i \,, \quad r_{\beta i} = \nu_{\beta i}/k \,, \end{aligned} \tag{3.17}$$

where d_i is the thickness of layer i.

We now use the above formulation and the continuity conditions of ground motion and stress at the interfaces:

$$\mathbf{S}_{l,i+1}(z_i) = \mathbf{S}_{li}(z_i) \,, \quad i = n-1, n-2, \cdots, 1 \,. \tag{3.18}$$

In addition to these, we also use the vector $\mathbf{\Delta}_l$ of the ground motion and stress discontinuities due to the point source (Sect. 2.2.5, hereinafter referred to as the **discontinuity vector**); therefore we can relate the potential vector $\mathbf{\Phi}_{ln}(z_{n-1})$ at the upper interface of the halfspace to the motion-stress vector $S_{l1}(0)$ at the ground surface, as shown in the following list and equations. $\mathbf{T}_n$ is the matrix given by (3.15), and $\mathbf{G}_i$ ($i = n$-1, n-2, ..., 1) are the matrices given by (3.17). An item in the list and its corresponding equation in (3.19) are similarly numbered, as follows.

(1) The inverse of $\mathbf{T}_n$ links $\mathbf{\Phi}_{ln}(z_{n-1})$ to $\mathbf{S}_{ln}(z_{n-1})$.
(2) The continuity condition equates $\mathbf{S}_{ln}(z_{n-1})$ to $\mathbf{S}_{l,n-1}(z_{n-1})$.
(3) $\mathbf{G}_{n-1}$ links $\mathbf{S}_{l,n-1}(z_{n-1})$ to $\mathbf{S}_{l,n-1}(z_{n-2})$.
(4) The repetition of (2) and (3) links $\mathbf{S}_{l,n-1}(z_{n-2})$ to $\mathbf{S}_{l,s+1}(z_s)$.
(5) $\mathbf{S}_{l,s+1}(z_s)$ is equated to the sum of $\mathbf{S}_{ls}(z_s)$ and $\mathbf{\Delta}_l$.
(6) $\mathbf{G}_i$ links $\mathbf{S}_{ls}(z_s)$ to $\mathbf{S}_{ls}(z_{s-1})$.
(7) The repetition of (2) and (3) again links $\mathbf{S}_{ls}(z_s)$ to $\mathbf{S}_{l1}(0)$.

[7]The elements of a matrix with a subscript such as $\mathbf{G}_i$ are denoted by $(G_i)_{11}$, $(G_i)_{12}$, $(G_i)_{21}$, and $(G_i)_{22}$ following the notation of Kind [68].

$$\begin{aligned}
\mathbf{\Phi}_{ln}(z_{n-1}) &= \mathbf{T}_n^{-1}\mathbf{S}_{ln}(z_{n-1}) && (1)\\
&= \mathbf{T}_n^{-1}\mathbf{S}_{l,n-1}(z_{n-1}) && (2)\\
&= \mathbf{T}_n^{-1}\mathbf{G}_{n-1}\mathbf{S}_{l,n-1}(z_{n-2}) && (3)\\
&\vdots\\
&= \mathbf{T}_n^{-1}\mathbf{G}_{n-1}\mathbf{G}_{n-2}\cdots\mathbf{G}_{s+1}\mathbf{S}_{l,s+1}(z_s) && (4)\\
&= \mathbf{T}_n^{-1}\mathbf{G}_{n-1}\mathbf{G}_{n-2}\cdots\mathbf{G}_{s+1}(\mathbf{\Delta}_l+\mathbf{S}_{ls}(z_s)) && (5)\\
&= \mathbf{T}_n^{-1}\mathbf{G}_{n-1}\mathbf{G}_{n-2}\cdots\mathbf{G}_{s+1}(\mathbf{\Delta}_l+\mathbf{G}_s\mathbf{S}_{ls}(z_{s-1})) && (6)\\
&\vdots\\
&= \mathbf{T}_n^{-1}\mathbf{G}_{n-1}\mathbf{G}_{n-2}\cdots\mathbf{G}_{s+1}(\mathbf{\Delta}_l+\mathbf{G}_s\mathbf{G}_{s-1}\cdots\mathbf{G}_1\mathbf{S}_{l1}(0)) && (7)
\end{aligned} \tag{3.19}$$

All the matrices in (3.19) can be known from the properties of the horizontally layered structure. The discontinuity vectors are given from the source potential in (2.129) as

$$\mathbf{\Delta}_0 = \begin{pmatrix}0\\0\end{pmatrix}, \quad \mathbf{\Delta}_1 = \frac{\overline{M}_0(\omega)}{4\pi\rho_s\omega^2}\begin{pmatrix}-2k_{\beta s}^2\\0\end{pmatrix}, \quad \mathbf{\Delta}_2 = \frac{\overline{M}_0(\omega)}{4\pi\rho_s\omega^2}\begin{pmatrix}0\\-\mu_s k k_{\beta s}^2\end{pmatrix} \tag{3.20}$$

and can also be known; therefore, (3.19) is a system of two linear equations for $\mathbf{\Phi}_{ln}(z_{n-1})$ and $\mathbf{S}_{l1}(0)$.

As there is no interface in the halfspace, the ground motion, radiating from the point source to the halfspace, only propagates downwards and never returns upwards. This is called the **radiation boundary condition**, and is equivalent to setting

$$\chi_{ln}^-(z_{n-1}) = 0 \tag{3.21}$$

in $\mathbf{\Phi}_{ln}(z_{n-1})$. In (3.5), if ν_β is imaginary, its sign is taken to be $-$ so that the remaining potential χ_{ln}^+ does not diverge at $z = +\infty$.

In addition, the stress vector (Sect. 1.2.2) vanishes above the ground surface, because there is no elastic body above the ground surface and a stress vector is an internal force in an elastic body. If the stress vector is zero, its component, i.e., the stress, here $\tau_{z\theta}$, is also zero. This is called a **stress-free condition**, and the ground surface is sometimes called a **free surface**. When $\tau_{z\theta}$ is zero, $\bar{\tau}_{z\theta}$ is also zero and

$$\tilde{p}_{l1}(0) = 0 \tag{3.22}$$

from (3.13). Following this, two unknowns, $\chi_{ln}^+(z_{n-1})$ and $\tilde{v}_{l1}(0)$, remain in $\mathbf{\Phi}_{ln}(z_{n-1})$ and $\mathbf{S}_{l1}(0)$, and we can therefore solve the system of two linear equations (Eq. (3.19)) as follows. Defining

$$\begin{aligned}
\mathbf{M} &= \mathbf{T}_n^{-1}\mathbf{G}_{n-1}\mathbf{G}_{n-2}\cdots\mathbf{G}_{s+1}\mathbf{G}_s\mathbf{G}_{s-1}\cdots\mathbf{G}_1\,,\\
\mathbf{M}^h &= \mathbf{T}_n^{-1}\mathbf{G}_{n-1}\mathbf{G}_{n-2}\cdots\mathbf{G}_{s+1}
\end{aligned} \tag{3.23}$$

where the superscript h indicates the portion of the medium from the halfspace to the point source, we rewrite (3.19) as

$$\begin{pmatrix} 0 \\ \chi_{ln}^{+}(z_{n-1}) \end{pmatrix} = \mathbf{M}^h \begin{pmatrix} \Delta_{l1} \\ \Delta_{l2} \end{pmatrix} + \mathbf{M} \begin{pmatrix} \tilde{\upsilon}_{l1}(0) \\ 0 \end{pmatrix} . \tag{3.24}$$

We then obtain the solution

$$\tilde{\upsilon}_{l1}(0) = \frac{-1}{M_{11}} (M_{11}^h \Delta_{l1} + M_{12}^h \Delta_{l2}) \tag{3.25}$$

using the first equation of (3.24) and the elements of $\mathbf{M}^h$ and $\mathbf{M}$.

3.1.3 P Wave and SV Wave

For the P wave and SV wave, the formulation is much more complex than for the SH wave, due to their **coupling**. Substituting $\bar{\chi} = 0$ into (3.6), we have

$$\bar{u}_r = \frac{\partial \bar{\phi}}{\partial r} + \frac{\partial^2 \bar{\psi}}{\partial r \partial z} , \quad \bar{u}_\theta = \frac{1}{r} \frac{\partial \bar{\phi}}{\partial \theta} + \frac{1}{r} \frac{\partial^2 \bar{\psi}}{\partial \theta \partial z} , \quad \bar{u}_z = \frac{\partial \bar{\phi}}{\partial z} + \frac{\partial^2 \bar{\psi}}{\partial z^2} . \tag{3.26}$$

Again, according to the proof of Aki and Richards [3] (Sect. 3.1.1), we need two boundary conditions for ground motion and two more for stress; therefore, here we select those for u_r, u_z, τ_{zr}, and τ_{zz}. Using τ_{zr} and τ_{zz} in the definitions of stresses in the cylindrical coordinate system (Eq. (1.34)) and substituting (3.26) into its Fourier transform, we obtain

$$\begin{aligned} \bar{\tau}_{zr} &= 2\mu \left(\frac{\partial^2 \bar{\phi}}{\partial r \partial z} + \frac{\partial^3 \bar{\psi}}{\partial r \partial z^2} \right) + \mu k_\beta^2 \frac{\partial \bar{\psi}}{\partial r} , \\ \bar{\tau}_{zz} &= -\lambda k_\alpha^2 \bar{\phi} + 2\mu \left(\frac{\partial^2 \bar{\phi}}{\partial z^2} + \frac{\partial^3 \bar{\psi}}{\partial z^3} + k_\beta^2 \frac{\partial \bar{\psi}}{\partial z} \right) . \end{aligned} \tag{3.27}$$

In (3.27), all the terms in $\bar{u}_r$ and $\bar{\tau}_{zr}$ include the partial derivative operator with respect to r, $\partial/\partial r$, whereas all the terms in $\bar{u}_z$ and $\bar{\tau}_{zz}$ do not include it. Hence, using the partial derivative with respect to r of the superposition of the cylindrical harmonics for $\bar{\phi}_i$ and $\bar{\psi}_i$ in (3.5), we represent $\bar{u}_r$ and $\bar{\tau}_{zr}$ in layer i as[8]

[8]The factor $1/ik$ attached to the integrands absorbs the difference between the cylindrical harmonics and the two-dimensional Cartesian harmonics (Sect. 3.1.4).

$$\bar{u}_{ri} = \sum_{l=0}^{2} \bar{u}_{li} = \sum_{l=0}^{2} \Lambda_l(\theta) \int_0^\infty \tilde{u}_{li}(z) \frac{dJ_l(kr)}{dr} \frac{1}{\mathrm{i}k} dk\,,$$
$$\bar{\tau}_{zri} = \sum_{l=0}^{2} \bar{t}_{li} = \sum_{l=0}^{2} \Lambda_l(\theta) \int_0^\infty \tilde{t}_{li}(z) \frac{dJ_l(kr)}{dr} \frac{1}{\mathrm{i}k} dk\,. \tag{3.28}$$

For $\bar{u}_z$ and $\bar{\tau}_{zz}$ in layer i, we directly use the superposition of the cylindrical harmonics without partial differentiation, and represent them as

$$\bar{u}_{zi} = \sum_{l=0}^{2} \bar{w}_{li} = \sum_{l=0}^{2} \Lambda_l(\theta) \int_0^\infty \tilde{w}_{li}(z) J_l(kr) dk\,,$$
$$\bar{\tau}_{zzi} = \sum_{l=0}^{2} \bar{s}_{li} = \sum_{l=0}^{2} \Lambda_l(\theta) \int_0^\infty \tilde{s}_{li}(z) J_l(kr) dk\,. \tag{3.29}$$

$\tilde{u}_{li}(z)$ and $\tilde{w}_{li}(z)$ can be regarded as the $\omega - k$ spectra or **Fourier–Hankel transform** of the P and SV ground motions.

We define the **potential vector** (Sect. 3.1.2) as $\mathbf{\Phi}_{li} = \left(\phi_{li}^-, \psi_{li}^-, \phi_{li}^+, \psi_{li}^+\right)^{\mathrm{T}}$ and the **motion-stress vector** (Sect. 3.1.2) as $\mathbf{S}_{li} = \left(\tilde{u}_{li}, \tilde{w}_{li}, \tilde{s}_{li}, \tilde{t}_{li}\right)^{\mathrm{T}}$. We substitute (3.28) and (3.29) into (3.26) and (3.27), and using the vectors defined above, we then arrange the results as

$$\mathbf{S}_{li} = \mathbf{T}_i\,\mathbf{\Phi}_{li}\,,\quad \mathbf{T}_i = \begin{pmatrix} +\mathrm{i}k & -\mathrm{i}\nu_{\beta i} & +\mathrm{i}k & +\mathrm{i}\nu_{\beta i} \\ +\mathrm{i}\nu_{\alpha i} & +\mathrm{i}k & -\mathrm{i}\nu_{\alpha i} & +\mathrm{i}k \\ +\mu_i l_i & -2\mu_i k\nu_{\beta i} & +\mu_i l_i & +2\mu_i k\nu_{\beta i} \\ -2\mu_i k\nu_{\alpha i} & -\mu_i l_i & +2\mu_i k\nu_{\alpha i} & -\mu_i l_i \end{pmatrix},$$
$$\mathbf{T}_i^{-1} = \frac{1}{2\mu_i\nu_{\alpha i}\nu_{\beta i}k_{\beta i}^2} \begin{pmatrix} -2\mathrm{i}\mu_i k\nu_{\alpha i}\nu_{\beta i} & +\mathrm{i}\mu_i l_i\nu_{\beta i} & -\nu_{\alpha i}\nu_{\beta i} & -k\nu_{\beta i} \\ -\mathrm{i}\mu_i l_i\nu_{\alpha i} & -2\mathrm{i}\mu_i k\nu_{\alpha i}\nu_{\beta i} & -k\nu_{\alpha i} & +\nu_{\alpha i}\nu_{\beta i} \\ -2\mathrm{i}\mu_i k\nu_{\alpha i}\nu_{\beta i} & -\mathrm{i}\mu_i l_i\nu_{\beta i} & -\nu_{\alpha i}\nu_{\beta i} & -k\nu_{\beta i} \\ +\mathrm{i}\mu_i l_i\nu_{\alpha i} & -2\mathrm{i}\mu_i k\nu_{\alpha i}\nu_{\beta i} & +k\nu_{\alpha i} & +\nu_{\alpha i}\nu_{\beta i} \end{pmatrix},$$
$$l_i = 2k^2 - k_{\beta i}^2\,. \tag{3.30}$$

These are the results by Fuchs [34] (Eqs. (18) and (19) in the original paper). The complicated inverse matrix $\mathbf{T}_i^{-1}$ can be obtained from the formula including the **adjugate matrix** [82]

$$\mathbf{A}^{-1} = \frac{1}{|\mathbf{A}|} \begin{pmatrix} \tilde{a}_{11} & \tilde{a}_{21} & \tilde{a}_{31} & \tilde{a}_{41} \\ \tilde{a}_{12} & \tilde{a}_{22} & \tilde{a}_{32} & \tilde{a}_{42} \\ \tilde{a}_{13} & \tilde{a}_{23} & \tilde{a}_{33} & \tilde{a}_{43} \\ \tilde{a}_{14} & \tilde{a}_{24} & \tilde{a}_{34} & \tilde{a}_{44} \end{pmatrix},\quad \tilde{a}_{12} = (-1)^{1+2} \begin{vmatrix} A_{21} & A_{23} & A_{24} \\ A_{31} & A_{33} & A_{34} \\ A_{41} & A_{43} & A_{44} \end{vmatrix},\ etc. \tag{3.31}$$

and others.

Next, the relation between $\mathbf{\Phi}(z_i)$ at the lower interface and $\mathbf{\Phi}(z_{i-1})$ at the upper interface of layer i is easily found to be

$$\mathbf{\Phi}_{li}(z_i) = \mathbf{E}_i\ \mathbf{\Phi}_{li}(z_{i-1})\ ,$$
$$\mathbf{E}_i = \begin{pmatrix} e^{+i\nu_{\alpha i}(z_i - z_{i-1})} & 0 & 0 & 0 \\ 0 & e^{+i\nu_{\beta i}(z_i - z_{i-1})} & 0 & 0 \\ 0 & 0 & e^{-i\nu_{\alpha i}(z_i - z_{i-1})} & 0 \\ 0 & 0 & 0 & e^{-i\nu_{\beta i}(z_i - z_{i-1})} \end{pmatrix}. \tag{3.32}$$

Hence, $\mathbf{S}_{li}(z_i)$ at the lower interface is related to $\mathbf{S}_{li}(z_{i-1})$ as[9]

$$\mathbf{S}_{li}(z_i) = \mathbf{G}_i\ \mathbf{S}_{li}(z_{i-1})\,,\quad \mathbf{G}_i = \mathbf{T}_i \mathbf{E}_i \mathbf{T}_i^{-1} = \begin{pmatrix} (G_i)_{11} & (G_i)_{12} & (G_i)_{13} & (G_i)_{14} \\ (G_i)_{21} & (G_i)_{22} & (G_i)_{23} & (G_i)_{24} \\ (G_i)_{31} & (G_i)_{32} & (G_i)_{33} & (G_i)_{34} \\ (G_i)_{41} & (G_i)_{42} & (G_i)_{43} & (G_i)_{44} \end{pmatrix}, \tag{3.33}$$

$$\begin{aligned}
&(G_i)_{11} = (G_i)_{44} = -\gamma_i \cos P_i + (\gamma_i + 1)\cos Q_i\,,\\
&(G_i)_{12} = (G_i)_{34} = \mathrm{i}\left\{(\gamma_i + 1) r_{\alpha i}^{-1} \sin P_i + \gamma_i r_{\beta i} \sin Q_i\right\},\\
&(G_i)_{13} = (G_i)_{24} = \mathrm{i}(\rho_i c\omega)^{-1}(-\cos P_i + \cos Q_i)\,,\\
&(G_i)_{14} = (\rho_i c\omega)^{-1}(r_{\alpha i}^{-1} \sin P_i + r_{\beta i} \sin Q_i)\,,\\
&(G_i)_{21} = (G_i)_{43} = -\mathrm{i}\left\{\gamma_i r_{\alpha i} \sin P_i + (\gamma_i + 1) r_{\beta i}^{-1} \sin Q_i\right\},\\
&(G_i)_{22} = (G_i)_{33} = (\gamma_i + 1)\cos P_i - \gamma_i \cos Q_i\,,\\
&(G_i)_{23} = (\rho_i c\omega)^{-1}(r_{\alpha i} \sin P_i + r_{\beta i}^{-1} \sin Q_i)\,,\\
&(G_i)_{31} = (G_i)_{42} = -\mathrm{i}\rho_i c\omega\gamma_i(\gamma_i + 1)(\cos P_i - \cos Q_i)\,,\\
&(G_i)_{32} = -\rho_i c\omega\left\{(\gamma_i + 1)^2 r_{\alpha i}^{-1} \sin P_i + \gamma_i^2 r_{\beta i} \sin Q_i\right\},\\
&(G_i)_{41} = -\rho_i c\omega\left\{\gamma_i^2 r_{\alpha i} \sin P_i + (\gamma_i + 1)^2 r_{\beta i}^{-1} \sin Q_i\right\},\\
&d_i = z_i - z_{i-1}\,,\ P_i = \nu_{\alpha i} d_i\,,\ Q_i = \nu_{\beta i} d_i\,,\\
&r_{\alpha i} = \nu_{\alpha i}/k\,,\ r_{\beta i} = \nu_{\beta i}/k\,,\ \gamma_i = -2k^2/k_{\beta i}^2\,,\ \rho_i c\omega = \mu_i k_{\beta i}^2/k\,,
\end{aligned}$$

where $c \equiv \omega/k$ is a **phase velocity** (Sect. 1.2.6). From the source potentials in (2.129), we have the **discontinuity vector**

$$\mathbf{\Delta}_0 = \frac{\overline{M}_0(\omega)}{4\pi\rho_s\omega^2}\begin{pmatrix} 0 \\ 4kk_{\alpha s}^2 \\ 0 \\ 2\mathrm{i}\mu_s k^2 \cdot (4k_{\alpha s}^2 - 3k_{\beta s}^2) \end{pmatrix},\quad \mathbf{\Delta}_1 = \frac{\overline{M}_0(\omega)}{4\pi\rho_s\omega^2}\begin{pmatrix} -2\mathrm{i}kk_{\beta s}^2 \\ 0 \\ 0 \\ 0 \end{pmatrix},\quad \mathbf{\Delta}_2 = \frac{\overline{M}_0(\omega)}{4\pi\rho_s\omega^2}\begin{pmatrix} 0 \\ 0 \\ 0 \\ -2\mathrm{i}\mu_s k^2 k_{\beta s}^2 \end{pmatrix}. \tag{3.34}$$

Using the above matrices and vectors, the potential vector $\mathbf{\Phi}_{ln}(z_{n-1})$ in the halfspace can be linked to the motion-stress vector $\mathbf{S}_{l1}(0)$ on the ground surface by (3.19), similar to the SH wave. However, the system of equations (3.19) is expanded from

[9] The notation of Fuchs [34] is used after correcting typographical errors in the original paper.

two linear equations to four linear equations by the coupling of P and SV waves. Here, the **radiation boundary condition** (Sect. 3.1.2) are

$$\phi_{ln}^{-}(z_{n-1}) = 0\,, \quad \psi_{ln}^{-}(z_{n-1}) = 0\,, \tag{3.35}$$

and the **stress-free condition** (Sect. 3.1.2) are

$$\tilde{s}_{l1}(0) = 0\,, \quad \tilde{t}_{l1}(0) = 0\,. \tag{3.36}$$

There remain four unknowns, which are $\bar{\phi}_{ln}^{+}(z_{n-1})$, $\bar{\psi}_{ln}^{+}(z_{n-1})$, $\bar{u}_{l1}(0)$, and $\bar{w}_{l1}(0)$. We can therefore solve the system of four linear equations and obtain the ground motions $\bar{u}_{l1}(0)$ and $\bar{w}_{l1}(0)$ at the ground surface. In (3.5), if ν_α or ν_β is imaginary, its sign is taken to be $-$ so that the remaining potential ϕ_{ln}^{+} or ψ_{ln}^{+} does not diverge at $z = +\infty$.

Although the system of equations is formally the same as (3.19) for the SH wave, it has been expanded to four equations; therefore, the vectors are expanded from two rows to four rows, and the matrices are also expanded from two rows and two columns to four rows and four columns. Equation (3.24) for the SH wave has become

$$\begin{pmatrix} 0 \\ 0 \\ \phi_{ln}^{+}(z_{n-1}) \\ \psi_{ln}^{+}(z_{n-1}) \end{pmatrix} = \mathbf{M}^h \begin{pmatrix} \Delta_{l1} \\ \Delta_{l2} \\ \Delta_{l3} \\ \Delta_{l4} \end{pmatrix} + \mathbf{M} \begin{pmatrix} \tilde{u}_{l1}(0) \\ \tilde{w}_{l1}(0) \\ 0 \\ 0 \end{pmatrix}, \tag{3.37}$$

and its solution has also been extended as follows:

$$\begin{aligned}
\tilde{u}_{l1}(0) &= \frac{1}{\hat{M}_{11}} \{-M_{22}(M_{11}^h\Delta_{l1} + M_{12}^h\Delta_{l2} + M_{13}^h\Delta_{l3} + M_{14}^h\Delta_{l4}) \\
&\qquad + M_{12}(M_{21}^h\Delta_{l1} + M_{22}^h\Delta_{l2} + M_{23}^h\Delta_{l3} + M_{24}^h\Delta_{l4})\}\,, \\
\tilde{w}_{l1}(0) &= \frac{1}{\hat{M}_{11}} \{+M_{21}(M_{11}^h\Delta_{l1} + M_{12}^h\Delta_{l2} + M_{13}^h\Delta_{l3} + M_{14}^h\Delta_{l4}) \\
&\qquad - M_{11}(M_{21}^h\Delta_{l1} + M_{22}^h\Delta_{l2} + M_{23}^h\Delta_{l3} + M_{24}^h\Delta_{l4})\}\,, \\
\hat{M}_{11} &= M_{11}M_{22} - M_{12}M_{21}\,.
\end{aligned} \tag{3.38}$$

Not only does the solution become complicated, but difficulties in numerical computations also arise. As previously mentioned, $\nu_{\alpha i}$ and $\nu_{\beta i}$, defined in (3.5) as wavenumbers in the depth direction, have a $-$ sign if they are imaginary, so that the downward potentials in the halfspace do not diverge at $z = +\infty$. However, since the upward potentials exist in the layers other than the halfspace, in this case in layer i, the matrix $\mathbf{G}_i$ includes the exponentially growing terms $e^{+|\nu_{\alpha i}|d_i}$ and $e^{+|\nu_{\beta i}|d_i}$. In particular, for the P and SV waves, as the solution of (3.38) includes the product of the elements of $\mathbf{M}$ and $\mathbf{M}^h$, $e^{+2|\nu_{\alpha i}|d_i}$ and $e^{+2|\nu_{\beta i}|d_i}$ also appear. Hence, **overflow** easily occurs if k or d_i is large. In other words, as $k = 2\pi f/c$, ground motions with high frequencies and low phase velocities passing through thick layers cannot be computed.

$\hat{M}_{11}$ in (3.38) is the determinant of the 2-by-2 submatrix made by extracting elements M_{11}, M_{12}, M_{21}, and M_{22} from $\mathbf{M}$. The determinant of a submatrix, such as $\hat{M}_{11}$, is hereinafter termed **subdeterminant**. Dunkin [31] proved the following theorem for subdeterminants. The subdeterminant made by extracting elements A_{pr}, A_{ps}, A_{qr}, and A_{qs} from the matrix $\mathbf{A}$ is represented by

$$\hat{A}_{kl} = A\Big|^{pq}_{rs} = A_{pr}\, A_{qs} - A_{ps}\, A_{qr} \tag{3.39}$$

using the notation of Červený [17], where the subscripts are rewritten as $12 \to 1$, $13 \to 2$, $14 \to 3$, $23 \to 4$, $24 \to 5$, and $34 \to 6$. In this representation and **summation convention** (Sect. 1.3.2), the theorem

$$\hat{B}_{kl} = (\hat{A}_1)_{kj_1} (\hat{A}_2)_{j_1 j_2} \cdots (\hat{A}_m)_{j_{m-1} l} \tag{3.40}$$

holds for $\mathbf{B} = \mathbf{A}_1 \mathbf{A}_2 \cdots \mathbf{A}_m$. Applying this theorem to $\mathbf{M}$ and $\mathbf{M}^h$ in (3.23), we have

$$\begin{aligned} \hat{M}_{kl} &= (\hat{\mathrm{T}}_n^{-1})_{kj_n} (\hat{G}_{n-1})_{j_n j_{n-1}} \cdots (\hat{G}_s)_{j_{s+1} j_s} (\hat{G}_{s-1})_{j_s j_{s-1}} \cdots (\hat{G}_1)_{j_2 l} \,, \\ \hat{M}^h_{kl} &= (\hat{\mathrm{T}}_n^{-1})_{kj_n} (\hat{G}_{n-1})_{j_n j_{n-1}} \cdots (\hat{G}_s)_{j_{s+1} l} \,. \end{aligned} \tag{3.41}$$

Fuchs [34] and Kind [68] gave explicit representations of $(\hat{\mathrm{T}}_i^{-1})_{1l}$ and $(\hat{G}_i)_{kl}$,[10] but with errors in the factors of $(\hat{\mathrm{T}}_i^{-1})_{1l}$ that were subsequently corrected by Baumgardt [9]. Parts of the corrected $(\hat{\mathrm{T}}_i^{-1})_{1l}$ and $(\hat{G}_i)_{kl}$ are listed in (3.163). In the above $(\hat{G}_i)_{kl}$, $e^{+2|\nu_{\alpha i}|d_i}$ and $e^{+2|\nu_{\beta i}|d_i}$ are canceled out and not included, so that the overflow problem in $\hat{M}_{11}$ should be about the same as that of the SH wave due to $e^{+|\nu_{\alpha i}|d_i}$ and $e^{+|\nu_{\beta i}|d_i}$.

Removing $e^{+2|\nu_{\alpha i}|d_i}$ and $e^{+2|\nu_{\beta i}|d_i}$ from parts other than $\hat{M}_{11}$ in (3.38) appears very difficult, because $\mathbf{M}$ contains $\mathbf{M}^h$; however, Kind [69] made it possible using

$$\mathbf{G}^l = \mathbf{G}_s \mathbf{G}_{s-1} \cdots \mathbf{G}_1 \tag{3.42}$$

as follows. The superscript l in (3.42) indicates a portion shallower than the point source. Cleverly using $\mathbf{M} = \mathbf{M}^h \mathbf{G}^l$ and the symmetry of $\mathbf{G}^l$, the parts other than $\hat{M}_{11}$ in (3.38), we obtain

[10] Because of the reason described in Sect. 3.1.5, we use a Roman font for subdeterminants of $\mathbf{T}_i$.

$$\tilde{u}_{l1}(0) = \frac{1}{\hat{M}_{11}} \left(\hat{M}^h_{11}\ \hat{M}^h_{12}\ \hat{M}^h_{13}\ \hat{M}^h_{15}\ \hat{M}^h_{16} \right) \begin{pmatrix} -G^l_{22} & G^l_{12} & 0 & 0 \\ -G^l_{32} & 0 & G^l_{12} & 0 \\ -G^l_{42} & -G^l_{32} & G^l_{22} & G^l_{12} \\ 0 & -G^l_{42} & 0 & G^l_{22} \\ 0 & 0 & -G^l_{42} & G^l_{32} \end{pmatrix} \begin{pmatrix} \Delta_{l1} \\ \Delta_{l2} \\ \Delta_{l3} \\ \Delta_{l4} \end{pmatrix},$$

$$\tilde{w}_{l1}(0) = \frac{-1}{\hat{M}_{11}} \left(\hat{M}^h_{11}\ \hat{M}^h_{12}\ \hat{M}^h_{13}\ \hat{M}^h_{15}\ \hat{M}^h_{16} \right) \begin{pmatrix} -G^l_{21} & G^l_{11} & 0 & 0 \\ -G^l_{31} & 0 & G^l_{11} & 0 \\ -G^l_{41} & -G^l_{31} & G^l_{21} & G^l_{11} \\ 0 & -G^l_{41} & 0 & G^l_{21} \\ 0 & 0 & -G^l_{41} & G^l_{31} \end{pmatrix} \begin{pmatrix} \Delta_{l1} \\ \Delta_{l2} \\ \Delta_{l3} \\ \Delta_{l4} \end{pmatrix}. \quad (3.43)$$

In (3.43), the subdeterminants of $\mathbf{M}^h$ are fully separated from the elements of $\mathbf{G}^l$, so that $\mathbf{G}_i$ of a layer is not included in both $\mathbf{M}^h$ and $\mathbf{G}^l$. Hence, $e^{+2|\nu_{\alpha i}|d_i}$ and $e^{+2|\nu_{\beta i}|d_i}$ never appear there.

In addition, Harvey [43] noted that the denominator $\hat{M}_{11}$ and $\Sigma \hat{M}^h_{1n} G^l_{n'1}$ in the numerator of (3.43) similarly include $e^{+|\nu_{\alpha i}|d_i}$ and $e^{+|\nu_{\beta i}|d_i}$. We can therefore normalize $e^{+|\nu_{\alpha i}|d_i}$ and $e^{+|\nu_{\beta i}|d_i}$ without affecting the result of (3.43). Through this normalization, it is possible to relieve the failure in relation to $e^{+|\nu_{\alpha i}|d_i}$ and $e^{+|\nu_{\beta i}|d_i}$, which not only exist in the P wave and the SV wave but also in the SH wave, even after the countermeasure of Dunkin and Kind is carried out. Wang [121] proposed another method using orthonormalization.

3.1.4 Haskell Matrix

So far, we have considered the cylindrical wave expansion (Sect. 2.2), which can conveniently represent the ground motion from the point source. However, when the source is sufficiently distant, such as for the teleseismic body wave (Sect. 3.1.11), plane wave expansion is sufficient. The plane wave expansion has been studied for longer than the cylindrical wave expansion. Thomson [112] showed in 1950 that matching the boundary conditions of the plane wave in a horizontally layered structure was possible through matrix operations, and Haskell [44] introduced this into seismology. This matrix was later shown to be a type of more general **propagator matrix** [39], but is often referred to as the **Haskell matrix** in the reference to the researcher who introduced it. Since the concept of the propagator matrix also applies to the cylindrical wave expansion, the matrix $\mathbf{G}_i$ that appears in Sect. 3.1.2 and Sect. 3.1.3 is also called a propagator matrix.

When the Cartesian coordinate system is set as shown by x and z in Fig. 3.1, the ground motion is a plane wave (Sect. 1.2.5) that is oscillatory in the x- and z-directions and does not change in the y-direction, perpendicular to the figure. Hence, $\partial/\partial y = 0$ and the Fourier transform of (1.44) give

$$\bar{u}_x = \frac{\partial \bar{\phi}}{\partial x} + \frac{\partial^2 \bar{\psi}}{\partial x \partial z}, \quad \bar{u}_y = -\frac{\partial \bar{\chi}}{\partial x}, \quad \bar{u}_z = \frac{\partial \bar{\phi}}{\partial z} - \frac{\partial^2 \bar{\psi}}{\partial x^2} \quad (3.44)$$

if there is no source. In this equation, $\bar{\phi}$ and $\bar{\psi}$ only appear in $\bar{u}_x$ and $\bar{u}_z$, and $\bar{\chi}$ only appears in $\bar{u}_y$, such that the P and SV waves are completely separated from the SH wave. The Fourier transforms of the displacement potentials in layer i can be represented by the superposition of the two-dimensional harmonics

$$e^{-\mathrm{i}kx}e^{\mp \mathrm{i}\nu_\upsilon z}\,, \tag{3.45}$$

which is one of the special solutions of the Helmholtz equation, that is the Fourier transform of the wave represented by (1.45) with no source, as

$$\begin{aligned}
&\bar{\phi}_i = \int_{-\infty}^{\infty} \tilde{\phi} e^{-\mathrm{i}kx} dk\,, \quad \frac{\partial \bar{\psi}_i}{\partial x} = \bar{\psi}_i' = \int_{-\infty}^{\infty} \tilde{\psi}_i' e^{-\mathrm{i}kx} dk\,, \quad \bar{\chi}_i = \int_{-\infty}^{\infty} \tilde{\chi}_i e^{-\mathrm{i}kx} dk \\
&\tilde{\phi}_i = A_i e^{+\mathrm{i}\nu_{\alpha i} z} + B_i e^{-\mathrm{i}\nu_{\alpha i} z} = \phi_i^- + \phi_i^+\,, \quad \tilde{\psi}_i = C_i e^{+\mathrm{i}\nu_{\beta i} z} + D_i e^{-\mathrm{i}\nu_{\beta i} z} = \psi_i^- + \psi_i^+ \\
&\tilde{\chi}_i = E_i e^{+\mathrm{i}\nu_{\beta i} z} + F_i e^{-\mathrm{i}\nu_{\beta i} z} = \chi_i^- + \chi_i^+ \\
&\nu_{\alpha i} = \begin{cases} (k_{\alpha i}^2 - k^2)^{\frac{1}{2}} \; k_{\alpha i} = \omega/\alpha_i \geq k \\ -\mathrm{i}(k^2 - k_{\alpha i}^2)^{\frac{1}{2}} \; k_{\alpha i} = \omega/\alpha_i < k \end{cases}, \quad \nu_{\beta i} = \begin{cases} (k_{\beta i}^2 - k^2)^{\frac{1}{2}} \; k_{\beta i} = \omega/\beta_i \geq k \\ -\mathrm{i}(k^2 - k_{\beta i}^2)^{\frac{1}{2}} \; k_{\beta i} = \omega/\beta_i < k \end{cases}.
\end{aligned} \tag{3.46}$$

Similarly, the Fourier transforms of the three components of elastic displacement and the three components of stress in the $x - y$ plane are represented using the two-dimensional harmonics, as

$$\begin{aligned}
\bar{u}_{xi} &= \int_{-\infty}^{\infty} \tilde{u}_i e^{-\mathrm{i}kx} dk\,, \quad \bar{u}_{yi} = \int_{-\infty}^{\infty} \tilde{\upsilon}_i e^{-\mathrm{i}kx} dk\,, \quad \bar{u}_{zi} = \int_{-\infty}^{\infty} \tilde{w}_i e^{-\mathrm{i}kx} dk\,, \\
\bar{\tau}_{zxi} &= \int_{-\infty}^{\infty} \tilde{t}_i e^{-\mathrm{i}kx} dk\,, \quad \bar{\tau}_{zyi} = \int_{-\infty}^{\infty} \tilde{p}_i e^{-\mathrm{i}kx} dk\,, \quad \bar{\tau}_{zzi} = \int_{-\infty}^{\infty} \tilde{s}_i e^{-\mathrm{i}kx} dk\,.
\end{aligned} \tag{3.47}$$

As $\bar{\psi}_i'$ is newly defined in (3.46), we rewrite (3.44) as

$$\bar{u}_x = \frac{\partial \bar{\phi}}{\partial x} + \frac{\partial \bar{\psi}'}{\partial z}\,, \quad \bar{u}_y = -\frac{\partial \bar{\chi}}{\partial x}\,, \quad \bar{u}_z = \frac{\partial \bar{\phi}}{\partial z} - \frac{\partial \bar{\psi}'}{\partial x}\,, \tag{3.48}$$

and obtain

$$\bar{\tau}_{zx} = \mu \left(\frac{\partial \bar{u}_x}{\partial z} + \frac{\partial \bar{u}_z}{\partial x} \right), \quad \bar{\tau}_{zy} = \mu \frac{\partial \bar{u}_y}{\partial z}\,, \quad \bar{\tau}_{zz} = \lambda \left(\frac{\partial \bar{u}_x}{\partial x} + \frac{\partial \bar{u}_z}{\partial z} \right) + 2\mu \frac{\partial \bar{u}_z}{\partial z} \tag{3.49}$$

using the stress definitions (1.24). We define the **potential vector** and **motion-stress vector** as

$$\mathbf{\Phi}_i = \left(\chi_i^-, \chi_i^+\right)^{\mathrm{T}}, \quad \mathbf{S}_i = \left(\mathrm{i}k\tilde{\upsilon}_i, \tilde{p}_i\right)^{\mathrm{T}}. \tag{3.50}$$

Using (3.46), (3.48), and (3.49), we then obtain

$$\mathbf{S}_i = \mathbf{T}_i\,\mathbf{\Phi}_i\,,\ \mathbf{T}_i = \mathrm{i}k\begin{pmatrix} +\mathrm{i}k & +\mathrm{i}k \\ +\mathrm{i}\mu_i\nu_{\beta i} & -\mathrm{i}\mu_i\nu_{\beta i}\end{pmatrix},\ \mathbf{T}_i^{-1} = \frac{1}{-2\mathrm{i}k^2\mu_i\nu_{\beta i}}\begin{pmatrix} +\mathrm{i}\mu_i\nu_{\beta i} & +\mathrm{i}k \\ +\mathrm{i}\mu_i\nu_{\beta i} & -\mathrm{i}k\end{pmatrix}. \tag{3.51}$$

For $\mathbf{\Phi}_i(z_{i-1})$ at the upper boundary of layer i and $\mathbf{\Phi}_i(z_i)$ at the lower boundary of layer i, we have the relation

$$\mathbf{\Phi}_i(z_i) = \mathbf{E}_i\,\mathbf{\Phi}_i(z_{i-1})\,,\quad \mathbf{E}_i = \begin{pmatrix} e^{+\mathrm{i}\nu_{\beta i}(z_i - z_{i-1})} & 0 \\ 0 & e^{-\mathrm{i}\nu_{\beta i}(z_i - z_{i-1})}\end{pmatrix}. \tag{3.52}$$

From (3.51) and (3.52), we have the Haskell matrix

$$\begin{aligned}
&\mathbf{G}_i = \mathbf{T}_i\mathbf{E}_i\mathbf{T}_i^{-1} = \begin{pmatrix} (G_i)_{11} & (G_i)_{12} \\ (G_i)_{21} & (G_i)_{22}\end{pmatrix},\\
&(G_i)_{11} = (G_i)_{22} = \cos Q_i\,,\ \ d_i = z_i - z_{i-1}\,,\\
&(G_i)_{12} = \mathrm{i}(\mu_i r_{\beta i})^{-1}\sin Q_i\,,\ \ Q_i = \nu_{\beta i} d_i\,,\\
&(G_i)_{21} = \mathrm{i}\mu_i r_{\beta i}\sin Q_i\,,\qquad r_{\beta i} = \nu_{\beta i}/k
\end{aligned} \tag{3.53}$$

for the SH wave [44].

Comparing this Haskell matrix with the propagator matrix (3.17) obtained in Sect. 3.1.2 shows that they are almost identical, even though the latter is for the cylindrical wave. This indicates that they result in similar matrix operations for both plane and cylindrical waves, if the potential and the motion-stress vectors can be properly defined in a horizontally layered structure. However, Haskell [44] defines only the ground motion element of the motion-stress vector as dimensionless with $\tilde{\dot{v}}_i/c$. As $\dot{v}$ indicates $\partial v/\partial t$ according to the formula in Table 4.3, the ground motion element is $\tilde{\dot{v}}_i/c = \mathrm{i}\omega\tilde{v}_i/c = \mathrm{i}k\tilde{v}_i$ as in (3.50). As a result of this nondimensionalization, $(G_i)_{12}$ in (3.53) is $\mathrm{i}k$ times that in (3.17), and $(G_i)_{21}$ is $1/\mathrm{i}k$ of that in (3.17). In addition, the factor $\mathrm{i}k$ drops from $\mathbf{T}_i$ because the original Haskell matrix [44] uses $\tilde{\chi}_i \equiv \tilde{v}_i$, but the difference in $\mathbf{T}_i$ does not affect $\mathbf{G}_i$ (as shown in (3.53), $\mathbf{T}_i$ and $\mathbf{T}_i^{-1}$ are included in $\mathbf{G}_i$ so that the influence is canceled).

The derivation for the P and SV waves is omitted, and only the result of Haskell [44] is shown below. The definitions of the parameters contained in the elements are the same as in (3.33), except for γ_i.

$$\begin{aligned}
&\mathbf{G}_i = \Big((G_i)_{jm}\Big),\quad \gamma_i = +2k^2/k_{\beta i}^2\,,\\
&(G_i)_{11} = (G_i)_{44} = \gamma_i\cos P_i - (\gamma_i - 1)\cos Q_i\,,\\
&(G_i)_{12} = (G_i)_{34} = \mathrm{i}\left\{(\gamma_i - 1)r_{\alpha i}^{-1}\sin P_i + \gamma_i r_{\beta i}\sin Q_i\right\},\\
&(G_i)_{13} = (G_i)_{24} = -(\rho_i c^2)^{-1}(\cos P_i - \cos Q_i)\,,\\
&(G_i)_{14} = \mathrm{i}(\rho_i c^2)^{-1}(r_{\alpha i}^{-1}\sin P_i + r_{\beta i}\sin Q_i)\,,\\
&(G_i)_{21} = (G_i)_{43} = -\mathrm{i}\left\{\gamma_i r_{\alpha i}\sin P_i + (\gamma_i - 1)r_{\beta i}^{-1}\sin Q_i\right\},\\
&(G_i)_{22} = (G_i)_{33} = -(\gamma_i - 1)\cos P_i + \gamma_i\cos Q_i\,,
\end{aligned}$$

$$
\begin{aligned}
(G_i)_{23} &= \mathrm{i}(\rho_i c^2)^{-1}(r_{\alpha i}\sin P_i + r_{\beta i}^{-1}\sin Q_i)\,,\\
(G_i)_{31} &= (G_i)_{42} = \rho_i c^2\gamma_i(\gamma_i - 1)(\cos P_i - \cos Q_i)\,,\\
(G_i)_{32} &= \mathrm{i}\rho_i c^2\left\{(\gamma_i - 1)^2 r_{\alpha i}^{-1}\sin P_i + \gamma_i^2 r_{\beta i}\sin Q_i\right\},\\
(G_i)_{41} &= \mathrm{i}\rho_i c^2\left\{\gamma_i^2 r_{\alpha i}\sin P_i + (\gamma_i - 1)^2 r_{\beta i}^{-1}\sin Q_i\right\}.
\end{aligned}
\tag{3.54}
$$

When we compare the above Haskell matrix with the propagator matrix (3.33) in Sect. 3.1.3, paying attention to the difference in the sign of γ_i, and we find them to be almost identical. Similar to the SH wave, the ground motion elements are defined as $\tilde{\dot{u}}_i/c$ and $\tilde{\dot{w}}_i/c$ with no dimensionality, so that $(G_i)_{13}$, $(G_i)_{14}$, $(G_i)_{23}$, and $(G_i)_{24}$ in (3.54) are ik times those in (3.33), and $(G_i)_{31}$, $(G_i)_{32}$, $(G_i)_{41}$, and $(G_i)_{42}$ in (3.54) are $(\mathrm{i}k)^{-1}$ of those in (3.33). The propagator matrix in (3.33) was originally proposed by Fuchs [34] for the plane P and SV waves. Kohketsu [70] then demonstrated that the same matrix could be used for the cylindrical P and SV waves, as shown in (Sect. 3.1.3). As Fuchs [34] used

$$
\bar{u}_x = \frac{\partial\bar{\phi}}{\partial x} - \frac{\partial\bar{\psi}'}{\partial z}\,,\quad \bar{u}_z = \frac{\partial\bar{\phi}}{\partial z} + \frac{\partial\bar{\psi}'}{\partial x}\,,
\tag{3.55}
$$

which are slightly different from (3.48), and used the two-dimensional harmonics

$$
e^{+\mathrm{i}kx}e^{\mp\mathrm{i}\nu_\upsilon z}\,,
\tag{3.56}
$$

which are also slightly different from (3.45); the signs of some elements are inverted. Summarizing the above differences and distinguishing the Haskell matrix (3.54) from the propagator matrix (3.33) of Fuchs using the superscript H, the following differences exist between them:

$$
\mathbf{G}_i^{\mathrm{H}} = \begin{pmatrix}
(G_i)_{11} & -(G_i)_{12} & \mathrm{i}k(G_i)_{13} & \mathrm{i}k(G_i)_{14}\\
-(G_i)_{21} & (G_i)_{22} & \mathrm{i}k(G_i)_{23} & \mathrm{i}k(G_i)_{24}\\
-(\mathrm{i}k)^{-1}(G_i)_{31} & (\mathrm{i}k)^{-1}(G_i)_{32} & (G_i)_{33} & -(G_i)_{34}\\
(\mathrm{i}k)^{-1}(G_i)_{41} & -(\mathrm{i}k)^{-1}(G_i)_{42} & -(G_i)_{43} & (G_i)_{44}
\end{pmatrix}.
\tag{3.57}
$$

Harkrider [42] showed that the Haskell matrix (3.54) can be used as it is, as well as for the cylindrical P and SV waves. Several other types of propagator matrices have been proposed since then. However, these matrices differ only slightly, depending on the potential vector and the motion-stress vector. Kennett [64, 65] studied the propagator matrix when the velocity and density change in the z-direction in a layer.

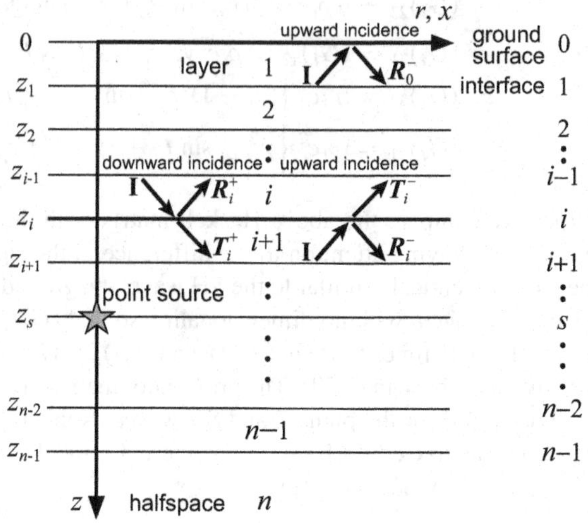

Fig. 3.2 Reflection and transmission matrices when ground motion is incident downwards (left) or upwards (right) on interface i of a horizontally layered structure (Fig. 3.1), and the reflection matrix when ground motion is incident upwards (top) on the ground surface (modified from Koketsu [72] with permission of Kindai Kagaku)

3.1.5 *Reflection/Transmission Matrix I*[11]

The problem of overflow in the propagator matrix and its countermeasures are described in detail in Sect. 3.1.3. However, as long as the propagator matrix is used, even if countermeasures are taken, there is still a small risk. In order to avoid this, the definition of the matrix itself must be changed. Kennett [64] separated the propagator matrix into parts related to reflection and transmission at interfaces and phase terms representing propagation in a layer. In addition, by dividing the operation into the part related to the upward potential and the part related to the downward potential, he succeeded in perfectly avoiding $e^{+|\nu_{\alpha i}|d_i}$ and $e^{+|\nu_{\beta i}|d_i}$ by always advancing the operation in the forward direction.

Consider that interface i is located at $z = z_i$ in the horizontally layered structure as shown in Fig. 3.1, and the ground motion of an SH wave of unit intensity is incident downwards from layer i (Fig. 3.2 (left)). Reflection and transmission occur at this interface, so that the **reflected wave** propagates upward in layer i and the **transmitted wave** propagates downward to layer i+1. As the Bessel function gives outward propagation (Sect. 3.1.1), the direction of incidence and propagation is obliquely downward or upward (arrows in the figure). In our formulation, the downward and upward directions correspond to the positive (+) and negative (−) directions along the z-axis, respectively. We therefore denote the reflection and transmission coefficients for the downward incidence at interface i as R_i^+ and T_i^+. The ground motion in layer i has downward and upward components with intensities of 1 and R_i^+, and the ground motion in layer i+1 has a downward component with intensity of T_i^+. The

[11]The descriptions in Sects. 3.1.5 and 3.1.6 are based on Kennett [64, 65], but the definitions and derivation of the reflection/transmission matrices use the results of Fuchs [34] and Červený [17].

potential vector in Sect. 3.1.2 is therefore $\mathbf{\Phi}_{li}(z_i) = \left(R_i^+, 1\right)^{\mathrm{T}}$ on the side of layer i, and $\mathbf{\Phi}_{l,i+1}(z_i) = \left(0, T_i^+\right)^{\mathrm{T}}$ on the side of layer i+1.

When these are substituted into (3.15) and the continuity condition (3.18),

$$\begin{pmatrix} 0 \\ T_i^+ \end{pmatrix} = \mathbf{Q}_i \begin{pmatrix} R_i^+ \\ 1 \end{pmatrix}, \quad \mathbf{Q}_i = \begin{pmatrix} (Q_i)_{11} & (Q_i)_{12} \\ (Q_i)_{21} & (Q_i)_{22} \end{pmatrix} = \mathbf{T}_{i+1}^{-1}\mathbf{T}_i \,. \tag{3.58}$$

From $\mathbf{T}_i$ and $\mathbf{T}_i^{-1}$ in (3.15), we obtain

$$(Q_i)_{11} = (Q_i)_{22} = \frac{\mu_{i+1}\nu_{\beta,i+1} + \mu_i \nu_{\beta i}}{2\mu_{i+1}\nu_{\beta,i+1}}, \quad (Q_i)_{12} = (Q_i)_{21} = \frac{\mu_{i+1}\nu_{\beta,i+1} - \mu_i \nu_{\beta i}}{2\mu_{i+1}\nu_{\beta,i+1}} \,. \tag{3.59}$$

Using these the simultaneous equations (3.58) can be solved and

$$R_i^+ = -\frac{(Q_i)_{12}}{(Q_i)_{11}} = \frac{\mu_i \nu_{\beta i} - \mu_{i+1}\nu_{\beta,i+1}}{\mu_i \nu_{\beta i} + \mu_{i+1}\nu_{\beta,i+1}}, \quad T_i^+ = (Q_i)_{21} R_i^+ + (Q_i)_{22} = \frac{2\mu_i \nu_{\beta i}}{\mu_i \nu_{\beta i} + \mu_{i+1}\nu_{\beta,i+1}} \tag{3.60}$$

is obtained.[12] With respect to R_i^- and T_i^- for upward incidence, the simultaneous equations are

$$\begin{pmatrix} 1 \\ R_i^- \end{pmatrix} = \mathbf{Q}_i \begin{pmatrix} T_i^- \\ 0 \end{pmatrix}. \tag{3.61}$$

Solving (3.61) we obtain

$$T_i^- = \frac{1}{(Q_i)_{11}} = \frac{2\mu_{i+1}\nu_{\beta,i+1}}{\mu_{i+1}\nu_{\beta,i+1} + \mu_i \nu_{\beta i}}, \quad R_i^- = (Q_i)_{21} T_i^- = \frac{\mu_{i+1}\nu_{\beta,i+1} - \mu_i \nu_{\beta i}}{\mu_{i+1}\nu_{\beta,i+1} + \mu_i \nu_{\beta i}} \,. \tag{3.62}$$

For the SH wave, the scalar quantities $R_i^\pm$ and $T_i^\pm$ representing reflection and transmission are termed **reflection/transmission coefficient**. However, when they are collectively referred to together with those for the P and SV waves, they are also called the reflection/transmission "matrices", each of which has only the scalar quantity as a single element.

When the P wave is incident downwards, the coupling between the P wave and the SV wave makes the formulation much more complicated. The potential vector is $\mathbf{\Phi}_{li}(z_i) = \left(R_i^{\mathrm{PP}+}, R_i^{\mathrm{PS}+}, 1, 0\right)^{\mathrm{T}}$ on the side of layer i, and $\mathbf{\Phi}_{l,i+1}(z_i) = \left(0, 0, T_i^{\mathrm{PP}+}, T_i^{\mathrm{PS}+}\right)^{\mathrm{T}}$ on the side of layer i+1. When these are substituted into (3.30) and the continuity condition (3.18),

[12] This agrees with Eq. (6.6.4) of Satô [104] if $\zeta \to \nu_\beta$.

$$\begin{pmatrix} 0 \\ 0 \\ T_i^{\text{PP}+} \\ T_i^{\text{PS}+} \end{pmatrix} = \mathbf{Q}_i \begin{pmatrix} R_i^{\text{PP}+} \\ R_i^{\text{PS}+} \\ 1 \\ 0 \end{pmatrix}, \quad \mathbf{Q}_i = \begin{pmatrix} (Q_i)_{11} & (Q_i)_{12} & (Q_i)_{13} & (Q_i)_{14} \\ (Q_i)_{21} & (Q_i)_{22} & (Q_i)_{23} & (Q_i)_{24} \\ (Q_i)_{31} & (Q_i)_{32} & (Q_i)_{33} & (Q_i)_{34} \\ (Q_i)_{41} & (Q_i)_{42} & (Q_i)_{43} & (Q_i)_{44} \end{pmatrix} = \mathbf{T}_{i+1}^{-1}\mathbf{T}_i . \tag{3.63}$$

We then sum (the first equation) $\times \{(-(Q_i)_{22}\}$ and (the second equation) $\times\{-(Q_i)_{11}\}$, obtaining the reflection coefficients

$$\begin{aligned} R_i^{\text{PP}+} &= \frac{-(Q_i)_{13}\{-(Q_i)_{22}\} - (Q_i)_{23}(Q_i)_{12}}{(Q_i)_{11}\{-(Q_i)_{22}\} + (Q_i)_{21}(Q_i)_{12}} = \frac{-(Q_i)\,\big|^{12}_{23}}{-(Q_i)\,\big|^{12}_{12}} = \frac{(\hat{Q}_i)_{14}}{(\hat{Q}_i)_{11}}, \\ R_i^{\text{PS}+} &= \frac{-(Q_i)_{13}(Q_i)_{21} - (Q_i)_{23}\{-(Q_i)_{11}\}}{(Q_i)_{12}(Q_i)_{21} + (Q_i)_{22}\{-(Q_i)_{11}\}} = \frac{(Q_i)\,\big|^{12}_{13}}{-(Q_i)\,\big|^{12}_{12}} = \frac{-(\hat{Q}_i)_{12}}{(\hat{Q}_i)_{11}}, \end{aligned} \tag{3.64}$$

where we use the **subdeterminant** (Sect. 3.1.3) of $\mathbf{Q}_i$.[13]

From the third and fourth equations of (3.63), the transmission coefficients

$$\begin{aligned} T_i^{\text{PP}+} &= (Q_i)_{31}R_i^{\text{PP}+} + (Q_i)_{32}R_i^{\text{PS}+} + (Q_i)_{33} = \frac{1}{(\hat{Q}_i)_{11}} \begin{vmatrix} (Q_i)_{11} & (Q_i)_{12} & (Q_i)_{13} \\ (Q_i)_{21} & (Q_i)_{22} & (Q_i)_{23} \\ (Q_i)_{31} & (Q_i)_{32} & (Q_i)_{33} \end{vmatrix} = \frac{(\tilde{Q}_i)_{44}}{(\hat{Q}_i)_{11}}, \\ T_i^{\text{PS}+} &= (Q_i)_{41}R_i^{\text{PP}+} + (Q_i)_{42}R_i^{\text{PS}+} + (Q_i)_{43} = \frac{1}{(\hat{Q}_i)_{11}} \begin{vmatrix} (Q_i)_{11} & (Q_i)_{12} & (Q_i)_{13} \\ (Q_i)_{21} & (Q_i)_{22} & (Q_i)_{23} \\ (Q_i)_{41} & (Q_i)_{42} & (Q_i)_{43} \end{vmatrix} = \frac{(\tilde{Q}_i)_{43}}{(\hat{Q}_i)_{11}} \end{aligned} \tag{3.65}$$

are obtained.[14] The second halves of both equations in (3.65) not only include the subdeterminant $(\hat{Q}_i)_{11}$ but also $(\tilde{Q}_i)_{44}$ and $(\tilde{Q}_i)_{43}$, which are the determinants of 3 by 3 submatrices of $\mathbf{Q}_i$. The submatrix for $(\tilde{Q}_i)_{kl}$ is constructed by removing the kth row and lth column from $\mathbf{Q}_i$ [17]. Conversely, when the SV wave is incident downwards, the potential vector is $\mathbf{\Phi}_{li}(z_i) = \left(R_i^{\text{SP}+}, R_i^{\text{SS}+}, 0, 1\right)^{\text{T}}$ on the side of layer i, and $\mathbf{\Phi}_{l,i+1}(z_i) = \left(0, 0, T_i^{\text{SP}+}, T_i^{\text{SS}+}\right)^{\text{T}}$ on the side of layer i+1. Hence, (3.63) yields

$$\begin{pmatrix} 0 \\ 0 \\ T_i^{\text{SP}+} \\ T_i^{\text{SS}+} \end{pmatrix} = \mathbf{Q}_i \begin{pmatrix} R_i^{\text{SP}+} \\ R_i^{\text{SS}+} \\ 0 \\ 1 \end{pmatrix}. \tag{3.66}$$

By solving these simultaneous equations, the **reflection/transmission coefficient**

$$R_i^{\text{SP}+} = \frac{(\hat{Q}_i)_{15}}{(\hat{Q}_i)_{11}}, \quad R_i^{\text{SS}+} = \frac{-(\hat{Q}_i)_{13}}{(\hat{Q}_i)_{11}}, \quad T_i^{\text{SP}+} = \frac{(\tilde{Q}_i)_{43}}{(\hat{Q}_i)_{11}}, \quad T_i^{\text{SS}+} = \frac{(\tilde{Q}_i)_{33}}{(\hat{Q}_i)_{11}} \tag{3.67}$$

[13]These agree with Eqs. (54) and (55) of Fuchs [34] if $m \to (Q_i)$.

[14]The first halves of the both equations agree with Eqs. (56) and (57) of Fuchs [34] if $m \to (Q_i)$. The second halves agree with Eq. (5) of Červený [17] if $D \to (\hat{Q}_i)_{11}$ and $\tilde{H} \to (\tilde{Q}_i)$.

are obtained.[15] To summarize the above, for the downward incidence of the P or SV wave, the reflection and transmission at interface i can be represented by the **reflection/transmission matrix**

$$\boldsymbol{R}_i^+ = \begin{pmatrix} R_i^{\mathrm{PP}+} & R_i^{\mathrm{PS}+} \\ R_i^{\mathrm{SP}+} & R_i^{\mathrm{SS}+} \end{pmatrix}, \quad \boldsymbol{T}_i^+ = \begin{pmatrix} T_i^{\mathrm{PP}+} & T_i^{\mathrm{PS}+} \\ T_i^{\mathrm{SP}+} & T_i^{\mathrm{SS}+} \end{pmatrix}, \tag{3.68}$$

which are 2 by 2 matrices.[16] Their elements are given in (3.64), (3.65), and (3.67). While the subdeterminants of $\mathbf{Q}_i$ in the elements can be analytically obtained from the subdeterminant of $\mathbf{T}_{i+1}^{-1}$, that of $\mathbf{T}_i$, and Dunkin's theorem [31] in (3.40), the explicit representations of the elements are given by Červený et al. [20].

When the P wave is incident upwards on interface i (Fig. 3.2, right), the potential vector is $\boldsymbol{\Phi}_{li}(z_i) = \left(T_i^{\mathrm{PP}-}, T_i^{\mathrm{PS}-}, 0, 0\right)^{\mathrm{T}}$ on the side of interface i, and $\boldsymbol{\Phi}_{l,i+1}(z_i) = \left(1, 0, R_i^{\mathrm{PP}-}, R_i^{\mathrm{PS}-}\right)^{\mathrm{T}}$ on the side of layer $i+1$. Conversely, when the SV wave is incident upwards, the potential vector is $\boldsymbol{\Phi}_{li}(z_i) = \left(T_i^{\mathrm{SP}-}, T_i^{\mathrm{SS}-}, 0, 0\right)^{\mathrm{T}}$ on the side of interface i, and $\boldsymbol{\Phi}_{l,i+1}(z_i) = \left(1, 0, R_i^{\mathrm{SP}-}, R_i^{\mathrm{SS}-}\right)^{\mathrm{T}}$ on the side of layer $i+1$. Hence, the simultaneous equations

$$\begin{pmatrix} 1 \\ 0 \\ R_i^{\mathrm{PP}-} \\ R_i^{\mathrm{PS}-} \end{pmatrix} = \mathbf{Q}_i \begin{pmatrix} T_i^{\mathrm{PP}-} \\ T_i^{\mathrm{PS}-} \\ 0 \\ 0 \end{pmatrix}, \quad \begin{pmatrix} 0 \\ 1 \\ R_i^{\mathrm{SP}-} \\ R_i^{\mathrm{SS}-} \end{pmatrix} = \mathbf{Q}_i \begin{pmatrix} T_i^{\mathrm{SP}-} \\ T_i^{\mathrm{SS}-} \\ 0 \\ 0 \end{pmatrix}, \quad \mathbf{Q}_i = \mathbf{T}_{i+1}^{-1}\mathbf{T}_i \tag{3.69}$$

must be solved. We then sum (the first equation) $\times \{(-(Q_i)_{22}\}$ and (the second equation) $\times \{-(Q_i)_{11}\}$, obtaining the reflection coefficients. We sum (the first equation) $\times (Q_i)_{22}$ and (the second equation) $\times \{-(Q_i)_{12}\}$, then sum (the first equation) $\times (Q_i)_{21}$ and (the second equation) $\times \{-(Q_i)_{11}\}$, obtaining the transmission coefficients[17]

[15] The reflection coefficients agree with Eq. (8) of Červený [17] if $W \to (\hat{Q}_i)$, and the transmission coefficients agree with Eq. (5) of Červený [17] if $D \to (\hat{Q}_i)_{11}$ and $\tilde{H} \to (\tilde{Q}_i)$.

[16] The reflection and transmission matrices should be in bold according to the notation used in this book, but in this case the transmission matrix is indistinguishable from $\mathbf{T}_i$ in (3.15) and (3.30), so we use italic bold.

[17] These agree with Eq. (5) of Červený [17] if $D \to (\hat{Q}_i)_{11}$ and $\tilde{H} \to (\tilde{Q}_i)$.

$$
\begin{aligned}
T_i^{\mathrm{PP}-} &= \frac{(Q_i)_{22}}{(Q_i)_{11}(Q_i)_{22}-(Q_i)_{21}(Q_i)_{12}} = \frac{(Q_i)_{22}}{(\hat{Q}_i)_{11}}, \\
T_i^{\mathrm{PS}-} &= \frac{(Q_i)_{21}}{(Q_i)_{12}(Q_i)_{21}-(Q_i)_{22}(Q_i)_{11}} = \frac{-(Q_i)_{21}}{(\hat{Q}_i)_{11}}, \\
T_i^{\mathrm{SP}-} &= \frac{-(Q_i)_{12}}{(Q_i)_{11}(Q_i)_{22}-(Q_i)_{21}(Q_i)_{12}} = \frac{-(Q_i)_{12}}{(\hat{Q}_i)_{11}}, \\
T_i^{\mathrm{SS}-} &= \frac{(Q_i)_{11}}{(Q_i)_{12}(Q_i)_{21}-(Q_i)_{22}(Q_i)_{11}} = \frac{(Q_i)_{11}}{(\hat{Q}_i)_{11}}.
\end{aligned}
\tag{3.70}
$$

Similarly, from the third and fourth equations, we obtain the reflection coefficients[18]

$$
\begin{aligned}
R_i^{\mathrm{PP}-} &= (Q_i)_{31}T_i^{\mathrm{PP}-} + (Q_i)_{32}T_i^{\mathrm{PS}-} = \frac{(Q_i)_{31}(Q_i)_{22}-(Q_i)_{32}(Q_i)_{21}}{(\hat{Q}_i)_{11}} = \frac{-(\hat{Q}_i)_{41}}{(\hat{Q}_i)_{11}}, \\
R_i^{\mathrm{PS}-} &= (Q_i)_{41}T_i^{\mathrm{PP}-} + (Q_i)_{42}T_i^{\mathrm{PS}-} = \frac{(Q_i)_{41}(Q_i)_{22}-(Q_i)_{42}(Q_i)_{21}}{(\hat{Q}_i)_{11}} = \frac{-(\hat{Q}_i)_{51}}{(\hat{Q}_i)_{11}}, \\
R_i^{\mathrm{SP}-} &= (Q_i)_{31}T_i^{\mathrm{SP}-} + (Q_i)_{32}T_i^{\mathrm{SS}-} = \frac{-(Q_i)_{31}(Q_i)_{22}+(Q_i)_{32}(Q_i)_{11}}{(\hat{Q}_i)_{11}} = \frac{(\hat{Q}_i)_{21}}{(\hat{Q}_i)_{11}}, \\
R_i^{\mathrm{SS}-} &= (Q_i)_{41}T_i^{\mathrm{SP}-} + (Q_i)_{42}T_i^{\mathrm{SS}-} = \frac{-(Q_i)_{41}(Q_i)_{22}+(Q_i)_{42}(Q_i)_{11}}{(\hat{Q}_i)_{11}} = \frac{(\hat{Q}_i)_{31}}{(\hat{Q}_i)_{11}}.
\end{aligned}
\tag{3.71}
$$

To summarize the above, for the upward incidence of the P or SV wave, the reflection and transmission at interface i can be represented by the **reflection/transmission matrix**

$$
\boldsymbol{R}_i^- = \begin{pmatrix} R_i^{\mathrm{PP}-} & R_i^{\mathrm{PS}-} \\ R_i^{\mathrm{SP}-} & R_i^{\mathrm{SS}-} \end{pmatrix}, \quad \boldsymbol{T}_i^- = \begin{pmatrix} T_i^{\mathrm{PP}-} & T_i^{\mathrm{PS}-} \\ T_i^{\mathrm{SP}-} & T_i^{\mathrm{SS}-} \end{pmatrix}
\tag{3.72}
$$

which are 2 by 2 matrices. Their elements are given in (3.70) and (3.71). While the subdeterminants of $\mathbf{Q}_i$ in the elements can be analytically obtained, such as for (3.68), the explicit representations of the elements are given by Červený et al. [20].

In addition, (3.63), (3.66), (3.68), (3.69), and (3.72) are summarized in a matrix form as

$$
\begin{pmatrix} \mathbf{0} \\ \boldsymbol{T}_i^+ \end{pmatrix} = \mathbf{Q}_i \begin{pmatrix} \boldsymbol{R}_i^+ \\ \mathbf{I} \end{pmatrix}, \quad \begin{pmatrix} \mathbf{I} \\ \boldsymbol{R}_i^- \end{pmatrix} = \mathbf{Q}_i \begin{pmatrix} \boldsymbol{T}_i^- \\ \mathbf{0} \end{pmatrix}, \quad \mathbf{0} = \begin{pmatrix} 0 & 0 \\ 0 & 0 \end{pmatrix}, \quad \mathbf{I} = \begin{pmatrix} 1 & 0 \\ 0 & 1 \end{pmatrix}.
\tag{3.73}
$$

The first and second equations in (3.73) correspond to the left and right diagrams in Fig. 3.2, and to (3.58) and (3.61) for the SH wave.

As a special case, we consider the reflection/transmission matrices at the ground surface ($i = 0$). However, since there is no ground motion in the upper air, only upward incidence from layer 1 exists, and no transmitted wave exists even in the case of upward incidence (Fig. 3.2 (upper)). First, in the case of the SH wave, we substitute the stress-free condition (3.22) and the potential vector for upward inci-

---

[18] These agree with Eq. (8) of Červený [17] if $W \to (\hat{Q}_i)$.

dence $\mathbf{\Phi}_{l1}(0) = \left(1, R_0^-\right)^{\mathrm{T}}$ into (3.15) at the ground surface, and obtain

$$\begin{pmatrix} \tilde{v}_{l1}(0) \\ 0 \end{pmatrix} = \begin{pmatrix} -1 & -1 \\ -\mathrm{i}\mu_1 \nu_{\beta 1} & +\mathrm{i}\mu_1 \nu_{\beta 1} \end{pmatrix} \begin{pmatrix} 1 \\ R_0^- \end{pmatrix} . \tag{3.74}$$

From the second equation of (3.74) we have

$$R_0^- = \frac{+\mathrm{i}\mu_1 \nu_{\beta 1}}{+\mathrm{i}\mu_1 \nu_{\beta 1}} = 1 . \tag{3.75}$$

Equation (3.75) indicates that the intensities of the reflected and incident waves are equal, so that **total reflection** occurs. In addition, the first equation of (3.74) yields

$$\tilde{v}_{l1}(0) = -1 - R_0^- = -2 . \tag{3.76}$$

The ground motion due to the incident wave of unit intensity can be regarded as an **amplification factor** [64, 65], which is denoted as W_0. Hence,

$$W_0 = -2 . \tag{3.77}$$

For the SH wave, W_0 is scalar, but when this is collectively referred to along with that for the P and SV waves, this is also called the **amplification factor matrix**, which has only the scalar quantity as a single element.

Secondly, in the case of the P and SV waves, we substitute the stress-free condition (3.36) and the potential vector for upward P wave incidence $\mathbf{\Phi}_{l1}(0) = \left(1, 0, R_0^{\mathrm{PP}-}, R_0^{\mathrm{PS}-}\right)^{\mathrm{T}}$ or the potential vector for upward SV wave incidence $\mathbf{\Phi}_{l1}(0) = \left(0, 1, R_0^{\mathrm{SP}-}, R_0^{\mathrm{SS}-}\right)^{\mathrm{T}}$ into (3.30) at the ground surface, and obtain

$$\begin{pmatrix} \tilde{u}_{l1}^{\mathrm{P}}(0) \\ \tilde{w}_{l1}^{\mathrm{P}}(0) \\ 0 \\ 0 \end{pmatrix} = \mathbf{T}_1 \begin{pmatrix} 1 \\ 0 \\ R_0^{\mathrm{PP}-} \\ R_0^{\mathrm{PS}-} \end{pmatrix} , \quad \begin{pmatrix} \tilde{u}_{l1}^{\mathrm{S}}(0) \\ \tilde{w}_{l1}^{\mathrm{S}}(0) \\ 0 \\ 0 \end{pmatrix} = \mathbf{T}_1 \begin{pmatrix} 0 \\ 1 \\ R_0^{\mathrm{SP}-} \\ R_0^{\mathrm{SS}-} \end{pmatrix} . \tag{3.78}$$

For confirmation, simultaneous equations are made from the third and fourth equations of (3.78), and (3.30) is substituted into them. Then,

$$\begin{aligned} \begin{pmatrix} \mu_1 l_1 & 2\mu_1 k \nu_{\beta 1} \\ 2\mu_1 k \nu_{\alpha 1} & -\mu_1 l_1 \end{pmatrix} \begin{pmatrix} R_0^{\mathrm{PP}-} \\ R_0^{\mathrm{PS}-} \end{pmatrix} &= \begin{pmatrix} -\mu_1 l_1 \\ 2\mu_1 k \nu_{\alpha 1} \end{pmatrix} , \\ \begin{pmatrix} \mu_1 l_1 & 2\mu_1 k \nu_{\beta 1} \\ 2\mu_1 k \nu_{\alpha 1} & -\mu_1 l_1 \end{pmatrix} \begin{pmatrix} R_0^{\mathrm{SP}-} \\ R_0^{\mathrm{SS}-} \end{pmatrix} &= \begin{pmatrix} 2\mu_1 k \nu_{\beta 1} \\ \mu_1 l_1 \end{pmatrix} \end{aligned} \tag{3.79}$$

is obtained.[19]

We return to (3.78), and denote the elements of $\mathbf{T}_i$ there as $(\mathrm{T}_i)_{kl}$, using Roman font to distinguish them from the transmission coefficients. We sum (the third equation) × $(\mathrm{T}_1)_{44}$ and (the fourth equation) × $\{-(\mathrm{T}_1)_{34}\}$. We also sum (the third equation) × $(\mathrm{T}_1)_{43}$ and (the fourth equation) × $\{-(\mathrm{T}_1)_{33}\}$. We then obtain the reflection coefficients

$$
\begin{aligned}
R_0^{\mathrm{PP}-} &= \frac{-(\mathrm{T}_1)_{31}(\mathrm{T}_1)_{44} - (\mathrm{T}_1)_{41}\{-(\mathrm{T}_1)_{34}\}}{(\mathrm{T}_1)_{33}(\mathrm{T}_1)_{44} + (\mathrm{T}_1)_{43}\{-(\mathrm{T}_1)_{34}\}} = \frac{-(\mathrm{T}_1)\left|^{34}_{14}\right.}{(\mathrm{T}_1)\left|^{34}_{34}\right.} = \frac{-(\hat{\mathrm{T}}_1)_{63}}{(\hat{\mathrm{T}}_1)_{66}}\,,\\
R_0^{\mathrm{PS}-} &= \frac{-(\mathrm{T}_1)_{31}(\mathrm{T}_1)_{43} - (\mathrm{T}_1)_{41}\{-(\mathrm{T}_1)_{33}\}}{(\mathrm{T}_1)_{34}(\mathrm{T}_1)_{43} + (\mathrm{T}_1)_{44}\{-(\mathrm{T}_1)_{33}\}} = \frac{-(\mathrm{T}_1)\left|^{34}_{13}\right.}{-(\mathrm{T}_1)\left|^{34}_{34}\right.} = \frac{(\hat{\mathrm{T}}_1)_{62}}{(\hat{\mathrm{T}}_1)_{66}}\,,\\
R_0^{\mathrm{SP}-} &= \frac{-(\mathrm{T}_1)_{32}(\mathrm{T}_1)_{44} - (\mathrm{T}_1)_{42}\{-(\mathrm{T}_1)_{34}\}}{(\mathrm{T}_1)_{33}(\mathrm{T}_1)_{44} + (\mathrm{T}_1)_{43}\{-(\mathrm{T}_1)_{34}\}} = \frac{-(\mathrm{T}_1)\left|^{34}_{24}\right.}{(\mathrm{T}_1)\left|^{34}_{34}\right.} = \frac{-(\hat{\mathrm{T}}_1)_{65}}{(\hat{\mathrm{T}}_1)_{66}}\,,\\
R_0^{\mathrm{SS}-} &= \frac{-(\mathrm{T}_1)_{32}(\mathrm{T}_1)_{43} - (\mathrm{T}_1)_{42}\{-(\mathrm{T}_1)_{33}\}}{(\mathrm{T}_1)_{34}(\mathrm{T}_1)_{43} + (\mathrm{T}_1)_{44}\{-(\mathrm{T}_1)_{33}\}} = \frac{-(\mathrm{T}_1)\left|^{34}_{23}\right.}{-(\mathrm{T}_1)\left|^{34}_{34}\right.} = \frac{(\hat{\mathrm{T}}_1)_{64}}{(\hat{\mathrm{T}}_1)_{66}}\,,
\end{aligned}
\tag{3.80}
$$

where $(\hat{\mathrm{T}}_1)_{kl}$ in Roman font indicates the elements of the subdeterminants of $\mathbf{T}_1$, and their explicit representations are given by Fuchs [34] and Kind [68] (Sect. 3.1.3). These reflection coefficients constitute the reflection matrix

$$
\boldsymbol{R}_0^- = \begin{pmatrix} R_0^{\mathrm{PP}-} & R_0^{\mathrm{PS}-} \\ R_0^{\mathrm{SP}-} & R_0^{\mathrm{SS}-} \end{pmatrix}. \tag{3.81}
$$

In addition, by solving the first and second equations of (3.78), we obtain

$$
\begin{aligned}
\tilde{u}^{\mathrm{P}}_{l1}(0) &= (\mathrm{T}_1)_{11} + (\mathrm{T}_1)_{13}R_0^{\mathrm{PP}-} + (\mathrm{T}_1)_{14}R_0^{\mathrm{PS}-} = \frac{1}{(\hat{\mathrm{T}}_1)_{66}}\begin{vmatrix} (\mathrm{T}_1)_{11} & (\mathrm{T}_1)_{13} & (\mathrm{T}_1)_{14} \\ (\mathrm{T}_1)_{31} & (\mathrm{T}_1)_{33} & (\mathrm{T}_1)_{34} \\ (\mathrm{T}_1)_{41} & (\mathrm{T}_1)_{43} & (\mathrm{T}_1)_{44} \end{vmatrix} = \frac{(\tilde{\mathrm{T}}_1)_{22}}{(\hat{\mathrm{T}}_1)_{66}}\,,\\
\tilde{w}^{\mathrm{P}}_{l1}(0) &= (\mathrm{T}_1)_{21} + (\mathrm{T}_1)_{23}R_0^{\mathrm{PP}-} + (\mathrm{T}_1)_{24}R_0^{\mathrm{PS}-} = \frac{1}{(\hat{\mathrm{T}}_1)_{66}}\begin{vmatrix} (\mathrm{T}_1)_{21} & (\mathrm{T}_1)_{23} & (\mathrm{T}_1)_{24} \\ (\mathrm{T}_1)_{31} & (\mathrm{T}_1)_{33} & (\mathrm{T}_1)_{34} \\ (\mathrm{T}_1)_{41} & (\mathrm{T}_1)_{43} & (\mathrm{T}_1)_{44} \end{vmatrix} = \frac{(\tilde{\mathrm{T}}_1)_{12}}{(\hat{\mathrm{T}}_1)_{66}}\,,\\
\tilde{u}^{\mathrm{S}}_{l1}(0) &= (\mathrm{T}_1)_{12} + (\mathrm{T}_1)_{13}R_0^{\mathrm{SP}-} + (\mathrm{T}_1)_{14}R_0^{\mathrm{SS}-} = \frac{1}{(\hat{\mathrm{T}}_1)_{66}}\begin{vmatrix} (\mathrm{T}_1)_{12} & (\mathrm{T}_1)_{13} & (\mathrm{T}_1)_{14} \\ (\mathrm{T}_1)_{32} & (\mathrm{T}_1)_{33} & (\mathrm{T}_1)_{34} \\ (\mathrm{T}_1)_{42} & (\mathrm{T}_1)_{43} & (\mathrm{T}_1)_{44} \end{vmatrix} = \frac{(\tilde{\mathrm{T}}_1)_{21}}{(\hat{\mathrm{T}}_1)_{66}}\,,\\
\tilde{w}^{\mathrm{S}}_{l1}(0) &= (\mathrm{T}_1)_{22} + (\mathrm{T}_1)_{23}R_0^{\mathrm{PP}-} + (\mathrm{T}_1)_{24}R_0^{\mathrm{PS}-} = \frac{1}{(\hat{\mathrm{T}}_1)_{66}}\begin{vmatrix} (\mathrm{T}_1)_{22} & (\mathrm{T}_1)_{23} & (\mathrm{T}_1)_{24} \\ (\mathrm{T}_1)_{32} & (\mathrm{T}_1)_{33} & (\mathrm{T}_1)_{34} \\ (\mathrm{T}_1)_{42} & (\mathrm{T}_1)_{43} & (\mathrm{T}_1)_{44} \end{vmatrix} = \frac{(\tilde{\mathrm{T}}_1)_{11}}{(\hat{\mathrm{T}}_1)_{66}}\,.
\end{aligned}
\tag{3.82}
$$

Using the above, we can calculate the amplification factor matrix

[19] This agrees with Eq. (7.1.6) of Satô [104] if $\xi \to k, c_P \to \alpha_1, c_S \to \beta_1, A'/A = R_0^{\mathrm{PP}-}, B'/A = R_0^{\mathrm{PS}-}, A'/B = R_0^{\mathrm{SP}-}, B'/B = R_0^{\mathrm{SS}-}$, and the both sides are multiplied by μ_1.

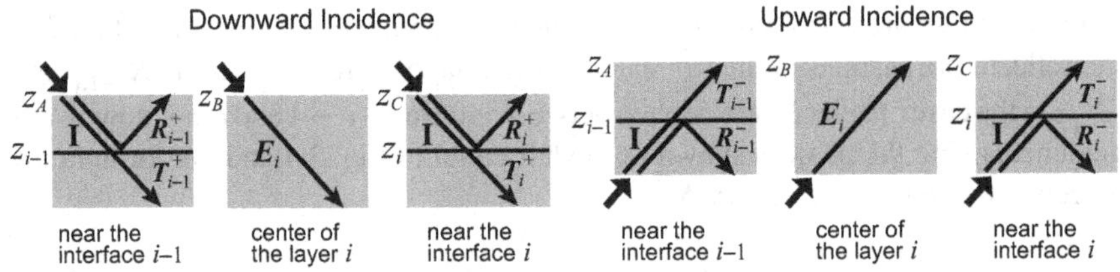

Fig. 3.3 Propagation effects of the three subregions in layer i on the downward incident wave (left) and the upward incident wave (right). Modified from Koketsu [72] with permission of Kindai Kagaku

$$\mathbf{W}_0 = \begin{pmatrix} W_0^{\mathrm{RP}} & W_0^{\mathrm{RS}} \\ W_0^{\mathrm{ZP}} & W_0^{\mathrm{ZS}} \end{pmatrix} = \begin{pmatrix} \tilde{u}_{l1}^{\mathrm{P}}(0) & \tilde{u}_{l1}^{\mathrm{S}}(0) \\ \tilde{w}_{l1}^{\mathrm{P}}(0) & \tilde{w}_{l1}^{\mathrm{S}}(0) \end{pmatrix} = \frac{1}{(\hat{\mathrm{T}}_1)_{66}} \begin{pmatrix} (\tilde{\mathrm{T}}_1)_{22} & (\tilde{\mathrm{T}}_1)_{21} \\ (\tilde{\mathrm{T}}_1)_{12} & (\tilde{\mathrm{T}}_1)_{11} \end{pmatrix}. \tag{3.83}$$

$(\tilde{\mathrm{T}}_1)_{kl}$ is the determinant of a 3 by 3 submatrix, which is made from $\mathbf{T}_1$ by excluding the kth row and the lth column [17]. This is the same as defined for (3.65).

3.1.6 Reflection/Transmission Matrix II

The **reflection/transmission matrix** defined in Sect. 3.1.5 is extended from a single interface to a layer including two upper and lower interfaces. If this layer is layer i, the entire layer is divided into three subregions from the top, namely the "near interface $i-1$", "center of layer i", and "near interface i" subregions (Fig. 3.3). Each subregion has its own characteristic propagation effect on the incident wave. In the "near interface i" subregion, the incident wave is guided to the interface, and reflection and transmission are caused according to the matrix obtained in Sect. 3.1.5. As the "center of layer i" subregion is homogeneous (Sect. 3.1.1), only forward propagation occurs there. This propagation causes only a phase delay, therefore the transmission matrix is the phase matrix[20]

$$E_i = e^{-\mathrm{i}\nu_{\beta i} d_i} \text{ (SH wave)}, \quad \boldsymbol{E}_i = \begin{pmatrix} e^{-\mathrm{i}\nu_{\alpha i} d_i} & 0 \\ 0 & e^{-\mathrm{i}\nu_{\beta i} d_i} \end{pmatrix} \text{ (P and SV waves)}, \tag{3.84}$$

and the reflection matrix is zero. Compared to $\mathbf{E}_i$ in (3.16) and (3.32) of the propagator matrix, this phase matrix does not include $e^{+\mathrm{i}\nu_{\alpha i} d_i}$ and $e^{+\mathrm{i}\nu_{\beta i} d_i}$, which is an advantage of the reflection/transmission matrix for the **overflow** problem (Sect. 3.1.3).

The propagation of the ground motion in layer i is then a combination of the propagation effects shown in Fig. 3.3. As the layer is sandwiched between the upper and lower interfaces, **reverberation** that repeat reflections between the interfaces

[20] If the bold face is used according to the notation in this book, it is indistinguishable from $\mathbf{E}_i$ in (3.16) and (3.32). Therefore, we use the italic-bold face. In the case of the SH wave, it is also called the phase "matrix" with a single element.

occur, and these are also combinations of the propagation effects shown in Fig. 3.4. The reflection/transmission matrix extended to a layer is represented as $\boldsymbol{R}^{\pm}_{i-1,i}$ and $\boldsymbol{T}^{\pm}_{i-1,i}$ in the case of layer i, sandwiched between interfaces $i-1$ and i. For downward incidence, using the process shown in the left panel of Fig. 3.4 and the formula for the geometric series of a matrix $\mathbf{A}$[21]

$$(\mathbf{I}-\mathbf{A})^{-1}=\mathbf{I}+\mathbf{A}+\mathbf{A}^2+\cdots \tag{3.85}$$

where $\mathbf{I}$ is the unit matrix defined in (3.73), we obtain[22]

$$\begin{aligned}
\boldsymbol{R}^{+}_{i-1,i} &= \boldsymbol{R}^{+}_{i-1}+\boldsymbol{T}^{-}_{i-1}\boldsymbol{E}_i\boldsymbol{R}^{+}_{i}\boldsymbol{E}_i\boldsymbol{T}^{+}_{i-1}+\boldsymbol{T}^{-}_{i-1}\boldsymbol{E}_i\boldsymbol{R}^{+}_{i}\boldsymbol{E}_i\boldsymbol{R}^{-}_{i-1}\boldsymbol{E}_i\boldsymbol{R}^{+}_{i}\boldsymbol{E}_i\boldsymbol{T}^{+}_{i-1}+\cdots\\
&= \boldsymbol{R}^{+}_{i-1}+\boldsymbol{T}^{-}_{i-1}\boldsymbol{E}_i\boldsymbol{R}^{+}_{i}\boldsymbol{E}_i\left(\mathbf{I}+\boldsymbol{R}^{-}_{i-1}\boldsymbol{E}_i\boldsymbol{R}^{+}_{i}\boldsymbol{E}_i+\cdots\right)\boldsymbol{T}^{+}_{i-1}\\
&= \boldsymbol{R}^{+}_{i-1}+\boldsymbol{T}^{-}_{i-1}\boldsymbol{E}_i\boldsymbol{R}^{+}_{i}\boldsymbol{E}_i\left(\mathbf{I}-\boldsymbol{R}^{-}_{i-1}\boldsymbol{E}_i\boldsymbol{R}^{+}_{i}\boldsymbol{E}_i\right)^{-1}\boldsymbol{T}^{+}_{i-1}\\
\boldsymbol{T}^{+}_{i-1,i} &= \boldsymbol{T}^{+}_{i}\boldsymbol{E}_i\boldsymbol{T}^{+}_{i-1}+\boldsymbol{T}^{+}_{i}\boldsymbol{E}_i\boldsymbol{R}^{-}_{i-1}\boldsymbol{E}_i\boldsymbol{R}^{+}_{i}\boldsymbol{E}_i\boldsymbol{T}^{+}_{i-1}+\cdots\\
&= \boldsymbol{T}^{+}_{i}\boldsymbol{E}_i\left(\mathbf{I}+\boldsymbol{R}^{-}_{i-1}\boldsymbol{E}_i\boldsymbol{R}^{+}_{i}\boldsymbol{E}_i+\cdots\right)\boldsymbol{T}^{+}_{i-1}=\boldsymbol{T}^{+}_{i}\boldsymbol{E}_i\left(\mathbf{I}-\boldsymbol{R}^{-}_{i-1}\boldsymbol{E}_i\boldsymbol{R}^{+}_{i}\boldsymbol{E}_i\right)^{-1}\boldsymbol{T}^{+}_{i-1}\,.
\end{aligned} \tag{3.86}$$

Similarly, for upward incidence, using the process shown in the right panel of Fig. 3.4 and the formula in (3.85), we obtain

$$\begin{aligned}
\boldsymbol{R}^{-}_{i-1,i} &= \boldsymbol{R}^{-}_{i}+\boldsymbol{T}^{+}_{i}\boldsymbol{E}_i\boldsymbol{R}^{-}_{i-1}\boldsymbol{E}_i\boldsymbol{T}^{-}_{i}+\boldsymbol{T}^{+}_{i}\boldsymbol{E}_i\boldsymbol{R}^{-}_{i-1}\boldsymbol{E}_i\boldsymbol{R}^{+}_{i}\boldsymbol{E}_i\boldsymbol{R}^{-}_{i-1}\boldsymbol{E}_i\boldsymbol{T}^{-}_{i}+\cdots\\
&= \boldsymbol{R}^{-}_{i}+\boldsymbol{T}^{+}_{i}\boldsymbol{E}_i\boldsymbol{R}^{-}_{i-1}\left(\mathbf{I}+\boldsymbol{E}_i\boldsymbol{R}^{+}_{i}\boldsymbol{E}_i\boldsymbol{R}^{-}_{i-1}+\cdots\right)\boldsymbol{E}_i\boldsymbol{T}^{-}_{i}\\
&= \boldsymbol{R}^{-}_{i}+\boldsymbol{T}^{+}_{i}\boldsymbol{E}_i\boldsymbol{R}^{-}_{i-1}\left(\mathbf{I}-\boldsymbol{E}_i\boldsymbol{R}^{+}_{i}\boldsymbol{E}_i\boldsymbol{R}^{-}_{i-1}\right)^{-1}\boldsymbol{E}_i\boldsymbol{T}^{-}_{i}\\
\boldsymbol{T}^{-}_{i-1,i} &= \boldsymbol{T}^{-}_{i-1}\boldsymbol{E}_i\boldsymbol{T}^{-}_{i}+\boldsymbol{T}^{-}_{i-1}\boldsymbol{E}_i\boldsymbol{R}^{+}_{i}\boldsymbol{E}_i\boldsymbol{R}^{-}_{i-1}\boldsymbol{E}_i\boldsymbol{T}^{-}_{i}+\cdots\\
&= \boldsymbol{T}^{-}_{i-1}\left(\mathbf{I}+\boldsymbol{E}_i\boldsymbol{R}^{+}_{i}\boldsymbol{E}_i\boldsymbol{R}^{-}_{i-1}+\cdots\right)\boldsymbol{E}_i\boldsymbol{T}^{-}_{i}=\boldsymbol{T}^{-}_{i-1}\left(\mathbf{I}-\boldsymbol{E}_i\boldsymbol{R}^{+}_{i}\boldsymbol{E}_i\boldsymbol{R}^{-}_{i-1}\right)^{-1}\boldsymbol{E}_i\boldsymbol{T}^{-}_{i}\,.
\end{aligned} \tag{3.87}$$

If the upper ends $z_{i-1}-$, $z_{i-1}+$, and z_i- of the three subregions of layer i are denoted as z_A, z_B, and z_C, respectively (Fig. 3.3), we can write $\boldsymbol{R}^{\pm}_{i-1,i+1}=\boldsymbol{R}^{\pm}_{AC}$ and $\boldsymbol{T}^{\pm}_{i-1,i+1}=\boldsymbol{T}^{\pm}_{AC}$. From the definitions in Sect. 3.1.5 and the illustrations in Fig. 3.4, we have

$$\begin{aligned}
&\boldsymbol{R}^{+}_{AB}=\boldsymbol{R}^{+}_{i-1}\,,\quad \boldsymbol{T}^{+}_{AB}=\boldsymbol{T}^{+}_{i-1}\,,\quad \boldsymbol{R}^{-}_{AB}=\boldsymbol{R}^{-}_{i-1}\,,\quad \boldsymbol{T}^{-}_{AB}=\boldsymbol{T}^{-}_{i-1}\,,\\
&\boldsymbol{R}^{+}_{BC}=\boldsymbol{E}_i\boldsymbol{R}^{+}_{i}\boldsymbol{E}_i\,,\quad \boldsymbol{T}^{+}_{BC}=\boldsymbol{T}^{+}_{i}\boldsymbol{E}_i\,,\quad \boldsymbol{R}^{-}_{BC}=\boldsymbol{R}^{-}_{i}\,,\quad \boldsymbol{T}^{-}_{BC}=\boldsymbol{E}_i\boldsymbol{T}^{-}_{i}\,.
\end{aligned} \tag{3.88}$$

Substituting these into (3.86) and (3.87) yields the **addition rule**[23] [64, 65]

---

[21] In order for this formula to hold, the absolute value of the eigenvalue of $\mathbf{A}$ must be less than 1.

[22] These agree with Eqs. (14.1.14) and (14.1.15) of Kennett [65] if $A \to i-1$, $B \to i$, $D \to +$, and $U \to -$]. However, there are some typographical errors in the original book.

[23] These agree with Eqs. (6.3) and (6.4) of Kennett [64] if $D \to +$ and $U \to -$.

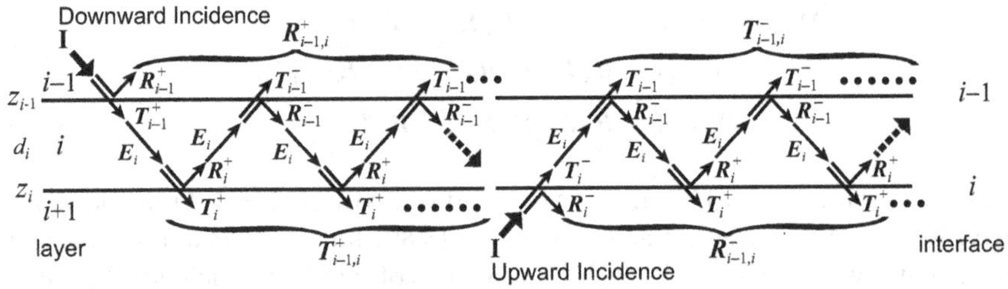

Fig. 3.4 Reflection/transmission matrices for downward incidence (left) and upward incidence (right), extended from the single interface i (Fig. 3.2) to layer i between interfaces $i-1$ and i (modified from Koketsu [72] with permission of Kindai Kagaku)

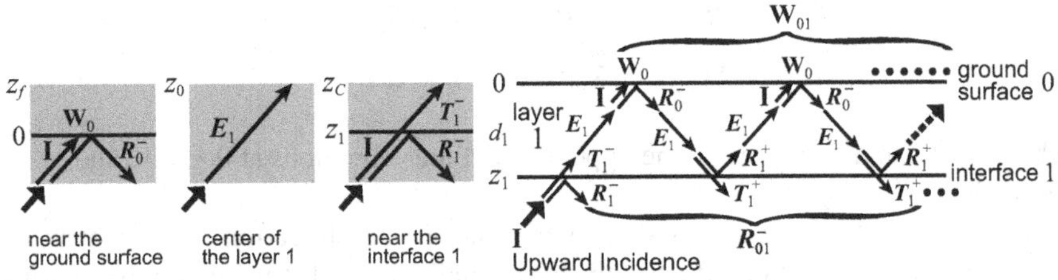

Fig. 3.5 Left: Propagation effects of the three subregions in layer 1 on the upward incident wave. Right: The reflection matrix and amplification factor matrix of layer 1, sandwiched between the ground surface and interface 1. Modified from Koketsu [72] with permission of Kindai Kagaku

$$\begin{aligned}
\boldsymbol{R}^+_{AC} &= \boldsymbol{R}^+_{AB} + \boldsymbol{T}^-_{AB}\boldsymbol{R}^+_{BC}\left(\mathbf{I} - \boldsymbol{R}^-_{AB}\boldsymbol{R}^+_{BC}\right)^{-1}\boldsymbol{T}^+_{AB}\,,\\
\boldsymbol{T}^+_{AC} &= \boldsymbol{T}^+_{BC}\left(\mathbf{I} - \boldsymbol{R}^-_{AB}\boldsymbol{R}^+_{BC}\right)^{-1}\boldsymbol{T}^+_{AB}\,,\\
\boldsymbol{R}^-_{AC} &= \boldsymbol{R}^-_{BC} + \boldsymbol{T}^+_{BC}\boldsymbol{R}^-_{AB}\left(\mathbf{I} - \boldsymbol{R}^+_{BC}\boldsymbol{R}^-_{AB}\right)^{-1}\boldsymbol{T}^-_{BC}\,,\\
\boldsymbol{T}^-_{AC} &= \boldsymbol{T}^-_{AB}\left(\mathbf{I} - \boldsymbol{R}^+_{BC}\boldsymbol{R}^-_{AB}\right)^{-1}\boldsymbol{T}^-_{BC}\,.
\end{aligned} \tag{3.89}$$

Here, it is shown that the addition rules hold for a series of subregions, but Kennett [64] showed that they hold for a series of layers or layer groups.

Similar to Sect. 3.1.5, as a special case we consider the addition rules for layer 1, where the upper interface is the ground surface. As there is no ground motion in the upper air, only upward incidence exists. Even for the upward incidence, the transmission matrix $\boldsymbol{T}^-_0$ does not exist, but the amplification factor matrix $\mathbf{W}_0$ (3.77) or (3.83) does exist. As in Fig. 3.3, layer 1 is divided into the three subregions, which are "near the ground surface", "center of layer 1", and "near interface 1". The propagation effects through these subregions are shown in the left panel of Fig. 3.5. The reverberations in layer 1 are represented in the right panel of Fig. 3.5, based on these propagation effects. R^-_{01} in this panel is identical to $R^1_{i-1,i}$ in the right panel of Fig. 3.4, if $i = 1$. For the amplification factor matrix, using the right panel of Fig. 3.5 and the formula (3.85), we obtain

$$\begin{aligned}\mathbf{W}_{01} &= \mathbf{W}_0 \boldsymbol{E}_1 \boldsymbol{T}_1^- + \mathbf{W}_0^- \boldsymbol{E}_1 \boldsymbol{R}_1^+ \boldsymbol{E}_1 \boldsymbol{R}_0^- \boldsymbol{E}_1 \boldsymbol{T}_1^- + \cdots \\ &= \mathbf{W}_0 \left(\mathbf{I} + \boldsymbol{E}_1 \boldsymbol{R}_1^+ \boldsymbol{E}_1 \boldsymbol{R}_0^- + \cdots\right) \boldsymbol{E}_1 \boldsymbol{T}_1^- \\ &= \mathbf{W}_0 \left(\mathbf{I} - \boldsymbol{E}_1 \boldsymbol{R}_1^+ \boldsymbol{E}_1 \boldsymbol{R}_0^-\right)^{-1} \boldsymbol{E}_1 \boldsymbol{T}_1^- .\end{aligned} \tag{3.90}$$

This agrees with $\boldsymbol{T}_{01}^-$ in (3.87) if $\boldsymbol{T}_0^-$ is replaced with $\mathbf{W}_0$ and $i = 1$. From the above agreements, we find that the addition rule in (3.89) for the reflection matrix and upward incidence holds as it is, even when a series of subregions include the ground surface. The addition rule for the transmission matrix and upward incidence also holds if the transmission matrix is replaced with the amplification factor matrix. We can therefore write these addition rules as

$$\begin{aligned}\boldsymbol{R}_{fC}^- &= \boldsymbol{R}_{0C}^- + \boldsymbol{T}_{0C}^+ \boldsymbol{R}_0^- \left(\mathbf{I} - \boldsymbol{R}_{0C}^+ \boldsymbol{R}_0^-\right)^{-1} \boldsymbol{T}_{0C}^- , \\ \mathbf{W}_{fC} &= \mathbf{W}_0 \left(\mathbf{I} - \boldsymbol{R}_{0C}^+ \boldsymbol{R}_0^-\right)^{-1} \boldsymbol{T}_{0C}^- ,\end{aligned} \tag{3.91}$$

where z_f, z_0, and z_C represent the upper ends of the subregions at $0-$, $0+$, and z_1-, respectively.[24]

Using the reflection/transmission matrix and the amplification factor matrix in the above, the propagation effect of the horizontally layered structure on the ground motion from the point source can be calculated, and the ground motion vector[25] at the ground surface

$$w_{l0} = \tilde{v}_{l1}(0) \text{ (SH wave)}, \quad \mathbf{w}_{l0} = \begin{pmatrix} \tilde{u}_{l1}(0) \\ \tilde{w}_{l1}(0) \end{pmatrix} \text{ (P and SV waves)} \tag{3.92}$$

can be obtained as shown in Fig. 3.6. Within this, the downgoing and upgoing wave vectors $\boldsymbol{\Sigma}_l^{\pm}$ are obtained from the source potentials modified in the form of (3.5), which are $\phi_{lS}^{\pm}$, $\psi_{lS}^{\pm}$, and $\chi_{lS}^{\pm}$, as

$$\Sigma_l^{\pm} = \chi_{lS}^{\pm} \text{ (SH wave)}, \quad \boldsymbol{\Sigma}_l^{\pm} = \begin{pmatrix} \phi_{lS}^{\pm} \\ \psi_{lS}^{\pm} \end{pmatrix} \text{ (P and SV waves)} . \tag{3.93}$$

In addition, the effect of propagation in the region between the point source and the halfspace is represented by the reflection matrix $\boldsymbol{R}_{s,n-1}^+$, which is obtained from the reflection/transmission matrix of the interface s, the phase matrix of the s+1 layer, and the addition rules in (3.88) and (3.89), or (3.86). The effect of propagation in the fs region between the ground surface and the point source is represented by the reflection matrix $\boldsymbol{R}_{fs}^-$ and the amplification factor matrix $\mathbf{W}_{fs}$. These are calculated from the amplification factor matrix at the ground surface, the reflection and transmission matrices at the interfaces 1 to s, the phase matrices in layers 1 to s,

[24] These agree with Eqs. (14.3.5) and (14.3.6) of Kennett [65] if $D \to +$, $U \to -$, and $F \to 0$.

[25] This is a scalar for the SH wave, but is also collectively called a "vector" with a single element, similar to the reflection/transmission and amplification factor matrices.

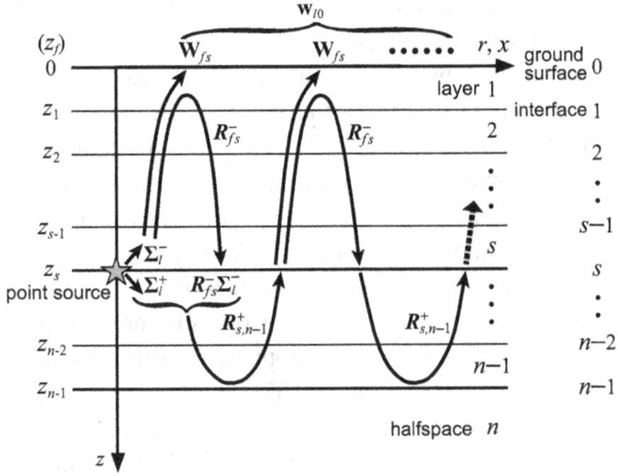

Fig. 3.6 Schematic diagram of the process in which the ground motion is generated by the reflection/transmission matrices and the amplification factor matrix acting on the downgoing and upgoing waves radiated from the point source (modified from Koketsu [72] with permission of Kindai Kagaku)

and the addition rules in (3.88) and (3.89), or (3.87). As reverberations occur in the entire f, n-1 region, including $\boldsymbol{\Sigma}_l^{\pm}$, from the illustration in Fig. 3.6 we obtain

$$\mathbf{w}_{l0} = \mathbf{W}_{fs}\Sigma_l^- + \mathbf{W}_{fs}\left(\mathbf{I} - \boldsymbol{R}^+_{s,n-1}\boldsymbol{R}^-_{fs}\right)^{-1}(\boldsymbol{\Sigma}_l^+ + \boldsymbol{R}^-_{fs}\boldsymbol{\Sigma}_l^-)\,. \tag{3.94}$$

3.1.7 Wavenumber Integration (Approximate)

In Sects. 3.1.2 – 3.1.6, when the Fourier transforms of the ground motions in the horizontally layered medium are represented in the form of the **cylindrical wave expansion** as

$$\begin{aligned}\bar{u}_{r0} &= \sum_{l=0}^{2}\Lambda_l(\theta)\int_0^{\infty}\tilde{u}_{l1}(0)\frac{dJ_l(kr)}{dr}\frac{dk}{ik}\,,\\ \bar{u}_{\theta 0} &= \sum_{l=0}^{2}\frac{d\Lambda_l(\theta)}{d\theta}\int_0^{\infty}\tilde{v}_{l1}(0)\frac{dJ_l(kr)}{dr}\,dk\,,\\ \bar{u}_{z0} &= \sum_{l=0}^{2}\Lambda_l(\theta)\int_0^{\infty}\tilde{w}_{l1}(0)J_l(kr)\,dk\,,\end{aligned} \tag{3.95}$$

we show that the integrands $\tilde{u}_{l1}(0)$, $\tilde{v}_{l1}(0)$, and $\tilde{w}_{l1}(0)$ can be calculated by applying the matrix operations shown in (3.25), (3.43), and (3.94) to the propagator matrices, the reflection/transmission matrices, etc. As a result, the Fourier transforms of the ground motions cannot be obtained without performing **wavenumber integra-**

tion, even if the integrands are obtained from the propagator matrices, the reflection/transmission matrices, etc.

While the integrands can be calculated by introducing the matrix operations, it is difficult to perform the wavenumber integration analytically. In the simplest velocity structure, which is a semi-infinite medium (a medium composed of only a halfspace with the upper end at the ground surface), the analytical integration was first studied by Lamb [83]. It was then formulated by Cagniard [15], refined by de Hoop [28], and Kawasaki et al. [61] obtained a final solution. As obtaining a solution for even the simplest case of a semi-infinite medium required 70 years, it is necessary to approximate the integrand or use a numerical method for a horizontally layered structure. In this subsection, approximations to the integrand are explained following the development of Kennett [65].

The formula in (3.85) is used in reverse for the integrand $\mathbf{w}_{l0}$ in (3.94), based on the reflection/transmission matrices of the horizontally layered structure, so that $\mathbf{w}_{l0}$ is decomposed as

$$\mathbf{w}_{l0} = \sum_I \mathbf{g}_{lI}\, e^{-\mathrm{i}\omega\tau_I}, \quad \mathbf{g}_{lI} = \mathbf{f}_{lI} \prod_j T_j \prod_k R_k\,,$$
$$\mathbf{f}_{lI} = \mathbf{W}_0 \boldsymbol{\Sigma}_l^{\pm}, \quad \omega\tau_I = \sum_m \nu_m d_m\,. \tag{3.96}$$

The expanded terms in (3.96) correspond to the **ray** (Sect. 1.2.5), which represent the trajectory of the ground motion, as shown by the example of a series of arrows in Fig. 3.4. However, these not only include body waves that can be calculated using the ray theory (Sect. 3.2.2) that is described later, but also non-body waves such as **head wave**, which are described in this subsection. Each of the terms was therefore referred to as a **generalized ray** [19], and the computation method for these was termed the **generalized ray theory** (hereinafter abbreviated as GRT) by Helmberger [45].

First, according to Cagniard [15], de Hoop [28], and Helmberger [45], we change the integration variable from the wavenumber k to the **slowness** $p \equiv k/\omega$,[26] so that the independent variables of the GRT are now ω and p. In (3.96), T_j and R_k represent the elements of the reflection and transmission matrices at the interface through which the generalized ray passes. For the SH wave, the single elements of the transmission/reflection matrices in (3.60) and (3.62) are rewritten as

$$R_i^+ = \frac{\mu_i q_{\beta i} - \mu_{i+1} q_{\beta,i+1}}{\mu_i q_{\beta i} + \mu_{i+1} q_{\beta,i+1}}\,, \quad T_i^+ = \frac{2\mu_i q_{\beta i}}{\mu_i q_{\beta i} + \mu_{i+1} q_{\beta,i+1}}\,, \quad R_i^- = \frac{\mu_{i+1} q_{\beta,i+1} - \mu_i q_{\beta i}}{\mu_{i+1} q_{\beta,i+1} + \mu_i q_{\beta i}}\,,$$
$$T_i^- = \frac{2\mu_{i+1} q_{\beta,i+1}}{\mu_{i+1} q_{\beta,i+1} + \mu_i q_{\beta i}}\,, \quad q_{\beta i} = \frac{\nu_{\beta i}}{\omega} = \begin{cases} \sqrt{\beta_i^{-2} - p^2}\,, & \beta_i^{-1} \ge p \\ -\mathrm{i}\sqrt{p^2 - \beta_i^{-2}}\,, & \beta_i^{-1} < p \end{cases} \tag{3.97}$$

after both their denominators and numerators are divided by ω. These are dependent only on p and independent of ω. It is assumed that $\mathbf{f}_{lI}$ is $\mathbf{W}_0 \Sigma_l^{\pm}$, representing the

[26] Because this is the reciprocal of a phase "velocity" $c \equiv \omega/k$, this is termed a "slowness".

influence of the point source and the ground surface [64], and is divided into the ω-dependent part $\bar{\mathsf{M}}(\omega)/\omega$ and the remaining part. As $e^{-\mathrm{i}\omega\tau_I}$ is the product of the elements of the phase matrices $\boldsymbol{E}_i$ in (3.84) for the layers through which the ground motion passes, $\omega\tau_I$ represents the sum of the phase delays caused by the z-direction propagation in the layers. Hence,

$$\tau_I = \sum_m \frac{\nu_m}{\omega} d_m = \sum_m q_m d_m \tag{3.98}$$

represents the sum of the time delays caused by the z-direction propagation in the layers, depending only on p but not on ω. Equations (3.97) and (3.98) can be valid for the P and SV waves [65], if q is replaced with

$$q_m = (\upsilon_m^{-2} - p^2)^{\frac{1}{2}} = \begin{cases} \sqrt{\upsilon_m^{-2} - p^2}, \ \upsilon_m^{-1} \geq p \\ -\mathrm{i}\sqrt{p^2 - \upsilon_m^{-2}}, \ \upsilon_m^{-1} < p \end{cases}, \quad \upsilon_m = \alpha_m \ \text{or} \ \beta_m, \tag{3.99}$$

which is defined similarly to ν_υ in the Sommerfeld integral (2.76). υ_m is the P wave velocity α_m or S wave velocity β_m of a layer in which the segment m of the generalized ray is located, depending on the wave type of the segment. As ν_υ in (2.76) represents the wavenumber in the depth direction, q_m of the segment m represents the slowness in the depth direction.

For example, if the influence of the point source is not dependent on the azimuth θ, only the term $l = 0$ is taken from the cylindrical harmonics (3.4).[27] We substitute $\mathbf{g}_{0I} = (\bar{\mathsf{M}}(\omega)/\omega)(-pF_I, -p\,G_I)^{\mathrm{T}}$ and (3.96) into (3.95) and perform the inverse Fourier transform, obtaining

$$\begin{aligned} u_{z0} &= \sum_I \frac{1}{2\pi}\int_{-\infty}^{\infty} e^{\mathrm{i}\omega t} d\omega \int_0^{\infty} \frac{\bar{\mathsf{M}}(\omega)(-p)G_I}{\omega} e^{-\mathrm{i}\omega\tau_I} J_0(kr)\, dk \\ &= \sum_I \frac{1}{2\pi}\int_{-\infty}^{\infty} e^{\mathrm{i}\omega t} d\omega\, \bar{\mathsf{M}}(\omega) \left\{ \int_0^{\infty} G_I\, e^{-\mathrm{i}\omega\tau_I} J_0(\omega p r)(-p) dp \right\} \\ &= \sum_I \mathsf{M}(t) * \frac{1}{2\pi}\int_{-\infty}^{\infty} e^{\mathrm{i}\omega t} d\omega \int_0^{\infty} G_I\, e^{-\mathrm{i}\omega\tau_I} J_0(\omega p r)(-p) dp = \mathsf{M}(t) * \sum_I u_{zI} \end{aligned} \tag{3.100}$$

for the z-component of the ground motion at the ground surface.[28] As shown in (3.100), two integrations, namely the inverse Fourier transform and the wavenumber integration or **slowness integration**, must be performed to obtain the time domain solution of the ground motion. In the GRT, the inverse Fourier transform and the slowness integration are performed in this order. Methods where the integrations are

[27] The problem of undersea explosion, which Helmberger [45] dealt with, corresponds to this.

[28] Compared with Eq. (16.1.22) in Kennett [65], the signs in $e^{-\mathrm{i}\omega\tau_I}$ and $(-p)$ are different because of the difference in the definition of the Fourier transform. For the same reason, $\mp$ in (3.101) is changed to $\pm$.

performed in this order are termed the **slowness method**, while methods such as described in Sect. 3.1.8, where the wavenumber integration and the inverse Fourier transform are performed in the reverse order, are referred to as the **spectral method** [64]. The inverse Fourier transform of u_{zI} is first performed according to the slowness method. As mentioned above, only $e^{-\mathrm{i}\omega\tau_I}$ and $J_0(\omega pr)$ are ω-dependent terms in u_{zI}. The term $e^{-\mathrm{i}\omega\tau_I}$ causes a time shift of τ_I, as shown in Table 4.3.

The latter term, the Bessel function $J_0(\omega pr)$, can be decomposed into the Hankel functions as in (3.8), so that u_{zI} in (3.100) yields

$$u_{zI} = \frac{1}{2\pi}\int_{-\infty}^{\infty} e^{\mathrm{i}\omega t} d\omega \int_0^{\infty} G_I\, e^{-\mathrm{i}\omega\tau_I} J_0(\omega pr)(-p)dp \tag{3.101}$$

$$= u_{zI-} + u_{zI+}\,, \quad u_{zI\mp} = \frac{1}{2\pi}\int_{-\infty}^{\infty} e^{\mathrm{i}\omega t} d\omega \int_0^{\infty} G_I\, e^{-\mathrm{i}\omega\tau_I} \frac{1}{2} H_0^{(1),(2)}(\omega pr)(-p)dp\,,$$

where the minus and plus signs in $\mp$ correspond to (1) and (2), respectively. No suitable formula has been found for the Fourier transform of the Hankel function, but for the **Laplace transform**

$$F_I(s) = \int_0^{\infty} f(t)e^{-st}dt\,, \tag{3.102}$$

we have[29]

$$H(t-a)(t^2-a^2)^{\nu} \Leftrightarrow F_I(s) = \frac{\Gamma(\nu+1)}{\sqrt{\pi}}\left(\frac{2a}{s}\right)^{\nu+1/2} K_{\nu+1/2}(as)\,. \tag{3.103}$$

$H(t)$ is a **step function** (Sect. 4.2.2), and $K_\nu(s)$ is the modified Bessel function. Between the Laplace transform $F_I(s)$ and the Fourier transform $F(\omega)$ for the same $f(t)$, the relation

$$F_I(s) = F(\omega)\,, \quad \omega = \frac{s}{\mathrm{i}} \tag{3.104}$$

holds [94]. From (3.103) and (3.104) with $\nu = -1/2$ and $a = pr$, $K_\nu(x) = \frac{\pi}{2}\mathrm{i}^{\nu+1}$ $H_\nu^{(1)}(\mathrm{i}x)$ (Eq. (11.17) of Arfken and Weber [5]), and (3.10),

$$\frac{H(t-pr)}{(t^2-p^2r^2)^{1/2}} \Leftrightarrow K_0(\mathrm{i}\omega pr) = \frac{\pi\mathrm{i}}{2}H_0^{(1)}(-\omega pr) = \frac{-\pi\mathrm{i}}{2}H_0^{(2)}(\omega pr)\,. \tag{3.105}$$

Let the third equation on the right-hand side of (3.105) be $F(\omega)$. Using the formula $\left\{H_0^{(2)}(x)\right\}^* = H_0^{(1)}(x)$ for the real x,[30] $\left\{H_0^{(2)}(x)\right\}^* = H_0^{(1)}(x)$, and (3.10), we obtain

[29]This formula is published on page 288 of Moriguchi et al. [91].

[30]This is written on page 659 of Arfken and Weber [5].

$$
\begin{aligned}
F(-\omega) &= \frac{-\pi \mathrm{i}}{2} H_0^{(2)}((-\omega)pr) = \frac{\pi \mathrm{i}}{2} H_0^{(1)}(\omega pr)\,, \\
\{F(\omega)\}^* &= \frac{\pi \mathrm{i}}{2} \left\{ H_0^{(2)}(\omega pr) \right\}^* = \frac{\pi \mathrm{i}}{2} H_0^{(1)}(\omega pr)\,,
\end{aligned}
\tag{3.106}
$$

which indicates that $H(t - pr)/(t^2 - p^2 r^2)^{1/2}$ on the left-hand side of (3.105) is a real function because of $F(-\omega) = \{F(\omega)\}^*$ in Table 4.3. Furthermore, using (3.105) and the time shift by $e^{-\mathrm{i}\omega\tau_I}$, we perform the inverse Fourier transform of u_{zI+} in (3.101) and obtain[31]

$$
u_{zI+} = \frac{1}{\pi \mathrm{i}} \int_0^\infty \frac{H(t - \tau_I - pr)}{\{(t-\tau_I)^2 - p^2 r^2\}^{1/2}} G_I \, p \, dp\,. \tag{3.107}
$$

In addition, applying $\{f(t)\}^* \Leftrightarrow \{F(-\omega)\}^*$ in Table 4.3 to the second equation on the right-hand side of (3.105) and using the fact that the left-hand side of (3.105) is a real function, we obtain

$$
\begin{aligned}
&\left\{ \frac{H(t-pr)}{(t^2 - p^2 r^2)^{1/2}} \right\}^* = \frac{H(t-pr)}{(t^2 - p^2 r^2)^{1/2}} \\
\Leftrightarrow &\left\{ \frac{\pi \mathrm{i}}{2} H_0^{(1)}(-(-\omega)pr) \right\}^* = \frac{-\pi \mathrm{i}}{2} \left\{ H_0^{(1)}(\omega pr) \right\}^*.
\end{aligned}
\tag{3.108}
$$

Using (3.108), the time shift by $e^{-\mathrm{i}\omega\tau_I}$, and the formula for the real integration variable x[32]

$$
\left\{ \int f(x)\, dx \right\}^* = \int \{f(x)\}^* dx\,, \tag{3.109}
$$

we perform the inverse Fourier transform of the complex conjugate of u_{zI-} and obtain

$$
\{u_{zI-}\}^* = \frac{1}{\pi \mathrm{i}} \int_0^\infty \frac{H(t - \tau_I - pr)}{\{(t-\tau_I)^2 - p^2 r^2\}^{1/2}} G_I \, p \, dp \tag{3.110}
$$

if $\{G_I\}^* = G_I$. As shown in (3.97), when $p \le \upsilon_i^{-1}, \upsilon_{i+1}^{-1}$ or $p \ge \upsilon_i^{-1}, \upsilon_{i+1}^{-1}$, T_j and R_k are real, therefore G_I composed of T_j and R_k is also real, then the condition $\{G_I\}^* = G_I$ is satisfied. Furthermore, the comparison of (3.110) with (3.107) results in $u_{zI+} = \{u_{zI-}\}^*$ (Kennett [65] noted that this relation could also be obtained by the **Schwarz reflection principle**). When $u_{zI+} = \{u_{zI-}\}^*$ and the integral part in (3.107) and (3.110) is represented as $R_I + \mathrm{i}I_I$,

$$
u_{zI+} = \frac{1}{\pi \mathrm{i}}(R_I + \mathrm{i}I_I) = \frac{1}{\pi}(I_I - \mathrm{i}R_I)\,, \quad u_{zI-} = \{u_{zI+}\}^* = \frac{1}{\pi}(I_I + \mathrm{i}R_I)\,, \tag{3.111}
$$

[31] This almost matches Eq. (16.1.24) of Kennett [65], but the factor is different by 1/2.

[32] This is suggested on page 251 of Saito [103].

and (3.101) then yields[33]

$$u_{zI} = u_{zI+} + u_{zI-} = \frac{2}{\pi} I_I = \frac{2}{\pi} \,\mathrm{Im} \int_0^\infty \frac{H(t - \tau_I - pr)}{\{(t - \tau_I)^2 - p^2 r^2\}^{1/2}} G_I \, p \, dp \,. \quad (3.112)$$

Next, the slowness integration is performed. As the first step, the integration variable is changed from p to[34]

$$\theta = \tau_I + pr = \sum_m q_m d_m + pr = \sum_m (v_m^{-2} - p^2)^{\frac{1}{2}} d_m + pr \,. \quad (3.113)$$

Cagniard [15] derived this from the argument of the exponential function obtained by replacing $J_0(kr)$ in the integrand of (3.100) with its integral representation (*ibid.*, page 55). He also showed that the right-hand side of (3.100), which underwent this change of variables, can be transformed into the form of the Laplace transform in (3.102) with $t = \theta$. de Hoop [28] showed that Cagniard's result also holds for (3.100), modified in the form of the **plane wave expansion** (Sect. 3.1.4). After the change of variables, (3.112) yields

$$u_{zI} = \frac{2}{\pi} \,\mathrm{Im} \int_{\tau_{0I}}^\infty \frac{H(t - \theta)}{\{(t - \theta)(t - \theta + 2pr)\}^{1/2}} G_I \, p \frac{\partial p}{\partial \theta} d\theta \,, \quad (3.114)$$

$$\frac{\partial p}{\partial \theta} = \left(\frac{\partial \theta}{\partial p}\right)^{-1} = \left\{ r - \sum_m \frac{p \, d_m}{(v_m^{-2} - p^2)^{\frac{1}{2}}} \right\}^{-1} , \quad \tau_{0I} = \tau_I \big|_{p=0} = \sum_m \frac{d_m}{v_m} \,.$$

As Cagniard [15] and de Hoop [28] suggested, θ is a quantity corresponding to time, and because time must be real, the integration path is along the real axis (Fig. 3.7 (left)). For simplicity, considering a generalized ray with only a segment of $m = 1$, which corresponds to a direct wave in a one-layer model, (3.113) yields

$$\theta = q_1 d_1 + pr = (v_1^{-2} - p^2)^{\frac{1}{2}} d_1 + pr \,. \quad (3.115)$$

By solving the quadratic equation modified from (3.115), we obtain

$$R^2 p = \theta r - \mathrm{i} d_1 \left(\theta^2 - \frac{R^2}{v_1^2}\right)^{\frac{1}{2}} , \quad R^2 = r^2 + d_1^2 \,. \quad (3.116)$$

This solution is based on Saito [103], however the sign before the square root is negative, following $e^{-\mathrm{i}\omega\tau_I}$ in (3.96). Along with this square root, a **branch point**

[33] This agrees with Eq. (A3) in Helmberger [45] if $\eta_1^{-1} \to G_I$ and the integration contour is moved from the imaginary axis to the real axis.

[34] In this book, since the inverse Fourier transform has already been performed in (3.107) and (3.110), the change of variables is a necessary condition from (3.105) rather than a sufficient condition for performing the inverse Fourier transform.

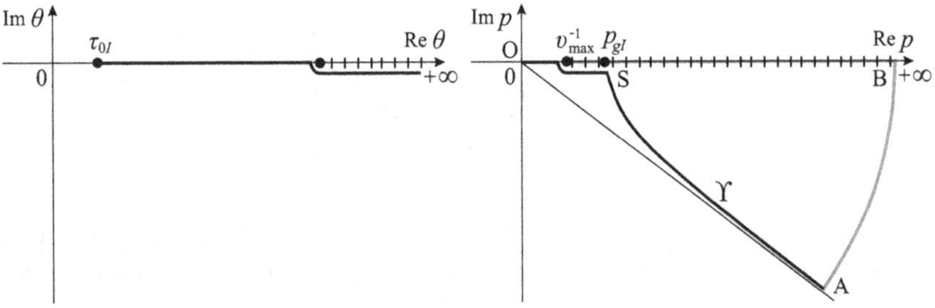

Fig. 3.7 Left: Path of the θ integration (based on Saito [103]). Right: The Cagniard path Υ of the slowness integration (based on Kennett [65], reprinted from Koketsu [72] with permission of Kindai Kagaku)

exists at $\theta = R/\upsilon_1$ as a singular point. From this point, the **branch cut** extends towards $\theta = +\infty$ (solid line with short vertical lines in the left panel of Fig. 3.7). When $p = 0$, $\theta = d_1/\upsilon_1$ from (3.114), and

$$\frac{d_1 r}{\upsilon_1} = \mathrm{i} d_1 \left(\frac{d_1^2}{\upsilon_1} - \frac{R^2}{\upsilon_1^2} \right)^{\frac{1}{2}} \tag{3.117}$$

from (3.116). The Riemann sheet where the imaginary part of the square root is negative must therefore be selected, and the path of the θ integration must extend along the fourth quadrant side of the real axis, as shown by the thick black line in the left panel of Fig. 3.7.[35]

The **Cagniard path** is obtained by remapping the path of the θ integration onto the complex plane of slowness p according to (3.113). Again, for simplicity, considering a generalized ray with only a segment of $m = 1$, the path of the slowness integration is determined by (3.116), so that θ is real and the square root is negative imaginary at $\theta \geq R/\upsilon_1$. For $d_1/\upsilon_1 \leq \theta < R/\upsilon_1$, since the square root is imaginary, from (3.116) p is found to be real (thick straight line in black in the right panel of Fig. 3.7). For $\theta \geq R/\upsilon_1$, the square root is real. Using $p_\mathrm{R} = \mathrm{Re}\, p$ and $p_\mathrm{I} = \mathrm{Im}\, p$ and comparing the real parts or the imaginary parts of both sides of (3.116), we obtain

$$R^2 p_\mathrm{R} = \theta r\,, \quad R^2 p_\mathrm{I} = -d_1 \sqrt{\theta^2 - \frac{R^2}{\upsilon^2}}\,. \tag{3.118}$$

Eliminating θ from both the equations gives the hyperbolic equation

$$\frac{p_\mathrm{R}^2}{a^2} - \frac{p_\mathrm{I}^2}{b^2} = 1\,, \quad a = \frac{r}{R\upsilon_1}\,, \ b = \frac{d_1}{R\upsilon_1}\,. \tag{3.119}$$

[35]This path is based on page 251 of Saito [103]. However, since the minus sign is selected in (3.116), we do not use the first quadrant, but the fourth quadrant.

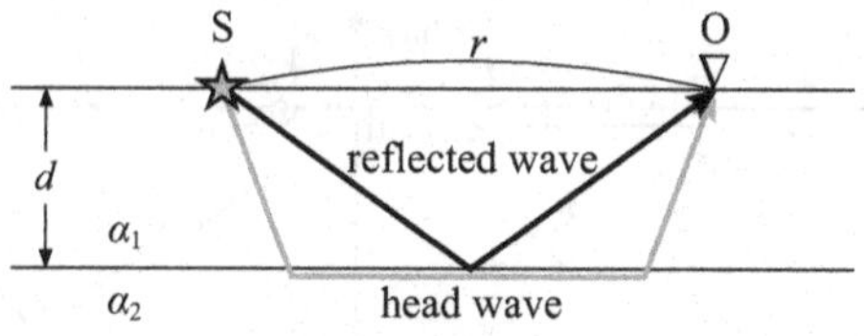

Fig. 3.8 Generalized ray for the reflected P wave in a two-layer structure. Both the reflected wave and the head wave are included (reprinted from Koketsu [72] with permission of Kindai Kagaku)

In the p complex plane, there is a branch point associated with the square root of q_1 in (3.99) on the real axis, and the branch cut then extends towards $p = +\infty$ (solid line with short vertical lines in the right panel in Fig. 3.7).[36] As the Riemann sheet with Im $p < 0$ is selected for the square root in (3.99) so that $e^{-i\omega\tau_I}$ does not diverge at the branch cut, the integration path must extend into the fourth quadrant.[37] The portion in the fourth quadrant of the hyperbola of (3.119) is therefore a part of the integration path for this generalized ray (thick black curve in the right panel of Fig. 3.7), which asymptotically approaches the line $p_{\mathrm{I}} = -(b/a)p_{\mathrm{R}}$ (diagonal solid line in the same panel). Substituting the approximation of the Hankel function (3.9) into u_{zI+} in (3.101), the vibrational part of u_{zI+} yields $e^{-i(\omega\tau_I + kr - \pi/4)} = e^{-i\omega\theta + i\pi/4}$. Accordingly, the **saddle point** of the **steepest descent path** (Sect. 3.1.9) is given by $\partial\theta/\partial p = 0$. We solve this equation using (3.113) and obtain the saddle point $p_{g1} = r/Rv_1$, which coincides with the vertex a of the hyperbola in (3.119).

In order to extend the above to multiple segments, for example, we assume the generalized ray for a reflected P wave in a two-layer structure as shown in Fig. 3.8, and set $\alpha_2 = v_{\max}$ in accordance with usual velocity structures. Although the structure has only two layers, we use $\alpha_2 = v_{\max}$ for generality for future uses. This generalized ray consists of two segments $m = 1, 2$, and $d_1 = d_2 = d$, $v_1 = v_2 = \alpha_1$, therefore from (3.113) we have

$$\theta = 2(\alpha_1^{-2} - p)^{\frac{1}{2}} d + pr \; . \tag{3.120}$$

Comparing (3.120) with (3.115), we find that the two agree if $2d \leftrightarrow d_1$ and $\alpha_1 \leftrightarrow v_1$. The integration path for the reflected P wave can therefore be obtained by applying these replacements to the integration path of the direct wave. However, R_1^+ in G_I not only includes $q_{\alpha 1}$ (corresponding to q_1 of the direct wave) but also $q_{\alpha 2}$, the branch point of which appears at $p = v_{\max}^{-1} = 1/\alpha_2$, and the branch cut extends from this to $p = +\infty$. When r is large, the saddle point $p_{gI} = r/(R\alpha_1) = r/\left(\alpha_1\sqrt{r^2 + 4d^2}\right)$ is located beyond this branch point, therefore the branch cut appears in section

[36] This branch cut is originally used for the Laplace transform [103], and it should not be used for the Fourier transform, but we here assume that it can be used, because the inverse Fourier transform has already been performed.

[37] Because Cagniard [15] and Aki and Richards [3] define the phase delay term differently from this book, the integration path extends to the first quadrant in Fig. 13 of Cagniard and Fig. 6.14 of Aki and Richards.

$\upsilon_{\max} \leq p \leq p_{gI}$ of the Cagniard path. This is therefore placed on the fourth quadrant side along the real axis (segment $\upsilon_{\max}^{-1}$S of the thick black line in the right panel of Fig. 3.7). The integration near $\upsilon_{\max}^{-1}$ in this segment generates a **head wave**[38] (thick gray line in Fig. 3.8). Conversely, if r is small, $p_{gI} < \upsilon_{\max}^{-1}$ and the Cagniard path does not pass around $\upsilon_{\max}^{-1}$, and no head wave is generated. If there are three or more layers, we cannot obtain a simple hyperbolic equation, but as each segment is a transmitted wave (direct wave in the middle of a generalized ray) or a reflected wave (Sect. 3.1.5), the Cagniard path should be a complex shape consisting of various hyperbolas.[39] In addition, the validity of changing the path of the slowness integration from along the real axis to the Cagniard path is confirmed as follows. The circular integration along the closed curve OSABO in the right panel of Fig. 3.7:

$$\oint_{\mathrm{OSABO}} F(p)dp = \int_{\Upsilon} F(p)dp + \int_{\mathrm{AB}} F(p)dp - \int_{0}^{\infty} F(p)dp \tag{3.121}$$

is zero because of **Cauchy's integral theorem** [82], and $\int_{\mathrm{AB}} F(p)dp$ on the right-hand side of (3.121) is also zero due to **Jordan's lemma** [5], therefore $\int_{\Upsilon} F(p)dp = \int_0^{\infty} F(p)dp$. It can also be noted that the Cagniard path is an interpretation of the real axis integration in the complex θ plane and the GRT does not actually use it to perform the slowness integration.

Although this subsection is entitled "Wavenumber Integration (Approximated)", only the inverse Fourier transform is analytically performed. The wavenumber integration (slowness integration) is converted into the time integration (θ integration) and remains. In the GRT, the time integration is performed numerically or is approximated into a standard integral form. In the case of the numerical time integration, as the curved part of the Cagniard path is adjacent to the steepest descent path as mentioned above, the integration is expected to converge rapidly and not to continue for a lengthy time period. The method of numerical integration is described in detail in Helmberger [45]. Alternatively, the time integration can be transformed into the form of a **convolution** (4.20) using approximations that yield standard integral forms. If r is large or the contribution at large t is negligible, the approximation $t - \theta + 2pr \approx 2pr$ holds. Substituting this into (3.114) and using $H(t) = 1$ in the integration range,

$$\begin{aligned} u_{zI} &= \frac{2}{\pi} \mathrm{Im} \int_{\tau_{0I}}^{\infty} \frac{H(t-\theta)}{(t-\theta)^{1/2}\sqrt{2pr}} G_I p \frac{\partial p}{\partial \theta} d\theta = \frac{2}{\pi} \frac{1}{\sqrt{t}} * \psi(t) , \\ \psi(t) &= \mathrm{Im} \frac{\sqrt{p}}{\sqrt{2r}} G_I \frac{\partial p}{\partial t} \end{aligned} \tag{3.122}$$

[38] This is often called a "refracted wave" because it is a type of seismic wave used in refraction exploration, but we do not use this term because it can be misunderstood as a transmitted wave with refraction at an interface.

[39] This is schematically shown in Fig. 9.2 of Aki and Richards [3].

is obtained,[40] and is termed the **first-motion approximation**. The convolutionary integration (4.20) remains in (3.122), but standard computer codes that can quickly compute the convolutionary integration exist.

Up to this point, each layer of the horizontally layered structure has been homogeneous, but we now consider extending the high-frequency GRT to the case where properties change in the z-direction. In this case, the separation of the SH wave from the P and SV waves (Sect. 3.1.1) is maintained, but the **wave equation** are modified from (1.42) because $\nabla\lambda$ and $\nabla\mu$ are not zero. If the equations obtained by Richards [100] in the spherical coordinate system hold in the cylindrical coordinate system, we have

$$\nabla^2\phi + \frac{\omega^2}{\alpha^2(z)}\phi = O\left(\frac{|\mathbf{u}|}{\omega}\right), \quad \nabla^2\boldsymbol{\psi} + \frac{\omega^2}{\beta^2(z)}\boldsymbol{\psi} = O\left(\frac{|\mathbf{u}|}{\omega}\right). \tag{3.123}$$

At high frequencies where ω is large, (3.123) agrees with (1.42) into which $\Phi = 0$, $\boldsymbol{\Psi} = \mathbf{0}$, $\partial^2\phi/\partial t^2 = -\omega^2\phi$, and $\partial^2\boldsymbol{\psi}/\partial t^2 = -\omega^2\boldsymbol{\psi}$ are substituted, except that α and β are functions of z. We solve (3.123) for large ω by variable separation and obtain the equation with respect to z

$$\frac{d^2Z}{dz^2} + \nu_\upsilon^2(z)Z = 0\,, \tag{3.124}$$

which also agrees with the equation from (1.42) [104] except that $\nu_\upsilon = \omega q_\upsilon$ in (2.76) and (3.99) is a function of z. When the **WKBJ approximation**[41] is applied to the second-order ordinary differential equation in the form of (3.124), Jeffreys [57] showed that the solution for large ω is

$$Z(z) = \frac{1}{|q_\upsilon|}\exp\left(\mp i\omega\int q_\upsilon(z)dz\right). \tag{3.125}$$

Equation (3.125) indicates that the phase in the depth direction is expressed by the z-direction integral of q_υ in the velocity structure that varies piecewise continuously in the z-direction (depth direction). If this is applied to u_{z0} for the source of $l = 0$ shown previously, the τ_I in (3.98) should be replaced with

$$\tau_I = \int_{z_1}^{z_2} q_\upsilon(z)dz\,. \tag{3.126}$$

---

[40]This agrees with Eq. (A17) of Helmberger [45] if $\mathcal{R}(p)\mathcal{T}(p)/\eta_1 \rightarrow G_I$.

[41]This is the same as the WKB approximation in quantum mechanics. However, while three papers by Wentzel, Kramer, and Brillouin were all published in 1926 for the Schrödinger equation, the paper by Jeffreys was published in 1925 for more general second-order ordinary differential equations. "Jeffreys" should therefore be included as in the WKBJ approximation because of his greatest contribution.

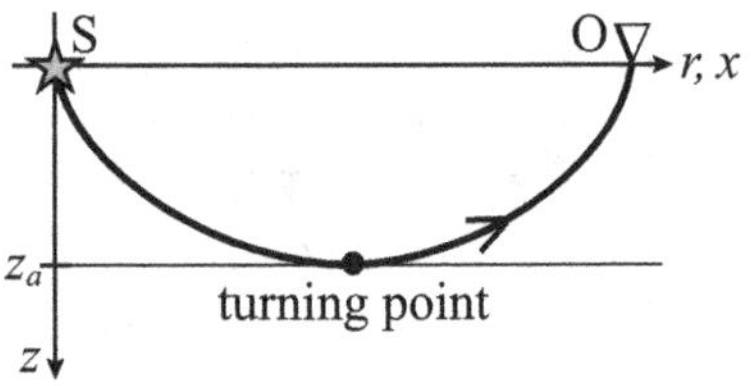

Fig. 3.9 Generalized ray in a semi-infinite medium with continuously increasing velocity in the depth direction. The black dot denotes the turning point (based on Chapman [22], reprinted from Koketsu [72] with permission of Kindai Kagaku)

Although the integration range $[z_1, z_2]$ is different for the generalized ray, in the semi-infinite medium in which the velocity continuously increases in the depth direction, the generalized ray turns as shown in Fig. 3.9, and (3.126) yields

$$\tau_I = 2\int_0^{z_a} q_v(z)dz \tag{3.127}$$

using the depth z_a of the turning point. According to Chapman [22], the inverse Fourier–Hankel transform of (3.101) into which (3.125) is incorporated, can be analytically performed using the formulas in (Table 4.3), etc. The result of this represents the first-motion approximation of the ground motion, and is called the **WKBJ seismogram**.

This method can calculate ground motions of a piecewise continuous velocity structure as fast as the GRT with the first-motion approximation. However, special considerations are required near the turning point, and an additional term is required in the formulation to include surface waves, etc. [64]. The method where the WKBJ approximation in the WKBJ seismogram is replaced with the Langer approximation is called by the slightly strange name "**full wave theory**". This eliminates the need for the special considerations near the turning point [25]. Computer programs for the GRT and related methods have been published in *Seismological Algorithms* [30]. In these methods, the computation is rapid and body waves as well as non-body waves, such as the head waves, are included. In addition surface waves can be considered by superimposing multiply reflected waves (**reverberation**, Sect. 3.1.6). However, as the ground motion is evaluated for each generalized ray and a subjective decision is required to determine which generalized rays are to be evaluated, accurate seismograms cannot be obtained if generalized rays with large contributions are omitted.

3.1.8 Wavenumber Integration (Numerical)

When it is difficult to perform the wavenumber integration analytically, the thought of performing it numerically follows naturally. However, this could be viewed as

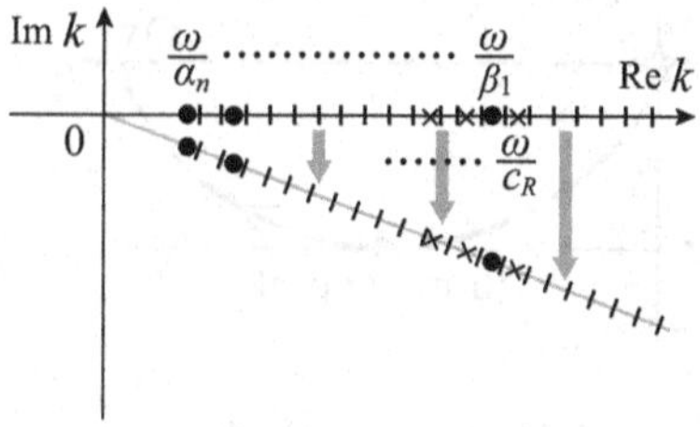

Fig. 3.10 In the k complex plane, the surface wave poles ($\times$), the branch points ($\bullet$), and the branch cuts (short vertical lines) are on the real axis, but when $-i\omega_I$ is introduced into ω, they move to the fourth quadrant (based on Phinney [96], reprinted from Koketsu [72] with permission of Kindai Kagaku)

"Columbus' egg"-type solution, and should be related to the advance of computers. Phinney [96] performed this numerical integration for the first time for a simple two-layer structure, obtaining synthetic seismograms in the time domain. Fuchs [35] then applied this to a multi-layer structure using the **cylindrical wave expansion**. Following them, Fuchs and Müller [36] reported the latter in detail, including a comparison with the seismograms from the GRT. As their aim was the structural analysis of the crust and upper mantle (Sect. 2.1.1), the same approximation as in (3.100) was applied to the shallow part of the velocity structure. As a result, their integrand is interpreted as the reflectivity (generalized reflection coefficient) from the target of the analysis, therefore they named their method the **reflectivity method**. The approximation in the shallow part was removed by Kind [69], and complete synthetic seismograms can now be calculated using this method. In the above, the propagator matrix is used to calculate the integrand (Kind [69] used the Haskell matrix; Kohketsu [70] used the matrix of Fuchs [34]), while Kennett [63] used the reflection/transmission matrix. Later, Zhu and Rivera [130] proposed a method that is stable even at $\omega \sim 0$ using the Haskell matrix (Sect. 3.1.12), and Hisada [49] proposed a method that is stable even for $r \sim 0$ and $h \sim 0$ using the reflection/transmission matrix.

The integrand has the following singular points. First, **branch point** exist, corresponding to the square roots of the wavenumbers in the depth direction $\nu_{\alpha i}$, $\nu_{\beta i}$ in (3.5), located at $k = \omega/\upsilon_i$ ($\upsilon = \alpha$ or β, $i = 1, 2, \cdots, n$) on the real axis. Since $\alpha_i > \beta_i$, and $\upsilon_{\max} = \upsilon_n$, $\upsilon_{\min} = \upsilon_1$ in a common velocity structure, the branch point closest to the origin is located at ω/α_n and the farthest branch point is located at ω/β_1 (black dots in Fig. 3.10). The definitions of $\nu_{\alpha i}$ and $\nu_{\beta i}$ in (3.5) originate from the **Sommerfeld integral** (Sect. 2.2.1). The signs before the square roots arise from the condition Im $\nu_{\upsilon n} \leq 0$, so that the potentials of the halfspace (layer n) $\phi_{ln}^{+} = B_{ln}e^{-i\nu_{\alpha n}z}$, $\psi_{ln}^{+} = D_{ln}e^{-i\nu_{\beta n}z}$, and $\chi_{ln}^{+} = F_{ln}e^{-i\nu_{\beta n}z}$, which remains unquenched by the radiation boundary condition, do not diverge at $z \to +\infty$. From (3.5), $\nu_{\upsilon n}$ is real if $\omega/\upsilon_n \geq k$, and this condition is automatically satisfied, so that there is no branch cut. If $\omega/\upsilon_n \leq k$, $\nu_{\upsilon n}$ is imaginary and there are cases in which the condition is satisfied and cases in which the condition is not satisfied depending on the selection of the Riemann sheet, therefore a branch cut is required from the branch point $k = \omega/v_n$ towards $+\infty$ along the real axis (solid line with short vertical lines in Fig. 3.10). Im $\nu_{vn} = 0$ on the branch cut, from which the Riemann sheet of Im $\nu_{vn} < 0$ is selected.[42] Similar branch cuts are set for ν_{vi} in the layers other than layer n.

If the ground motion is expanded to generalized rays as in Sect. 3.1.7, there are no singular points other than the branch points. However, if not, there are multiple **poles** related to surface waves on the real axis, in addition to the branch points. The farthest from the origin is related to the **Rayleigh wave** (Sect. 3.1.10). As its phase velocity c_R is slower than β_1, the farthest pole is located to the right of ω/β_1. The other poles are aligned from this pole towards the origin ($\times$ in Fig. 3.10). The integrands diverge at these **surface wave poles** (Sect. 3.1.9); therefore, they represent obstacles in the numerical wavenumber integration. Phinney [96] identified a method to work around the obstacles. When a negative imaginary number is added to the angular frequency ω, which is real, as

$$\omega \to \omega - \mathrm{i}\omega_{\mathrm{I}} , \tag{3.128}$$

the wavenumbers of the singular points yield those such as $\omega/c_R \to \omega/c_R - \mathrm{i}\omega_{\mathrm{I}}/c_R$, so that all the poles and branch points move to the fourth quadrant and lie on a straight line with slope $-\omega_{\mathrm{I}}/\omega$ (solid gray line in Fig. 3.10). As ω becomes complex, the definitions of $\nu_{\alpha i}$ and $\nu_{\beta i}$ in (3.5) are replaced by

$$\nu_{\alpha i} = \begin{cases} \sqrt{k_{\alpha i}^2 - k^2}\,, & |k_{\alpha i}| \geq |k| \\ -\mathrm{i}\sqrt{k^2 - k_{\alpha i}^2}\,, & |k_{\alpha i}| < |k| \end{cases}, \quad \nu_{\beta i} = \begin{cases} \sqrt{k_{\beta i}^2 - k^2}\,, & |k_{\beta i}| \geq |k| \\ -\mathrm{i}\sqrt{k^2 - k_{\beta i}^2}\,, & |k_{\beta i}| < |k| \end{cases} \tag{3.129}$$

where $k_{vi} = \omega/v_i$ [62]. Since Im $\nu_{vi} = 0$ is equivalent to $k = \omega/v_i - \mathrm{i}\omega_{\mathrm{I}}/v_i$, the branch cuts also move to the fourth quadrant and extend along the line of slope $-\omega_{\mathrm{I}}/\omega$. As a result, neither singular points nor branch cuts exist on the real axis; therefore, the wavenumber integration along the real axis is possible.

If the formula of frequency shift (Table 4.4)

$$f(t)e^{\mathrm{i}\omega_0 t} \Leftrightarrow F(\omega - \omega_0) \tag{3.130}$$

is valid even for an imaginary shift, we can substitute $\omega_0 = \mathrm{i}\omega_{\mathrm{I}}$ into (3.130) to obtain

$$f(t)e^{-\omega_{\mathrm{I}} t} \Leftrightarrow F(\omega - \mathrm{i}\omega_{\mathrm{I}}) . \tag{3.131}$$

The effect of $-\omega_{\mathrm{I}}$ can therefore only be removed by multiplying the obtained ground motion by $e^{+\omega_{\mathrm{I}} t}$. However, as this operation may unnecessarily enlarge the computational error in the later portion of $f(t)$, it is empirically proven that $\omega_{\mathrm{I}} \sim \pi/T$ is appropriate when the duration T is required for $f(t)$. A similar effect can also be obtained by giving **Q** of **intrinsic attenuation** to the medium and making the velocities complex (Sect. 1.2.6) [69]. In other words, as the medium constituting the Earth has more or less attenuation property, the problem of the surface wave poles

[42] Since Phinney [96] defined $\nu = \left\{k^2 - (\omega/v_n)^2\right\}^{1/2}$, the Riemann sheet of Re $\nu > 0$ is selected.

can be solved naturally only by considering this. However, this consideration should be incorporated because cases without attenuation may occasionally be necessary.

The range of the wavenumber integration is formally from 0 to ∞ as shown in (3.95), but the range of the phase velocity $[c_1, c_2]$ is set so as to sufficiently cover the phase velocities of all the body and surface waves. As the integrands are small outside the corresponding wavenumber range $[\omega/c_2, \omega/c_1]$, it is sufficient to perform the integration within this range. Our experiences have shown that c_1 is required to be 90% of β_1, which corresponds to the **Rayleigh wave velocity** c_R (Sect. 3.1.10) as mentioned above, and c_2 is required to be greater than the largest among $\alpha_1, \alpha_2, \cdots, \alpha_n$. However, if a layer other than the first layer gives the minimum S wave velocity $\beta_{\min}$ and this is less than about 90% of β_1, c_1 must be less than $\beta_{\min}$.

The numerical implementation of the wavenumber integration for the plane wave expansion in (3.47) using the **trapezoidal rule** is equivalent to the approximation

$$\int_{-\infty}^{+\infty} \tilde{u}_i(k, z, \omega) e^{-\mathrm{i}kx} dk \approx \sum_{n=-N}^{N-1} \frac{\Delta k}{2} \left\{ \tilde{u}(k_n, z, \omega) e^{-\mathrm{i}k_n x} + \tilde{u}(k_{n+1}, z, \omega) e^{-\mathrm{i}k_{n+1} x} \right\}$$

$$= \Delta k \sum_{n=-N}^{N} \epsilon_n \tilde{u}_i(k_n, z, \omega) e^{-\mathrm{i}k_n x}, \ k_n = n\Delta k\,,\ \epsilon_n = \begin{cases} 1/2 \text{ if } n = -N, N \\ 1 \quad \text{otherwise} \end{cases}. \quad (3.132)$$

If we write $\Delta k = 2\pi/L$, this approximation corresponds to the "discrete wavenumber representation" of Bouchon and Aki [12]. Although the factors at both ends differ by 1/2, the difference does not matter if N is set sufficiently large so that correct seismograms can be obtained. Similarly, the trapezoidal approximation of the wavenumber integration in the cylindrical wave expansion agrees with the discrete wavenumber representation of Bouchon [13]. The **discrete wavenumber method** is therefore exactly equivalent to the reflectivity method where the wavenumber integration is performed numerically with the trapezoidal rule.

As the inverse Fourier transform of a spectrum by the **FFT** (Sect. 4.2.4) brings about such aliasing in the time domain as the repetition of a finite time history (Sect. 4.2.3), the discretization of the wavenumber integration in (3.132) brings about aliasing in the space domain. The physical interpretation of the **aliasing** in the space domain gives rise to the repetition of sources with the interval L along the x-axis [12] as shown in (3.132), as well as the repetition of sources with the interval L along the r-axis, connecting the true source and observation points [13] for the numerical integration with the cylindrical wave expansion (Fig. 3.11). Just as the repetition period of the time history is adjusted to avoid aliasing in the time domain, the repetition interval of the source can be adjusted through Δk so that a neighboring virtual source does not affect the correct seismograms.

If the wavenumber integration is performed through the above numerical scheme, the same result should be obtained no matter what matrix is used for the integrands. In Fig. 3.12, the velocity seismograms for a dip-slip point source in the western Taiwan velocity structure were computed using the method of Kohketsu [70] with the matrix

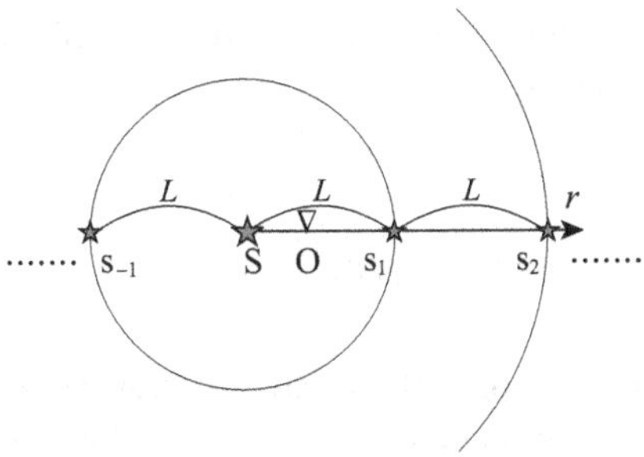

Fig. 3.11 Aliasing due to the numerical wavenumber integration for the cylindrical wave expansion (based on Bouchon [13]). The sources $\cdots$, s_{-1}, (S), s_1, s_2, $\cdots$ are repeatedly distributed along the r-axis connecting the true source S and the observation point O. Reprinted from Koketsu [72] with permission of Kindai Kagaku

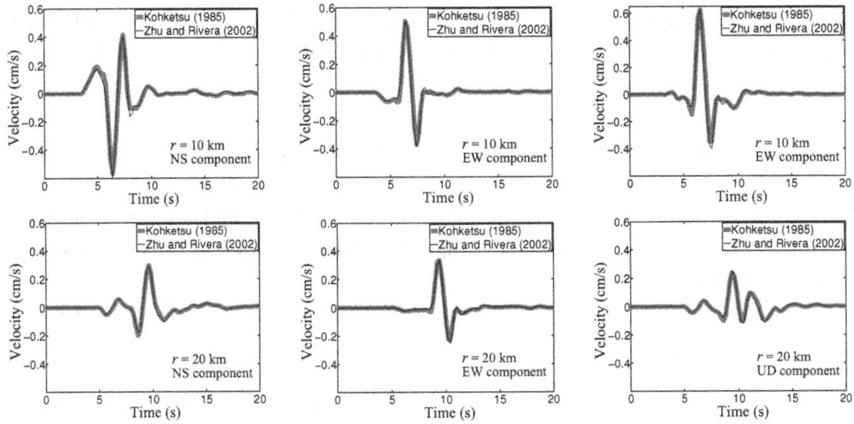

Fig. 3.12 Comparison of the synthetic velocity seismograms due to a dip-slip point source in the velocity structure for western Taiwan. The thick gray traces and thin black traces were computed using the methods of Kohketsu [70] and Zhu and Rivera [130], respectively (modified from the Electronic Supplement of Diao et al. [29])

of Fuchs [34] and the method of Zhu and Rivera [130] with the Haskell matrix. A comparison shows that the two methods agree well. Furthermore, similar seismograms should be not only obtained when comparing different numerical wavenumber integration methods but also when comparing a numerical wavenumber integration method and an approximate wavenumber integration method, because the aim is the same. The above-mentioned comparison between the reflectivity method and the **generalized ray theory** [36], and the comparison between the reflectivity method and the **full wave theory** [24] were performed and good agreement was generally found.

Finally, in the case of the numerical wavenumber integration method, as what is obtained after the wavenumber integration is the Fourier transform of the ground motion, the inverse Fourier transform is still required, but this is not special; therefore, it will be described in Chap. 4.

3.1.9 Surface Wave (Love Wave)[43]

Surface waves are phenomenologically "a wave that travels along a (ground) surface" [117], however physically they have the following meaning. If the energy is conserved in a system, there exist **normal oscillation**, and any vibration of the system can be represented by the superposition of the normal oscillations [84]. The velocity structure can be viewed as a system consisting of an elastic medium, and its normal oscillations are termed **normal mode** [3]. In the case of a 1-D velocity structure as shown in Fig. 3.1, layer n is a halfspace that is open toward $z = +\infty$. When the displacement potentials there ϕ_{ln}^+, ψ_{ln}^+, and χ_{ln}^+ (ϕ_{ln}^-, ψ_{ln}^-, and χ_{ln}^- vanish due to the radiation boundary condition) are the equations in (3.5) for $k \le k_{\alpha_n} < k_{\beta n}$ ($k_{\alpha n} = \omega/\alpha_n$, $k_{\beta n} = \omega/\beta_n$)

$$B_{ln}e^{-\mathrm{i}\sqrt{k_{\alpha n}^2-k^2}z}\,,\quad D_{ln}e^{-\mathrm{i}\sqrt{k_{\beta n}^2-k^2}z}\,,\quad F_{ln}e^{-\mathrm{i}\sqrt{k_{\beta n}^2-k^2}z}\,, \tag{3.133}$$

the ground motions are **sine wave** propagating to $z = +\infty$; therefore, the energy is not conserved. The displacement potentials for the normal modes must therefore be the equations in (3.5) for

$$k > k_{\beta n} > k_{\alpha_n}\,, \tag{3.134}$$

which are

$$B_{ln}e^{-\sqrt{k^2-k_{\alpha n}^2}z}\,,\quad D_{ln}e^{-\sqrt{k^2-k_{\beta n}^2}z}\,,\quad F_{ln}e^{-\sqrt{k^2-k_{\beta n}^2}z}\,. \tag{3.135}$$

As these exponentially decay towards $z = +\infty$, most of the energy of the normal modes is concentrated near the ground surface at $z = 0$; therefore, the ground motions of the normal modes can be termed **surface wave.**

The normal oscillation is the solution of **free oscillation**[44] when the wave equation is regarded as the equation of oscillation, and this is also termed **characteristic oscillation**. In the ground motion solution (3.25) obtained from the wave equation of the SH wave for the cylindrical wave expansion, because of "free oscillation" the discontinuous vector is zero ($\Delta_{l1} = \Delta_{l2} = 0$), therefore

$$\tilde{v}_{l1}(0) = \frac{-1}{M_{11}}(M_{11}^h\Delta_{l1} + M_{12}^h\Delta_{l2}) \Rightarrow M_{11}\tilde{v}_{l1}(0) = 0\,. \tag{3.136}$$

[43] For more information on Sects. 3.1.9 and 3.1.10, see Saito [103].

[44] In seismology, the normal oscillation of the Earth is often referred to as "free oscillation" (e.g., Sect. 2.3.4). However, in general physics, this signifies "oscillation performed by internal forces of the system without external force" [98], and we use it here in this sense.

$M_{11} = 0$ is required for the right-hand equation to have a solution other than the trivial solution $\tilde{v}_{l1}(0) = 0$. For simplicity, if we first consider a single-layer structure (assuming $n = 1$ in Fig. 3.1), that is, a semi-infinite medium, $\mathbf{M} = \mathbf{T}_1^{-1}$, and $M_{11} = (\mathrm{T}_1^{-1})_{11}$ is equivalently 1/2, so that $M_{11} = 0$ can never be realized. There is therefore no SH type surface wave in a semi-infinite medium. However, in the case of a two-layer structure with $n = 2$, $\mathbf{M} = \mathbf{T}_2^{-1}\mathbf{G}_1$, and therefore from (3.15) and (3.17)[45] we have

$$M_{11} = (\mathrm{T}_2^{-1})_{11}(G_1)_{11} + (\mathrm{T}_2^{-1})_{12}(G_1)_{21} = \frac{1}{2}\left(-\cos Q_1 + \frac{-1}{\mathrm{i}\mu_2\nu_{\beta 2}}(-\mu_1\nu_{\beta 1})\sin Q_1\right) = 0$$

$$\Rightarrow \tan Q_1 = \tan \nu_{\beta 1} d_1 = \frac{\mathrm{i}\mu_2\nu_{\beta 2}}{\mu_1\nu_{\beta 1}} . \tag{3.137}$$

If, in addition to (3.134), the condition that the ground motion is oscillatory in layer 1 is imposed, then $k_{\beta 2} < k < k_{\beta 1}$. Hence, using (3.5) and the **phase velocity** $c \equiv \omega/k$, the **characteristic equation**

$$\tan\left(\omega d_1\sqrt{\frac{1}{\beta_1^2} - \frac{1}{c^2}}\right) = \frac{\mu_2}{\mu_1}\left(\sqrt{\frac{1}{\beta_1^2} - \frac{1}{c^2}}\right)^{-1}\sqrt{\frac{1}{c^2} - \frac{1}{\beta_2^2}} \tag{3.138}$$

is obtained.[46]

The above condition that the ground motion is oscillatory in layer 1 indicates that the surface wave can be decomposed into body waves. The characteristic equation (3.138) can be obtained from the condition in which constructive interference occurs among the SH waves composing the reverberations [103]. As described in Sect. 3.1.7, surface waves are computed through the **generalized ray theory** using this property. When the ground motions come from a deep part of the medium, that is, when the source is deep, the reflected waves on the ground surface travel back near vertically and constructive interference is unlikely to occur. Surface waves are therefore less likely to develop when the source is deep.

The characteristic equation is now solved with respect to c under the condition $\beta_1 < c < \beta_2$, which is derived from the wavenumber condition $k_{\beta 2} < k < k_{\beta 1}$ divided by ω. As the argument of tan is complicated, we first perform the change of variables $x = d_1\sqrt{1/\beta_1^2 - 1/c^2}$, and then we have $\tan \omega x$ on the left-hand side and $C/x \cdot \sqrt{A - x^2}$ on the right-hand side, where $C = \mu_2/\mu_1$ and $A = d_1^2(\beta_1^{-2} - \beta_2^{-2})$. As shown in Fig. 3.13, the solution of the characteristic equation can therefore be obtained graphically from the plots of $y = \tan \omega x$ and $y = C/x \cdot \sqrt{A - x^2}$. The search range for the solution is from $x = 0$ corresponding to $c = \beta_1$, to $x = d_1\sqrt{1/\beta_1^2 - 1/\beta_2^2}$ corresponding to $c = \beta_2$. If $\beta_1 < \beta_2$, one or more solutions surely exist, and this type of surface wave is named a **Love wave**, because A. E. H.

---

[45]The same result can be obtained using (3.51) and (3.53) for the Haskell matrix.

[46]This is written in many documents. For example, this agrees with Eq. (6.4.8) of Satô [104] if $0 \to 2$ and $c\xi \to \omega$, and with Eq. (7.6) of Aki and Richards [3] if $H \to d_1$.

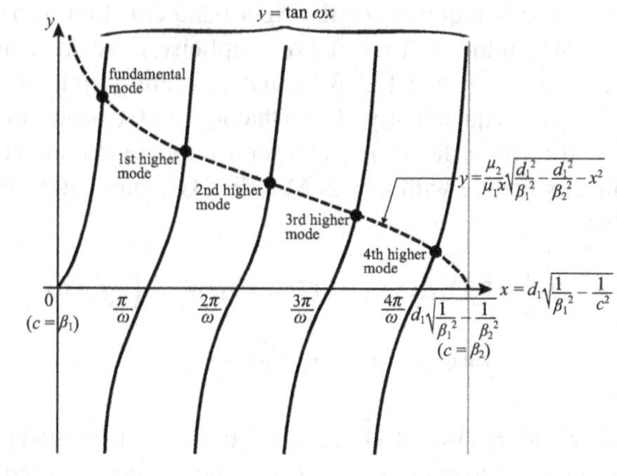

Fig. 3.13 Graphical solution of the characteristic equation of the Love wave in a two-layer structure, in the case of $4\pi/\omega \leq d_1\sqrt{1/\beta_1^2 - 1/\beta_2^2} < 5\pi/\omega$ (based on Aki and Richards [3], reprinted from Koketsu [72] with permission of Kindai Kagaku)

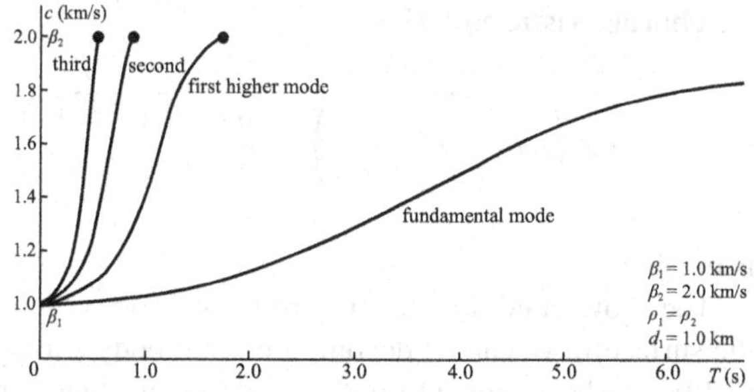

Fig. 3.14 Dispersion curves with respect to the period T in the two-layer structure. The parameters of the structure are shown in the lower right (based on Satô [104], reprinted from Koketsu [72] with permission of Kindai Kagaku)

Love demonstrated it theoretically for the first time in 1911 [86].[47] The number of solutions depends on how many $y = \tan \omega x$ curves can be plotted in the search range. The case of $4\pi/\omega \leq d_1\sqrt{1/\beta_1^2 - 1/\beta_2^2} < 5\pi/\omega$ is shown in Fig. 3.13, for which five solutions exist. The Love wave for the solution with the smallest c is termed the **fundamental mode**, and the Love waves for other solutions are successively termed the first-order **higher mode**, the second-order higher mode, etc., in sequence of c. For the same ω, c therefore increases as the mode order increases (the fundamental mode is the zeroth order). In this way, if a two-layer structure with $n = 2$ is used, the graphical solution is possible in principle, but the numerical solution of an equation is generally used (this will be explained in Sect. 3.2.3). In a multi-layer structure with $n \geq 3$, there is no choice but to use the numerical solution.

As ω is included as a parameter on the left-hand side of (3.138), the phase velocity solution c is a function of the angular frequency ω or the period $T = 2\pi/\omega$. Such a phenomenon in which the phase velocity depends on the period is termed **dispersion**. This property is different from a body wave that propagates only at the P wave velocity or S wave velocity of the medium regardless of the period. In Fig. 3.13, each mode corresponds to one of the $y = \tan \omega x$ curves, and the curves move towards the origin

[47] This is written on page 6 of Utsu [117].

of the x-axis when ω increases. Conversely, the curve $y = C/x \cdot \sqrt{A - x^2}$ does not include ω; therefore, it does not move even if ω increases. The solution, which is the intersection of the two curves, therefore also moves towards the origin of the x-axis, the solution x becomes smaller, and c also becomes smaller. As ω decreases, T increases. When c is calculated and plotted for various Ts, it therefore becomes a monotonically increasing curve as shown in Fig. 3.14 [104]. This curve is termed a **dispersion curve**. In the dispersion curves of all modes, c is decreasing and always approaches β_1 when T is decreasing (ω is increasing). However, c is increasing when T is increasing (ω is decreasing), and the curves are cut off if c reaches the upper limit of the search range. Along the curve $y = C/x \cdot \sqrt{A - x^2}$ in Fig. 3.13, $c = \beta_2$ only at the point $\left(d_1\sqrt{1/\beta_1^2 - 1/\beta_2^2}, 0\right)$ on the x-axis. In addition, the curve $y = \tan \omega x$ of the m-th mode intersects at the point $(m\pi/\omega, 0)$. When the dispersion curve is cut off, these two points coincide so that the **cut-off frequency** is[48]

$$\omega_{cm} = \frac{m\pi}{d_1}\left(\sqrt{\frac{1}{\beta_1^2} - \frac{1}{\beta_2^2}}\right)^{-1} . \tag{3.139}$$

The black dots in Fig. 3.14 represent $2\pi/\omega_{cm}$.

Using the above, we can now obtain the phase velocity c or the wavenumber $k \equiv \omega/c$ of the Love wave. Since this k gives $M_{11} = 0$ in (3.137) and M_{11} appears in the denominator of the ground motion solution (3.25) (Sect. 3.1.2), this k denotes a **pole** of the ground motion solution and is termed the **surface wave pole** (Sect. 3.1.8). In order to obtain the ground motion due to the Love waves only,[49] it is therefore possible to evaluate the wavenumber integration in (3.95) with the **residue** at the surface wave poles. Harkrider [42] gives a concrete computation method for this. In addition, Harvey [43] showed that the normal mode solution from the residue computations at the surface wave poles gives a good approximation of the complete ground motion, including the body waves, when a virtual high-velocity layer is introduced in the lowermost halfspace.

Consider that the ground motion is composed of the Love waves. The Love waves are smoothly dispersed as shown in Fig. 3.14, so that the components with slightly different ω's have slightly different c's. In such a situation, if the components with slightly different ω's are taken out, a phenomenon similar to the **beat** of a sound wave occurs.[50] For simplicity, we consider only $l = 1$ for $\bar{u}_{\theta 0}$ in (3.95) ($l = 0$ is always zero for the SH wave as shown in Sect. 2.2.5), and the phase velocity is obtained as $c(\omega)$ through the method described in this subsection or other methods. The normal mode solutions are also obtained as $\bar{v}_{10}^L(0)H_1^{(2)}(kr)$ through the residue computations based on Harkrider [42] or others. If the product of this and $\partial\Lambda_1/\partial\theta$

---

[48]This agrees with Eq. (7.8) of Aki and Richards [3] if $n \to m$ and $H \to d$.

[49]The ground motion including all body waves and surface waves can be calculated using the method in Sect. 3.1.8.

[50]However, for a sound wave, the "beat" is often explained as a phenomenon only in the time domain (e.g., *Physics Dictionary* [98]).

is represented as $\bar{v}_{10}(0)H_1^{(2)}(kr)$, $\bar{u}_{\theta 0} = \bar{v}_{10}(0)H_1^{(2)}(kr)$. The ground motion is the inverse Fourier transform of the above, therefore

$$u_{\theta 0} = \frac{1}{2\pi}\int_{-\infty}^{+\infty} \bar{v}_{10}(0)H_1^{(2)}(kr)e^{\mathrm{i}\omega t}d\omega\,. \tag{3.140}$$

We then introduce the approximation of the Hankel function in (3.9), and for removing the components with slightly different ω's we limit the integration range to $[\omega_0 - \Delta\omega/2, \omega_0 + \Delta\omega/2]$ and obtain

$$u_{\theta 0} = \frac{1}{2\pi}\int_{\omega_0-\Delta\omega/2}^{\omega_0+\Delta\omega/2} V(\omega)e^{\mathrm{i}(\omega t - kr)}d\omega\,,\quad V(\omega) = \sqrt{\frac{2}{\pi kr}}e^{3\pi\mathrm{i}/4}\bar{v}_{10}(0)\,. \tag{3.141}$$

If we can assume that $V(\omega)$ does not vary much around ω_0, $V(\omega) \sim V(\omega_0)$ [103], and $k(\omega) = \omega/c(\omega)$ is approximated by the first order Taylor expansion around ω_0:

$$k(\omega) = k_0 + k_0'(\omega - \omega_0)\,,\quad k_0 = k(\omega_0)\,,\ k_0' = \left.\frac{dk}{d\omega}\right|_{\omega=\omega_0} = \left.\frac{dk}{d\omega}\right|_0\,. \tag{3.142}$$

In the following, $\left.\right|_{\omega=\omega_0}$ is abbreviated with $\left.\right|_0$ as in (3.142) [103]. We then perform the change of variables $w = \omega - \omega_0$ and obtain[51]

$$\begin{aligned} u_{\theta 0} &= \frac{1}{2\pi}V(\omega_0)\int_{-\Delta\omega/2}^{+\Delta\omega/2} e^{\mathrm{i}(wt+\omega_0 t-(k_0+k_0' w)r)}dw \\ &= \frac{\Delta\omega}{2\pi}\frac{\sin(t-k_0' r)\Delta\omega/2}{(t-k_0' r)\Delta\omega/2}V(\omega_0)e^{\mathrm{i}(\omega_0 t - k_0 r)}\,. \end{aligned} \tag{3.143}$$

Since

$$k_0' = \frac{d}{d\omega}\left.\left(\frac{\omega}{c(\omega)}\right)\right|_0 = \frac{1}{c_0}\left(1 - \frac{\omega_0}{c_0}\left.\frac{dc}{d\omega}\right|_0\right), \tag{3.144}$$

k_0' has the reciprocal dimension of velocity; therefore, we can let k_0' be $1/U_0$. We substitute $k_0' = 1/U_0$ and $k_0 = \omega_0/c(\omega_0) = \omega_0/c_0$ into (3.143) obtaining

$$u_{\theta 0} = \frac{\Delta\omega}{2\pi}\frac{\sin(t - r/U_0)\Delta\omega/2}{(t - r/U_0)\Delta\omega/2}V(\omega_0)e^{\mathrm{i}\omega_0(t - r/c_0)}\,. \tag{3.145}$$

In addition, for a more general expression we make the replacements $\omega_0 \to \omega$, $c_0 \to c$, and $U_0 \to U$ and obtain

[51] This almost agrees with the equation at the top of page 141 of Saito [103] if $x \to r$, but the sign of the exponential argument is inverted because the definition of the Fourier transform is different from this book.

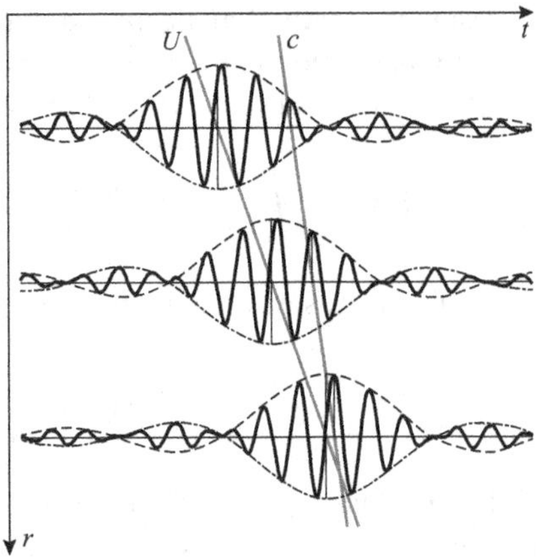

Fig. 3.15 Example of the Love wave ground motion in a two-layer structure. Seismograms of the ground motions at multiple r (thick solid lines) and $F(t)$ of (3.146) (dashed lines) are drawn in the form of record section. The envelope of the ground motion, which is a combination of $F(t)$ and its mirror image (dashed-dotted lines), the phase velocity c, and the group velocity U are also shown (modified from Saito [103] with permission of the author)

$$u_{\theta 0} = F\left(t - \frac{r}{U}\right) V(\omega) e^{\mathrm{i}\omega(t-r/c)}, \ \frac{1}{U} = \frac{1}{c}\left(1 - \frac{\omega}{c}\frac{dc}{d\omega}\right), \ F(t) = \frac{\Delta\omega}{2\pi}\frac{\sin \Delta\omega t/2}{\Delta\omega t/2}. \tag{3.146}$$

In $u_{\theta 0}$ of (3.146), $V(\omega)e^{\mathrm{i}\omega(t-r/c)} = V(\omega)e^{\mathrm{i}(\omega t - kr)}$ corresponds to the **spectrum** (Sect. 4.2.2) of $u_{\theta 0}$ in (3.141), and thus represents the spectrum of the ground motion of the Love wave at an angular frequency ω, i.e., the **sine wave** of amplitude $V(\omega)$, angular frequency ω, and phase velocity c. The remaining part of $u_{\theta 0}$, $F(t)$, has the form $A \sin at/at$, the main part of which is a smooth chevron in $[-\pi/a, \pi/a]$ centering around $t = 0$.[52] As $a = \Delta\omega/2$a is small in $F(t)$, its waveform is conspicuous with the wide chevron centered at $t = 0$, as shown by the broken line in Fig. 3.15. Moreover, since (3.146) includes $F(t - r/U)$, this chevron propagates at the velocity U.

$u_{\theta 0}$, which is a combination of both of the above, is composed of wavetrains[53] in which several cycles of the former are contained in the chevron of the latter. U is termed the **group velocity** because U is the velocity at which the wavetrain propagates. An example is shown in Fig. 3.15, where the Love wave ground motions at the three sites are drawn with black solid lines in the form of a **record section** in which each seismogram is plotted at the vertical position corresponding to an epicentral distance r, with a common time axis t. Hence, if the points of a specific phase of the ground motion, for example the points for the peaks, are connected, the slope of the connecting line is the phase velocity c (gray line on the right in the figure). In addition, if the centers of the chevrons are connected, the slope of this connecting line is the group velocity U (gray line on the left in the figure). As described above, in the case of the two-layer structure, c is monotonically increasing with respect to T, that is, monotonically decreasing with respect to ω, so that $U < c$ because $dc/d\omega < 0$ and (3.146).

[52]For example, the graph of $\sin at/\pi t$ was drawn in Figs. 2–10 of Papoulis [94].

[53]This term is used in Aki and Richards [3] and Kennett [65].

Using the **method of steepest descent**, it is possible to obtain an approximation of the ground motion even in the general case where the integration range of the inverse Fourier transform is not limited to $[\omega_0 - \Delta\omega/2, \omega_0 + \Delta\omega/2]$ as in (3.141) but

$$u_{\theta 0} = \frac{1}{2\pi}\int_{-\infty}^{+\infty} V(\omega)e^{\mathrm{i}(\omega t - kr)}d\omega = \int_{-\infty}^{+\infty}\frac{V(\omega)}{2\pi}e^{\mathrm{i}(\omega t - kr)}d\omega\,. \tag{3.147}$$

According to this method, in the complex integration

$$I(s) = \int_C g(z)e^{sf(z)}dz\,, \tag{3.148}$$

if the integration path C is an open curve or a closed curve at the ends of which the integrand is zero, s is a large real number, and $g(z)$ is "dominated"[54] by the exponential term, the approximation

$$I(s) \sim \frac{\sqrt{2\pi}\,g(z_0)e^{sf(z_0)}e^{\mathrm{i}\alpha}}{\left|sf''(z_0)\right|^{1/2}}\,, \quad f'(z_0) = 0\,, \quad \alpha = \arg(z - z_0) \tag{3.149}$$

holds and z_0 is named the **saddle point** [5]. If $z \to \omega$, $s \to r$, $f(z) \to \mathrm{i}(\omega t/r - k)$, and $g(z) \to V(\omega)/2\pi$, the integrand and integration variable in (3.148) coincide with those in (3.147). The integration path C is a closed curve that starts at $-\infty$ on the real axis of the ω complex plane, extends to $+\infty$ along the real axis, then wraps around the upper half plane, and returns to $-\infty$. According to **Jordan's lemma**, the portion in the upper half plane is zero; therefore, $I(s)$ in (3.148) is identical to $u_{\theta 0}$ in (3.147). The method of steepest descent can be applied to (3.147), as r is a large real number because of distant observation points and the change of V is already assumed to be small in (3.142). Using

$$\begin{aligned} &z_0 \to \omega_0\,, \quad f'(z_0) = 0 \to \mathrm{i}r\left(\frac{t}{r} - \left.\frac{dk}{d\omega}\right|_0\right) = 0 \Rightarrow U(\omega_0) = \frac{t}{r} \\ &sf''(z_0) \to \mathrm{i}r\left.\frac{d}{d\omega}\left(\frac{t}{r} - \frac{dk}{d\omega}\right)\right|_0 = -\mathrm{i}r\left.\frac{d^2k}{d\omega^2}\right|_0 \end{aligned} \tag{3.150}$$

from (3.142) and (3.144), we obtain[55]

[54]This expression is used by Arfken and Weber [5], and means that the variation in $g(z)$ is much smaller than the variation in the exponential term.

[55]This almost agrees with Eq. (18.2.10) of Satô [104] if $f^* \to V$, $\xi \to k$, and $x \to r$, but his inverse Fourier transform does not have the factor $1/2\pi$ and his result is multiplied by 2π. This also almost agrees with Eq. (6.1.10) of Saito [103] and Eq. (7.18) of Aki and Richards [3] if $x \to r$, but the sign of the exponential argument is inverted because the definition of the Fourier transform is different from this book.

$$u_{\theta 0} \sim \frac{V(\omega_0)}{\sqrt{2\pi r \left| d^2k/d\omega^2|_0 \right|}} \exp\left\{i(\omega_0 t - k_0 r \mp \frac{\pi}{4})\right\}, \tag{3.151}$$

where a minus or plus sign in $\mp$ is taken according to whether $d^2k/d\omega^2|_0$ is positive or negative. Equation (3.151) indicates that the **geometrical spreading** of the Love wave is in inverse proportion to the square root of the **epicentral distance** r.

In the case $d^2k/d\omega^2|_0 \sim 0$, (3.151) has a maximum and a large wavetrain appears. However, (3.151) in this form diverges if $d^2k/d\omega^2|_0 = 0$. The reason for this problem is that the method of steepest descent only uses the second-order Taylor expansion [5], such as

$$f(z) \sim f(z_0) + f'(z_0)(z - z_0) + \frac{1}{2} f''(z_0)(z - z_0)^2 . \tag{3.152}$$

We therefore use the third-order Taylor expansion instead, and also use $f'(z_0) = f''(z_0) = 0$ from (3.150), obtaining [104]

$$2\pi\, u_{\theta 0} \sim \frac{2V(\omega_0)}{\sqrt[3]{(r/6)\, d^3k/d\omega^3|_0}} \exp\{i(\omega_0 t - k_0 r)\} \int_0^\infty \cos(y^3) dy . \tag{3.153}$$

However, for the same reason as the footnote for (3.151), 2π is added at the beginning of the result of Satô [104]. We substitute $x = 3^{1/3} y$ into the formula for the Airy function Ai(x)[56]

$$\mathrm{Ai}(0) = \frac{1}{\pi} \int_0^\infty \cos\left(\frac{x^3}{3}\right) dx = \frac{3^{-2/3}}{\Gamma(2/3)} \tag{3.154}$$

and we have the integral in (3.153). We therefore obtain[57]

$$u_{\theta 0} \sim \frac{V(\omega_0)}{\sqrt[3]{(r/6)\, d^3k/d\omega^3|_0}} \frac{1}{3\Gamma(2/3)} \exp\{i(\omega_0 t - k_0 r)\} . \tag{3.155}$$

As the formula of the Airy function (3.154) is used to obtain (3.155), the large wavetrain is termed the **Airy phase**.

---

[56] Equation (G.10) of Ben-Menahem and Singh [10].

[57] This almost agrees with the equation at the bottom of page 146 of Saito [103] if $x \to r$, but the factor is different by $1/2\pi$.

3.1.10 Surface Wave (Rayleigh Wave)

In the same way that the Love wave is obtained from the free oscillation solution of the wave equation for the SH wave, Lord Rayleigh[58] theoretically showed the existence of the surface waves related to the free vibration solutions of the wave equations of P and SV waves for a uniform halfspace in his 1885 paper [99], earlier than the 1911 book by Love (from Utsu [117]). These surface waves are therefore named **Rayleigh wave**. In the ground motion solutions (3.38) obtained from the wave equations of the P and SV waves for the cylindrical wave expansion, because of "free oscillation" the discontinuous vector is zero ($\Delta_{l1} = \Delta_{l2} = \Delta_{l3} = \Delta_{l4} = 0$), therefore

$$\begin{aligned}
\tilde{u}_{l1}(0) &= \frac{1}{\hat{M}_{11}} \{-M_{22}(M^h_{11}\Delta_{l1} + M^h_{12}\Delta_{l2} + M^h_{13}\Delta_{l3} + M^h_{14}\Delta_{l4}) \\
&\quad + M_{12}(M^h_{21}\Delta_{l1} + M^h_{22}\Delta_{l2} + M^h_{23}\Delta_{l3} + M^h_{24}\Delta_{l4})\} \Rightarrow \hat{M}_{11}\tilde{u}_{l1}(0) = 0\,, \\
\tilde{w}_{l1}(0) &= \frac{1}{\hat{M}_{11}} \{+M_{21}(M^h_{11}\Delta_{l1} + M^h_{12}\Delta_{l2} + M^h_{13}\Delta_{l3} + M^h_{14}\Delta_{l4}) \\
&\quad - M_{11}(M^h_{21}\Delta_{l1} + M^h_{22}\Delta_{l2} + M^h_{23}\Delta_{l3} + M^h_{24}\Delta_{l4})\} \Rightarrow \hat{M}_{11}\tilde{w}_{l1}(0) = 0\,.
\end{aligned} \tag{3.156}$$

$\hat{M}_{11} = 0$ is required for the right-hand equations to have solutions other than the trivial solutions $\tilde{u}_{l1}(0) = \tilde{w}_{l1}(0) = 0$. In a single-layer structure (assuming $n = 1$ in Fig. 3.1), that is, a semi-infinite medium, from (3.30) we have [9]

$$\hat{M}_{11} = (\hat{\mathrm{T}}_1^{-1})_{11} = -\frac{\beta_1^4}{4\omega^4}\left(4k^2 + \frac{l_1^2}{\nu_{\alpha 1}\nu_{\beta 1}}\right) = 0\,. \tag{3.157}$$

l_1 is defined in (3.30). As this equation is not a monomial equation such as for the SH wave, it can be solved under the condition $k > k_{\beta 1} > k_{\alpha 1}$ in (3.134). Similar to the Love wave, using (3.5) and the phase velocity $c \equiv \omega/k$, the characteristic equation of the Rayleigh waves in a semi-infinite medium can be obtained as[59]

[58] His real name was John William Strutt, third Baron Rayleigh. However, he used "Lord Rayleigh" as for the authorship of his papers, listed in the bibliography. In addition to his scientific career at Cambridge University and the Royal Society, he was the third generation of Baron Rayleigh's family, hence "third Baron" (Therefore, the title Lord is used in the author name.). His younger brother, who was entrusted with the management of the territory in Essex, ran a dairy farm and opened several directly-managed shops of dairy products, called "Lord Rayleigh's Dairies" in the city of London, so he was also well known among the British public [53].

[59] Equation (3.158) agrees with Eq. (3.85) of Utsu [117] if $q \to k$, $k \to k_{\beta 1}$, $a = \sqrt{k^2 - k_{\alpha 1}^2}$, and $b = \sqrt{k^2 - k_{\beta 1}^2}$. Equation (3.159) agrees with Eq. (6.3.7) of Saito [103].

$$
\begin{aligned}
&4k^2 + \frac{l_1^2}{\nu_{\alpha 1}\nu_{\beta 1}} = 0 \\
&\Rightarrow (2k^2 - k_{\beta 1}^2)^2 = 4k^2\sqrt{k^2 - k_{\alpha 1}^2}\sqrt{k^2 - k_{\beta 1}^2} \qquad (3.158) \\
&\Rightarrow \left(\frac{2}{c^2} - \frac{1}{\beta_1^2}\right)^4 = \frac{16}{c^4}\left(\frac{1}{c^2} - \frac{1}{\alpha_1^2}\right)\left(\frac{1}{c^2} - \frac{1}{\beta_1^2}\right) \\
&\Rightarrow (2x - 1)^4 = 16x^2(x - a^2)(x - 1)\,, \quad x = \frac{\beta_1^2}{c^2},\ a = \frac{\beta_1}{\alpha_1} \\
&\Rightarrow 16(1 - a^2)x^3 + (16a^2 - 24)x^2 + 8x - 1 = 0\,. \qquad (3.159)
\end{aligned}
$$

The phase velocity c_R obtained by solving this is termed the **Rayleigh wave velocity** [103]. Below, we shall follow Saito [103]. We first assume **Poisson's ratio** ν (Sect. 1.2.3) to be the typical value 1/4, so that $\lambda_1 = \mu_1$ (Sect. 1.2.3) and $a = \beta_1/\alpha_1 = \sqrt{\mu_1/(\lambda_1 + 2\mu_1)} = 1/\sqrt{3}$. Substituting this into (3.159), we have

$$
\frac{32}{3}x^3 - \frac{56}{3}x^2 + 8x - 1 = (4x - 1)\left(\frac{8}{3}x^2 - 4x + 1\right) = 0\,. \qquad (3.160)
$$

Of the solutions of this equation, only $x = (3 + \sqrt{3})/4$ satisfies the condition (3.134) and $c < \beta_1$ or $x > 1$. Thus, in the semi-infinite medium where $\nu = 1/4$, the Rayleigh wave velocity is given as

$$
c_R = \frac{2}{\sqrt{3 + \sqrt{3}}}\,\beta_1 \approx 0.919402\,\beta_1\,. \qquad (3.161)
$$

This does not include ω; hence, there is no dispersion.

Next, considering the Rayleigh waves in a two-layer structure such as for the Love wave, from (3.40) we have

$$
\hat{M}_{11} = \sum_{j=1}^{6}(\hat{\mathrm{T}}_2^{-1})_{1j}(\hat{G}_1)_{j1} = 0\,. \qquad (3.162)
$$

$(\hat{\mathrm{T}}_2^{-1})_{1j}$ and $(\hat{G}_1)_{j1}$ here are given from (3.30) and (3.33) as [9, 34]

$$
\begin{aligned}
&(\hat{\mathrm{T}}_2^{-1})_{11} = -\frac{\beta_2^4}{4\omega^4}\left(4k^2 + \frac{l_2^2}{\nu_{\alpha 2}\nu_{\beta 2}}\right), \quad (\hat{\mathrm{T}}_2^{-1})_{12} = \frac{\mathrm{i}\beta_2^2}{4\mu_2\nu_{\beta 2}\omega^2}, \\
&(\hat{\mathrm{T}}_2^{-1})_{13} = (\hat{\mathrm{T}}_2^{-1})_{14} = -\frac{\mathrm{i}\beta_2^4}{4\mu_2\omega^3 c}\left(2 + \frac{l_2}{\nu_{\alpha 2}\nu_{\beta 2}}\right), \\
&(\hat{\mathrm{T}}_2^{-1})_{15} = -\frac{\mathrm{i}\beta_2^2}{4\mu_2\nu_{\alpha 2}\omega^2}, \quad (\hat{\mathrm{T}}_2^{-1})_{16} = -\frac{1}{4\rho_2^2\omega^4}\left(1 + \frac{k^2}{\nu_{\alpha 2}\nu_{\beta 2}}\right), \\
&(\hat{G}_1)_{11} = -2\gamma_1(\gamma_1 + 1) + \{2\gamma_1(\gamma_1 + 1) + 1\}\cos P_1 \cos Q_1
\end{aligned}
$$

$$
\begin{aligned}
&\quad - \left\{(\gamma_1+1)^2 W_1 Y_1 + \gamma_1^2 X_1 Z_1\right\}, \\
(\hat{G}_1)_{21} &= -\rho_1 c\omega \left\{\gamma_1^2 Z_1 \cos P_1 + (\gamma_1+1)^2 W_1 \cos Q_1\right\}, \\
(\hat{G}_1)_{31} &= (\hat{G}_1)_{41} = \mathrm{i}\rho_1 c\omega\,\gamma_1\big[(\gamma_1+1)(2\gamma_1+1)(1-\cos P_1 \cos Q_1) \\
&\quad + \left\{(\gamma_1+1)^3 W_1 Y_1 + \gamma_1^3 X_1 Z_1\right\}\big], \\
(\hat{G}_1)_{51} &= \rho_1 c\omega \left\{(\gamma_1+1)^2 \cos P_1\, Y_1 + \gamma_1^2 X_1 \cos Q_1\right\}, \\
(\hat{G}_1)_{61} &= -(\rho_1 c\omega)^2 \big\{2\gamma_1^2(\gamma_1+1)^2(1-\cos P_1 \cos Q_1) \\
&\quad + (\gamma_1+1)^4 W_1 Y_1 + \gamma_1^4 X_1 Z_1\big\},
\end{aligned}
\tag{3.163}
$$

$$
\begin{aligned}
&W_1 = r_{\alpha 1}^{-1} \sin P_1\,,\ \ Y_1 = r_{\beta 1}^{-1} \sin Q_1\,,\ \ X_1 = r_{\alpha 1} \sin P_1\,,\ \ Z_1 = r_{\beta 1} \sin Q_1\,, \\
&\gamma_1 = -2k^2/k_{\beta 1}^2\,,\ r_{\alpha 1} = \nu_{\alpha 1}/k\,,\ r_{\beta 1} = \nu_{\beta 1}/k\,,\ P_1 = \nu_{\alpha 1} d_1\,,\ Q_1 = \nu_{\beta 1} d_1\,.
\end{aligned}
$$

Similar to (3.158) for $n = 1$, we extract $-\beta_2^4/(4\omega^4)$ from each term of $\hat{M}_{11}$ as a common factor and obtain the characteristic equation

$$
\begin{aligned}
\Delta_R &= \sum_{j=1}^{6} (\hat{\mathrm{T}}_2^{-1})'_{1j} (\hat{G}_1)_{j1} = 0\,, \\
(\hat{\mathrm{T}}_2^{-1})'_{11} &= 4k^2 + \frac{l_2^2}{\nu_{\alpha 2}\nu_{\beta 2}}\,,\quad (\hat{\mathrm{T}}_2^{-1})'_{12} = \frac{\omega^2}{\mathrm{i}\mu_2 \nu_{\beta 2}\beta_2^2}\,, \\
(\hat{\mathrm{T}}_2^{-1})'_{13} &= (\hat{\mathrm{T}}_2^{-1})'_{14} = \frac{\mathrm{i}\omega}{\mu_2 c}\left(2 + \frac{l_2}{\nu_{\alpha 2}\nu_{\beta 2}}\right), \\
(\hat{\mathrm{T}}_2^{-1})'_{15} &= \frac{\mathrm{i}\omega^2}{\mu_2 \nu_{\alpha 2}\beta_2^2}\,,\quad (\hat{\mathrm{T}}_2^{-1})'_{16} = \frac{1}{\rho_2^2\beta_2^4}\left(1 + \frac{k^2}{\nu_{\alpha 2}\nu_{\beta 2}}\right).
\end{aligned}
\tag{3.164}
$$

First, if the Rayleigh wave has low frequencies and ω is small, and if the P and SV waves are oscillatory in layer 1, we have $k_{\alpha 2} < k_{\beta 2} < k < k_{\alpha 1} < k_{\beta 1}$ and $\beta_1 < \alpha_1 < c < \beta_2 < \alpha_2$ in addition to (3.134). Hence, from (3.5)

$$
\begin{aligned}
\nu_{\alpha 1} &= \omega\sqrt{\alpha_1^{-2} - c^{-2}}\,,\ \nu_{\beta 1} = \omega\sqrt{\beta_1^{-2} - c^{-2}}\,, \\
\nu_{\alpha 2} &= -\mathrm{i}\omega\sqrt{c^{-2} - \alpha_2^{-2}}\,,\ \nu_{\beta 2} = -\mathrm{i}\omega\sqrt{c^{-2} - \beta_2^{-2}}\,.
\end{aligned}
\tag{3.165}
$$

Since ω is small and $\omega \to 0$, $(\hat{\mathrm{T}}_2^{-1})'_{12}$, $(\hat{\mathrm{T}}_2^{-1})'_{13}$, $(\hat{\mathrm{T}}_2^{-1})'_{14}$, $(\hat{\mathrm{T}}_2^{-1})'_{15}$, $(\hat{G}_1)_{61} \to 0$. In addition, since $\nu_{\alpha i}$, $\nu_{\beta i}$, P_1, $Q_1 \to 0$, $(\hat{G}_1)_{11} \to 1$. The characteristic equation then yields

$$
\Delta_R \to (\hat{\mathrm{T}}_2^{-1})'_{11} = 4k^2 + \frac{l_2^2}{\nu_{\alpha 2}\nu_{\beta 2}} = 0\,.
\tag{3.166}
$$

This agrees with (3.158) with $1 \to 2$ and is the characteristic equation of the Rayleigh wave in a semi-infinite medium composed of only layer 2. However, since in the above we impose the condition of being oscillatory in layer 1, this corresponds to the Rayleigh wave composed of the reverberations in layer 1. The phase velocity of this Rayleigh wave is denoted by c_{R2}.

We next consider the case in which the Rayleigh wave has high frequencies and ω is large, and $c < \beta_1 < \alpha_1$. From (3.5), $\nu_{\alpha 1} = -\mathrm{i}\sqrt{k^2 - \omega^2/\alpha_1^2} = -\mathrm{i}\omega\sqrt{1/c^2 - 1/\alpha_1^2}$, $\nu_{\beta 1} = -\mathrm{i}\sqrt{k^2 - \omega^2/\beta_l^2} = -\mathrm{i}\omega\sqrt{1/c^2 - 1/\beta_1^2}$, and $\cos z = \cosh \mathrm{i}z$, $\sin z = \mathrm{i}^{-1}\sinh \mathrm{i}z$ [91]. Hence,

$$\cos P_1 = \cos \nu_{\alpha 1} d_1 = \cosh \mathrm{i}\nu_{\alpha 1} d_1 \,, \tan P_1 = \mathrm{i}^{-1} \tanh \mathrm{i}\nu_{\alpha 1} d_1 \,.$$

Using the formulas for the hyperbolic functions [91], $\lim_{x\to\infty} \cosh x = \infty$ and $\lim_{x\to\infty} \tanh x = 1$, we have $\cos P_1 \to \infty$, $\tan P_1 \to -\mathrm{i}$ in the case of $\omega \to \infty$. In addition, $\cos Q_1 \to \infty$ and $\tan Q_1 \to -\mathrm{i}$. From these and (3.163), we obtain

$$\begin{aligned}
(\hat{G}_1)_{11} &= \cos P_1 \cos Q_1 \left[\frac{-2\gamma_1(\gamma_1+1)}{\cos P_1 \cos Q_1} + \{2\gamma_1(\gamma_1+1)+1\} \right. \\
&\quad \left. - \left\{ (\gamma_1+1)^2 \frac{\tan P_1 \tan Q_1}{r_{\alpha 1} r_{\beta 1}} + \gamma_1^2 r_{\alpha 1} r_{\beta 1} \tan P_1 \tan Q_1 \right\} \right] \\
&\to \frac{\cos P_1 \cos Q_1}{r_{\alpha 1} r_{\beta 1}} \left[r_{\alpha 1} r_{\beta 1} \{2\gamma_1(\gamma_1+1)+1\} + \{(\gamma_1+1)^2 + \gamma_1^2 r_{\alpha 1}^2 r_{\beta 1}^2\} \right] \\
&= \frac{\cos P_1 \cos Q_1}{r_{\alpha 1} r_{\beta 1}} (1 + r_{\alpha 1} r_{\beta 1}) \{(\gamma_1+1)^2 + \gamma_1^2 r_{\alpha 1} r_{\beta 1}\} \,, \\
(\hat{G}_1)_{21} &= \cos P_1 \cos Q_1 (-\rho_1 c\omega) \left\{ \gamma_1^2 r_{\beta 1} \tan Q_1 + (\gamma_1+1)^2 r_{\alpha 1}^{-1} \tan P_1 \right\} \\
&\to \frac{\cos P_1 \cos Q_1}{r_{\alpha 1} r_{\beta 1}} (-\rho_1 c\omega) \{ \gamma_1^2 r_{\alpha 1} r_{\beta 1}^2 (-\mathrm{i}) + (\gamma_1+1)^2 r_{\beta 1}(-\mathrm{i}) \} \\
&= \frac{\cos P_1 \cos Q_1}{r_{\alpha 1} r_{\beta 1}} \mathrm{i}\rho_1 c\omega\, r_{\beta 1} \{(\gamma_1+1)^2 + \gamma_1^2 r_{\alpha 1} r_{\beta 1}\} \,, \\
(\hat{G}_1)_{31} &= \frac{\cos P_1 \cos Q_1}{r_{\alpha 1} r_{\beta 1}} \mathrm{i}\rho_1 c\omega \left[\gamma_1(\gamma_1+1)(2\gamma_1+1) \left(\frac{1}{\cos P_1 \cos Q_1} - 1 \right) r_{\alpha 1} r_{\beta 1} \right. \\
&\quad \left. + \left\{ (\gamma_1+1)^3 \tan P_1 \tan Q_1 + \gamma_1^3 \tan P_1 \tan Q_1 r_{\alpha 1}^2 r_{\beta 1}^2 \right\} \right] \\
&\to \frac{\cos P_1 \cos Q_1}{r_{\alpha 1} r_{\beta 1}} \mathrm{i}\rho_1 c\omega \left[-\gamma_1(\gamma_1+1)(2\gamma_1+1) r_{\alpha 1} r_{\beta 1} \right. \\
&\quad \left. - \{(\gamma_1+1)^3 + \gamma_1^3 r_{\alpha 1}^2 r_{\beta 1}^2\} \right] \\
&= \frac{\cos P_1 \cos Q_1}{r_{\alpha 1} r_{\beta 1}} (-\mathrm{i}\rho_1 c\omega) \left[(\gamma_1 + 1 + \gamma_1 r_{\alpha 1} r_{\beta 1}) \{(\gamma_1+1)^2 + \gamma_1^2 r_{\alpha 1} r_{\beta 1}\} \right] , \\
(\hat{G}_1)_{51} &= \cos P_1 \cos Q_1 \rho_1 c\omega \left\{ (\gamma_1+1)^2 r_{\beta 1}^{-1} \tan Q_1 + \gamma_1^2 r_{\alpha 1} \tan P_1 \right\} \\
&\to \frac{\cos P_1 \cos Q_1}{r_{\alpha 1} r_{\beta 1}} \rho_1 c\omega \{ (\gamma_1+1)^2 r_{\alpha 1}(-\mathrm{i}) + \gamma_1^2 r_{\alpha 1}^2 r_{\beta 1}(-\mathrm{i}) \}
\end{aligned}$$

$$
\begin{aligned}
&= \frac{\cos P_1 \cos Q_1}{r_{\alpha 1} r_{\beta 1}} (-\mathrm{i}\rho_1 c\omega r_{\alpha 1}) \left\{ (\gamma_1+1)^2 + \gamma_1^2 r_{\alpha 1} r_{\beta 1} \right\}, \\
(\hat{G}_1)_{61} &= \cos P_1 \cos Q_1 \left\{ -(\rho_1 c\omega)^2 \right\} \left[\frac{2\gamma_1^2(\gamma_1+1)^2}{\cos P_1 \cos Q_1} - 2\gamma_1^2(\gamma_1+1)^2 \right. \\
&\quad \left. + \left\{ (\gamma_1+1)^4 \frac{\tan P_1 \tan Q_1}{r_{\alpha 1} r_{\beta 1}} + \gamma_1^4 r_{\alpha 1} r_{\beta 1} \tan P_1 \tan Q_1 \right\} \right] \\
&\to \frac{\cos P_1 \cos Q_1}{r_{\alpha 1} r_{\beta 1}} (\rho_1 c\omega)^2 \left[r_{\alpha 1} r_{\beta 1} \left\{ 2\gamma_1^2(\gamma_1+1)^2 \right\} - \left\{ -(\gamma_1+1)^4 - \gamma_1^4 r_{\alpha 1}^2 r_{\beta 1}^2 \right\} \right] \\
&= \frac{\cos P_1 \cos Q_1}{r_{\alpha 1} r_{\beta 1}} (\rho_1 c\omega)^2 \left\{ (\gamma_1+1)^2 + \gamma_1^2 r_{\alpha 1} r_{\beta 1}) \right\}^2 . \qquad (3.167)
\end{aligned}
$$

It can be noted that $\cos P_1 \cos Q_1/(r_{\alpha 1} r_{\beta 1})\left\{(\gamma_1+1)^2 + \gamma_1^2 r_{\alpha 1} r_{\beta 1}\right\}$ is included as a common factor in all $\hat{G}_{j1}$ in (3.167). However, $(\hat{\mathrm{T}}_2)_{1j}$ in (3.164) has no common factor. Hence, the characteristic equation is in the form[60]

$$
\begin{aligned}
\Delta_R &= \sum_{j=1}^{6} (\hat{\mathrm{T}}_2^{-1})'_{1j} (\hat{G}_1)_{j1} = 0 \\
&\to \frac{\cos P_1 \cos Q_1}{r_{\alpha 1} r_{\beta 1}} \left\{ (\gamma_1+1)^2 + \gamma_1^2 r_{\alpha 1} r_{\beta 1} \right\} F(c) = 0 . \qquad (3.168)
\end{aligned}
$$

The solution of the characteristic equation now yields a solution of the equation

$$
(\gamma_1+1)^2 + \gamma_1^2 r_{\alpha 1} r_{\beta 1} = 0 \qquad (3.169)
$$

which is the second factor of (3.168). Substituting $r_{\alpha i} = \nu_{\alpha i}/k$, $r_{\beta i} = \nu_{\beta i}/k$, and $\gamma_i = -2k^2/k_{\beta i}^2$ (these are in (3.33)), and $l_1 = 2k^2 - k_{\beta 1}^2$ in (3.157), and $k_{\beta i} = \omega/\beta_i$ in (3.5) into (3.169), we obtain

$$
\left(-\frac{2k^2}{k_{\beta 1}^2} + 1 \right)^2 + \frac{4k^4}{k_{\beta 1}^4} \frac{\nu_{\alpha 1}}{k} \frac{\nu_{\beta 1}}{k} = 0 \Rightarrow 4k^2 + \frac{l_1^2}{\nu_{\alpha 1} \nu_{\beta 1}} = 0 . \qquad (3.170)
$$

The right-hand side of (3.170) agrees with the characteristic equation (3.158) of the Rayleigh wave in a semi-infinite medium composed of only layer 1, therefore the phase velocity is equal to the Rayleigh wave velocity c_R in (3.161). This is the slowest phase velocity, and corresponds to the **fundamental mode** of the Rayleigh wave.

Finally, consider **higher mode** of the Rayleigh wave at high frequencies where ω is large but with phase velocities c faster than the fundamental mode and $\beta_1 < c < \alpha_1$. Since, from $\beta_1 < c < \alpha_1$ and (3.5), $\nu_{\alpha 1} = -\mathrm{i}\sqrt{k^2 - \omega^2/\alpha_1^2} = -\mathrm{i}\omega\sqrt{1/c^2 - 1/\alpha_1^2}$,

---

[60]This agrees with the equation on page 164 of Saito [103] if $C_\alpha, C_\beta \to \cos P_1, \cos Q_1, \gamma_1 \to -\gamma_1$, $\mathrm{i}\xi_1 \to r_{\alpha 1}/c$, and $\mathrm{i}\eta_1 \to r_{\beta 1}/c$. An actual expression of F is also given there.

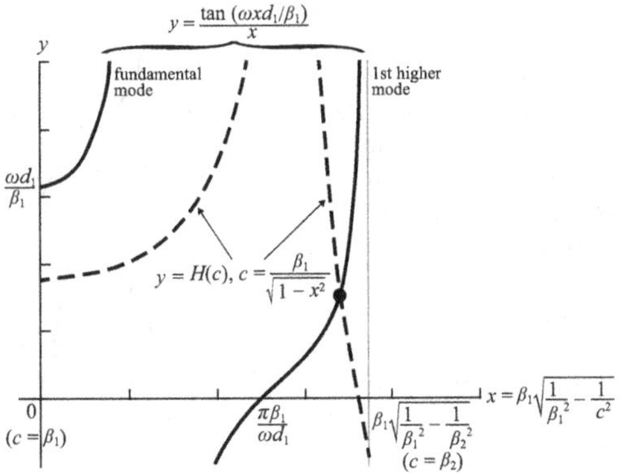

Fig. 3.16 Graphical solution of the characteristic equation of the Rayleigh wave in a two-layer structure, in the case of $\pi/\omega \cdot \beta_1/d_1 \le \beta_1\sqrt{1/\beta_1^2 - 1/\beta_2^2} < 2\pi/\omega \cdot \beta_1/d_1$ (modified from Saito [103] with permission of the author)

and $\nu_{\beta 1} = \sqrt{\omega^2/\beta_1^2 - k^2} = \omega\sqrt{1/\beta_1^2 - 1/c^2}$, we have

$$\lim_{\omega\to\infty} \cos P_1 = \infty\,, \quad \lim_{\omega\to\infty} \tan P_1 = -\mathrm{i}\,. \tag{3.171}$$

However, $\cos Q_1$ and $\tan Q_1$ remain as trigonometric functions even for $\omega \to \infty$. Hence, although the characteristic equation remains very complicated, $\cos Q_1$ can be extracted as a common factor from all $(\hat{G}_1)_{j1}$ as in the fundamental mode. The characteristic equation is then rewritten as (3.168), and its core part contains only $\tan Q_1$, therefore the equation

$$\tan Q_1 = \frac{\nu_{\beta 1}}{k_{\beta 1}} H(c)\,, \tag{3.172}$$

which is similar to (3.137) for the Love wave, can be obtained [103]. Furthermore, by substituting $Q_1 = \nu_{\beta 1} d_1$ in (3.163) and $\nu_{\beta 1} = \omega\sqrt{1/\beta_1^2 - 1/c^2}$ above, we obtain

$$\tan\left(\omega d_1\sqrt{\frac{1}{\beta_1^2} - \frac{1}{c^2}}\right) = \beta_1\sqrt{\frac{1}{\beta_1^2} - \frac{1}{c^2}}\,H(c)\,, \tag{3.173}$$

which is similar to (3.138) for the Love wave. Transforming (3.173) to

$$\frac{\tan(\omega x d_1/\beta_1)}{x} = H(c)\,,\ c = \frac{\beta_1}{\sqrt{1-x^2}} \tag{3.174}$$

with $x = \beta_1\sqrt{1/\beta_1^2 - 1/c^2}$ and plotting y = left-hand side and y = right-hand side similarly to the case of the Love wave, we can graphically obtain the intersection of the two curves as a solution of the characteristic equation (3.173), as shown in Fig. 3.16. The search range of the solution is from $x = 0$ corresponding to $c =$

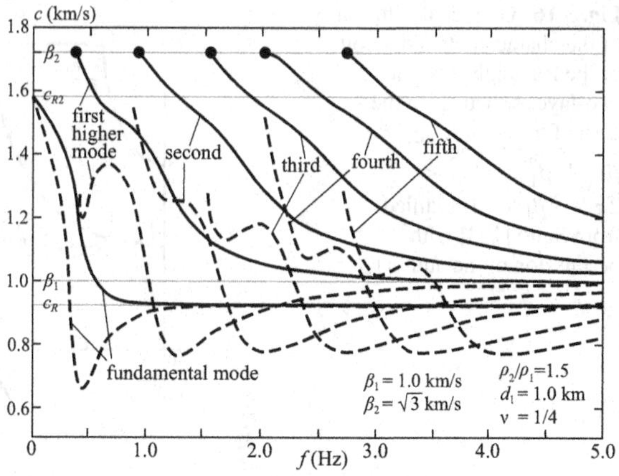

Fig. 3.17 Dispersion curves with respect to the frequency f for the Rayleigh waves in a two-layer structure. The black dots indicate the cut-off frequencies. In addition to the phase velocities shown by the solid lines, the group velocities are shown as broken lines. The parameters of the two-layer structure are shown in the lower right corner (modified from Saito [103] with permission of the author)

β_1, to $x = \beta_1\sqrt{1/\beta_1^2 - 1/\beta_2^2}$ corresponding to $c = \beta_2$, based on $c < \beta_2 < \alpha_2$ and $\beta_1 < c < \alpha_1$, which are the conditions that the ground motion is a higher mode of the surface wave. However, if the P wave velocity of layer 1 is very slow and $\alpha_1 < \beta_2$, the upper limit of c must be α_1. The number of solutions (number of higher modes) depends on the number of curves $y = \tan(\omega x d_1/\beta_1)/x$ in the search range. In Fig. 3.16, the case where $\pi/\omega \cdot \beta_1/d_1 \le \beta_1\sqrt{1/\beta_1^2 - 1/\beta_2^2} < 2\pi/\omega \cdot \beta_1/d_1$ is depicted, and there is one solution of the higher mode. The x-intercept $n\beta_1/(\omega d_1)$ of $y = \tan(\omega x d_1/\beta_1)/x$ approaches the lower limit of the range as ω increases, and the solution approaches the lower limit $c = \beta_1$. Conversely, as ω decreases, the x-intercept approaches the upper limit, and the solution approaches the upper limit $c = \beta_2$. Because of $\lim_{\theta\to\infty}(\tan\theta)/\theta = 1$, $y = \tan(\omega x d_1/\beta_1)/x$ for the fundamental mode does not pass through the origin as in the Love wave, so this does not intersect $y = H(c)$ in the range. As described above, the phase velocity of the fundamental mode, c_R, is smaller than β_1. Including the fundamental mode as the zero-th order mode, c increases with increasing mode order for the same ω, which is the same as for the Love wave. Further, similar to the Love wave, at present, the numerical solution of an equation is used rather than the graphical solution with the high frequency approximation explained here.

Figure 3.17 shows the **dispersion curve** of the Rayleigh wave in a two-layer structure obtained by numerical solution of the dispersion equation. The parameters of the two-layer structure are given in the lower right corner of the figure. Although c in the fundamental mode is equal to c_{R2} at $f = 0$ ($\omega = 0$), c rapidly approaches c_R when f or ω increases even slightly; therefore, in a horizontally layered structure it is generally appropriate that the phase velocity of the fundamental mode is approximated by the **Rayleigh wave velocity** c_R of the one-layer structure with the parameters of layer 1. The higher modes start at the upper limit β_2 of the range (the frequencies at this limit are also termed **cut-off frequency**) and asymptotically approach the lower limit β_1 of the range as f or ω increases. These characteristics of the dispersion curves are in agreement with the results in the previous paragraphs, which were examined approximately for the limits of $\omega \to 0$ and $\omega \to \infty$.

The phase velocity c or the wavenumber $k \equiv \omega/c$ of the Rayleigh wave can be obtained as explained above. As in the Love wave, the **group velocity** U is also obtained from the $dc/d\omega$ of a dispersion curve and (3.146), and is illustrated in Fig. 3.17. The weakness for a deep source, geometrical spreading in inverse proportion to $\sqrt{r}$, the **Airy phase**, etc. are also similar to the Love wave. The obtained k gives $\hat{M}_{11} = 0$ in (3.162) and $\hat{M}_{11}$ appears in the denominator of the ground motion solution in (3.38), therefore a **surface wave pole** is located at this k as in the Love wave. Similarly to the Love wave, the ground motion of only the Rayleigh wave is represented by the solution of the **normal mode**, the normal mode solution is obtained through the residue computation, and a virtual high-velocity layer in the lowermost halfspace gives a good approximation for the complete ground motion.

3.1.11 Teleseismic Body Wave

According to the USGS glossary [116], "teleseismic" is an adjective meaning "pertaining to earthquakes at distances greater than 1,000 km from the measurement site", and "teleseisms" indicate ground motions due to such distant earthquakes [65], which are termed **teleseismic earthquake**. The **teleseismic body wave** is therefore the body wave component of the ground motion caused by a teleseismic earthquake. A global network of broadband seismometers has been constructed, and observed teleseisms have sufficient detection capability that the moment tensor can be determined by a **CMT inversion** (Sect. 2.3.4) for an earthquake occurring anywhere in the world with an M_w of about 5.0 or greater. Moreover, as within a few hours following an earthquake waveform data from the data management center of **IRIS** (Incorporated Research Institutions for Seismology) are open via the Internet, these represent essential data for the **source inversion** (Sect. 2.3.6) of overseas earthquakes. Even for domestic earthquakes, although the resolution is inferior to other data because they come from distant stations, they are often useful to stabilize a solution of the source inversion (Sect. 2.3.6).

Figure 3.18 shows an example of the usefulness of teleseismic body wave data. The result of the joint inversion in the lower right shows that the source process of the Kobe earthquake is characterized by a deep large slip near the rupture initiation point ($x = 0$, $y = 0$ km), a shallow large slip ($x = 4 \sim 8$, $y = 12$ km) on the Awaji Island side (A in the left panel of Fig. 2.35), and a rather large slip near the central part ($x = 20$, $y = 4$ km) on the Kobe side (B in the left panel of Fig. 2.35). However, in the result of the single inversion of the strong motion data in the upper right, the second feature, that is the shallow large slip on the Awaji Island side, is not very clear. In addition, in the result of the single inversion of the teleseismic body wave data in the upper left, the third feature, that is the rather large slip near the central part on the Kobe side, is not visible due to poor resolution, but the second feature

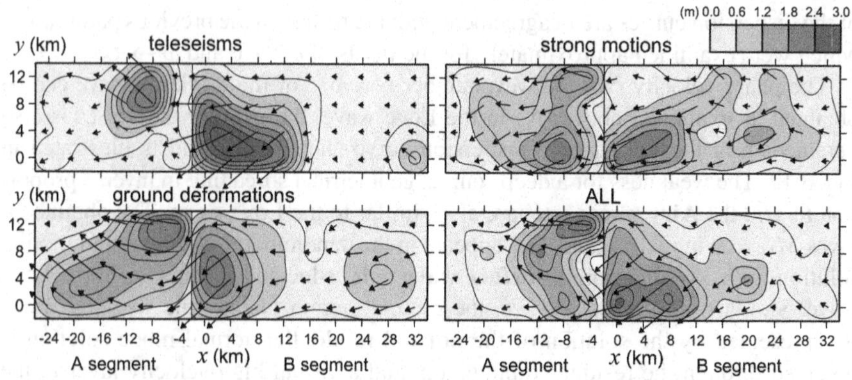

Fig. 3.18 All results of Yoshida et al.'s [127] source inversions for the Kobe earthquake (slip vectors and slip distributions). The results of the single inversions of teleseismic body wave data, strong motion data, and crustal deformation data are shown in the upper left, upper right, and lower left, respectively. The lower right is the result of the joint inversion of all three datasets. The two on the right-hand side are the same as in Fig. 2.35. Reprinted from Koketsu [72] with permission of Kindai Kagaku

is clearly visible. The joint inversion including both the strong motion data and the teleseismic body wave data can therefore reveal both the second and third features.

As the method of Kikuchi and Kanamori [67] is the most widely used for single inversions of teleseismic body wave data, the computation of the Green's functions used there is explained here. The computation of the Green's functions in Yoshida et al. [127] is also similar to this, but the effects of velocity structures near the source and the observation point are excluded. The Earth forms a sphere, hence the **spherical coordinate system** (R, θ, ϕ) in Fig. 2.23 with the origin at the Earth's center is used. We assume that the Earth has a 1-D velocity structure as in the title of this section (Sect. 3.1), and this assumption indicates that the structure is in spherical symmetry varying only in the R direction, and that the properties (Sect. 1.2.4) are functions of R only. The **ray theory** (Sect. 3.2.2), which is the most approximate method for computing the effect of propagation, is applied to this spherically symmetric structure. In addition, since the earthquake is distant, the source can be assumed to be a point source and the body wave emitted from the point source can be represented only by the **far-field term** (Sect. 2.1.4) in (2.56), and their elastic displacements are given in (2.162).

First, consider **ray tracing** (Sect. 3.2.3) in the general spherical coordinate system in Fig. 3.19. Since

$$\nabla = \left(\frac{\partial}{\partial R}, \frac{1}{R}\frac{\partial}{\partial \theta}, \frac{1}{R\sin\theta}\frac{\partial}{\partial \phi} \right) \tag{3.175}$$

in the spherical coordinate system, the vector representing the direction of the ray, $\mathbf{p} = \nabla\tau$ in (3.244) (Fig. 3.22) yields

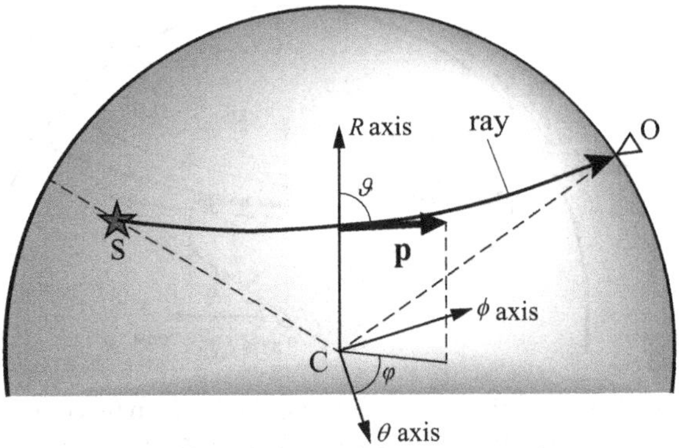

Fig. 3.19 The azimuthal angle φ, the takeoff angle ϑ, and the direction vector **p** (thick arrow) of the ray (thick solid line) for the teleseismic body wave in the general spherical coordinate system placed at the center of the Earth. Reprinted from Koketsu [72] with permission of Kindai Kagaku

$$\mathbf{p} = \left(p_R, \frac{p_\theta}{R}, \frac{p_\phi}{R\sin\theta}\right), \quad p_R = \frac{\partial\tau}{\partial R}, \quad p_\theta = \frac{\partial\tau}{\partial\theta}, \quad p_\phi = \frac{\partial\tau}{\partial\phi}. \tag{3.176}$$

The eikonal equation (3.237) yields

$$p_R^2 + \frac{p_\theta^2}{R^2} + \frac{p_\phi^2}{R^2\sin^2\theta} = s^2, \quad s = \alpha^{-1} \text{ or } \beta^{-1}. \tag{3.177}$$

The ray equation (3.243) yields[61]

$$\begin{aligned} &\frac{dR}{d\lambda} = \frac{p_R}{s}, \quad \frac{d\theta}{d\lambda} = \frac{p_\theta}{sR^2}, \quad \frac{d\phi}{d\lambda} = \frac{p_\phi}{sR^2\sin^2\theta}, \\ &\frac{dp_R}{d\lambda} = \frac{\partial s}{\partial R} + \frac{p_\theta^2}{sR^3} + \frac{p_\phi^2}{sR^3\sin^2\theta}, \quad \frac{dp_\theta}{d\lambda} = \frac{\partial s}{\partial\theta} + \frac{\cot\theta}{sR^2\sin^2\theta}p_\phi^2, \quad \frac{dp_\phi}{d\lambda} = \frac{\partial s}{\partial\phi}. \end{aligned} \tag{3.178}$$

Next, consider the **azimuthal angle** φ and the **takeoff angle** ϑ (Sect. 3.2.3). φ is defined to be an angle between the $\theta-\phi$ component of **p** and the θ-axis, and ϑ is defined to be an angle between **p** and the R-axis (Fig. 3.19). Using these and (3.177), and following (3.277), we obtain[62]

$$p_R = s\cos\vartheta, \quad \frac{p_\theta}{R} = s\sin\vartheta\cos\varphi, \quad \frac{p_\phi}{R\sin\theta} = s\sin\vartheta\sin\varphi. \tag{3.179}$$

The subscript S indicates these coordinates and parameters at the source.

As the slowness s in (3.177) is a property, s depends only on R being independent of θ and ϕ. $dp_\phi/d\lambda = 0$ from the sixth equation in (3.178); therefore, p_ϕ does not change from the initial value $p_{\phi S}$ at the point source, that is $p_\phi \equiv p_{\phi S}$. In this case, since the directions of the θ and ϕ axes of the spherical coordinate system do not affect the results if they are orthogonal to each other [19], the axes are oriented such

[61] Equation (3.177) agrees with Eq. (3.5.28) of Červený [19] if $T \to \tau$ and $T_j \to p_j$. Equation (3.178) agrees with Eq. (3.5.31) of Červený [19] if $n = 1$, $u \to \lambda$, and $T_j \to p_j$.

[62] This agrees with Eq. (15.206) of Dahlen and Tromp [27] if $r \to R$, $i \to \vartheta$, and $\zeta \to \varphi$.

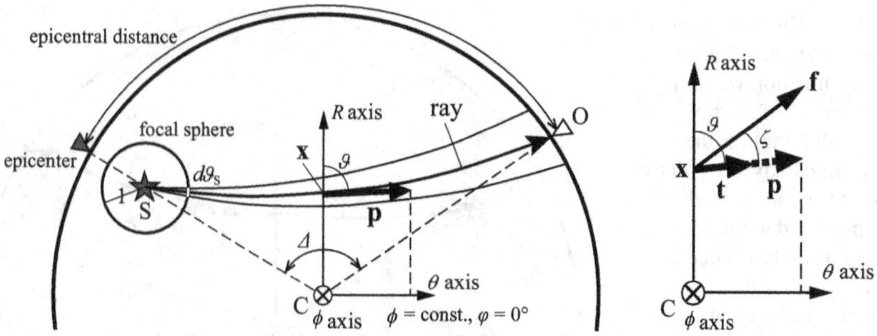

Fig. 3.20 Ray (thick solid line), direction vector **p** (thick arrow), azimuthal angle φ, and takeoff angle ϑ of the teleseismic body wave in the cylindrical coordinate system. Δ stands for the epicentral distance between the point source S and the observation point O. The right diagram shows an enlarged view illustrating the relationship of the vectors near **x**. Reprinted from Koketsu [72] with permission of Kindai Kagaku

that the azimuthal angle at the point source φ_S is 0°, and then $p_\phi \equiv p_{\phi S} = 0$. $p_\phi = 0$ and the third equation in (3.179) leads to $\sin\varphi = 0$ or $\sin\vartheta = 0$. As from the latter $\sin\vartheta = 0$, we only have the trivial solution of $\mathbf{p} = (s, 0, 0)$, we obtain $\varphi = 0°$ from the former $\sin\varphi = 0$ as long as we limit rays emitted forward. Substituting this into the third equation in (3.178), $d\phi/d\lambda = 0$ is obtained, and the ray remains in the plane of $\phi =$ constant. Then, substituting $p_\phi = 0$ and $\partial s/\partial\theta = 0$ into the fifth equation in (3.178)), $dp_\theta/d\lambda = 0$ is obtained similarly, and $p_\theta =$ constant. Finally, substituting $\varphi = 0°$ and $s = \upsilon^{-1}$ into the second equation of (3.179) and letting the constant value of p_θ be p, **Snell's law** in the spherical coordinate system

$$p_\theta = \frac{R\sin\vartheta}{\upsilon} = \text{constant} = p \tag{3.180}$$

is obtained.[63] p is called the **ray parameter**.

In global-scale problems such as teleseismic body waves, the **epicentral distance** is not shown in km but with the angle of the great circle connecting the epicenter and the observation point with the center of the Earth (Δ in Fig. 3.20). This angle corresponds to the ϑ of the observation point O viewed from the point source S, and here $\vartheta = \theta$ from $\varphi = 0°$, and using (3.178)

$$\Delta = \int_S^O \frac{d\theta}{d\lambda} d\lambda = \int_{R_S}^{R_O} \frac{d\theta}{dR} dR\,, \quad \frac{d\theta}{dR} = \frac{d\theta}{d\lambda}\frac{d\lambda}{dR} = \frac{p_\theta}{R^2 p_R}\,. \tag{3.181}$$

From (3.177) and $p_\phi \equiv 0$, we have

[63] This agrees with Eq. (4.6) of Utsu [117] if $r \to R$ and $i \to \vartheta$. In the Cartesian coordinate system, if s is a function of only z, p_x in (3.277) is constant, and Snell's law is $\sin\vartheta/\upsilon =$ constant.

$$p_R = \left(s^2 - \frac{p_\theta^2}{R^2}\right)^{1/2} = \left(\frac{1}{\upsilon^2} - \frac{p^2}{R^2}\right)^{1/2}, \quad \upsilon = \alpha \text{ or } \beta \tag{3.182}$$

and obtain

$$\frac{d\theta}{dR} = \frac{p_\theta}{R^2 p_R} = \frac{p}{R^2\left(1/\upsilon^2 - p^2/R^2\right)^{1/2}} = \frac{p}{R\left\{(R/\upsilon)^2 - p^2\right\}^{1/2}}. \tag{3.183}$$

If the point source is relatively shallow and can be regarded as close to the ground surface, a turning point appears at the midpoint of the ray as shown in Fig. 3.9, and the ray becomes symmetrical with respect to that point. Hence, we let the Rs of the ground surface and the turning point be R_0 and R_m, and we have[64]

$$\Delta = 2\int_{R_m}^{R_0} \frac{d\theta}{dR} dR = 2\int_{R_m}^{R_0} \frac{p\,dR}{R\left\{(R/\upsilon)^2 - p^2\right\}^{1/2}}. \tag{3.184}$$

It is worth noting that the time T for which the ground motion propagates from the point source to the observation point, that is, the travel time (Sect. 3.2.2), has the derivative $dT/d\lambda = s$ (Sect. 3.2.3); hence,

$$\frac{dT}{dR} = \frac{dT}{d\lambda}\frac{d\lambda}{dR} = \frac{s^2}{p_R} = \frac{1}{\upsilon^2\left(1/\upsilon^2 - p^2/R^2\right)^{1/2}} = \frac{R}{\upsilon\left(R^2 - \upsilon^2 p^2\right)^{1/2}}. \tag{3.185}$$

Similar to Δ, in the case of a point source close to the ground surface, we have[65]

$$T = 2\int_{R_m}^{R_0} \frac{dT}{dR} dR = 2\int_{R_m}^{R_0} \frac{R\,dR}{\upsilon\left(R^2 - \upsilon^2 p^2\right)^{1/2}}. \tag{3.186}$$

In the general case where the point source is located at a depth h,

$$\begin{aligned} \Delta &= 2\int_{R_m}^{R_0-h} \frac{p\,dR}{R\left\{(R/\upsilon)^2 - p^2\right\}^{1/2}} + \int_{R_0-h}^{R_0} \frac{p\,dR}{R\left\{(R/\upsilon)^2 - p^2\right\}^{1/2}} \\ T &= 2\int_{R_m}^{R_0-h} \frac{R\,dR}{\upsilon\left(R^2 - \upsilon^2 p^2\right)^{1/2}} + \int_{R_0-h}^{R_0} \frac{R\,dR}{\upsilon\left(R^2 - \upsilon^2 p^2\right)^{1/2}}. \end{aligned} \tag{3.187}$$

If no turning point appears, only the second term is required. Equation (3.187) indicates that the integrals of R replace the ray tracing in the spherically symmetric structure. A spherically symmetric velocity structure $\upsilon(R)$ must be given for the integration, but the Green's functions used for the source inversion do not need to

[64]Equations (3.183) and (3.184) respectively agree with Eqs. (4.11) and (4.12) of Utsu [117] if $r \to R$.

[65]Equations (3.185) and (3.186) respectively agree with Eqs. (3.7.33) and (3.7.37) of Červený [19].

be as precise. For example, in Kikuchi and Kanamori [67], the classical velocity structure of Jeffreys and Bullen [58] is the default.

We next consider dynamic ray tracing (Sect. 3.2.3) in a spherically symmetric structure. Comparing Figs. 3.20 and 3.22, the coordinate system $\mathbf{x}$ for the position of the ray in Sect. 3.2.2 corresponds to the spherical coordinate system (R, θ, ϕ) here, and γ_1, γ_2 of the ray-centered coordinate system $\boldsymbol{\lambda}$ in Sect. 3.2.2 corresponds to the coordinates along the ϑ and φ axes here. For later convenience, we set $\gamma_1 = \vartheta_S$ and $\gamma_2 = \varphi_S$. The **ray Jacobian** $\mathcal{J}$ is defined in Sect. 3.2.2 for the Cartesian coordinate system, and this is transformed into the spherical coordinate system as

$$\mathcal{J} = \frac{D(x, y, z)}{D(\vartheta_S, \varphi_S, \lambda)} = \frac{D(x, y, z)}{D(R, \theta, \phi)} \frac{D(R, \theta, \phi)}{D(\vartheta_S, \varphi_S, \lambda)}, \quad \frac{D(R, \theta, \phi)}{D(\vartheta_S, \varphi_S, \lambda)} = \begin{vmatrix} \dfrac{\partial R}{\partial \vartheta_S} & \dfrac{\partial R}{\partial \varphi_S} & \dfrac{dR}{\partial \lambda} \\ \dfrac{\partial \theta}{\partial \vartheta_S} & \dfrac{\partial \theta}{\partial \varphi_S} & \dfrac{d\theta}{d\lambda} \\ \dfrac{\partial \phi}{\partial \vartheta_S} & \dfrac{\partial \phi}{\partial \varphi_S} & \dfrac{d\phi}{d\lambda} \end{vmatrix} . \tag{3.188}$$

The Jacobian $D(x, y, z)/D(R, \theta, \phi)$ between the Cartesian and spherical coordinate systems is $R^2 \sin\theta$ [55]. In addition, since ϕ = constant and $\varphi = 0°$ as described above, $\partial\phi/\partial\vartheta_S = 0$ and $\partial\phi/\partial\varphi_S = 1$ [19], and $d\phi/d\lambda = 0$. Substituting these into (3.188) and performing the **cofactor expansion**, we obtain

$$\mathcal{J} = R^2 \sin\theta \begin{vmatrix} \dfrac{\partial R}{\partial \vartheta_S} & \dfrac{dR}{d\lambda} \\ \dfrac{\partial \theta}{\partial \vartheta_S} & \dfrac{d\theta}{d\lambda} \end{vmatrix} = R \sin\theta \left| \left(\frac{\partial R}{\partial \vartheta_S}, R\frac{\partial \theta}{\partial \vartheta_S}, 0 \right) \times \left(\frac{dR}{d\lambda}, R\frac{d\theta}{d\lambda}, 0 \right) \right| . \tag{3.189}$$

From (3.176), (3.178), and $p_\phi = 0$, we find that

$$\left(\frac{dR}{d\lambda}, R\frac{d\theta}{d\lambda}, 0 \right) = \frac{1}{s} \left(p_R, \frac{p_\theta}{R}, \frac{p_\phi}{R \sin\theta} \right) = \frac{\mathbf{p}}{s} = \mathbf{t} \tag{3.190}$$

is a unit vector in the tangential direction of the ray (Fig. 3.22). Using the formula for the vector product $|\mathbf{a} \times \mathbf{b}| = |\mathbf{a}||\mathbf{b}| \sin\eta$ (η is the angle between $\mathbf{a}$ and $\mathbf{b}$) and the configuration shown in the right panel of Fig. 3.20, we obtain

$$\mathcal{J} = R \sin\theta \, |\,\mathbf{f}\,| \sin\zeta \,, \quad \mathbf{f} = \left(\frac{\partial R}{\partial \vartheta_S}, R\frac{\partial \theta}{\partial \vartheta_S}, 0 \right) . \tag{3.191}$$

From here, the takeoff angle and related derivatives are evaluated on spheres of R = constant [19] (the ground surface is also a sphere of R = constant). Since $\partial R/\partial\vartheta_S = 0$, $\mathbf{f}$ has only the θ-component and is parallel to the θ-axis, and $\zeta = 90° - \vartheta$. We therefore obtain the ray Jacobian:

$$\mathcal{J} = R^2 \sin\theta \cos\vartheta \frac{\partial \theta}{\partial \vartheta_S} . \tag{3.192}$$

Using (3.267), the amplitude $u_{1\mathrm{O}}^{(0)}$ of the P wave ground motion at the observation point O is related to the amplitude $u_{1\mathrm{S}}^{(0)}$ at the point source as

$$u_{1\mathrm{O}}^{(0)} = \frac{1}{\mathcal{L}} u_{1\mathrm{S}}^{(0)}, \quad \frac{1}{\mathcal{L}} = \left(\frac{\rho_\mathrm{S} \alpha_\mathrm{S} \mathcal{J}_\mathrm{S}}{\rho_\mathrm{O} \alpha_\mathrm{O} \mathcal{J}_\mathrm{O}} \right)^{1/2}, \tag{3.193}$$

where $1/\mathcal{L}$ is termed **geometrical spreading**. In Sect. 1.2.5, the phenomenon of decreasing amplitude due to the expansion of the wavefront is also named "geometrical spreading", as well as the mathematical expression of the amplitude decrease, such as $1/\mathcal{L}$. The latter is sometimes called a "geometrical spreading factor" to distinguish it [27, 67]. As described above, $u_{i\mathrm{S}}^{(0)}$ is given by (2.162), but the term $1/R$, where R indicates the distance from the point source, diverges at the point source, hence a **focal sphere** (Sect. 2.3.2) with a radius of 1 is set around the point source as shown in Fig. 3.20. As the takeoff angle is evaluated in the plane of $R =$ constant, its infinitesimal change (Fig. 3.20) is given as

$$d\vartheta_\mathrm{S} = \cos\vartheta_\mathrm{S} d\theta_\mathrm{S} \,. \tag{3.194}$$

We use this and $\theta_\mathrm{O} = \Delta$ from (3.181), obtaining[66]

$$\frac{1}{\mathcal{L}} = \left(\frac{\rho_\mathrm{S} \alpha_\mathrm{S} R_\mathrm{S}^2 \sin\theta_\mathrm{S} \cos\vartheta_\mathrm{S} \dfrac{\partial\theta_\mathrm{S}}{\partial\vartheta_\mathrm{S}}}{\rho_\mathrm{O} \alpha_\mathrm{O} R_\mathrm{O}^2 \sin\theta_\mathrm{O} \cos\vartheta_\mathrm{O} \dfrac{\partial\theta_\mathrm{O}}{\partial\vartheta_\mathrm{S}}} \right)^{1/2} = \frac{1}{R_\mathrm{O}} \left(\frac{\rho_\mathrm{S} \alpha_\mathrm{S} \sin\theta_\mathrm{S}}{\rho_\mathrm{O} \alpha_\mathrm{O} \sin\Delta \cos\vartheta_\mathrm{O}} \frac{d\vartheta_\mathrm{S}}{d\Delta} \right)^{1/2}. \tag{3.195}$$

For the amplitudes $u_2^{(0)}$ and $u_3^{(0)}$ of the S waves, from (3.276) we similarly obtain

$$u_{2\mathrm{O}}^{(0)} = \frac{1}{\mathcal{L}} u_{2\mathrm{S}}^{(0)}, \quad u_{3\mathrm{O}}^{(0)} = \frac{1}{\mathcal{L}} u_{3\mathrm{S}}^{(0)}, \quad \frac{1}{\mathcal{L}} = \frac{1}{R_\mathrm{O}} \left(\frac{\rho_\mathrm{S} \beta_\mathrm{S} \sin\theta_\mathrm{S}}{\rho_\mathrm{O} \beta_\mathrm{O} \sin\Delta \cos\vartheta_\mathrm{O}} \frac{d\vartheta_\mathrm{S}}{d\Delta} \right)^{1/2}. \tag{3.196}$$

In a spherically symmetric structure, the ray remains in the plane containing the point source S, the observation point O, and the Earth's center C (Fig. 3.20). The vertical vector from C to O and the normal vector **n** of the ray (Fig. 3.22) lie in this plane; therefore, the amplitude $u_2^{(0)}$ in (3.250) for the **n**-component of the S wave ground motion represents the **SV wave** (Sect. 1.2.4). In addition, since the binormal vector **b** is perpendicular to this plane, the amplitude $u_3^{(0)}$ of the **b**-component represents the **SH wave** (Sect. 1.2.4). Equations (3.195) and (3.196) indicate that dynamic ray tracing is not necessary in the spherically symmetric structure, and the amplitudes can be calculated only from Δ and T obtained from the integrations on R and physical properties at the point source and observation point.

[66] This agrees with Eq. (8.22) of Kikuchi [66] if $h, 0 \to \mathrm{S}, \mathrm{O}$, also with Eq. (4.20) of Utsu [117] if $r \to R$ and $\sin e_0 \to \cos\vartheta_\mathrm{O}$, and with equations in many other documents. While the equations were explained by approximate geometry in these documents, the physics behind (3.195) is explained in this subsection, Sects. 3.2.2, and 3.2.3.

As most of the ray of a teleseismic body wave is located in the **mantle** (Sect. 2.1.1) within the Earth's interior, and the spatial changes of physical properties in the mantle are gentle, it is sufficient to apply the ray theory. However, it is desirable to use the methods explained in Sects. 3.1.1 – 3.1.8, because the segments of the ray under the observation point and near the source of a shallow earthquake are in the uppermost part of the Earth, i.e., the **crust** (Sect. 2.1.1), where the spatial changes of physical properties are large. The ground motions that propagate from the point source to the lowest part of the crustal structure just under the observation point, spreading into the plane waves, are calculated using the ray theory. The results are then converted into displacement potentials and the Fourier transforms are performed on the potentials. As the source does not exist in the crustal structure beneath the observation point, the equations are solved by substituting the Fourier transforms into $\chi_{ln}^-(z_{n-1})$ for the SH waves, $\phi_{ln}^-(z_{n-1})$ for the P waves, or $\psi_{ln}^-(z_{n-1})$ for SV waves, in (3.24) or (3.37) with $\boldsymbol{\Delta}_{li} = 0$. The effect of the crustal structure near the point source is similarly obtained using the **reciprocity theorem** (Sect. 1.3.2) to interchange the observation point with the point source. The ground motion spectra including the geometrical spreading can now be obtained through the method explained above. Kikuchi and Kanamori [67] used the **Haskell matrix** (Sect. 3.1.4) to evaluate the effects of the crustal structures.

Finally, **intrinsic attenuation** must be considered in order to compute realistic teleseismic body waves. However, the method used in Sects. 3.1.1 – 3.1.8, in which the **Q**'s of the intrinsic attenuation are given to the seismic wave velocities by making them complex, cannot be simply applied. This is because the formulation of ray tracing and dynamic ray tracing cannot be established when the velocities are complex. The simplest way to make the method applicable is to approximate the Earth as being locally homogeneous. The 1-D ground motion in a homogeneous medium with velocity υ and Q yields a **damped oscillation** (Sect. 1.2.6), and its amplitude decays according to

$$e^{-\eta t}, \quad \eta = \frac{\omega}{2Q} \tag{3.197}$$

which are derived from (1.61) and (1.65). The ground motions according to the ray theory can be regarded as 1-D ground motions along the ray, and if the effect in (3.197) is continuous along the ray, ground motions at the observation point can be approximated to decay according to

$$e^{-\omega Y}, \quad Y = \int_0^T \frac{1}{2Q} d\tau = \int_S^O \frac{1}{2Q} \frac{d\lambda}{\upsilon} . \tag{3.198}$$

Further, $\overline{Q}$, which is the average Q, is defined as $T/2\overline{Q} = Y$, and t^* is defined as $t^* = T/\overline{Q}$, and we then obtain

$$e^{-\omega Y} = e^{-\omega T/2\overline{Q}} = e^{-\omega t^*/2}. \tag{3.199}$$

The standard values of t^* in the mantle are known to be 1 s for the P wave and 4 s for the S wave [67]. These standard values are often used instead of performing the integration in (3.198). In addition, we add a term to prevent (3.199) from violating the causality of the ground motion as shown in Sect. 1.2.6, and obtain[67]

$$D(\omega) = \exp\left(-\frac{1}{2}\omega t^* + \mathrm{i}\omega\frac{t^*}{\pi}\ln\frac{\omega}{\omega_r}\right). \tag{3.200}$$

$D(\omega)$ is termed a **dissipation filter** [19]. Multiplying this by the ground motion spectrum with the geometrical spreading and performing the inverse Fourier transform, we obtain the ground motion including both the geometrical spreading and the intrinsic attenuation.

3.1.12 Crustal Deformation

crustal deformation, which was observed by surveying in the past and in recent years has been observed by satellite technology such as **GNSS** (Global Navigation Satellite System)[68] and **InSAR** (Interferometric Synthetic Aperture Radar), is also referred to as static displacement, permanent displacement, etc., and is often thought of as different from ground "motion". However, crustal deformation is the part of ground motion with a frequency component of $\omega = 0$, because this is a result of the effect of the flat parts after the rise time of the moment time function and the slip time function, e.g., the $t > \tau$ part of the ramp function in Fig. 2.25.

In the source inversion, although crustal deformation data contain no information on the time history, they contribute the same stabilizing effect as teleseismic body wave data. In particular, they represent the optimum data to eliminate the **tradeoff** (Sect. 2.3.6) between time and position of fault rupture, because they have no time information. In the case of the Kobe earthquake in Fig. 3.18, the single inversion of strong motion data shows a rather deep large slip on the Awaji Island side, while the single inversion of crustal deformation data shows a correct large slip near the Nojima fault, which appears on the ground surface. As a result, the source process model with the correct slip positions and time information can be obtained from the joint inversion.

We can compute crustal deformation in a 1-D velocity structure, by applying $\omega \to 0$ to the methods in Sects. 3.1.1 – 3.1.8 as follows [130]. Here, the result of applying $\omega \to 0$ to $f(\omega)$ is represented as $\lim_{\omega\to 0} f(\omega) = (f)_0$. Since the $\omega - k$ spectrum of the ground motion of the SH wave, $\tilde{v}_{l1}(0)$, is computed from the substitution of (3.23) for

---

[67]The exponent in this agrees with the exponent in Eq. (8.12) of Kikuchi [67] if $T/\overline{Q} \to t^*$ and $2\pi f \to \omega$. This also agrees with Eq. (5.5.19) of Červený [19], though the sign of $\mathrm{i}\omega$ is inverted due to the difference in the definition of the Fourier transform.

[68]Though this is often called **GPS** (Global Positioning System), it should be called this because GPS is a proper noun of the observation network by the USA.

(3.25), $\tilde{v}_{l1}(0)$ is composed of the discontinuity vector $\mathbf{\Delta}_l$ and the propagator matrices $\mathbf{G}_i$ and $\mathbf{T}_n^{-1}$ (Sect. 3.1.2). For $\mathbf{\Delta}_l$ in (3.20), the square of $k_{\beta i} = \omega/\beta_i$ in (3.5) cancels ω^2 in the denominator, therefore

$$(\mathbf{\Delta}_0)_0 = \begin{pmatrix} 0 \\ 0 \end{pmatrix}, \quad (\mathbf{\Delta}_1)_0 = \frac{\overline{M}_0(0)}{4\pi\rho_s}\begin{pmatrix} -\dfrac{2}{\beta_s^2} \\ 0 \end{pmatrix}, \quad (\mathbf{\Delta}_2)_0 = \frac{\overline{M}_0(0)}{4\pi\rho_s}\begin{pmatrix} 0 \\ -\dfrac{\mu_s k}{\beta_s^2} \end{pmatrix}. \tag{3.201}$$

Applying the definition $\nu_{\beta i} = -\mathrm{i}\sqrt{k^2 - k_{\beta i}^2}$ for $k_{\beta i} < k$ in (3.5) and $(\nu_{\beta i})_0 = -ik$ by $(k_{\beta i})_0 = 0$ to (3.17) and (3.15), we obtain

$$(\mathbf{G}_i)_0 = \begin{pmatrix} \cosh kd_i & (\mu_i k)^{-1}\sinh kd_i \\ \mu_i k \sinh kd_i & \cosh kd_i \end{pmatrix}, \quad (\mathbf{T}_n^{-1})_0 = \frac{1}{2}\begin{pmatrix} -1 & \dfrac{-1}{\mu_n k} \\ -1 & \dfrac{+1}{\mu_n k} \end{pmatrix}. \tag{3.202}$$

Equation (3.202) almost agrees with Eq. (A10) of Zhu and Rivera [130]. However, due to Haskell's [44] nondimensionalization of $\tilde{v}_{l1}(0)$ (Sect. 3.1.4) and the difference between Eq. (A5) of Zhu and Rivera [130] and Eq. (9.6) of Haskell [44] (or (3.53) in this book), $((G_i)_{12})_0$ becomes $1/(\mathrm{i}^2 k)$ times and $((G_i)_{21})_0$ becomes $\mathrm{i}^2 k$ times. By substituting the obtained $(\mathbf{\Delta}_l)_0$, $(\mathbf{G}_i)_0$, and $(\mathbf{T}_n^{-1})_0$ into (3.23) and (3.25), we can compute the SH type of crustal deformation $(\tilde{v}_{l1}(0))_0$.

For the P and SV types of crustal deformation, the $\omega - k$ spectra of the ground motions of the P and SV waves, $\tilde{u}_{l1}(0)$ and $\tilde{w}_{l1}(0)$, are composed of $\mathbf{\Delta}_l$, $\mathbf{G}_i$, and $\mathbf{T}_n^{-1}$, which are given in (3.34), (3.33), and (3.30). For $\mathbf{\Delta}_l$ in (3.34), similarly to the SH type, the square of $k_{\alpha i} = \omega/\alpha_i$ or $k_{\beta i} = \omega/\beta_i$ in (3.5) cancels ω^2 in the denominator, therefore

$$(\mathbf{\Delta}_0)_0 = \frac{\overline{M}_0(0)}{4\pi\rho_s}\begin{pmatrix} 0 \\ 4k/\alpha_s^2 \\ 0 \\ 2\mathrm{i}\mu_s k^2(4/\alpha_s^2 - 3/\beta_s^2) \end{pmatrix}, \quad (\mathbf{\Delta}_1)_0 = \frac{\overline{M}_0(0)}{4\pi\rho_s}\begin{pmatrix} -2\mathrm{i}k/\beta_s^2 \\ 0 \\ 0 \\ 0 \end{pmatrix},$$
$$(\mathbf{\Delta}_2)_0 = \frac{\overline{M}_0(0)}{4\pi\rho_s}\begin{pmatrix} 0 \\ 0 \\ 0 \\ -2\mathrm{i}\mu_s k^2/\beta_s^2 \end{pmatrix}. \tag{3.203}$$

In contrast to this, $(\mathbf{G}_i)_0$ requires complex derivation, but if we use **L'Hôpital's rule** [5][69]

$$\lim_{x\to x_0} f(x) = 0, \lim_{x\to x_0} g(x) = 0 \Rightarrow \lim_{x\to x_0}\frac{f(x)}{g(x)} = \lim_{x\to x_0}\frac{f'(x)}{g'(x)}, \tag{3.204}$$

[69] Many works including Arfken and Weber [5] write "L'Hôpital" as "L'Hospital", an older spelling.

where "'" represents the first-order derivative, we can derive them [130]. Since $(\cos P_i - \cos Q_i)_0 = 0$, $(\gamma_i^{-1})_0 = 0$ (P_i, Q_i, γ_i are given in (3.33)), from this rule we have

$$\left(\frac{\cos P_i - \cos Q_i}{\gamma_i^{-1}}\right)_0 = \left(\frac{-(P_i)' \sin P_i + (Q_i)' \sin Q_i}{(\gamma_i^{-1})'}\right)_0 . \tag{3.205}$$

From the definition $\nu_{\alpha i} = -\mathrm{i}\sqrt{k^2 - k_{\alpha i}^2}$ for $k_{\alpha i} < k$ in (3.5), we find that the factor of the first term on the right-hand side of (3.205) is

$$\left(\frac{(P_i)'}{(\gamma_i^{-1})'}\right)_0 = \lim_{\omega\to 0} \frac{-\mathrm{i}}{2}\left(k^2 - \frac{\omega^2}{\alpha_i^2}\right)^{-1/2} \frac{-2\omega}{\alpha_i^2} d_i \left(\frac{-2\omega}{2k^2\beta_i^2}\right)^{-1} = -\mathrm{i}kd_i \frac{\beta_i^2}{\alpha_i^2} . \tag{3.206}$$

Similarly, the factor of the second term is $\left((Q_i)'/(\gamma_i^{-1})'\right)_0 = -\mathrm{i}kd_i$. Using $(\sin P_i)_0 = (\sin Q_i)_0 = \mathrm{i}^{-1}\sinh kd_i$, we then obtain[70]

$$\left(\gamma_i(\cos P_i - \cos Q_i)\right)_0 = -kd_i(1 - \xi_i)\sinh kd_i \,, \quad \xi_i = \frac{\beta_i^2}{\alpha_i^2} . \tag{3.207}$$

Subsequently, applying the L'Hôpital's rule to $\left(r_{\alpha i}^{-1} \sin P_i + r_{\beta i} \sin Q_i\right)_0 = 0$ ($r_{\alpha i}$ and $r_{\beta i}$ are given in (3.33)) and $(\gamma_i^{-1})_0 = 0$ yields

$$\left(\frac{r_{\alpha i}^{-1} \sin P_i + r_{\beta i} \sin Q_i}{\gamma_i^{-1}}\right)_0$$
$$= \left(\frac{(r_{\alpha i}^{-1})' \sin P_i + r_{\alpha i}^{-1}(P_i)' \cos P_i + (r_{\beta i})' \sin Q_i + r_{\beta i}(Q_i)' \cos Q_i}{(\gamma_i^{-1})'}\right)_0 . \tag{3.208}$$

Substituting (3.206), etc., and $\left(r_{\alpha i}^{-1}\right)_0 = (k/\nu_{\alpha i})_0 = \mathrm{i}$, $\left(r_{\beta i}\right)_0 = \left(\nu_{\beta i}/k\right)_0 = -\mathrm{i}$, and

$$\left(\frac{\left(r_{\alpha i}^{-1}\right)'}{(\gamma_i^{-1})'}\right)_0 = \lim_{\omega\to 0} \frac{-1}{r_{\alpha i}^2}\frac{-\mathrm{i}}{2}\left(k^2 - \frac{\omega^2}{\alpha_i^2}\right)^{-1/2} \frac{-2\omega}{\alpha_i^2}\frac{1}{k}\left(\frac{-2\omega}{2k^2\beta_i^2}\right)^{-1} = -\mathrm{i}\frac{\beta_i^2}{\alpha_i^2} = -\mathrm{i}\xi_i \,,$$
$$\left(\frac{\left(r_{\beta i}\right)'}{(\gamma_i^{-1})'}\right)_0 = \lim_{\omega\to 0} \frac{-\mathrm{i}}{2}\left(k^2 - \frac{\omega^2}{\beta_i^2}\right)^{-1/2} \frac{-2\omega}{\beta_i^2}\frac{1}{k}\left(\frac{-2\omega}{2k^2\beta_i^2}\right)^{-1} = -\mathrm{i} \,, \tag{3.209}$$

etc. into (3.208) and using $(\cos P_i)_0 = (\cos Q_i)_0 = \cosh kd_i$, etc., we obtain

[70] This agrees with Eq. (28) of Zhu and Rivera [130] except for the sign inversion. The sign inversion is because the sign of γ_i is inverted from that in page 18 of Haskell [44] (Eq. (3.54) in this book) which was used by Zhu and Rivera, to that in page 409 of Fuchs [27] (Eq. (3.17) in this book).

$$\left(\gamma_i\left(r_{\alpha i}^{-1}\sin P_i + r_{\beta i}\sin Q_i\right)\right)_0 = -(1+\xi_i)\sinh kd_i - (1-\xi_i)kd_i\cosh kd_i\ . \tag{3.210}$$

In addition, applying L'Hôpital's rule to $\left(r_{\alpha i}\sin P_i + r_{\beta i}^{-1}\sin Q_i\right)_0 = 0$ and $(\gamma_i^{-1})_0 = 0$ yields

$$\left(\frac{r_{\alpha i}\sin P_i + r_{\beta i}^{-1}\sin Q_i}{\gamma_i^{-1}}\right)_0 = \left(\frac{(r_{\alpha i})'\sin P_i + r_{\alpha i}(P_i)'\cos P_i + (r_{\beta i}^{-1})'\sin Q_i + r_{\beta i}^{-1}(Q_i)'\cos Q_i}{(\gamma_i^{-1})'}\right)_0 . \tag{3.211}$$

Similar to (3.208), substituting $(r_{\alpha i})_0 = (\nu_{\alpha i}/k)_0 = -\mathrm{i}$, $\left(r_{\beta i}^{-1}\right)_0 = \left(k/\nu_{\beta i}\right)_0 = \mathrm{i}$, and

$$\begin{aligned}
\left(\frac{(r_{\alpha i})'}{(\gamma_i^{-1})'}\right)_0 &= \lim_{\omega\to 0}\frac{-\mathrm{i}}{2}\left(k^2-\frac{\omega^2}{\alpha_i^2}\right)^{-1/2}\frac{-2\omega}{\alpha_i^2}\frac{1}{k}\left(\frac{-2\omega}{2k^2\beta_i^2}\right)^{-1} = -\mathrm{i}\frac{\beta_i^2}{\alpha_i^2} = -\mathrm{i}\xi_i\ ,\\
\left(\frac{\left(r_{\beta i}^{-1}\right)'}{(\gamma_i^{-1})'}\right)_0 &= \lim_{\omega\to 0}\frac{-1}{r_{\beta i}^2}\frac{-\mathrm{i}}{2}\left(k^2-\frac{\omega^2}{\beta_i^2}\right)^{-1/2}\frac{-2\omega}{\beta_i^2}\frac{1}{k}\left(\frac{-2\omega}{2k^2\beta_i^2}\right)^{-1} = -\mathrm{i}\frac{\beta_i^2}{\beta_i^2} = -\mathrm{i}\ ,
\end{aligned} \tag{3.212}$$

etc. into (3.211) we obtain

$$\left(\gamma_i\left(r_{\alpha i}\sin P_i + r_{\beta i}^{-1}\sin Q_i\right)\right)_0 = -(1+\xi_i)\sinh kd_i + (1-\xi_i)kd_i\cosh kd_i\ . \tag{3.213}$$

For the propagator matrix $\mathbf{G}_i$ in (3.33), we let $(\rho_i c\omega)^{-1} = -\gamma_i/2k\mu_i$ take the limit $\omega \to 0$, and substitute (3.207), (3.210), and (3.213). We then obtain $(\mathbf{G}_i)_0 = \left(((G_i)_{jm})_0\right)$ as

$$\begin{aligned}
((G_i)_{11})_0 &= ((G_i)_{44})_0 = -(\gamma_i(\cos P_i - \cos Q_i))_0 + (\cos Q_i)_0 = xS + C\ ,\\
((G_i)_{12})_0 &= ((G_i)_{34})_0 = \mathrm{i}\left(\gamma_i\left(r_{\alpha i}^{-1}\sin P_i + r_{\beta i}\sin Q_i\right)\right)_0 + \mathrm{i}\left(r_{\alpha i}^{-1}\sin P_i\right)_0\\
&= -\mathrm{i}(1+\xi_i)S - \mathrm{i}(1-\xi_i)kd_iC + \mathrm{i}S = \mathrm{i}^{-1}(\xi_iS + xC)\ ,\\
((G_i)_{13})_0 &= ((G_i)_{24})_0 = \frac{-\mathrm{i}}{2k\mu_i}(\gamma_i(\cos P_i - \cos Q_i))_0 = \frac{1}{\mathrm{i}k}\frac{-xS}{2\mu_i}\ ,\\
((G_i)_{14})_0 &= \frac{-1}{2k\mu_i}\left(\gamma_i\left(r_{\alpha i}^{-1}\sin P_i + r_{\beta i}\sin Q_i\right)\right)_0 = \frac{1}{-k}\frac{-(1+\xi_i)S - xC}{2\mu_i}\ ,\\
((G_i)_{21})_0 &= ((G_i)_{43})_0 = -\mathrm{i}\left(\gamma_i\left(r_{\alpha i}\sin P_i + r_{\beta i}^{-1}\sin Q_i\right)\right)_0 - \mathrm{i}\left(r_{\beta i}^{-1}\sin Q_i\right)_0\\
&= -\mathrm{i}\{-(1+\xi_i)S + (1-\xi_i)kd_iC\} - \mathrm{i}S = \mathrm{i}(\xi_iS - xC)\ ,\\
((G_i)_{22})_0 &= ((G_i)_{33})_0 = (\gamma_i(\cos P_i - \cos Q_i))_0 + (\cos P_i)_0 = -xS + C\ ,\\
((G_i)_{23})_0 &= \frac{-1}{2k\mu_i}\left(\gamma_i\left(r_{\alpha i}\sin P_i + r_{\beta i}^{-1}\sin Q_i\right)\right)_0 = \frac{1}{-k}\frac{-(1+\xi_i)S + xC}{2\mu_i}\ ,\\
((G_i)_{31})_0 &= ((G_i)_{42})_0 = 2\mathrm{i}k\mu_i\left\{(\gamma_i(\cos P_i - \cos Q_i))_0\right.
\end{aligned}$$

$$
\begin{aligned}
&\quad + (\cos P_i - \cos Q_i)_0\} = -\mathrm{i}k \cdot 2\mu_i x S\,,\\
((G_i)_{32})_0 &= 2k\mu_i \left\{ \left(2 + \frac{1}{\gamma_i}\right)_0 \left(r_{\alpha i}^{-1} \sin P_i\right)_0 + \left(\gamma_i \left(r_{\alpha i}^{-1} \sin P_i + r_{\beta i} \sin Q_i\right)\right)_0 \right\}\\
&= 2k\mu_i \{2S - (1+\xi_i)S - xC\} = -k \cdot 2\mu_i \{-(1-\xi_i)S + xC\}\,,\\
((G_i)_{41})_0 &= 2k\mu_i \left\{ \left(\gamma_i \left(r_{\alpha i} \sin P_i + r_{\beta i}^{-1} \sin Q_i\right)\right)_0 + \left(2 + \frac{1}{\gamma_i}\right)_0 \left(r_{\beta i}^{-1} \sin Q_i\right)_0 \right\}\\
&= 2k\mu_i \{-(1+\xi_i)S + xC + 2S\} = -k \cdot (-2\mu_i)\{(1-\xi_i)S + xC\}\,,
\end{aligned}
\tag{3.214}
$$

where $S = \sinh kd_i$, $C = \cosh kd_i$, and $x = kd_i(1-\xi_i)$. Equation (3.214) generally agrees with Eq. (27) of Zhu and Rivera [130], but there are subtle differences, which are caused by the differences in the formulation of the propagator matrices. The formulation of Zhu and Rivera [130] is almost the same as that of Haskell [44], but their propagator matrix $\mathbf{a}_i$ differs from the Haskell matrix $\mathbf{G}_i^{\mathrm{H}}$ in (3.54) as

$$
\mathbf{a}_i = \begin{pmatrix}
(G_i^{\mathrm{H}})_{11} & \mathrm{i}^{-1}(G_i^{\mathrm{H}})_{12} & (G_i^{\mathrm{H}})_{13} & \mathrm{i}(G_i^{\mathrm{H}})_{14}\\
\mathrm{i}(G_i^{\mathrm{H}})_{21} & (G_i^{\mathrm{H}})_{22} & \mathrm{i}(G_i^{\mathrm{H}})_{23} & -(G_i^{\mathrm{H}})_{24}\\
(G_i^{\mathrm{H}})_{31} & \mathrm{i}^{-1}(G_i^{\mathrm{H}})_{32} & (G_i^{\mathrm{H}})_{33} & \mathrm{i}(G_i^{\mathrm{H}})_{34}\\
\mathrm{i}^{-1}(G_i^{\mathrm{H}})_{41} & -(G_i^{\mathrm{H}})_{42} & -\mathrm{i}(G_i^{\mathrm{H}})_{43} & (G_i^{\mathrm{H}})_{44}
\end{pmatrix}
\tag{3.215}
$$

so that i does not appear explicitly in $\mathbf{a}_i$. Furthermore, the propagator matrix $\mathbf{G}_i$ in (3.33), on which (3.214) is based, has differences from the Haskell matrix as shown in (3.57); therefore, the relationship

$$
\mathbf{a}_i = \begin{pmatrix}
(G_i)_{11} & \mathrm{i}(G_i)_{12} & \mathrm{i}k(G_i)_{13} & -k(G_i)_{14}\\
-\mathrm{i}(G_i)_{21} & (G_i)_{22} & -k(G_i)_{23} & -\mathrm{i}k(G_i)_{24}\\
-(\mathrm{i}k)^{-1}(G_i)_{31} & -k^{-1}(G_i)_{32} & (G_i)_{33} & \mathrm{i}^{-1}(G_i)_{34}\\
-k^{-1}(G_i)_{41} & (\mathrm{i}k)^{-1}(G_i)_{42} & \mathrm{i}(G_i)_{43} & (G_i)_{44}
\end{pmatrix}
\tag{3.216}
$$

exists between $\mathbf{a}_i$ and $\mathbf{G}_i$. This is also valid between Eq. (27) of Zhu and Rivera [130] and (3.214).

The last element $\mathbf{T}_n^{-1}$, which represents the $\omega - k$ spectra $\tilde{u}_{l1}(0)$ and $\tilde{w}_{l1}(0)$ of the ground motions of the P and SV waves, diverges when $\omega \to 0$, because $k_{\beta n}^2 = \omega^2/\beta_n^2$ is included in the denominator as a common factor, as shown in (3.30). However, in (3.43) for $\tilde{u}_{l1}(0)$ and $\tilde{w}_{l1}(0)$, $\hat{M}_{11}$ in the denominator and $\hat{M}_{1j'}^h$, $j' = 1, 2, \cdots, 6$ in the numerator include these in common in the form of a **subdeterminant** (Sect. 3.1.3). Hence, if the diverging factors can be grouped together from the subdeterminants, they are canceled in the denominator and the numerator [130]. The subdeterminants $(\hat{\mathrm{T}}_n^{-1})_{1j}$ are given by (3.163) in Sect. 3.1.9, where $n = 2$ but (3.163) holds for a generic n. Extracting the common factor $1/2\mu_n \nu_{\alpha n} \nu_{\beta n} k_{\beta i}^2$ in (3.30) from $(\hat{\mathrm{T}}_n^{-1})_{1j}$ is known to result in [9]

$$(\hat{\mathrm{T}}_n^{-1})'_{11} = -\frac{\beta_n^4 \rho_n}{2\omega^2}\left(4k^2 \nu_{\alpha n}\nu_{\beta n} + l_n^2\right), \quad (\hat{\mathrm{T}}_n^{-1})'_{12} = \frac{\mathrm{i}\nu_{\alpha n}}{2},$$
$$(\hat{\mathrm{T}}_n^{-1})'_{13} = (\hat{\mathrm{T}}_n^{-1})'_{14} = -\frac{\mathrm{i}\beta_n^2}{2\omega c}\left(l_n + 2\nu_{\alpha n}\nu_{\beta n}\right),$$
$$(\hat{\mathrm{T}}_n^{-1})'_{15} = -\frac{\mathrm{i}\nu_{\beta n}}{2}, \quad (\hat{\mathrm{T}}_n^{-1})'_{16} = -\frac{1}{2\rho_n\omega^2}\left(\nu_{\alpha n}\nu_{\beta n} + k^2\right) \tag{3.217}$$

as described in section 6.3 of Fuchs [34]. If

$$\delta_n = \gamma_n\left(1 + \frac{\nu_{\alpha n}\nu_{\beta n}}{k^2}\right) + 1 = \gamma_n\left(1 + r_{\alpha n} r_{\beta n}\right) + 1 \tag{3.218}$$

is newly defined here, (3.217) is rewritten as

$$(\hat{\mathrm{T}}_n^{-1})'_{11} = k^2\mu_n\left(\delta_n + 1 - \gamma_n^{-1}\right), \quad (\hat{\mathrm{T}}_n^{-1})'_{12} = \frac{\mathrm{i}\nu_{\alpha n}}{2}, \tag{3.219}$$
$$(\hat{\mathrm{T}}_n^{-1})'_{13} = (\hat{\mathrm{T}}_n^{-1})'_{14} = \frac{\mathrm{i}k}{2}\delta_n, \quad (\hat{\mathrm{T}}_n^{-1})'_{15} = -\frac{\mathrm{i}\nu_{\beta n}}{2}, \quad (\hat{\mathrm{T}}_n^{-1})'_{16} = \frac{1}{4\mu_n}(\delta_n - 1).$$

Applying L'Hôpital's rule to $\left(1 + r_{\alpha n} r_{\beta n}\right)_0 = 0$ and $\left(\gamma_n^{-1}\right)_0 = 0$ and substituting (3.209) and (3.212) into the results, we obtain

$$\left(\frac{1 + r_{\alpha n} r_{\beta n}}{\gamma_n^{-1}}\right)_0 = \left(\frac{(r_{\alpha n})' r_{\beta n} + r_{\alpha n}\left(r_{\beta n}\right)'}{\left(\gamma_n^{-1}\right)'}\right)_0 = -\xi_n - 1 \tag{3.220}$$

and $(\delta_n)_0 = -\xi_n$. We then use this and $(\nu_{\alpha n})_0 = \left(\nu_{\beta n}\right)_0 = -\mathrm{i}k$, obtaining

$$\left((\hat{\mathrm{T}}_n^{-1})'_{11}\right)_0 = k^2\mu_n(1 - \xi_n), \quad \left((\hat{\mathrm{T}}_n^{-1})'_{12}\right)_0 = \frac{k}{2}, \tag{3.221}$$
$$\left((\hat{\mathrm{T}}_n^{-1})'_{13}\right)_0 = \left((\hat{\mathrm{T}}_n^{-1})'_{14}\right)_0 = \frac{-\mathrm{i}k\xi_n}{2}, \quad \left((\hat{\mathrm{T}}_n^{-1})'_{15}\right)_0 = -\frac{k}{2}, \quad \left((\hat{\mathrm{T}}_n^{-1})'_{16}\right)_0 = \frac{-1}{4\mu_n}(1 + \xi_n).$$

We finally remove the additional common factor $k/2$ and have

$$\left((\hat{\mathrm{T}}_n^{-1})'_{11}\right)_0 = 2k\mu_n(1 - \xi_n), \quad \left((\hat{\mathrm{T}}_n^{-1})'_{12}\right)_0 = 1, \tag{3.222}$$
$$\left((\hat{\mathrm{T}}_n^{-1})'_{13}\right)_0 = \left((\hat{\mathrm{T}}_n^{-1})'_{14}\right)_0 = -\mathrm{i}\xi_n, \quad \left((\hat{\mathrm{T}}_n^{-1})'_{15}\right)_0 = -1, \quad \left((\hat{\mathrm{T}}_n^{-1})'_{16}\right)_0 = \frac{-1}{2k\mu_n}(1 + \xi_n).$$

(3.222) almost agrees with the vector part of Eq. (34) of Zhu and Rivera [130] but there are subtle differences, such as between $\mathbf{a}_i$ [130] and $\mathbf{G}_i$ in (3.33) and these are reflected in (3.222).

From $((G_i)_{jm})_0$ in (3.214), $((G^l)_{jm})_0$ in (3.42) and $((\hat{G}_i)_{j'm'})_0$ are computed. Combining the latter with $\left((\hat{\mathrm{T}}_n^{-1})'_{1j'}\right)_0$ in (3.222), we compute $(\hat{M}_{11})_0$ and $(\hat{M}^h_{1m'})_0$ from (3.41). By substituting these results and (3.203) into (3.43), we obtain the P and SV types of crustal deformation $(\tilde{u}_{l1}(0))_0$ and $(\tilde{w}_{l1}(0))_0$. In addition, Zhu and Rivera [130] also reported that crustal deformation can be computed with sufficient accuracy only by giving a small imaginary constant to the angular frequency (for example, $\omega = 0$ and $\omega_I = 2\pi \times 0.01$ Hz for the angular frequency in (3.128)), while preserving the formulation in Sects. 3.1.1 – 3.1.8 without the formulation of $\omega \to 0$ explained in this subsection.

3.2 Propagation in 3-D Velocity Structures

3.2.1 3-D Velocity Structure

In this section we show how we can employ a more general and realistic structure than just a 1-D velocity structure that changes only in the depth direction. Working with a 2-D velocity structure ("2-D" stands for "two-dimensional") that adds a change in one of two horizontal directions was the first straightforward step, and theoretical research on ground motion has progressed on this course since around 1970. However, because the 2-D case is intermediate between a realistic **3-D velocity structure** ("3-D" stands for "three-dimensional") and a 1-D velocity structure, this remains theoretical research and is rarely used for actual ground motion analysis. This section is therefore entitled "Propagation in 3-D velocity structures". However, understanding 2-D velocity structure is indispensable for the extension to propagation in a 3-D velocity structure, therefore except for ray theory, theories and methods will be explained for **2-D velocity structure**.

First, the ray theory (Sects. 3.2.2 and 3.2.3) and the finite difference method (Sect. 3.2.4) are explained in the following subsections, as these represent the theory and method typically used for a realistic velocity structure. There exist various additional methods, such as the finite element method, the boundary element method, and the Aki-Larner method. These appear to be completely different methods at first glance, however looking carefully at their formulation reveals that they all depend on the same mathematical foundation, that is the **method of weighted residuals**. When there exist a partial differential equation $L(u) = 0$ and a boundary condition $B(u) = 0$ for a function $u(\mathbf{x})$ of the spatial coordinates $\mathbf{x}$ of the velocity structure, u is expanded as

$$u(\mathbf{x}) = \sum_{n=1}^{N} a_n \phi_n(\mathbf{x}) \tag{3.223}$$

using linearly independent functions $\phi_n(\mathbf{x})$. The method of weighted residuals is used to determine a_n so that the residuals of $L(\sum a_n \phi_n(\mathbf{x}))$ and/or $B(\sum a_n \phi_n(\mathbf{x}))$ are minimized [32, 33]. In this method, ϕ_n is termed a **trial function**. If the time term of the equation of motion in (1.30) is Fourier transformed and $u \to \bar{\mathbf{u}}$, the ground motion can be obtained by solving the equation with this method. The approach of the method of weighted residuals is very similar to the Rayleigh–Ritz method mentioned in deriving the formulation of the finite element method (especially when using the

Galerkin method described below). However, it is more general because it can be applied to problems in which the energy is not preserved and the variational principle does not exist. In fact, in a medium with a **radiation boundary** (a boundary on which the **radiation boundary condition** (Sect. 3.1.2) is imposed), energy conservation is not possible because body waves leak from there.

The method of weighted residuals can be classified into three types, the **domain method**, **boundary method**, and **mixed method**, according to the trial functions. In the domain method, the trial function that already satisfies the boundary condition is used, and the residual of the equation of motion in the internal domain is forced to zero. To the contrary, in the boundary method the trial function that already satisfies the equation of motion is used, and the residual is made to be zero on the boundary. However, it is not guaranteed that a function which satisfies the boundary condition or the equation of motion can always be found. In such a case, the sum of the residuals of the equation of motion and the boundary condition is made to be zero by using a trial function that does not satisfy both, and this is called the mixed method. The **finite element method** (Sect. 3.2.5), which is the typical domain method among the three types listed above, has a long history of study in many fields including **elastodynamics** since the mid-1950s [98], and is explained in Sect. 3.2.5.

The boundary element method, which represents the boundary method among the three types listed above, has also been in development since as early as the 1960s [98], but is rarely used in the field of seismology. Nevertheless, other boundary methods have been studied in the field of ground motion since around 1970, probably because the structure model used in the boundary method, in which homogeneous layers are separated by irregular interfaces, is close to the situation that many people have in mind for the velocity structure. In addition, the scale of the necessary computation appears to be smaller than for the domain method, because only the interfaces are treated, not the internal region that occupies a large part of the velocity structure. However, this advantage in the computation is lost as the irregularities of the interfaces become realistic. The **Aki-Larner method** [1], which is the best studied of the boundary methods and is specific to the field of ground motion seismology, is reviewed in Sect. 3.2.6.

In the method of weighted residuals, the residual is defined with **weighting function** w_k as

$$\epsilon_k = \int w_k(\mathbf{x}) L\left(\sum_{n=1}^{N} a_n \phi_n(\mathbf{x})\right) dV \tag{3.224}$$

for the domain method in domain V, and as

$$\epsilon_k = \int w_k(\mathbf{x}) B\left(\sum_{n=1}^{N} a_n \phi_n(\mathbf{x})\right) dS \tag{3.225}$$

for the boundary method at surface S. There are several kinds of weighting functions, as shown in Table 3.1. The least-squares method minimizes $\int L^2 dV$, but the integral can be partially differentiated by a_n and formally defined as the weighting function

Table 3.1 Classification of the method of weighted residuals by the weighting function

	Name	Weighting function
(a)	Collocation method	$w_k = \delta(\mathbf{x} - \mathbf{x}_k)$
(b)	Subdomain method	$w_k = \begin{cases} 1 \text{ in } D \\ 0 \text{ outside} \end{cases}$
(c)	Method of moments	$w_k = \mathbf{x}^{k-1}$
(d)	Galerkin method	$w_k = \phi_k(\mathbf{x})$
(e)	Least-squares method	$w_k = \partial L/\partial a_k$

(e) in Table 3.1. $w_1, w_2, \ldots$, and w_N are prepared, and the simultaneous equations of $\epsilon_1 = 0, \epsilon_2 = 0, \ldots, \epsilon_N = 0$ are made, so that a_n $(n = 1, 2, \cdots, N)$ can be obtained by solving them. For (a) to (c), we can make the number of w_k greater than N and obtain a solution using the least-squares method for data processing (Sect. 4.4). The appropriate weighting function depends on what the problem is, but generally the closer to (d) and (e) the better the accuracy ((d) and (e) are comparable), and the closer to (a) the easier the formulation. The main methods used are (a) the **collocation method**, (d) the **Galerkin method**, and (e) the least-squares method. The weighting functions and others used in the finite element method and the Aki-Larner method are discussed in Sects. 3.2.5 and 3.2.6, respectively.

3.2.2 Ray Theory

The most classical and well-studied method for computing ground motions in a 3-D velocity structure is to apply **geometrical optics** for the light, to ground motions. For the solution of the equation of motion (1.30) without the body force ($\rho\mathbf{f} = \mathbf{0}$), it is assumed that the Fourier-transformed time dependence is a single harmonic oscillation, the time delay through the propagation to the point $\mathbf{x}$ is $\tau(\mathbf{x})$, and the amplitude is expressed in a power series of i/ω, therefore[71]

$$\mathbf{u}(\mathbf{x}, t) = e^{\mathrm{i}\omega(t-\tau(\mathbf{x}))} \sum_{k=0}^{\infty} \mathbf{u}^{(k)}(\mathbf{x}) \left(\frac{\mathrm{i}}{\omega}\right)^k . \tag{3.226}$$

Since the higher order terms of i/ω can be ignored in (3.226) when ω is large, the operation of substituting this equation into the equation of motion and collecting only the lower order terms corresponds to the high frequency approximation. The

[71] This agrees with the definition of Červený and Ravindra [21]. However, compared to the definitions of Červený et al. [20] and Červený [19], the sign of $\mathrm{i}\omega$ is inverted because of the difference in the definition of the Fourier transform.

assumption of (3.226) is equivalent to applying the **WKBJ approximation** to the ground motion in a piecewise continuous 1-D velocity structure.[72]

The time delay τ_I in (3.96) is an integral of the slowness in the depth direction as in (3.126) in a 1-D velocity structure, but it is a complicated calculation in a 3-D velocity structure, as described below. In addition, although (3.226) contains only $e^{i\omega(t-\tau(\mathbf{x}))}$ as a phase term, the original solutions are the superposition of two-dimensional harmonics (plane wave expansion, Sect. 3.1.4) and the superposition of cylindrical harmonics (cylindrical wave expansion, Sect. 3.1.1). The superposition is represented by the wavenumber integral. Here, we do not perform the superposition, but consider only the phase term for **x** determined using the following method, where **x** corresponds to the ray defined in Sect. 1.2.5. The following method is therefore called the **ray theory**. Although the wavenumber integration is accompanied by theoretical difficulty even in 1-D velocity structures (Sects. 3.1.7 and Sect. 3.1.8), the ray theory does not include this integration, therefore it has a wide application range if only paying attention to the degree of approximation.

Substituting (3.226) into (1.30), arranging the result, and defining the operators as

$$\begin{aligned}
\mathbf{N}(\mathbf{u}^{(k)}) &= -\rho\mathbf{u}^{(k)} + (\lambda+\mu)\nabla\tau(\nabla\tau\cdot\mathbf{u}^{(k)}) + \mu(\nabla\tau)^2\mathbf{u}^{(k)} \\
\mathbf{M}(\mathbf{u}^{(k)}) &= (\lambda+\mu)\left\{\nabla\tau(\nabla\cdot\mathbf{u}^{(k)}) + \nabla(\nabla\tau\cdot\mathbf{u}^{(k)})\right\} + \mu\left\{(\nabla^2\tau)\mathbf{u}^{(k)} + 2(\nabla\tau\cdot\nabla)\mathbf{u}^{(k)}\right\} \\
&\quad + \nabla\lambda(\nabla\tau\cdot\mathbf{u}^{(k)}) + \nabla\mu\times(\nabla\tau\times\mathbf{u}^{(k)}) + 2(\nabla\mu\cdot\nabla\tau)\mathbf{u}^{(k)} \\
\mathbf{L}(\mathbf{u}^{(k)}) &= (\lambda+\mu)\nabla(\nabla\cdot\mathbf{u}^{(k)}) + \mu\nabla^2\mathbf{u}^{(k)} + \nabla\lambda(\nabla\cdot\mathbf{u}^{(k)}) \\
&\quad + \nabla\mu\times(\nabla\times\mathbf{u}^{(k)}) + 2(\nabla\mu\cdot\nabla)\mathbf{u}^{(k)}\,, \qquad (3.227)
\end{aligned}$$

we obtain

$$e^{i\omega(t-\tau)}\sum_{k=0}^{\infty}\left(\frac{i}{\omega}\right)^k\left\{\mathbf{N}(\mathbf{u}^{(k)}) - \mathbf{M}(\mathbf{u}^{(k-1)}) - \mathbf{L}(\mathbf{u}^{(k-2)})\right\} = \mathbf{0}\,,\ \mathbf{u}^{(-2)} \equiv \mathbf{u}^{(-1)} \equiv \mathbf{0}\,. \qquad (3.228)$$

From the linearly independent nature of the power series $(i/\omega)^k$, we can rewrite (3.228) to the recurrence equations for $\mathbf{u}^{(k)}$

$$\mathbf{N}(\mathbf{u}^{(k)}) - \mathbf{M}(\mathbf{u}^{(k-1)}) - \mathbf{L}(\mathbf{u}^{(k-2)}) = \mathbf{0}\,,\ \ \mathbf{u}^{(-2)} \equiv \mathbf{u}^{(-1)} \equiv \mathbf{0}\,, \qquad (3.229)$$

which must be satisfied for $k = 0, 1, 2, \cdots$.

When only the recurrence equation for $k = 0$ is taken out,

$$\mathbf{N}(\mathbf{u}^{(0)}) = -\rho\mathbf{u}^{(0)} + (\lambda+\mu)\nabla\tau(\mathbf{u}^{(0)}\cdot\nabla\tau) + \mu(\nabla\tau)^2\mathbf{u}^{(0)} = \mathbf{0} \qquad (3.230)$$

is obtained as a zeroth-order high frequency approximation. We further rewrite (3.230) as

[72] This was written in page 70 of Červený [19].

$$
\begin{aligned}
(3.230)\cdot\nabla\tau &= \left\{-\rho+(\lambda+2\mu)(\nabla\tau)^2\right\}(\mathbf{u}^{(0)}\cdot\nabla\tau)=0\,,\\
(3.230)\times\nabla\tau &= \left\{-\rho+\mu(\nabla\tau)^2\right\}(\mathbf{u}^{(0)}\times\nabla\tau)=\mathbf{0}\,.
\end{aligned} \tag{3.231}
$$

In order to satisfy the two equations in (3.231) simultaneously, we then have the two sets of simultaneous equations

$$
\begin{aligned}
&(\nabla\tau)^2=1/\alpha^2,\ \mathbf{u}^{(0)}\times\nabla\tau=\mathbf{0} \quad \text{or}\\
&(\nabla\tau)^2=1/\beta^2,\quad \mathbf{u}^{(0)}\cdot\nabla\tau=0\,.
\end{aligned} \tag{3.232}
$$

$\nabla\tau$ denotes the direction of a ray, i.e., the propagation direction (Sect. 1.2.5). In (3.232), the first set therefore represents a P wave with the propagation velocity α and the amplitude in the direction along the propagation (ray). The second set represents an S wave with the propagation velocity β and the amplitude in the direction perpendicular to the propagation (ray).

The first equations of the two sets:

$$
(\nabla\tau)^2=1/\alpha^2,\quad (\nabla\tau)^2=1/\beta^2 \tag{3.233}
$$

are nonlinear first-order partial differential equations, and are named the **eikonal equations**. In the Cartesian coordinate system $\mathbf{x}=(x_i)=(x,y,z)$, the nonlinear first-order partial differential equations

$$
H(x_i,p_i,\tau)=0\,,\quad p_i=\frac{\partial\tau}{\partial x_i} \tag{3.234}
$$

are transformed into the set of first-order ordinary differential equations

$$
\frac{dx_i}{\partial H/\partial p_i}=\frac{-dp_i}{\partial H/\partial x_i+p_i\partial H/\partial\tau}=\frac{d\tau}{\sum\limits_{j=1}^{3}p_j\partial H/\partial p_j}\,, \tag{3.235}
$$

which are called the characteristic differential equations [55]. If H does not contain τ explicitly and (3.235) is equated with $d\xi$, the first half of (3.235) yields

$$
\frac{dx_i}{d\xi}=\frac{\partial H}{\partial p_i}\,,\quad \frac{dp_i}{d\xi}=-\frac{\partial H}{\partial x_i}\,. \tag{3.236}
$$

These equations are in the same form as the **canonical equation** in analytical mechanics, where p_i is a **momentum** and H is the **Hamiltonian** [84]. However, as demonstrated later the independent variable here is the ray length, whereas in analytical mechanics the independent variable is time. The position coordinates (x_i) here are represented as the generalized coordinates (q_i) in analytical mechanics. The eikonal equations (3.233) and (3.234) yield

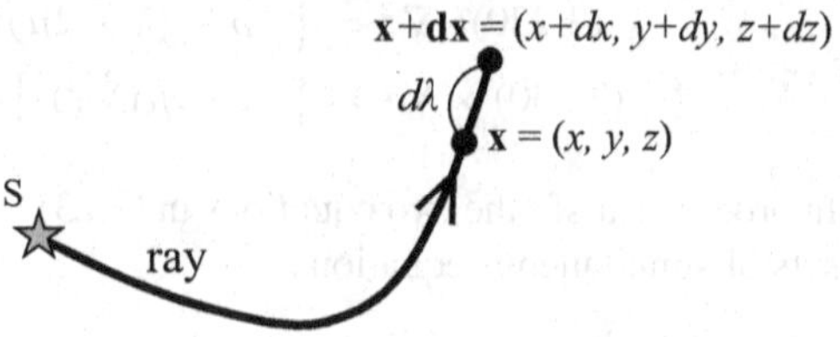

Fig. 3.21 The magnitude of the infinitesimal change in the position of the ray is equivalent to the infinitesimal change of the ray length and is denoted by $d\lambda$ (reprinted from Koketsu [72] with permission of Kindai Kagaku)

$$(\nabla\tau)^2 = \sum_{j=1}^{3} p_j^2 = s^2, \quad s = \alpha^{-1} \text{ or } \beta^{-1} . \tag{3.237}$$

Here, s is the reciprocal of the P wave velocity or S wave velocity and represents the slowness as a physical property of the medium. From (3.237), it is found that the Hamiltonian

$$H(x_i, p_i) = \sqrt{\sum_{j=1}^{3} p_j^2} - s = 0 \tag{3.238}$$

is possible. Substituting this into the first equation in (3.236), we obtain

$$dx_i = \frac{\partial H}{\partial p_i} d\xi = \frac{p_i}{s} d\xi . \tag{3.239}$$

As mentioned above, $\mathbf{x} = (x_i) = (x, y, z)$ obtained by solving the eikonal equation is the position of the ray, and the magnitude of its infinitesimal change $d\lambda$ is given from Fig. 3.21 as

$$d\lambda = \sqrt{dx^2 + dy^2 + dz^2} = \sqrt{\sum_{j=1}^{3} dx_j^2} . \tag{3.240}$$

From Fig. 3.21, we also find that this represents an infinitesimal change in the length of the ray measured from the point source S. Substituting (3.239) for (3.240) gives

$$d\lambda = \sqrt{\sum_{j=1}^{3} \frac{p_j^2}{s^2} d\xi^2} = d\xi , \tag{3.241}$$

so the independent variable ξ is the ray length λ under the Hamiltonian in (3.238).

The canonical equations yield

$$\frac{dx_i}{d\lambda} = \frac{p_i}{s}, \quad \frac{dp_i}{d\lambda} = \frac{\partial s}{\partial x_i} . \tag{3.242}$$

Using the momentum vector $\mathbf{p} = (p_i)$, we rewrite (3.242) in a vector form as

$$\frac{d\mathbf{x}}{d\lambda} = \frac{\mathbf{p}}{s}, \quad \frac{d\mathbf{p}}{d\lambda} = \nabla s. \tag{3.243}$$

These equations are called **ray equation.** From $(p_i) = \left(\frac{\partial \tau}{\partial x_i}\right)$ in (3.234), we have

$$\mathbf{p} = \nabla \tau, \tag{3.244}$$

therefore **p** represents the direction of the ray.

The Hamiltonian can also be set to

$$H(x_i, p_i) = \frac{1}{2}\left(\sum_{j=1}^{3} \frac{p_j^2}{s^2} - 1\right) = 0 \tag{3.245}$$

by modifying (3.238). When we substitute this into the latter half of (3.235), we obtain

$$d\xi = \frac{d\tau}{\sum_{j=1}^{3} p_j \partial H/\partial p_j} = \frac{d\tau}{\sum_{j=1}^{3} p_j^2/s^2} = d\tau, \tag{3.246}$$

therefore the independent variable ξ is now τ, which is time or time delay as in analytical mechanics. However, since the calculation to determine **x** is more stable to set the independent variable to λ (Sect. 3.2.3), only the case of λ is explained in this book.

Continuing the analogy to analytical mechanics, **Hamilton's principle** [98] in analytical mechanics

$$\delta \int L\, dt = 0, \quad L = \sum_{j=1}^{3} p_j \frac{dq_j}{dt} - H \tag{3.247}$$

yields

$$\delta \int L\, d\lambda = 0, \quad L = \sum_{j=1}^{3} p_j \frac{dx_j}{d\lambda} - H \tag{3.248}$$

in the ray theory through $t \to \lambda$ and $q_j \to x_j$. Substituting the eikonal equations in (3.237), the Hamiltonian in (3.238), and the ray equations in (3.242) into the **Lagrangian** L in (3.248), we obtain **Fermat's principle**

$$\delta \int s\, d\lambda = 0. \tag{3.249}$$

Since $s\, d\lambda = d\lambda/\alpha$ or $d\lambda/\beta$ in (3.249) represents the time during which the P or S wave propagates through the infinitesimal portion $d\lambda$, the integral $\int s\, d\lambda$ corresponds

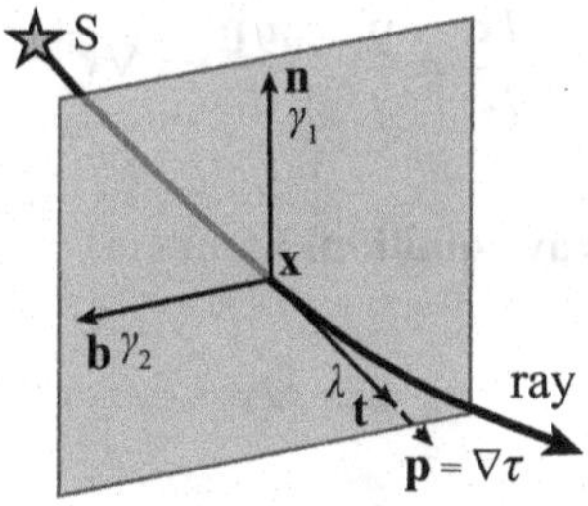

Fig. 3.22 The ray-centered coordinate system consists of the ray-tangential direction (unit vector **t**), the principal normal direction (unit vector **n**), and the binormal direction (unit vector **b**) at the position **x**. **t** and $\mathbf{p} = \nabla\tau$ overlap. The gray square is normal to **t** and includes **n** and **b**, therefore represents the normal plane or wavefront (based on Červený [19], reprinted from Koketsu [72] with permission of Kindai Kagaku)

to the time during which the ground motion propagates from the point source to a certain point, that is the **travel time** T. Thus, Fermat's principle indicates that the ray is the route from the point source to a certain point along which the travel time is the minimum ($\delta T = 0$).

From the above results, the formulation for calculating the position **x** of the ray is obtained. Since the travel time T is also calculated in the process of this calculation, in (3.226), which is a solution of the equation of motion, the phase term $e^{i\omega(t-\tau(\mathbf{x}))}$ can be obtained. Then, if the amplitude term $\mathbf{u}^{(0)}$ can be calculated, the zeroth-order high frequency approximation of the ground motion can be obtained. From (3.232)), $\mathbf{u}^{(0)}$ of the P wave has a component along the ray and $\mathbf{u}^{(0)}$ of the S wave has a component perpendicular to the ray, therefore the first step to calculate $\mathbf{u}^{(0)}$ is to set up the ray-centered coordinate system[73] as shown in Fig. 3.22. In this coordinate system, **t** is a unit vector of the coordinate tangential to the ray. **n** and **b** are unit vectors of the principal normal and binormal coordinates, which are each perpendicular to the ray. The ground motion is decomposed as

$$\mathbf{u}^{(0)} = u_1^{(0)}\mathbf{t} + u_2^{(0)}\mathbf{n} + u_3^{(0)}\mathbf{b}\,, \tag{3.250}$$

where $u_1^{(0)}\mathbf{t}$ represents the zeroth-order P wave ground motion, and the vector in the normal plane

$$\mathbf{u}_{\perp}^{(0)} = u_2^{(0)}\mathbf{n} + u_3^{(0)}\mathbf{b} \tag{3.251}$$

represents the zeroth-order S-wave ground motion. Incidentally, **t** and $\mathbf{p} = \nabla\tau$ in (3.244) overlap because $\nabla\tau$ indicates a direction along the propagation or the ray as shown in (3.232), and the normal plane can be called the **wavefront** because the wavefront is a plane perpendicular to the ray (Sect. 1.2.5).

[73]This is the definition by Červený et al. [20]. Červený [19] provided another definition, in which **n** and **b** are rotated on the normal plane (wavefront) such that the torsion T is 0, as described later.

As the phase term including τ in (3.226) is common regardless of the degree k of the power series, the eikonal equations (3.233) should not only hold for the zeroth-order approximation of $k = 0$ but also for $k \geq 1$.[74] Hence, as the recurrence equation for $k = 0$ $\mathbf{N}(\mathbf{u}^{(0)}) = \mathbf{0}$ in (3.230) yields (3.231), $\mathbf{N}(\mathbf{u}^{(1)}) = \mathbf{0}$ should yield

$$\begin{aligned} \mathbf{N}(\mathbf{u}^{(1)}) \cdot \nabla\tau &= \left\{-\rho + (\lambda + 2\mu)(\nabla\tau)^2\right\}(\mathbf{u}^{(1)} \cdot \nabla\tau) = 0 \\ \mathbf{N}(\mathbf{u}^{(1)}) \times \nabla\tau &= \left\{-\rho + \mu(\nabla\tau)^2\right\}(\mathbf{u}^{(1)} \times \nabla\tau) = \mathbf{0}\,. \end{aligned} \tag{3.252}$$

In addition, the recurrence equation for $k = 1$ in (3.229) is

$$\mathbf{N}(\mathbf{u}^{(1)}) - \mathbf{M}(\mathbf{u}^{(0)}) = \mathbf{0}\,. \tag{3.253}$$

We then take (3.253)$\cdot\nabla\tau$ and (3.253)$\times\nabla\tau$, and substitute the results into (3.252), obtaining

$$\begin{aligned} \mathbf{M}(\mathbf{u}^{(0)}) \cdot \nabla\tau &= \mathbf{N}(\mathbf{u}^{(1)}) \cdot \nabla\tau = 0 \\ \mathbf{M}(\mathbf{u}^{(0)}) \times \nabla\tau &= \mathbf{N}(\mathbf{u}^{(1)}) \times \nabla\tau = \mathbf{0}\,. \end{aligned} \tag{3.254}$$

When we consider only the P wave ground motion, from the description of (3.250) we find

$$\mathbf{u}^{(0)} = u_1^{(0)}\mathbf{t} = u_t^{(0)}\nabla\tau\,, \quad u_t^{(0)} = \alpha u_1^{(0)}\,. \tag{3.255}$$

Substituting (3.255) into the second equation for $k = 0$ in (3.227) and using $\nabla\tau \cdot \nabla\tau = (\nabla\tau)^2 = \alpha^{-2}$, $\nabla\tau \times \nabla\tau = \mathbf{0}$, $\nabla \cdot \nabla\tau = \nabla^2\tau$, $\nabla(u_t^{(0)}\nabla\tau) = \nabla u_t^{(0)}\nabla\tau + u_t^{(0)}\nabla(\nabla\tau)$, $\nabla \cdot u_t^{(0)}\nabla\tau = \nabla u_t^{(0)} \cdot \nabla\tau + u_t^{(0)}\nabla \cdot \nabla\tau$, and $(\nabla\tau \cdot \nabla)\nabla\tau = 1/2\nabla(\nabla\tau)^2 = 1/2\nabla\alpha^{-2}$,[75] we obtain

$$\begin{aligned} \mathbf{M}(\mathbf{u}^{(0)}) = \mathbf{M}(u_t^{(0)}\nabla\tau) &= (\lambda + \mu)\left\{\nabla\tau(\nabla \cdot u_t^{(0)}\nabla\tau) + \nabla(\nabla\tau \cdot u_t^{(0)}\nabla\tau)\right\} \\ &+ \mu\left\{(\nabla^2\tau)u_t^{(0)}\nabla\tau + 2(\nabla\tau \cdot \nabla)u_t^{(0)}\nabla\tau\right\} \\ &+ \nabla\lambda(\nabla\tau \cdot u_t^{(0)}\nabla\tau) + \nabla\mu \times (\nabla\tau \times u_t^{(0)}\nabla\tau) + 2(\nabla\mu \cdot \nabla\tau)u_t^{(0)}\nabla\tau \\ &= (\lambda + \mu)\left\{(\nabla u_t^{(0)} \cdot \nabla\tau + u_t^{(0)}\nabla^2\tau)\nabla\tau + \nabla u_t^{(0)}\alpha^{-2} + u_t^{(0)}\nabla\alpha^{-2}\right\} \\ &+ \mu\left\{u_t^{(0)}(\nabla^2\tau)\nabla\tau + 2\nabla\tau \cdot \nabla u_t^{(0)}\nabla\tau + u_t^{(0)}\nabla\alpha^{-2}\right\} \\ &+ u_t^{(0)}\alpha^{-2}\nabla\lambda + u_t^{(0)}(\nabla(2\mu) \cdot \nabla\tau)\nabla\tau\,. \end{aligned} \tag{3.256}$$

Next, substituting (3.256) into the first equation in (3.254) and using $(\lambda + 2\mu)/\rho = \alpha^2$, we obtain

[74] The footnote on page 20 of Červený and Ravindra [21] states this, but shows no proof.

[75] These are derived from the eikonal equations, formulas in vector geometry, Appendix of Červený and Ravindra [21], Appendix A of Ben-Menahem and Singh [10], and others.

$$
\begin{aligned}
\mathbf{M}(\mathbf{u}^{(0)}) \cdot \nabla\tau &= (\lambda+\mu)\left\{(\nabla u_t^{(0)} \cdot \nabla\tau + u_t^{(0)}\nabla^2\tau)\alpha^{-2} + \nabla u_t^{(0)} \cdot \nabla\tau\alpha^{-2} + u_t^{(0)}\nabla\alpha^{-2} \cdot \nabla\tau\right\} \\
&+ \mu\left\{u_t^{(0)}(\nabla^2\tau)\alpha^{-2} + 2\nabla\tau \cdot \nabla u_t^{(0)}\alpha^{-2} + u_t^{(0)}\nabla\alpha^{-2} \cdot \nabla\tau\right\} \\
&+ u_t^{(0)}\alpha^{-2}\nabla\lambda \cdot \nabla\tau + u_t^{(0)}(\nabla(2\mu) \cdot \nabla\tau)\nabla\tau \cdot \nabla\tau \\
&= (\lambda+2\mu)\left\{u_t^{(0)}\alpha^{-2}(\nabla^2\tau) + 2\nabla\tau \cdot \nabla u_t^{(0)}\alpha^{-2} + u_t^{(0)}\nabla\alpha^{-2} \cdot \nabla\tau\right\} \\
&+ u_t^{(0)}\alpha^{-2}\nabla(\lambda+2\mu) \cdot \nabla\tau \\
&= 2\rho\nabla\tau \cdot \nabla u_t^{(0)} + u_t^{(0)}\left\{\rho\nabla^2\tau + \nabla\tau \cdot \nabla\rho\right\} = 0\,. \qquad (3.257)
\end{aligned}
$$

$\nabla = (\partial/\partial x_i)$ from the definition, $(p_i) = \nabla\tau$ from (3.234), and τ, $\mathbf{u}^{(0)}$ in (3.226) are functions of $\mathbf{x} = (x_i)$ in (3.226), therefore the first equation of the ray equations (3.242) yields

$$
\nabla\tau \cdot \nabla u_t^{(0)} = \sum_{j=1}^{3} p_j \frac{\partial u_t^{(0)}}{\partial x_j} = \frac{1}{\alpha}\frac{du_t^{(0)}}{d\lambda}\,. \qquad (3.258)
$$

We can similarly transform $\nabla\tau \cdot \nabla\rho$ because the density ρ is also a function of $\mathbf{x} = (x_i)$, and finally obtain the equation for the amplitude $u_1^{(0)}$ of the P wave ground motion from (3.258) as[76]

$$
\frac{du_t^{(0)}}{d\lambda} + u_t^{(0)}\left(\frac{\alpha}{2}\nabla^2\tau + \frac{1}{2\rho}\frac{d\rho}{d\lambda}\right) = 0\,, \quad u_t^{(0)} = \alpha u_1^{(0)}\,. \qquad (3.259)
$$

This equation is termed the **transport equation.**

Equation (3.259) can be made into a first-order ordinary differential equation such as the ray equations, by replacing the included second-order partial derivative $\nabla^2\tau$ as follows. In the ray-centered coordinate system (Fig. 3.22), the coordinates in the principal normal and binormal directions are denoted by γ_1 and γ_2. The coordinate in the tangential direction is λ. "**Jacobian**" generally stands for the determinant of a transformation matrix between two coordinate systems [88], and the Jacobian between the spatial coordinate system for $\mathbf{x}$ and the ray-centered coordinate system $\boldsymbol{\lambda} = (\gamma_1, \gamma_2, \lambda)$ is specifically called the **ray Jacobian** [19]. If $\mathbf{x}$ is in the Cartesian coordinate system, the ray Jacobian is

$$
\mathcal{J} = \frac{D\mathbf{x}}{D\boldsymbol{\lambda}} = \frac{D(x,y,z)}{D(\gamma_1,\gamma_2,\lambda)} = \begin{vmatrix} \partial x/\partial\gamma_1 & \partial x/\partial\gamma_2 & \partial x/\partial\lambda \\ \partial y/\partial\gamma_1 & \partial y/\partial\gamma_2 & \partial y/\partial\lambda \\ \partial z/\partial\gamma_1 & \partial z/\partial\gamma_2 & \partial z/\partial\lambda \end{vmatrix} = \frac{\partial x}{\partial\gamma_1}\frac{\partial y}{\partial\gamma_2}\frac{\partial z}{\partial\lambda} \qquad (3.260)
$$

$$
+\frac{\partial y}{\partial\gamma_1}\frac{\partial z}{\partial\gamma_2}\frac{\partial x}{\partial\lambda} + \frac{\partial x}{\partial\gamma_2}\frac{\partial y}{\partial\lambda}\frac{\partial z}{\partial\gamma_1} - \frac{\partial x}{\partial\lambda}\frac{\partial y}{\partial\gamma_2}\frac{\partial z}{\partial\gamma_1} - \frac{\partial x}{\partial\gamma_2}\frac{\partial y}{\partial\gamma_1}\frac{\partial z}{\partial\lambda} - \frac{\partial x}{\partial\gamma_1}\frac{\partial y}{\partial\lambda}\frac{\partial z}{\partial\gamma_2}\,.
$$

According to **Smirnov's lemma** [111], in the case where the solution $\mathbf{x}$ of

---

[76]This agrees with Eq. (2.25') of Červený and Ravindra [21] if $W_0^{\tau} \to u_t^{(0)}$ and $s \to \lambda$.

$$\frac{d\mathbf{x}}{d\xi} = \mathbf{f}(\mathbf{x}) \tag{3.261}$$

is represented with the independent variable ξ and the two parameters η and ζ, we have

$$\frac{d}{d\xi} \ln\left(\frac{D\mathbf{x}}{D\boldsymbol{\xi}}\right) = \nabla \cdot \mathbf{f}\,, \quad \boldsymbol{\xi} = (\eta, \zeta, \xi)\,, \tag{3.262}$$

where $D\mathbf{x}/D\boldsymbol{\xi}$ is the Jacobian between $\mathbf{x}$ and $\boldsymbol{\xi}$. If $\xi \to \lambda$ and $\mathbf{f} \to \mathbf{p}/\alpha^{-1}$, (3.261) yields the first ray equation in (3.243) such that $\boldsymbol{\xi} = \boldsymbol{\lambda}$ and $D\mathbf{x}/D\boldsymbol{\xi} = \mathcal{J}$. Furthermore, using (3.244) and (3.262), we obtain

$$\frac{d}{d\lambda} \ln \mathcal{J} = \nabla \cdot \left(\frac{\nabla\tau}{\alpha^{-1}}\right) = \alpha\nabla^2\tau + \nabla\alpha \cdot \nabla\tau \;\Rightarrow\; \nabla^2\tau = \frac{1}{\alpha\mathcal{J}}\frac{d\mathcal{J}}{d\lambda} - \frac{1}{\alpha}\nabla\alpha \cdot \nabla\tau\,. \tag{3.263}$$

Since α is a function of $\mathbf{x} = (x_i)$, similarly to (3.258),

$$\nabla\tau \cdot \nabla\alpha = \frac{1}{\alpha}\frac{d\alpha}{d\lambda}\,. \tag{3.264}$$

The substitution of this into (3.263) results in[77]

$$\nabla^2\tau = \frac{1}{\alpha\mathcal{J}}\frac{d\mathcal{J}}{d\lambda} - \frac{1}{\alpha^2}\frac{d\alpha}{d\lambda} = \frac{1}{\mathcal{J}}\left\{\frac{1}{\alpha}\frac{d\mathcal{J}}{d\lambda} + \mathcal{J}\frac{d}{d\lambda}\left(\frac{1}{\alpha}\right)\right\} = \frac{1}{\mathcal{J}}\frac{d}{d\lambda}\left(\frac{\mathcal{J}}{\alpha}\right). \tag{3.265}$$

Further substituting (3.265) into the transport equation (3.259), we obtain

$$\begin{aligned} &\frac{du_t^{(0)}}{d\lambda} + u_t^{(0)}\left[\frac{\alpha}{2}\frac{1}{\mathcal{J}}\left\{\frac{1}{\alpha}\frac{d\mathcal{J}}{d\lambda} + \mathcal{J}\frac{d}{d\lambda}\left(\frac{1}{\alpha}\right)\right\} + \frac{1}{2\rho}\frac{d\rho}{d\lambda}\right] \\ &= \frac{du_t^{(0)}}{d\lambda} + u_t^{(0)}\left[\frac{1}{2\rho\alpha^{-1}\mathcal{J}}\left\{\rho\alpha^{-1}\frac{d\mathcal{J}}{d\lambda} + \rho\mathcal{J}\frac{d}{d\lambda}\left(\alpha^{-1}\right) + \alpha\mathcal{J}\frac{d\rho}{d\lambda}\right\}\right] \\ &= \frac{du_t^{(0)}}{d\lambda} + u_t^{(0)}\left\{\frac{1}{2\rho\alpha^{-1}\mathcal{J}}\frac{d}{d\lambda}\left(\rho\alpha^{-1}\mathcal{J}\right)\right\} = 0\,, \quad u_t^{(0)} = \alpha u_1^{(0)}\,. \end{aligned} \tag{3.266}$$

Using the derivative of a square root, (3.266) can be solved analytically as[78]

$$u_t^{(0)} = \left(\rho\alpha^{-1}\mathcal{J}\right)^{-1/2}\varphi_1 \;\Rightarrow\; u_1^{(0)} = \frac{u_t^{(0)}}{\alpha} = \left(\rho\alpha\mathcal{J}\right)^{-1/2}\varphi_1\,. \tag{3.267}$$

[77] This agrees with Eq. (3.10.30) of Červený [19], if $T \to \tau$, $s \to \lambda$, and $V \to \alpha$.

[78] This agrees with Eq. (2.30) of Červený et al. [20]. Although they used τ as the independent variable and the definition of $\mathcal{J}$ is different, their definition for τ is equivalent to our definition for λ, so that the agreement is established, as written in page 206 of Červený [19].

In (3.267), φ_1 is an integration constant independent of λ and is determined from the initial condition for $u_1^{(0)}$. However, since $\mathcal{J}$ can only be obtained numerically, $u_1^{(0)}$ follows a numerical solution similar to the position $\mathbf{x}$ of the ray (Sect. 3.2.3).

As the two components are related to the S wave ground motion as shown in (3.251), we modify the second equation in (3.252) to

$$\mathbf{N}(\mathbf{u}^{(1)}) \cdot \mathbf{n} = 0\,, \quad \mathbf{N}(\mathbf{u}^{(1)}) \cdot \mathbf{b} = 0\,, \tag{3.268}$$

and substitute (3.253)·**n** and (3.253)·**b** into (3.268), obtaining

$$\mathbf{M}(\mathbf{u}^{(0)}) \cdot \mathbf{n} = \mathbf{N}(\mathbf{u}^{(1)}) \cdot \mathbf{n} = 0\,, \quad \mathbf{M}(\mathbf{u}^{(0)}) \cdot \mathbf{b} = \mathbf{N}(\mathbf{u}^{(1)}) \cdot \mathbf{b} = 0\,. \tag{3.269}$$

We next modify the second equation in (3.227) for $k = 0$ by substituting (3.251) and using $\nabla\tau/\beta^{-1} = \mathbf{t}$, the eikonal equation $(\nabla\tau)^2 = \beta^{-2}$, and $d\mathbf{n}/d\lambda = \mathrm{T}\mathbf{b} - \mathrm{K}\mathbf{t}$, $d\mathbf{b}/d\lambda = -\mathrm{T}\mathbf{n}$ from **Frenet's formulas** [20]. The obtained $\mathbf{M}(\mathbf{u}^{(0)})$ and (3.269) yield [21]

$$\begin{aligned}
&\frac{du_2^{(0)}}{d\lambda} - \mathrm{T}u_3^{(0)} + \frac{\beta}{2\mu}u_2^{(0)}(\nabla \cdot \mu\nabla\tau) = 0\,,\\
&\frac{du_3^{(0)}}{d\lambda} + \mathrm{T}u_2^{(0)} + \frac{\beta}{2\mu}u_3^{(0)}(\nabla \cdot \mu\nabla\tau) = 0\,,
\end{aligned} \tag{3.270}$$

where T stands for the **torsion** (the change rate of the binormal direction) [110] of the ray. Similar to the P wave ground motion, we perform the change of variables, $u_n^{(0)} = \beta u_2^{(0)}$ and $u_b^{(0)} = \beta u_3^{(0)}$, and obtain

$$\begin{aligned}
&\frac{1}{\beta}\frac{du_n^{(0)}}{d\lambda} - \frac{1}{\beta^2}\frac{d\beta}{d\lambda}u_n^{(0)} - \mathrm{T}\frac{u_b^{(0)}}{\beta} + \frac{1}{2\mu}u_n^{(0)}(\nabla \cdot \mu\nabla\tau) = 0\,,\\
&\frac{1}{\beta}\frac{du_b^{(0)}}{d\lambda} - \frac{1}{\beta^2}\frac{d\beta}{d\lambda}u_b^{(0)} + \mathrm{T}\frac{u_n^{(0)}}{\beta} + \frac{1}{2\mu}u_b^{(0)}(\nabla \cdot \mu\nabla\tau) = 0\,.
\end{aligned} \tag{3.271}$$

Using the formula $\nabla \cdot a\mathbf{b} = \mathbf{b} \cdot \nabla a + a(\nabla \cdot \mathbf{b})$ [21], we have

$$\nabla \cdot \mu\nabla\tau = \nabla\tau \cdot \nabla\mu + \mu(\nabla \cdot \nabla\tau) = \nabla\tau \cdot \nabla\mu + \mu\nabla^2\tau\,. \tag{3.272}$$

In addition, as μ is a function of $\mathbf{x} = (x_i)$, similarly to (3.258),

$$\nabla\tau \cdot \nabla\mu = \sum_{j=1}^{3} p_j \frac{\partial\mu}{\partial x_j} = \frac{1}{\beta}\frac{d\mu}{d\lambda} = \beta\frac{d\rho}{d\lambda} + 2\rho\frac{d\beta}{d\lambda}\,. \tag{3.273}$$

We substitute (3.272) and (3.273) into (3.271) multiplied by β, include the above change of variables, and obtain

$$\frac{du_n^{(0)}}{d\lambda} - \mathrm{T}u_b^{(0)} + u_n^{(0)}\left(\frac{\beta}{2}\nabla^2\tau + \frac{1}{2\rho}\frac{d\rho}{d\lambda}\right) = 0\,, \quad u_n^{(0)} = \beta u_2^{(0)}\,,$$
$$\frac{du_b^{(0)}}{d\lambda} + \mathrm{T}u_n^{(0)} + u_b^{(0)}\left(\frac{\beta}{2}\nabla^2\tau + \frac{1}{2\rho}\frac{d\rho}{d\lambda}\right) = 0\,, \quad u_b^{(0)} = \beta u_3^{(0)}\,. \quad (3.274)$$

The two equations in (3.274) form the **transport equation** for the S wave ground motion. They are coupled to each other through cross terms with a coefficient of T. As they have the same form as the transport equation (3.259) of the P wave ground motion except for the cross terms, they are rewritten as

$$\frac{du_n^{(0)}}{d\lambda} - \mathrm{T}u_b^{(0)} + u_n^{(0)}\left\{\frac{1}{2\rho\beta^{-1}\mathcal{J}}\frac{d}{d\lambda}\left(\rho\beta^{-1}\mathcal{J}\right)\right\} = 0\,, \quad u_n^{(0)} = \beta u_2^{(0)},$$
$$\frac{du_b^{(0)}}{d\lambda} + \mathrm{T}u_n^{(0)} + u_b^{(0)}\left\{\frac{1}{2\rho\beta^{-1}\mathcal{J}}\frac{d}{d\lambda}\left(\rho\beta^{-1}\mathcal{J}\right)\right\} = 0\,, \quad u_b^{(0)} = \beta u_3^{(0)} \quad (3.275)$$

using the **ray Jacobian** $\mathcal{J}$ for the S wave. Equation (3.275) can be solved analytically in the same way as (3.266), and the solutions are[79]

$$u_2^{(0)} = \left(\rho\beta\mathcal{J}\right)^{-1/2}(\varphi_2\cos\Theta + \varphi_3\sin\Theta)\,,$$
$$u_3^{(0)} = \left(\rho\beta\mathcal{J}\right)^{-1/2}(\varphi_3\cos\Theta - \varphi_2\sin\Theta)\,, \quad \Theta = \int_S^O \mathrm{T}\,d\lambda\,. \quad (3.276)$$

As φ_1 in (3.267), φ_2 and φ_3 are integration constants independent of λ and are determined from the initial conditions for $u_2^{(0)}$ and $u_3^{(0)}$. The integration in Θ is performed along the ray from the point source S to the observation point O. Thus, the amplitudes of the S wave ground motion and the P wave ground motion result in the problem of obtaining $\mathcal{J}$ numerically. Specific numerical solutions are explained in Sect. 3.2.3.

As the degree of approximation in the ray theory is high, various problems arise in a 3-D velocity structure. For example, the ground motion cannot be calculated if no ray reaches the observation point (this is termed a **shadow**), or conversely, if innumerable rays are concentrated and the geometrical spreading diverges (this situation is termed **caustic**). Although specific countermeasures have been considered for individual cases of such problems, these cannot represent a general solution, because the positions in a structure at which the problems occur must be known beforehand in order to apply them. However, in the 1980s, methods were formulated to avoid these problems as extensions of the ray theory. In these extensions, the energy of ground motion is not simply approximated to be concentrated on the ray, but is considered to be distributed around the ray in a way specified by a certain function. The penetration of the energy by the distribution becomes a **diffracted wave** in the shadow. In the caustic situation, the ray becomes a wave tube with a width because of the penetration, and the geometrical spreading, which is the reciprocal

[79]This agrees with Eq. (2.32) of Červený et al. [20], and with Eq. (2.28) if $\beta d\tau \to d\lambda$.

of the width, can be calculated. As a Gaussian distribution is used for the energy distribution, this method is called the **Gaussian beam method** [18]. However, there is no theory regarding how much the width of the energy distribution should be taken into account, and it is a defect of the method that the width must be decided empirically. In addition, a method to obtain the slowness spectra was proposed, using the property that the amplitude term, Fourier-transformed in the spatial coordinate along the ray, does not cause the problems of shadows and caustic situations. The method then involves inverting the slowness spectra through the Maslov approximation. The solution of the ground motion through this method is called a **Maslov seismogram** [23]. Furthermore, the use of **Kirchhoff's integral**, which is a general representation of **Huygens' principle**, has been proposed to compensate for the defect of the ray theory [41].

3.2.3 Ray Tracing

The first step in the calculation of the ground motion based on the ray theory is to obtain the position $\mathbf{x} = (x, y, z)$ of the ray, by solving the ordinary differential equations of the ray equations in (3.242) under the given boundary conditions at the start point (point source) and the end point (observation point). This is termed **ray tracing**. The numerical solutions of boundary value problems of ordinary differential equations are roughly classified into two methods, namely, the shooting method and the relaxation method [97].

The shooting method calculates multiple solutions which satisfy the differential equation and also the boundary condition at the start point, and to search for the solution which satisfies the boundary condition at the end point. This belongs to the type of "boundary method" explained in Sect. 3.2.1. For ray tracing, it is also termed **shooting method**, because the ray from the point source is shot at the observation point, but may not hit without adjustment. The unknowns of the ray equations are formally defined as the three positions x, y, z and the three momenta p_x, p_y, p_z. However, from the condition $\sum_{j=1}^{3} p_j^2 = p_x^2 + p_y^2 + p_z^2 = s^2$ in (3.237), we have

$$p_x = s \sin\vartheta \cos\varphi\,, \quad p_y = s \sin\vartheta \sin\varphi\,, \quad p_z = s \cos\vartheta\,, \tag{3.277}$$

which indicate the change of variables from the momenta p_x, p_y, p_z to the **azimuthal angle** φ and **takeoff angle** ϑ[80] [20]. As the momentum vector $\mathbf{p} = (p_x, p_y, p_z)$ is in the direction of the ray (Sect. 3.2.2), ϑ is the angle between the $\mathbf{p}$ and the z-axis, and φ is the angle between the horizontal component of $\mathbf{p}$ and the x-axis (Fig. 3.23). Substituting (3.277) into the first equation of the ray equations in (3.242), we obtain

[80]The names of the angles are from Ben-Menahem and Singh [10].

$$\frac{dx}{d\lambda} = \cos\varphi \sin\vartheta\,, \quad \frac{dy}{d\lambda} = \sin\varphi \sin\vartheta\,, \quad \frac{dz}{d\lambda} = \cos\vartheta\,. \tag{3.278}$$

This substitution is also applied to the second equation, so that

$$\begin{aligned}
\frac{dp_x}{d\lambda} &= \frac{ds}{d\lambda}\cos\varphi\sin\vartheta - s\sin\varphi\frac{d\varphi}{d\lambda}\sin\vartheta + s\cos\varphi\cos\vartheta\frac{d\vartheta}{d\lambda} = \frac{\partial s}{\partial x}\,,\\
\frac{dp_y}{d\lambda} &= \frac{ds}{d\lambda}\sin\varphi\sin\vartheta + s\cos\varphi\frac{d\varphi}{d\lambda}\sin\vartheta + s\sin\varphi\cos\vartheta\frac{d\vartheta}{d\lambda} = \frac{\partial s}{\partial y}\,,\\
\frac{dp_z}{d\lambda} &= \frac{ds}{d\lambda}\cos\vartheta - s\sin\vartheta\frac{d\vartheta}{d\lambda} = \frac{\partial s}{\partial z}\,.
\end{aligned} \tag{3.279}$$

Therefore,

$$\begin{aligned}
&\frac{\partial s}{\partial x}\sin\varphi - \frac{\partial s}{\partial y}\cos\varphi = -s\frac{d\varphi}{d\lambda}\sin\vartheta\,,\\
&\left(\frac{\partial s}{\partial x}\cos\varphi + \frac{\partial s}{\partial y}\sin\varphi\right)\cos\vartheta - \frac{\partial s}{\partial z}\sin\vartheta = s\frac{d\vartheta}{d\lambda}
\end{aligned} \tag{3.280}$$

is obtained. The combination of (3.278) and (3.280) results in the ray equations[81]

$$\begin{aligned}
&\frac{dx}{d\lambda} = \cos\varphi\sin\vartheta\,, \quad \frac{dy}{d\lambda} = \sin\varphi\sin\vartheta\,, \quad \frac{dz}{d\lambda} = \cos\vartheta\,,\\
&\frac{d\varphi}{d\lambda} = \frac{-1}{s\sin\vartheta}\left(\frac{\partial s}{\partial x}\sin\varphi - \frac{\partial s}{\partial y}\cos\varphi\right),\\
&\frac{d\vartheta}{d\lambda} = \frac{1}{s}\left(\frac{\partial s}{\partial x}\cos\varphi + \frac{\partial s}{\partial y}\sin\varphi\right)\cos\vartheta - \frac{1}{s}\frac{\partial s}{\partial z}\sin\vartheta\,.
\end{aligned} \tag{3.281}$$

The equations in (3.281) are simultaneous first-order ordinary differential equations with five unknowns. In the shooting method, the boundary conditions at the start point are

$$x = x_S\,, \quad y = y_S\,, \quad z = z_S\,, \tag{3.282}$$

where $\mathbf{x}_S = (x_S, y_S, z_S)$ is the location of the point source as shown in Fig. 3.23. In addition to these, the remaining unknowns φ and ϑ are given tentative values, and the ray equations are solved numerically as an initial value problem. The numerical solution of the initial value problem of the first-order ordinary differential equation has long been considered, and the **Runge–Kutta method** is a typical example.[82] First, we consider advancing the solution x of the uncoupled ordinary differential equation

[81] These agree with Eq. (3.5) of Červený et al. [20] if $d\tau = s\,d\lambda$, which is obtained from (3.235) and (3.238), and $v = s^{-1}$.

[82] According to Kreyszig [82], the original papers were Runge (1895; *Math. Annalen*, **46**, 167–178) and Kutta (1901; *Zeitschr. Math. Phys.*, **46**, 435–453).

Fig. 3.23 The momentum vector **p**, the azimuthal angle φ, and the takeoff angle φ of the ray, shown in the Cartesian coordinate system (reprinted from Koketsu [72] with permission of Kindai Kagaku)

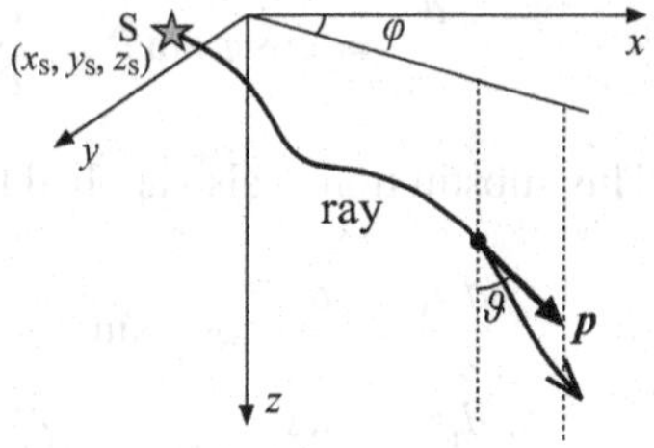

$$\frac{dx}{d\lambda} = f(\lambda, x) \tag{3.283}$$

from $x(\lambda)$ to $x(\lambda + h)$. The Runge–Kutta method is a method for computing the fourth-order Taylor expansion of $x(\lambda + h)$ using only the values of f without calculating the derivatives, and can be regarded as a type of explicit method of the fourth-order finite difference method (Sect. 3.2.4). The derivation process is omitted and only the result is shown as

$$x(\lambda + h) = x(\lambda) + \frac{h}{6}(k_1 + 2k_2 + 2k_3 + k_4), \quad k_1 = f(\lambda, x), \tag{3.284}$$
$$k_2 = f(\lambda + \frac{h}{2}, x + \frac{hk_1}{2}), \quad k_3 = f(\lambda + \frac{h}{2}, x + \frac{hk_2}{2}), \quad k_4 = f(\lambda + h, x + hk_3).$$

If we give h a small value and calculate $x(\lambda + h)$ successively from the initial values of λ and x using this formula, we can extend x sequentially. From the above, even in the case of simultaneous differential equations such as the ray equations (3.281), the solution of a single equation can be applied to each differential equation independently.

As only tentative initial values are given for φ and ϑ at this stage, it is rare that the ray shoots at the position $\mathbf{x}_O = (x_O, y_O, z_O)$ of the observation point. For example, even if the azimuthal angle and takeoff angle of the straight line connecting $\mathbf{x}_S$ and $\mathbf{x}_O$ are used as simple initial values φ_S and ϑ_S, the ray does not shoot to $\mathbf{x}_O$ because the ray is not a straight line in an inhomogeneous velocity structure (Fig. 3.24). The combination of φ_S and ϑ_S to shoot is therefore sought, but a search for two variables is generally difficult.[83] It is possible to extend the case of the following two-dimensional problem, but it does not always work well, therefore the relaxation method should be used instead of the shooting method when a general technique is constructed.

For a two-dimensional problem where $\partial/\partial y = 0$, since $p_y = \partial\tau/\partial y = 0$ and the momentum condition is $p_x^2 + p_z^2 = s^2$,

$$p_x = s \sin\vartheta, \quad p_z = s \cos\vartheta \tag{3.285}$$

[83]"In multidimensions, you can never be sure that the root is there at all until you have found it" (page 256 of Press et al. [97]).

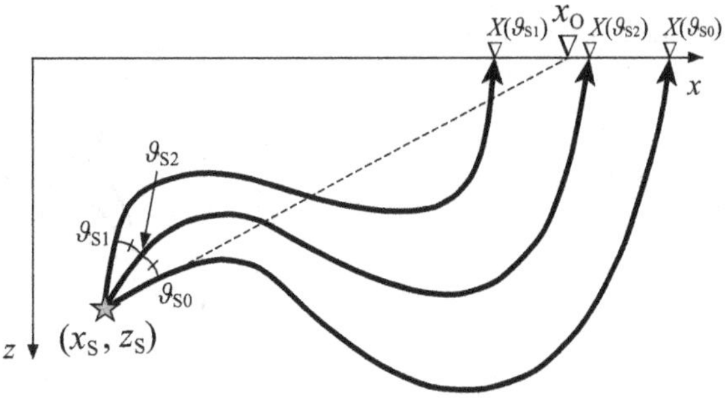

Fig. 3.24 Schematic illustration of the changes of the ray in the shooting method (reprinted from Koketsu [72] with permission of Kindai Kagaku)

and the unknown variable can be only the takeoff angle ϑ [20]. The ray equations yield[84]

$$\frac{dx}{d\lambda} = \sin\vartheta\,, \quad \frac{dz}{d\lambda} = \cos\vartheta\,,$$
$$\frac{d\vartheta}{d\lambda} = \frac{1}{s}\frac{\partial s}{\partial x}\cos\vartheta - \frac{1}{s}\frac{\partial s}{\partial z}\sin\vartheta\,. \tag{3.286}$$

Let $X(\vartheta_S)$ be the x-coordinate of the point where the ray reaches z_O, which is the z-coordinate of the observation point, after solving the Eqs. (3.286) with the Runge–Kutta method and the initial value of the takeoff angle ϑ_S at the point source $\mathbf{x}_S = (x_S, z_S)$. The two-dimensional shooting method consists in solving $f(\vartheta_S) = X(\vartheta_S) - x_O = 0$ using general numerical solutions of an equation. To do this, an interval of ϑ_S including the correct solution for $f(\vartheta_S) = 0$ must first be obtained (this process is called **bracketing** [97]). The takeoff angle ϑ_{SO} of the straight line connecting $\mathbf{x}_S$ and $\mathbf{x}_O$ is used as the initial value, and it is assumed that $X(\vartheta_{SO}) > x_O$, therefore $f(\vartheta_{SO}) > 0$, as shown in Fig. 3.24. Generally, if the takeoff angle is measured as shown in Fig. 3.23, the smaller the takeoff angle, the farther the ray extends. Here, the takeoff angle is therefore increased as ϑ_{S1}. However, unless $X(\vartheta_{S1}) < x_O$ or $f(\vartheta_{SO}) < 0$, the increment of the takeoff angle continues until $f(\vartheta_{SO}) < 0$ is realized. When the interval including the correct solution is obtained in this way, the interval is narrowed down to the correct solution using a well-known iteration method, such as the **bisection method**, the **false position method**, or the **secant method** [97]. Figure 3.24 shows an example using the bisection method, in which the interval is divided into two intervals by $\vartheta_{S2} = (\vartheta_{SO} + \vartheta_{S1})/2$, and when $X(\vartheta_{S2}) > x_O$ ($f(\vartheta_{S2}) > 0$) as shown in the figure, $[\vartheta_{S1}, \vartheta_{S2}]$ is set as a new interval. However, iteration methods such as Newton's method, which require a derivative of $f(\vartheta_S)$, cannot be used.

Of the two numerical solutions for boundary value problems of ordinary differential equations described at the beginning of this subsection, the shooting method is used for the ray tracing, as explained above. In addition, methods where bound-

[84] This agrees with Eq. (3.16) of Červený et al. [20] if $d\tau = s\,d\lambda$ and $\upsilon = s^{-1}$.

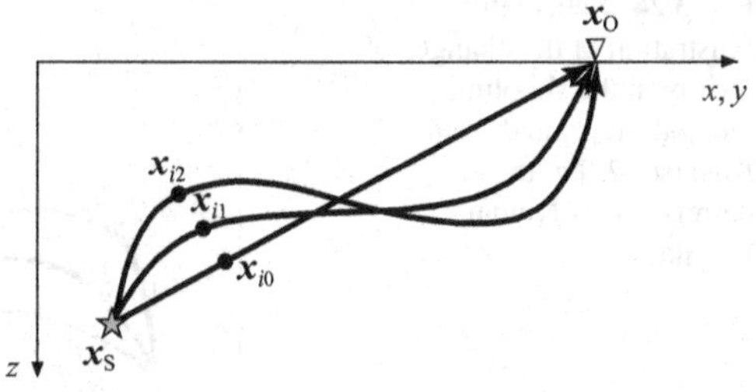

Fig. 3.25 Schematic illustration of the changes of the ray in the bending method (reprinted from Koketsu [72] with permission of Kindai Kagaku)

ary value problems of ordinary differential equations are solved using the **finite difference method** (Sect. 3.2.4) are generally termed the **relaxation method**. In ray tracing, the ray equations are discretized by arranging many points on the ray, therefore they become a large set of simultaneous equations with the position and momentum of each point as variables. Although the ray equations (3.242) in the Cartesian coordinate system are linear, such large simultaneous equations should be solved iteratively. In an orthogonal curvilinear coordinate system such as the spherical coordinate system, the ray equations are nonlinear and therefore must be solved iteratively. Assuming that the initial ray is a straight line connecting the point source and the observation point, as shown in Fig. 3.25, the ray is gradually bent in the process of an iterative solution, so the relaxation method in ray tracing is termed the **bending method**.

The points resulting from the discretization should be equally spaced on the ray for numerical computation, and this can be realized if the independent variable of the ray equation is the ray length λ. This is the reason why it is stated in Sect. 3.2.2 that "the calculation to determine $\mathbf{x}$ is more stable to set the independent variable to λ". However, the length of the interval cannot be determined before the discretization, because the total ray length L cannot be known in advance. This problem can be avoided by carrying out the change of variables of $l \equiv \lambda/L$ and adding the formal differential equation $dL/dl = 0$, so that the range of l is limited to [0, 1], and L is obtained with the same accuracy as the ray position $\mathbf{x}$. Similarly, the travel time T can be obtained with the same accuracy as $\mathbf{x}$ by adding the differential equation $dT/d\lambda = s$, which is derived from the differentiation of the Fermat's principle $T = \int s\, d\lambda$ with respect to λ. Applying this change of variables and the addition of the differential equations to (3.242), and using the fact that there is no λ derivative on the right-hand side of (3.242), we obtain

$$\frac{d\boldsymbol{\omega}}{dl} = \mathbf{f}(\boldsymbol{\omega}), \quad \boldsymbol{\omega} = (x, y, z, p_x, p_y, p_z, T, L)^{\mathrm{T}}, \tag{3.287}$$

where

$$\mathbf{f} = L\left(\frac{p_x}{s}, \frac{p_y}{s}, \frac{p_z}{s}, \frac{\partial s}{\partial x}, \frac{\partial s}{\partial y}, \frac{\partial s}{\partial z}, s, 0\right)^{\mathrm{T}}. \tag{3.288}$$

As the bending method is more numerical than the shooting method, it can be solved more stably without applying the change of variables in (3.277).

The discretization of (3.287) at N points yields

$$\frac{\boldsymbol{\omega}_{n+1}-\boldsymbol{\omega}_n}{\Delta l}=\frac{\mathbf{f}(\boldsymbol{\omega}_{n+1})+\mathbf{f}(\boldsymbol{\omega}_n)}{2}\,,\quad n=1,2,\cdots,N-1 \tag{3.289}$$

using $\Delta l = 1/(N-1)$. The left-hand side of (3.289) is the finite difference approximation at the midpoint between $\boldsymbol{\omega}_{n+1}$ and $\boldsymbol{\omega}_n$ of the derivative $d\boldsymbol{\omega}/dl$ on the left-hand side of (3.287). Hence, $f(\boldsymbol{\omega})$ on the right-hand side of (3.287) is more accurate when approximated by the mean at the midpoint of $f(\boldsymbol{\omega}_{n+1})$ and $f(\boldsymbol{\omega}_n)$, as in (3.289). However, in this case, both sides of (3.289) include $\boldsymbol{\omega}_{n+1}$, therefore the equation must be solved as described below. This solution is termed the **implicit method** within the category of the finite element method. There exists another solution in this category, termed the **explicit method** (Sect. 3.2.4), where $\boldsymbol{\omega}_{n+1}$ on the right-hand side is replaced with some known value $\boldsymbol{\omega}'_{n+1}$. For this, for example, we can use the estimate at a previous step of the iterative solution. Although the accuracy is lower than that of the implicit method, the right-hand side of (3.289) does not contain the unknown $\boldsymbol{\omega}_{n+1}$, and yields the recursive equation

$$\boldsymbol{\omega}_{n+1}=\boldsymbol{\omega}_n+\frac{\Delta l}{2}\left\{\mathbf{f}(\boldsymbol{\omega}'_{n+1})+\mathbf{f}(\boldsymbol{\omega}_n)\right\}\,,\quad n=1,2,\cdots N-1\,, \tag{3.290}$$

which requires only arithmetic calculations. Although the explicit method is not often used in ray tracing, the parameterized shooting method [107] is one of the few examples. The name of this method includes the word "shooting"; however, it corresponds to a combination of the relaxation method and the explicit method.

In the bending method, which is an implicit method of ray tracing, (3.289) is first changed to

$$\mathbf{h}(\boldsymbol{\omega}_n,\boldsymbol{\omega}_{n+1})=\mathbf{f}(\boldsymbol{\omega}_{n+1})+\mathbf{f}(\boldsymbol{\omega}_n)-\frac{2}{\Delta l}(\boldsymbol{\omega}_{n+1}-\boldsymbol{\omega}_n)=\mathbf{0}\,. \tag{3.291}$$

Equation (3.291) does not contain $p_x^2+p_y^2+p_z^2=s^2$ in (3.237), but this condition is automatically satisfied if it is included in the initial conditions at $n=1$ [95]. The boundary conditions for $n=1$ and $n=N$ are therefore given as

$$\begin{aligned}&\left(x_1-x_\mathrm{S},\ y_1-y_\mathrm{S},\ z_1-z_\mathrm{S},\ p_{x1}^2+p_{y1}^2+p_{z1}^2-s_1^2,\ T_1\right)^\mathrm{T}=\mathbf{g}_1(\boldsymbol{\omega}_1)=\mathbf{0}\,,\\&(x_N-x_\mathrm{O},\ y_N-y_\mathrm{O},\ z_N-z_\mathrm{O})=\mathbf{g}_N(\boldsymbol{\omega}_N)=\mathbf{0}\,.\end{aligned} \tag{3.292}$$

where $(x_\mathrm{S}, y_\mathrm{S}, z_\mathrm{S})$ and $(x_\mathrm{O}, y_\mathrm{O}, z_\mathrm{O})$ are the positions of the point source and the observation point ($\mathbf{x}_\mathrm{S}$ and $\mathbf{x}_\mathrm{O}$ in Fig. 3.25), and s_1 is the slowness at $\mathbf{x}_1=\mathbf{x}_\mathrm{S}$. The fifth equation of $\mathbf{g}_1$ is the initial condition specifying that the travel time T starts at 0 s. Combining (3.291) and (3.292), we define

Fig. 3.26 Schematic diagram of the pseudo-bending method (reprinted from Koketsu and Sekine [80] with permission of Blackwell Publishing)

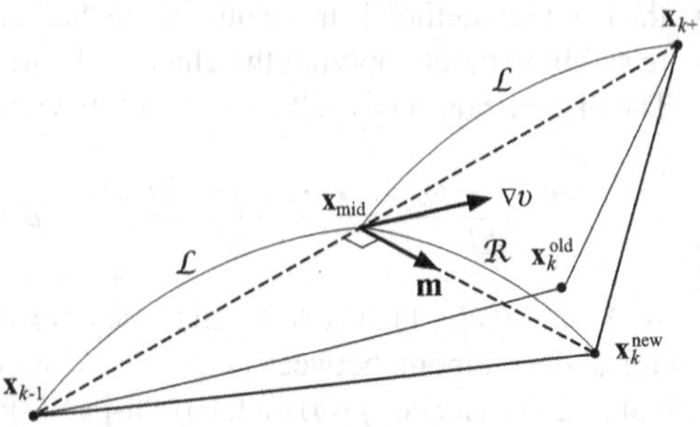

$$\mathbf{F}(\boldsymbol{\Omega}) = (\mathbf{g}_1(\boldsymbol{\omega}_1),\ \mathbf{h}(\boldsymbol{\omega}_1, \boldsymbol{\omega}_2),\ \mathbf{h}(\boldsymbol{\omega}_2, \boldsymbol{\omega}_3), \cdots,\ \mathbf{h}(\boldsymbol{\omega}_{N-1}, \boldsymbol{\omega}_N),\ \mathbf{g}_N(\boldsymbol{\omega}_N))^{\rm T} = \mathbf{0}\,,$$
$$\boldsymbol{\Omega} = (\boldsymbol{\omega}_1,\ \boldsymbol{\omega}_2,\ \cdots,\ \boldsymbol{\omega}_N)^{\rm T}\,. \tag{3.293}$$

Representing the estimate of $\boldsymbol{\Omega}$ at the mth step of the iterative solution with $\boldsymbol{\Omega}_m$, we have the approximation

$$\mathbf{F}(\boldsymbol{\Omega}_{m+1}) \approx \mathbf{F}(\boldsymbol{\Omega}_m) + \frac{\partial \mathbf{F}(\boldsymbol{\Omega}_m)}{\partial \boldsymbol{\Omega}} \delta \boldsymbol{\Omega}_m \approx \mathbf{0}\,, \tag{3.294}$$

and therefore, the simultaneous linear equations to be solved are

$$\frac{\partial \mathbf{F}(\boldsymbol{\Omega}_m)}{\partial \boldsymbol{\Omega}} \delta \boldsymbol{\Omega}_m = -\mathbf{F}(\boldsymbol{\Omega}_m)\,. \tag{3.295}$$

While (3.295) contains $8N$ unknowns, which are the elements of $\delta\boldsymbol{\Omega}_m$, we have $8(N-1)$ equations from (3.291). However, in addition, we also have 8 equations from the boundary conditions in (3.292), so that the total number of equations coincides with the number of unknowns. We can then solve the simultaneous linear equations (3.295) using general solving schemes such as the **Gaussian elimination** [82]. If the band structure of $\partial \mathbf{F}/\partial \boldsymbol{\Omega}$ is utilized, we can compute the solution efficiently.

Another relaxation method is the pseudo-bending method, which does not solve the ray equation but continuously applies the local **Fermat's principle** [115]. We take the three neighboring points on the ray, $\mathbf{x}_{k-1}$, $\mathbf{x}_k$, and $\mathbf{x}_{k+1}$, and apply the local Fermat's principle to move $\mathbf{x}_k$ from $\mathbf{x}_k^{\rm old}$ to the new point $\mathbf{x}_k^{\rm new}$ (Fig. 3.26). The direction of the vector from the midpoint $\mathbf{x}_{\rm mid}$ between $\mathbf{x}_{k-1}$ and $\mathbf{x}_{k+1}$ to $\mathbf{x}_k^{\rm new}$ should approximately match the ray bending direction, that is, the curvature direction. The curvature direction is opposite to the normal principal direction (unit vector $\mathbf{n}$ in Fig. 3.22), and this is represented by the unit vector $\mathbf{m}$. The length of the vector should be determined by Fermat's principle to minimize the time $\mathcal{T}$ during which the ground motion propagates along the ray as $\mathbf{x}_{k-1} \to \mathbf{x}_k^{\rm new} \to \mathbf{x}_{k+1}$.

As for the direction of the vector, since Ben-Menahem and Singh [10] etc. obtained the analytical solution of $\mathbf{n}$

$$\mathbf{n} = \frac{d\mathbf{t}}{d\lambda} \bigg/ \left|\frac{d\mathbf{t}}{d\lambda}\right|, \quad \frac{d\mathbf{t}}{d\lambda} = \frac{1}{s}\{\nabla s - (\mathbf{t}\cdot\nabla s)\,\mathbf{t}\} = -\frac{1}{\upsilon}\{\nabla\upsilon - (\mathbf{t}\cdot\nabla\upsilon)\,\mathbf{t}\}\,, \quad (3.296)$$

$\mathbf{m}$ is obtained from this solution with the inverted sign. Regarding the length of the vector, we first represent $\mathcal{T}$ from the geometry in Fig. 3.26 as

$$\mathcal{T} = \left(\mathcal{L}^2 + \mathcal{R}^2\right)^{\frac{1}{2}}\left\{s_k^{\text{new}} + \frac{s_{k-1} + s_{k+1}}{2}\right\}, \quad (3.297)$$

where $\mathcal{R}$ is the distance between the midpoint $\mathbf{x}_{\text{mid}}$ and $\mathbf{x}_k^{\text{new}}$, and $\mathcal{L}$ is the distance between the midpoint $\mathbf{x}_{\text{mid}}$ and $\mathbf{x}_{k-1}$ or $\mathbf{x}_{k+1}$. s_{k-1}, s_k^{new}, and s_{k+1} stand for the slowness at $\mathbf{x}_{k-1}$, $\mathbf{x}_k^{\text{new}}$, and $\mathbf{x}_{k+1}$, respectively. By solving $\partial\mathcal{T}/\partial\mathcal{R} = 0$, we can obtain a minimum of $\mathcal{T}$, which satisfies Fermat's principle. If we partially differentiate (3.297) by $\mathcal{R}$ and ignore higher-order derivatives, we obtain a quadratic equation, of which a solution is [115]

$$\hat{\mathcal{R}} = -\frac{c\upsilon_{\text{mid}} + 1}{4c\mathbf{n}\cdot\nabla\upsilon_{\text{mid}}} + \left\{\left(\frac{c\upsilon_{\text{mid}} + 1}{4c\mathbf{n}\cdot\nabla\upsilon_{\text{mid}}}\right)^2 + \frac{\mathcal{L}^2}{2c\upsilon_{\text{mid}}}\right\}^{\frac{1}{2}}, \quad (3.298)$$

where $c = (s_{k-1} + s_{k+1})/2$, and υ_{mid} stands for the velocity at $\mathbf{x}_{\text{mid}}$. Using the resulting $\mathbf{m}$ and $\hat{\mathcal{R}}$, $\mathbf{x}_k$ is moved sequentially from $k = 2$ to $k = N-1$. In the pseudo-bending method, these movements form one cycle, and the cycle is repeated until they converge. Although this method does not have the same strictness as a numerical solution compared to the bending method, this method benefits from a small amount of calculation and numerical instability rarely occurs.

So far, we have explained the ray tracing for position $\mathbf{x}$ in the Cartesian coordinate system. For the spherical coordinate system, the shooting method was described by Jacob [56], etc., and the bending and pseudo-bending methods are described by Koketsu and Sekine [80], etc.

In order to compute the ground motion in the ray theory, in addition to $\mathbf{x}$, the amplitude $\mathbf{u}^{(0)}$ has to be traced along the ray, and the latter is called **dynamic ray tracing** when distinguishing it from the former. In Sect. 3.2.2, it is shown that the dynamic ray tracing results in the tracing of the **ray Jacobian** $\mathcal{J}$. Using the ray equations (3.242) and (3.244), the relationship between $\mathbf{p}$ and $\mathbf{t}$ shown in Fig. 3.22 and the eikonal equation (3.232), $\mathcal{J}$ in (3.260) yields[85]

$$\mathcal{J} = \left(\frac{\partial y}{\partial\gamma_1}\frac{\partial z}{\partial\gamma_2} - \frac{\partial z}{\partial\gamma_1}\frac{\partial y}{\partial\gamma_2}\right)\alpha p_x + \left(\frac{\partial z}{\partial\gamma_1}\frac{\partial y}{\partial\gamma_2} - \frac{\partial x}{\partial\gamma_1}\frac{\partial z}{\partial\gamma_2}\right)\alpha p_y$$
$$+ \left(\frac{\partial x}{\partial\gamma_1}\frac{\partial y}{\partial\gamma_2} - \frac{\partial y}{\partial\gamma_1}\frac{\partial x}{\partial\gamma_2}\right)\alpha p_z = \mathbf{\Omega}\cdot\mathbf{t}, \quad \mathbf{\Omega} = \frac{\partial\mathbf{x}}{\partial\gamma_1}\times\frac{\partial\mathbf{x}}{\partial\gamma_2}. \quad (3.299)$$

---

[85]This agrees with Eq. (3.10.16) of Červený [19].

Hence, tracing $\mathcal{J}$ can be replaced with tracing $\mathbf{y} = (\partial\mathbf{x}/\partial\gamma_1, \partial\mathbf{x}/\partial\gamma_2)^{\mathrm{T}}$. However, as the ordinary differential equation to be satisfied by $\mathbf{y}$ is derived from the ray equation (3.243) and this includes $\mathbf{p}$, $\mathbf{z} = (\partial\mathbf{p}/\partial\gamma_1, \partial\mathbf{p}/\partial\gamma_2)^{\mathrm{T}}$ must also be added as an unknown. Partially differentiating (3.243) by γ_1 and γ_2, we obtain the simultaneous ordinary differential equations

$$\frac{d}{d\lambda}\left(\frac{\partial\mathbf{x}}{\partial\gamma_1}, \frac{\partial\mathbf{x}}{\partial\gamma_2}\right)^{\mathrm{T}} = \frac{\partial s^{-1}}{\partial\mathbf{x}}\left(\frac{\partial\mathbf{x}}{\partial\gamma_1}, \frac{\partial\mathbf{x}}{\partial\gamma_2}\right)^{\mathrm{T}}\mathbf{p} + s^{-1}\left(\frac{\partial\mathbf{p}}{\partial\gamma_1}, \frac{\partial\mathbf{p}}{\partial\gamma_2}\right)^{\mathrm{T}} \Rightarrow \frac{d\mathbf{y}}{d\lambda} = \frac{\partial s^{-1}}{\partial\mathbf{x}}\mathbf{y}\mathbf{p} + s^{-1}\mathbf{z}$$

$$\frac{d}{d\lambda}\left(\frac{\partial\mathbf{p}}{\partial\gamma_1}, \frac{\partial\mathbf{p}}{\partial\gamma_2}\right)^{\mathrm{T}} = \frac{\partial\nabla s}{\partial\mathbf{x}}\left(\frac{\partial\mathbf{x}}{\partial\gamma_1}, \frac{\partial\mathbf{x}}{\partial\gamma_2}\right)^{\mathrm{T}} \Rightarrow \frac{d\mathbf{z}}{d\lambda} = \frac{\partial\nabla s}{\partial\mathbf{x}}\mathbf{y}\,. \tag{3.300}$$

Solving (3.300) corresponds to the dynamic ray tracing, and since $\mathbf{y}$ and $\mathbf{z}$ both have six elements, those are simultaneous ordinary differential equations with twelve unknowns. Since the equations in (3.300) also contain $\mathbf{x}$ and $\mathbf{p}$, they must be solved simultaneously with the ray equations (3.243) for the ray tracing. As the ray equations are simultaneous ordinary differential equations with six unknowns, we have to solve the simultaneous ordinary differential equations with eighteen unknowns in order to obtain the amplitude $\mathbf{u}^{(0)}$ as well as the position $\mathbf{x}$ of the ray (we can reduce the number of unknowns and equations using the change of variables in (3.277), etc.).

The zeroth-order approximation of the effect of propagation $e^{i\omega(t-\tau(\mathbf{x}))}\mathbf{u}^{(0)}(\mathbf{x})$ in (3.226) is computed by executing the ray tracing and the dynamic ray tracing. The effect of the point source in (2.56), etc. is then convolved, and the ground motion can be obtained. Good program packages, which execute these in a 3-D velocity structure, are available, e.g., in *Seismological Algorithms* [30]. However, due to the use of the ray theory or its extension, the problems associated with the ray theory persist, such as the high frequency approximation, difficulties in the phase selection, etc. However, there are examples in which the ray theory and the Gaussian beam method give good results for the problem where components of low frequencies and reverberations are rather important, such as in Fig. 3.27. In addition, for the high frequency approximation such as the ray theory, there is a long history of study in optics and electromagnetics, which generally deal with higher frequencies than that of ground motion. All of the techniques mentioned here borrow the basic principles from these fields.

3.2.4 *Finite Difference Method*[86]

At the opposite pole of the "ray theory" is the **finite difference method**, which is the most classical numerical solution of partial differential equations. In the finite difference method, a lattice is set in the spatiotemporal region for the problem, and algebraic equations, which are obtained by replacing partial derivatives with **difference quotients**, are solved by a computer. In the ground motion problem, the partial differential equation is the equation of motion (1.30), and the unknown is the elas-

[86]For more information on Sect. 3.2.4, see Aki and Richards [4].

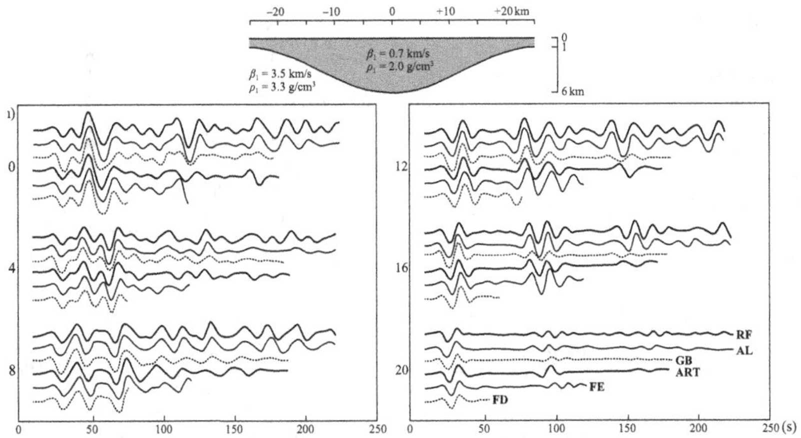

Fig. 3.27 Upper: Two-dimensional sedimentary basin. Lower: Comparison of the ground motions computed by various methods for the case where an SH plane wave is incident from the bottom upon the basin. ART uses the ray theory (Sect. 3.2.2), GB uses the Gaussian beam method (Sect. 3.2.2), FD uses the finite difference method (Sect. 3.2.4), FE uses the finite element method (Sect. 3.2.5), and AL and RF use the Aki-Lerner method and the 2-D reflectivity method (Sect. 3.2.6) (modified from Kohketsu [71] with permission of Blackwell Publishing)

tic displacement (ground motion). In order to explain the principle understandably, two-dimensional plane waves (Sect. 1.2.5) in a 2-D velocity structure are considered below.

Assuming that the ground motion and the velocity structure do not change in the y-direction of the Cartesian coordinate system and that no body forces exist because of the plane wave problem, $\partial/\partial y = 0$ and $f_x = f_y = f_z = 0$ are substituted for the equations of motion (1.27), (1.28), and (1.29), and we have

$$\rho\frac{\partial^2 u_x}{\partial t^2} = (\lambda+\mu)\frac{\partial}{\partial x}\left(\frac{\partial u_x}{\partial x}+\frac{\partial u_z}{\partial z}\right)+\mu\left(\frac{\partial^2}{\partial x^2}+\frac{\partial^2}{\partial z^2}\right)u_x$$
$$+\frac{\partial\lambda}{\partial x}\left(\frac{\partial u_x}{\partial x}+\frac{\partial u_z}{\partial z}\right)+\frac{\partial\mu}{\partial z}\frac{\partial u_z}{\partial x}+2\frac{\partial\mu}{\partial x}\frac{\partial u_x}{\partial x}+\frac{\partial\mu}{\partial z}\frac{\partial u_x}{\partial z}, \tag{3.301}$$

$$\rho\frac{\partial^2 u_y}{\partial t^2} = \mu\left(\frac{\partial^2}{\partial x^2}+\frac{\partial^2}{\partial z^2}\right)u_y+\frac{\partial\mu}{\partial z}\frac{\partial u_y}{\partial z}+\frac{\partial\mu}{\partial x}\frac{\partial u_y}{\partial x}, \tag{3.302}$$

$$\rho\frac{\partial^2 u_z}{\partial t^2} = (\lambda+\mu)\frac{\partial}{\partial z}\left(\frac{\partial u_x}{\partial x}+\frac{\partial u_z}{\partial z}\right)+\mu\left(\frac{\partial^2}{\partial x^2}+\frac{\partial^2}{\partial z^2}\right)u_z$$
$$+\frac{\partial\lambda}{\partial z}\left(\frac{\partial u_x}{\partial x}+\frac{\partial u_z}{\partial z}\right)+\frac{\partial\mu}{\partial x}\frac{\partial u_x}{\partial z}+\frac{\partial\mu}{\partial x}\frac{\partial u_z}{\partial x}+2\frac{\partial\mu}{\partial z}\frac{\partial u_z}{\partial z}. \tag{3.303}$$

Among these equations, (3.301) and (3.303) contain only u_x and u_z, and (3.302) contains only u_y. In addition, when $\partial/\partial y = 0$ is substituted into (1.44),

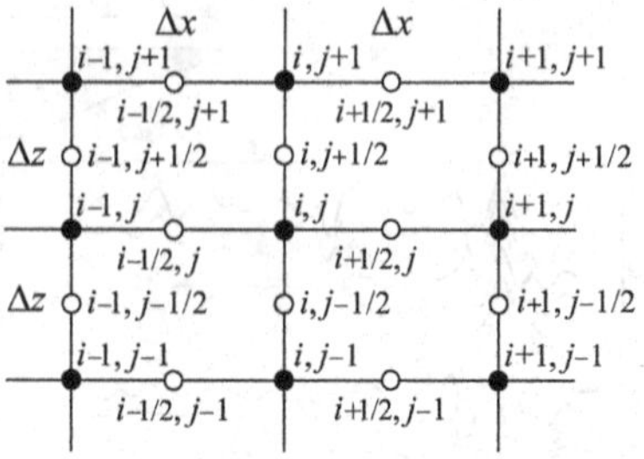

Fig. 3.28 Spatial lattice set in the 2-D velocity structure, where the ground motion is evaluated at the grid points ● and properties such as the rigidity are evaluated at the intermediate points ○ (reprinted from Koketsu [72] with permission of Kindai Kagaku)

$$u_x = \frac{\partial \phi}{\partial x} + \frac{\partial^2 \psi}{\partial x \partial z}, \quad u_y = -\frac{\partial \chi}{\partial x}, \quad u_z = \frac{\partial \phi}{\partial z} - \frac{\partial^2 \psi}{\partial x^2} \tag{3.304}$$

are obtained, therefore u_x and u_z are composed of only the displacement potentials ϕ for the P wave and ψ for the SV wave, and u_y is composed only of the displacement potential χ for the SH wave. In the two-dimensional plane wave problem in a 2-D velocity structure, the ground motion of the P and SV waves and the ground motion of the SH wave are therefore separated, as in a 1-D velocity structure.

The application of the finite difference method to the ground motion problem began in the late 1960s, and the results from that time to the early 1970s were reviewed by Boore [11]. The SH plane wave problem in a 2-D velocity structure is shown as an example in this review. The equation to be solved is (3.302), and setting $v = u_y$, this is changed to

$$\rho \frac{\partial^2 v}{\partial t^2} = \frac{\partial}{\partial x}\left(\mu \frac{\partial v}{\partial x}\right) + \frac{\partial}{\partial z}\left(\mu \frac{\partial v}{\partial z}\right). \tag{3.305}$$

Here, the lattice of Fig. 3.28 is set in a 2-D velocity structure, and as a difference quotient, the **central difference** of second-order accuracy is used for the first-order space derivatives and a second-order time derivative. Because the spatial differentiation is doubly imposed, the discrete spacing is half of the lattice spacing and we use

$$\frac{\partial f}{\partial x} \sim \frac{f^p_{i+1/2,j} - f^p_{i-1/2,j}}{\Delta x}, \quad \frac{\partial f}{\partial z} \sim \frac{f^p_{i,j+1/2} - f^p_{i,j-1/2}}{\Delta z}, \tag{3.306}$$

$$\frac{\partial^2 f}{\partial t^2} \sim \frac{f^{p+1}_{i,j} - 2 f^p_{i,j} + f^{p-1}_{i,j}}{(\Delta t)^2}, \tag{3.307}$$

where p represents the time step, Δt is the time interval, and $f^p_{i,j}$ is the value of the function f at the grid point (i, j) in the time step p.

First, we replace the first term on right-hand side of (3.305) with the first difference quotient in (3.306) for $f = \mu \partial v / \partial x$, and obtain

$$\frac{\partial}{\partial x}\left(\mu\frac{\partial v}{\partial x}\right) \sim \frac{1}{\Delta x}\left\{\left(\mu\frac{\partial v}{\partial x}\right)^p_{i+1/2,j} - \left(\mu\frac{\partial v}{\partial x}\right)^p_{i-1/2,j}\right\},$$

Furthermore, $\partial v/\partial x$ in this equation is replaced by the first difference quotient in (3.306) for $f = v$ centered at the midpoint $(i+1/2, j)$ or $(i-1/2, j)$, and we obtain

$$\begin{aligned}\frac{\partial}{\partial x}\left(\mu\frac{\partial v}{\partial x}\right) &\sim \frac{1}{\Delta x}\left\{\mu_{i+1/2,j}\frac{v^p_{i+1,j}-v^p_{i,j}}{\Delta x} - \mu_{i-1/2,j}\frac{v^p_{i,j}-v^p_{i-1,j}}{\Delta x}\right\}\\ &\sim \frac{\mu_{i+1/2,j}v^p_{i+1,j} - (\mu_{i+1/2,j}+\mu_{i-1/2,j})v^p_{i,j} + \mu_{i-1/2,j}v^p_{i-1,j}}{(\Delta x)^2},\end{aligned} \tag{3.308}$$

where $\mu_{i\pm1/2,j}$ are the rigidities at the midpoints. Similarly, the second term on right-hand side of (3.305) yields

$$\frac{\partial}{\partial z}\left(\mu\frac{\partial v}{\partial z}\right) \sim \frac{\mu_{i,j+1/2}v^p_{i,j+1} - (\mu_{i,j+1/2}+\mu_{i,j-1/2})v^p_{i,j} + \mu_{i,j-1/2}v^p_{i,j-1}}{(\Delta z)^2}. \tag{3.309}$$

Substituting (3.307) for $f = v$, (3.308), and (3.309) into (3.305), we obtain the finite difference equation

$$\begin{aligned}v^{p+1}_{i,j} &= 2v^p_{i,j} - v^{p-1}_{i,j}\\ &+ \frac{(\Delta t)^2}{\rho_{i,j}}\left\{\frac{\mu_{i+1/2,j}v^p_{i+1,j} - (\mu_{i+1/2,j}+\mu_{i-1/2,j})v^p_{i,j} + \mu_{i-1/2,j}v^p_{i-1,j}}{(\Delta x)^2}\right.\\ &+ \left.\frac{\mu_{i,j+1/2}v^p_{i,j+1} - (\mu_{i,j+1/2}+\mu_{i,j-1/2})v^p_{i,j} + \mu_{i,j-1/2}v^p_{i,j-1}}{(\Delta z)^2}\right\}.\end{aligned} \tag{3.310}$$

The algebraic calculations shown in this equation are recursively carried out under the initial and boundary conditions of the ground motion problem. These recursive calculations are stable if $\beta_{\max}\Delta t/\min(\Delta x, \Delta z) \le 1/\sqrt{n}$, where $\beta = \sqrt{\mu/\rho}$ and n, which is the number of dimensions, is 2 in this case [11].

For example, in the problem where SH plane waves are incident from below upon a two-dimensional sedimentary basin, the initial condition is this incidence. The ground surface is located in the computational domain and the **stress-free condition** is realized by setting $\mu = 0$ above it. In addition, **artificial reflection** occurs on artificial boundaries, which must be located in the circumference of the computational domain because the domain is limited, although an actual velocity structure spreads widely. In order to avoid the influence of the artificial reflection, the computation is stopped before the time at which this influence begins to appear, or boundary conditions similar to those imposed at **radiation boundary** are given to the artificial boundaries. For the latter case, a "viscous boundary", "transmission boundary", "combination boundary", etc. have been proposed, but none of them works perfectly. Buffer zones with large attenuation are therefore located in front of the artificial boundaries, giving damping terms $k\,\partial v/\partial t$ to the equation of motion and the finite difference equation.

The ground motion labeled **FD** in Fig. 3.27 was computed by Boore [11] using the method described above, in which the ground motion was terminated before the influence of the artificial reflection appeared.

If the formulation is started not from the equations of motion (1.27), (1.28), and (1.29) in which the stresses are eliminated, but from the equations of motion (1.26) in which the stresses remain and the definitions of the stresses (1.24), the accuracy of the boundary conditions increases, and this formulation is now the mainstream. As both elastic displacements and stresses must be stored, the required memory and disk space are doubled, but this issue was alleviated by the advance of computers. We substitute $\partial/\partial y = 0$ and $f_x = f_y = f_z = 0$ into (1.26) and the time derivative of (1.24), and perform the change of variables $\upsilon_x = \partial u_x/\partial t$, $\upsilon_y = \partial u_y/\partial t$, and $\upsilon_z = \partial u_z/\partial t$.[87] We then obtain

$$\rho\frac{\partial \upsilon_x}{\partial t} = \frac{\partial \tau_{xx}}{\partial x} + \frac{\partial \tau_{xz}}{\partial z},$$

$$\frac{\partial \tau_{xx}}{\partial t} = \lambda\left(\frac{\partial \upsilon_x}{\partial x} + \frac{\partial \upsilon_z}{\partial z}\right) + 2\mu\frac{\partial \upsilon_x}{\partial x}, \quad \frac{\partial \tau_{xz}}{\partial t} = \mu\left(\frac{\partial \upsilon_x}{\partial z} + \frac{\partial \upsilon_z}{\partial x}\right) \tag{3.311}$$

$$\rho\frac{\partial \upsilon_y}{\partial t} = \frac{\partial \tau_{xy}}{\partial x} + \frac{\partial \tau_{yz}}{\partial z}, \quad \frac{\partial \tau_{xy}}{\partial t} = \mu\frac{\partial \upsilon_y}{\partial x}, \quad \frac{\partial \tau_{yz}}{\partial t} = \mu\frac{\partial \upsilon_y}{\partial z}, \tag{3.312}$$

$$\rho\frac{\partial \upsilon_z}{\partial t} = \frac{\partial \tau_{xz}}{\partial x} + \frac{\partial \tau_{zz}}{\partial z},$$

$$\frac{\partial \tau_{xz}}{\partial t} = \mu\left(\frac{\partial \upsilon_x}{\partial z} + \frac{\partial \upsilon_z}{\partial x}\right), \quad \frac{\partial \tau_{zz}}{\partial t} = \lambda\left(\frac{\partial \upsilon_x}{\partial x} + \frac{\partial \upsilon_z}{\partial z}\right) + 2\mu\frac{\partial \upsilon_z}{\partial z}, \tag{3.313}$$

noting that τ_{xz} appears both in (3.311) and in (3.313). The ground motion velocity υ_y of the SH wave appears only in (3.312). The SH plane wave problem in a 2-D velocity structure is therefore solved algebraically by converting the simultaneous partial differential equations (3.312) into the simultaneous finite differential equations with the central differences. It is well known in the field of numerical analysis that in cases of multiple unknowns, it is more stable to use **staggered grids**, which are lattices shifted from each other, than to evaluate all unknowns at the same grid point [108]. In addition, the condition under which the computation is stably carried out is also relaxed to double [38].

Consequently, only υ and ρ are evaluated at the original grid points ● in Fig. 3.28, and the intermediate points ○ are newly defined as grid points for σ, τ, and μ. The intermediate steps $p \pm 1/2$ corresponding to each time step p are also provided. The difference equation thus yields

[87] We follow the notation of Virieux [119], so that here the definition υ is different from (3.305), etc.

$$
\begin{aligned}
\upsilon_{i,j}^{p+1/2} &= \upsilon_{i,j}^{p-1/2} + \frac{\Delta t}{\rho_{i,j}} \frac{\sigma_{i+1/2,j}^{p} - \sigma_{i-1/2,j}^{p}}{\Delta x} + \frac{\Delta t}{\rho_{i,j}} \frac{\tau_{i,j+1/2}^{p} - \sigma_{i,j-1/2}^{p}}{\Delta z} \\
\sigma_{i+1/2,j}^{p+1} &= \sigma_{i+1/2,j}^{p} + \Delta t \cdot \mu_{i+1/2,j} \frac{\upsilon_{i+1,j}^{p+1/2} - \upsilon_{i,j}^{p+1/2}}{\Delta x} \\
\tau_{i,j+1/2}^{p+1} &= \tau_{i,j+1/2}^{p} + \Delta t \cdot \mu_{i,j+1/2} \frac{\upsilon_{i,j+1}^{p+1/2} - \upsilon_{i,j}^{p+1/2}}{\Delta x} .
\end{aligned}
\tag{3.314}
$$

Virieux [119] used these equations to compute the ground motions shown in Fig. 3.27, and the results showed a good agreement with the results of the other methods over 180 s.

Numerical computations of the ground motion in 3-D velocity structures using the finite difference method are widely performed (e.g., Graves [40]), although here this method has been applied to 2-D velocity structures for the sake of clarity. Performing the numerical computation of the ground motion in a 3-D velocity structure using the finite difference method or the finite element method (Sect. 3.2.5) is therefore often referred to as **ground motion simulation**. In addition to the conventional finite difference method explained here, the **pseudo-spectral method** [37], where the Fourier transform and the **FFT** (Sect. 4.2.4) are used for spatial differentiation, has also been proposed, however this method is seldom used because it is difficult to gain a high parallelization efficiency. The introduction of anelastic damping (Sect. 1.2.6) and point sources is discussed in the next subsection Sect. 3.2.5.

3.2.5 *Finite Element Method*

If a **domain method** of the method of weighted residuals (Sect. 3.2.1) is applied to a domain with a complicated velocity structure, even the **Galerkin method** cannot provide an accurate solution without preparing a large number of trial functions. In addition, even if a large number of trial functions can be prepared, their properties tend to deteriorate when they are higher-order functions, and simultaneous equations tend to be ill conditioned when they are simply applied. In such a case, a function which has a value only in a small element within the domain is considered, and the element is variously changed to generate trial functions. This approach to the Galerkin method is termed the **computational Galerkin method** [33]. If elements are made smaller so that the structure does not change significantly within an element, we can use simple trial functions but the number of elements increases. Conversely, it is difficult to choose an appropriate function if the number of elements is reduced. The numerical Galerkin method usually takes the former approach and adopts a low-order polynomial such as a linear function and a spline. Although the equation of motion for the ground motion includes second-order derivatives, this can be avoided by executing the **integration by parts (weak formulation)**, so that a linear function can be used as a trial function. The **finite element method** is a domain method with the numerical Galerkin method.

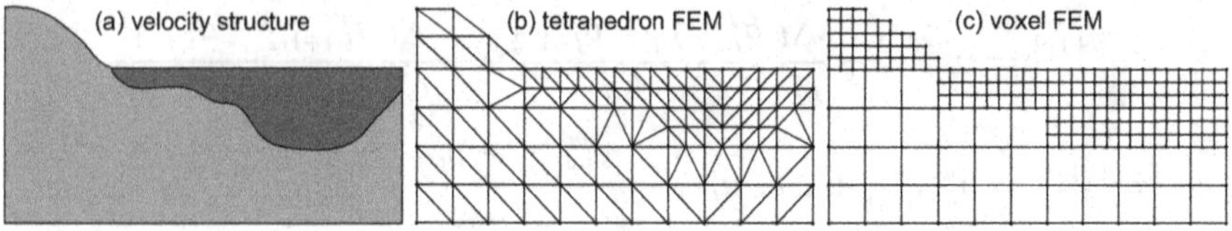

Fig. 3.29 (a) Complex velocity structure, (b) a tetrahedral mesh of the conventional finite element method, and (c) a voxel mesh of the voxel finite element method (reprinted from Koketsu et al. [73] with permission of Springer Nature)

The advantages of the finite element method over the finite difference method are: (1) the stress-free condition is automatically incorporated on the circumference of the domain, (2) the Courant condition for time integration is relaxed (this will be described later), and (3) a complex domain can be handled due to the availability of various element shapes. However, in return for these advantages the method contains defects such as large-scale computation and time-consuming preprocessing (element mesh generation, etc.). Here, these defects are solved by renouncing advantage (3), limiting the element shape to a "voxel" (a rectangular prism, or hexahedron), and applying the explicit method. Figure 3.29 schematically illustrates the above situation. For a complex velocity structure such as shown in panel (a), the conventional finite element method generates a mesh such as in panel (b) using tetrahedrons and other shapes that can flexibly fit into any structure. In contrast, the voxel finite element method uses only hexahedrons (voxels) to generate such a grid mesh, as in panel (c), which is similar to that of the finite difference method, but the goodness of fit to (a) is weaker than the grid in (b); therefore, a finer mesh is generated to compensate for this.

We assume that the general equation of motion derived from (1.81) and (1.82):

$$\rho \frac{\partial^2 u_i}{\partial t^2} = \frac{\partial}{\partial x_j} C_{ijkl} \frac{\partial u_k}{\partial x_l} + \rho f_i \tag{3.315}$$

is valid in the region V in Fig. 1.9. However, (3.315) does not include anelastic damping (Sect. 1.2.6); therefore, we include it now and there are three options for this inclusion. The first option is the most widely used, in which the damping term $c' dU/dt$ in the equation of motion (1.60) for the 1-D problem is given to (3.315), with $U \to u_i$ and the damping coefficient $c' = 2\pi f_0 \rho Q^{-1}$ in (1.66). f_0 is the frequency at which Q is measured. As the damping coefficient is proportional to ρ, the anelastic damping in this way is called **mass-proportional damping** as in the 1-D problem (Sect. 1.2.6), and C' stands for this proportional constant in the description below. In the second option, anelastic damping is included in (1.81) with the strain rate de_{kl}/dt as

$$\tau_{ij} = C_{ijkl} e_{kl} + \Gamma_{ijkl} \frac{de_{kl}}{dt}, \quad e_{kl} = \frac{\partial u_k}{\partial x_l}. \tag{3.316}$$

From this relationship, Γ_{ijkl} is considered to be approximately proportional to C_{ijkl} [54]; therefore, it is called **stiffness-proportional damping**[88] [8]. The final option is termed **Rayleigh damping** [8], and is a linear combination of the mass-proportional and stiffness-proportional damping. The damping spectrum (the distribution of damping coefficients for various f) in the mass-proportional damping is independent of f, and the damping spectrum in the stiffness-proportional damping is proportional to f^2. The Rayleigh damping which combines the two options can therefore correspond to various damping spectra, including the constant Q damping proportional to f [54]. If the weights of the mass-proportional damping and the stiffness-proportional damping in the combination are W_M and W_K, respectively, the damping terms added to the equation of motion (3.315) due to anelastic damping are

$$W_M\, C' \rho \,\frac{\partial u_i}{\partial t}\,, \quad W_K \Gamma_{ijkl}\,\frac{\partial}{\partial t}\frac{\partial u_k}{\partial x_l}\,. \tag{3.317}$$

As (3.317) represents the mass-proportional damping if $W_K = 0$, and the stiffness-proportional damping if $W_M = 0$, (3.317) covers all three options. These three options can be used even in the finite difference method, if (3.317) is added to the appropriate equation of motion (Eq. (3.301) – (3.303) in the two-dimensional case).

When region V in Fig. 1.9 is defined as the domain of the numerical Galerkin method and its circumference S is a **free surface** so that T_i is 0, applying the numerical Galerkin method to (3.315) with (3.317) added in, we obtain the weak form of the integral equation

$$\begin{aligned}
&\int N^n \rho \frac{\partial^2 u_i}{\partial t^2} dV + W_M C' \int N^n \rho \frac{\partial u_i}{\partial t} dV + W_K \int N^n \frac{\partial}{\partial x_j} \Gamma_{ijkl} \frac{\partial}{\partial t}\frac{\partial u_k}{\partial x_l} dV \\
&- \int N^n \frac{\partial}{\partial x_j} \Gamma_{ijkl} \frac{\partial u_k}{\partial x_l} dV = \int N^n \rho f_i dV, \quad n = 1, 2, \cdots\,, \\
&u_i(\mathbf{x}, t) = \sum_n N^n(\mathbf{x})\, u_i^n(t)\,.
\end{aligned} \tag{3.318}$$

$N^n(\mathbf{x})$ in the above is a trial function of the numerical Galerkin method (ϕ_n in (3.223)), and is also called a **shape function**. $u_i(\mathbf{x}, t)$ in each element that divides region V is approximated from the interpolation of its values $u_i^k(t)$ at the vertexes of each element, which are the **node** of the grid mesh. Hence, as the shape function has a value only in the vicinity of each element and is zero at S, the integration by parts can be applied to the third and fourth terms of the first equation in (3.318), and they yield

$$-W_K \int \frac{\partial N^n}{\partial x_j} \Gamma_{ijkl} \frac{\partial}{\partial t}\frac{\partial u_k}{\partial x_l} dV\,, \quad \int \frac{\partial N^n}{\partial x_j} C_{ijkl} \frac{\partial u_k}{\partial x_l} dV\,.$$

Then, substituting the second equation into the first equation in (3.318), we obtain

[88] This name is fixed in engineering, but it should be called "elasticity-proportional damping" in the terminology of this book. "stiffness" is another name for the elastic constant in engineering.

$$\mathbf{M}\frac{d^2\boldsymbol{\delta}}{dt^2}+\mathbf{C}\frac{d\boldsymbol{\delta}}{dt}+\mathbf{K}\boldsymbol{\delta}=\mathbf{f}\,,\quad \mathbf{C}=W_M C'\mathbf{M}+W_K\mathbf{G}\,, \tag{3.319}$$
$$\mathbf{M}=\int\mathbf{N}^{\mathrm{T}}\rho\mathbf{N}\,dV,\ \mathbf{K}=\int\mathbf{B}^{\mathrm{T}}\mathbf{D}\mathbf{B}\,dV,\ \mathbf{G}=\int\mathbf{B}^{\mathrm{T}}\boldsymbol{\Gamma}\mathbf{B}\,dV,\ \mathbf{f}=\int\mathbf{N}^{\mathrm{T}}\mathbf{F}\,dV\,,$$

where $\boldsymbol{\delta}$ and $\mathbf{F}$ are the vectors composed of u_i^n or ρf_i, respectively. **N**, **B**, **D**, and **G** are the matrices composed of N^n, $\partial N^n/\partial x_i$, C_{ijkl}, or Γ_{ijkl}. Comparing the second equation of (3.318) with (3.223), $\boldsymbol{\delta}$ is found to be an unknown vector because u_i^n corresponds to a_n, which is determined by minimizing the sum of weighted residuals in (3.224).

When we replace the second-order time derivative by the central difference, and the first-order time derivative by the **backward difference**, (3.319) yields

$$\mathbf{M}\frac{\boldsymbol{\delta}_{t+\Delta t}-2\boldsymbol{\delta}_t+\boldsymbol{\delta}_{t-\Delta t}}{(\Delta t)^2}+\mathbf{C}\frac{\boldsymbol{\delta}_t-\boldsymbol{\delta}_{t-\Delta t}}{\Delta t}+\mathbf{K}\boldsymbol{\delta}_t=\mathbf{f}_t\,,\quad \boldsymbol{\delta}_t=\boldsymbol{\delta}(t)\,,\ \mathbf{f}_t=\mathbf{f}(t)\,. \tag{3.320}$$

Here, it is assumed that the mass (density) of each element is discretely distributed at the nodes around the element (vertexes of the element). Under this assumption, which is referred to as the **lumped mass**, **M** in (3.319) becomes a diagonal matrix, so that **M** can be analytically inverted, and (3.320) yields a linear equation of $\boldsymbol{\delta}_{t+\Delta t}$, which can be solved using the **explicit method**. The **spectral element method** [81] is a finite element method which realizes the diagonalization of **M** by taking an irregular node arrangement using the Legendre polynomials, instead of the lumped mass assumption.

The shape function N^n and the matrices and vectors in (3.319) are now explained for the voxel finite element method. We consider the two-dimensional case $(\partial/\partial y=0)$ of the P and SV wave ground motions for visual clarity. In this case, the voxel elements extend infinitely in the y-direction, therefore we consider only the rectangles of their cross sections and refer to them as "voxel elements". As a layered structure is a good approximation to the velocity structure, the region is divided into several subregions by horizontal planes as shown in Fig. 3.29c, and the element sizes Δx and Δz are constant in each subregion. Within a subregion, elements of a certain area $\Delta x\Delta z$ are distributed, and the vertexes of the (i,j) element are the i,j node, the $i+1,j$ node, the $i,j+1$ node, and the $i+1,j+1$ node (Fig. 3.30). If we take only the components related to the (i,j) element from $\boldsymbol{\delta}$, we have

$$\boldsymbol{\delta}^{(i,j)}=\left(\mathbf{u}^{i,j},\mathbf{u}^{i+1,j},\mathbf{u}^{i,j+1},\mathbf{u}^{i+1,j+1}\right)^{\mathrm{T}},\quad \mathbf{u}^{k,l}=\left(u_x^{k,l},u_z^{k,l}\right)^{\mathrm{T}}\,. \tag{3.321}$$

Because of $\mathbf{M}^{(i,j)}=\rho\Delta x\Delta z/4\cdot\mathbf{I}$, the elements of $\mathbf{M}\boldsymbol{\delta}_t$ in (3.320) are given as

$$(\mathbf{M}\boldsymbol{\delta}_t)^{i,j}=\rho\Delta x\Delta z\,\mathbf{u}^{i,j}\,, \tag{3.322}$$

then (3.320) yields the recursive equation

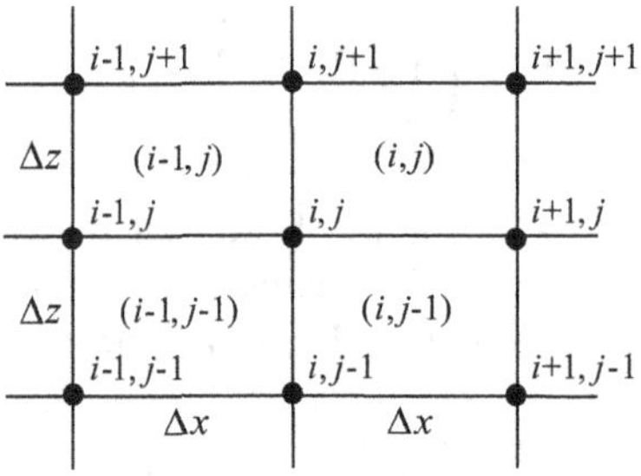

Fig. 3.30 The arrangement of voxel elements and nodes within a subregion where Δx and Δz are constant (reprinted from Koketsu et al. [73] with permission of Springer Nature)

$$\mathbf{u}_{t+\Delta t}^{i,j} = 2\mathbf{u}_t^{i,j} - \mathbf{u}_{t-\Delta t}^{i,j} + \frac{(\Delta t)^2}{\rho \Delta x \Delta z} \left\{ \mathbf{f}_t^{i,j} - (\mathbf{K}\boldsymbol{\delta}_t)^{i,j} - \frac{1}{\Delta t} (\mathbf{C}(\boldsymbol{\delta}_t - \boldsymbol{\delta}_{t-\Delta t}))^{i,j} \right\} . \tag{3.323}$$

If the distribution of elements is set so that the position of the point source coincides with the center of an element, $\mathbf{f}_t$ of the point source is delivered to the nodes around the element by the operation of $\mathbf{N}^{\mathrm{T}}\mathbf{F}$, and the other nodes have no source contribution. The portions delivered to the surrounding nodes are determined from the moment tensor of the point source [54]. This method is equivalent to that of Graves [40] in the finite difference method.

The finite element method has been well studied in engineering and there is an extensive literature. In order to remain consistent with those studies, the definition from engineering (Sect. 1.2.1, and footnote 3)

$$\gamma_{ij} = 2e_{ij} = \frac{\partial u_i}{\partial x_j} + \frac{\partial u_j}{\partial x_i} \tag{3.324}$$

is used for the **shear strain** in the rest of this subsection. As this is the two-dimensional case of the P and SV wave ground motions, and therefore $\partial/\partial y = 0$ and $u_y = 0$, the strains in (1.81) with the shear strains γ_{ij} are zero, except for e_{xx}, e_{zz}, and γ_{zx}. Assuming that the velocity structure is isotropic according to the policy of this book, $\mathbf{D} = \left(C_{ijkl}\right)$ yields the 3×3 matrix

$$\mathbf{D} = \begin{pmatrix} \lambda + 2\mu & \lambda & 0 \\ \lambda & \lambda + 2\mu & 0 \\ 0 & 0 & \mu \end{pmatrix} . \tag{3.325}$$

This is from (1.20), but $e_{zx} \rightarrow \gamma_{zx}$ results in the change of the (3,3) element from 2μ to μ. The matrix $\boldsymbol{\Gamma} = (\Gamma_{ijkl})$, where Γ_{ijkl} is in (3.316), has a similar form to (3.325) because this is proportional to C_{ijkl} in the stiffness-proportional damping. When the **Q** of the P wave, measured at frequency f_0, is Q_α and the Q of the S wave is Q_β, $\boldsymbol{\Gamma}$ is given as [6]

$$\boldsymbol{\Gamma} = \begin{pmatrix} \Gamma_\alpha & \Gamma_\lambda & 0 \\ \Gamma_\lambda & \Gamma_\alpha & 0 \\ 0 & 0 & \Gamma_\beta \end{pmatrix}, \; \Gamma_\alpha = \frac{\lambda + 2\mu}{2\pi f_0 Q_\alpha}, \; \Gamma_\beta = \frac{\mu}{2\pi f_0 Q_\beta}, \; \Gamma_\lambda = \Gamma_\alpha - 2\Gamma_\beta . \tag{3.326}$$

As examples of the trial function (shape function) of the numerical Galerkin method, we have already given linear functions, among which we here adopt the the functions of the **Lagrange interpolation**

$$N^1 = \frac{\Delta x - x}{\Delta x}\frac{\Delta z - z}{\Delta z}, \; N^2 = \frac{x}{\Delta x}\frac{\Delta z - z}{\Delta z}, \; N^3 = \frac{\Delta x - x}{\Delta x}\frac{z}{\Delta z}, \; N^4 = \frac{x}{\Delta x}\frac{z}{\Delta z} . \tag{3.327}$$

Since

$$\mathbf{u} = \mathbf{N}\boldsymbol{\delta}, \quad \mathbf{u} = (u_x, u_z)^{\mathrm{T}} \tag{3.328}$$

from the second equation of (3.318), in the case of $\boldsymbol{\delta}$ defined in (3.321), the matrix, which is the component of **N** related only to the (i, j) element, must have the form

$$\mathbf{N}^{(i,j)} = \begin{pmatrix} N^1 & 0 & N^2 & 0 & N^3 & 0 & N^4 & 0 \\ 0 & N^1 & 0 & N^2 & 0 & N^3 & 0 & N^4 \end{pmatrix} . \tag{3.329}$$

Meanwhile, since

$$\mathbf{e} = \mathbf{E}\,\mathbf{u}, \quad \mathbf{e} = (e_{xx}, e_{zz}, \gamma_{zx}), \quad \mathbf{E} = \begin{pmatrix} \frac{\partial}{\partial x} & 0 & \frac{\partial}{\partial z} \\ 0 & \frac{\partial}{\partial z} & \frac{\partial}{\partial x} \end{pmatrix}^{\mathrm{T}}, \tag{3.330}$$

the matrix, which is the component of $\mathbf{B} = \mathbf{E}\,\mathbf{N}$ related only to the (i, j) element, must have the form

$$\mathbf{B}^{(i,j)} = \begin{pmatrix} \frac{\partial N^1}{\partial x} & 0 & \frac{\partial N^2}{\partial x} & 0 & \frac{\partial N^3}{\partial x} & 0 & \frac{\partial N^4}{\partial x} & 0 \\ 0 & \frac{\partial N^1}{\partial z} & 0 & \frac{\partial N^2}{\partial z} & 0 & \frac{\partial N^3}{\partial z} & 0 & \frac{\partial N^4}{\partial z} \\ \frac{\partial N^1}{\partial z} & \frac{\partial N^1}{\partial x} & \frac{\partial N^2}{\partial z} & \frac{\partial N^2}{\partial x} & \frac{\partial N^3}{\partial z} & \frac{\partial N^3}{\partial x} & \frac{\partial N^4}{\partial z} & \frac{\partial N^4}{\partial x} \end{pmatrix} . \tag{3.331}$$

K, obtained from (3.329) and (3.331), is given in Washizu et al. [124] and others. When the substitutions $\lambda + 2\mu \rightarrow \Gamma_\alpha$, $\mu \rightarrow \Gamma_\beta$, and $\lambda \rightarrow \Gamma_\lambda$ are applied to **K**, we can obtain **G**. If C', W_M, and W_K are given from the expected attenuation spectra, **C** is obtained using **G** and **M** for the lumped mass, such that all coefficients in the recurrence equation (3.323) are determined, and $\mathbf{u}_{t+\Delta t}^{i,j}$ is then obtained from $\mathbf{u}_t^{i,j}$, $\mathbf{u}_{t-\Delta t}^{i,j}$, and $\boldsymbol{\delta}_t$, $\boldsymbol{\delta}_{t-\Delta t}$.

To estimate the accuracy of the voxel finite element method, we extract $\partial u_x/\partial x$ and $\partial u_z/\partial x$ from (3.315). However, the results are the same for $\partial u_x/\partial x$ and $\partial u_z/\partial x$, so we use u for both u_x and u_z below. From the same numerical Galerkin method as in (3.318) and the shape function in (3.327), $\partial u/\partial x$ in the (i, j) element is discretized as

$$\frac{\partial u}{\partial x} \sim \frac{1}{\Delta x \Delta z}\mathbf{J}\delta_t\,, \quad \mathbf{J}^{(i,j)} = \left(J^{mn}\right), \quad J^{mn} = \int_0^{\Delta x}\int_0^{\Delta z} N^m \frac{\partial N^n}{\partial x} dx\, dz\,. \tag{3.332}$$

Performing the integrations in J^{mn}, we obtain

$$\begin{aligned}(\mathbf{J}\delta_t)^{(i,j)} = &-\frac{\Delta z}{12}u^{i-1,j-1} + \frac{\Delta z}{12}u^{i+1,j-1} - \frac{\Delta z}{3}u^{i-1,j} \\ &+\frac{\Delta z}{3}u^{i+1,j} - \frac{\Delta z}{12}u^{i-1,j+1} + \frac{\Delta z}{12}u^{i+1,j+1}\,.\end{aligned} \tag{3.333}$$

When (3.333) is rearranged and the result is substituted into (3.332),

$$\begin{aligned}\frac{\partial u}{\partial x} \sim \frac{1}{3}\Bigg\{&2\cdot\frac{1}{2\Delta x}\left(u^{i+1,j} - u^{i-1,j}\right) \\ &+1\cdot\frac{1}{2\Delta x}\left(\frac{u^{i+1,j+1}+u^{i+1,j-1}}{2} - \frac{u^{i-1,j+1}+u^{i-1,j-1}}{2}\right)\Bigg\}\end{aligned} \tag{3.334}$$

is obtained. The first term of (3.334) is just the central difference for the first-order derivative $1/(2\Delta x)\cdot\left(u^{i+1,j} - u^{i-1,j}\right)$ in the finite difference method. In addition, in the second term, $\left(u^{i+1,j+1} + u^{i+1,j-1}\right)/2$ approximates $u^{i+1,j}$ and $\left(u^{i-1,j+1} + u^{i-1,j-1}\right)/2$ approximates $u^{i-1,j}$. That is, although the second term also gives the central difference, the central difference is given not from the left and right nodes along the x-axis but from the diagonally upper and lower nodes (Fig. 3.30). This second term achieves a higher accuracy in the voxel finite element method than in the finite difference method. For the same problem of the ground motion in a semi-infinite medium, the maximum value of C in the **Courant condition** [26] for this problem

$$\Delta t < C\frac{\min(\Delta x, \Delta z)}{v_{\max}}\,, \quad v_{\max} = \alpha \tag{3.335}$$

where numerical computation is not unstable, was compared between the voxel finite element method and finite difference method [73]. While the maximum C was 0.45 in the finite difference method, the maximum value was 0.80 in the voxel finite element method due to the high accuracy.

As already noted in this subsection, the first advantage of the finite element method is that the **stress-free condition** is automatically satisfied on the circumference of region V. This condition is satisfactory for the ground surface, however for the side boundaries and lower boundary, the radiation boundary conditions should be given. Consequently, the stress-free condition is automatically given to these boundaries and the problems of **artificial reflection** by the free surface occur at these boundaries. We therefore have to take the same measures as in the finite difference method. The

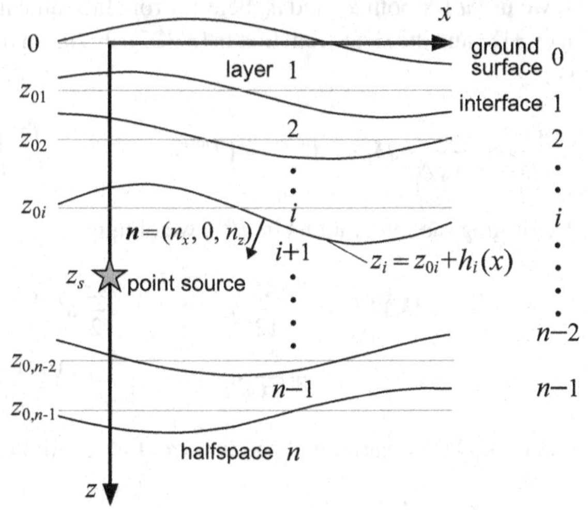

Fig. 3.31 A two-dimensional (symmetry in the y-direction) irregularly layered structure (modified from Koketsu [72] with permission of Kindai Kagaku)

ground motion computed by T. L. Hong and D. Kosloff using the finite element method, which is assumed to be the conventional finite element method, is shown in Fig. 3.27, and this result is in good agreement with the results of the other methods over 110 s.

3.2.6 Aki-Larner Method

For a velocity structure where homogeneous layers are separated by irregular interfaces that are not horizontal (an **irregularly layered structure**), the ground motions and stresses in a layer should be represented by those explained in Sects. 3.1.1 – 3.1.8. Using these as trial functions, it is therefore possible to solve the propagation problem in an irregularly layered structure, if the **boundary method** of the **method of weighted residuals** (Sect. 3.2.1) is applied so as to satisfy the boundary conditions at the irregular interfaces. The Cartesian coordinate system, which is shown by x and z in Fig. 3.1, is set in the irregularly layered structure. It is assumed that the depth z_i of the ith irregular interface separating the ith and $i+1$st homogeneous layers is z_{0i} on average, and varies with

$$z_i = z_{0i} + h_i(x) \tag{3.336}$$

where h_i is a function of x, as shown in Fig. 3.31.

In such a 2-D velocity structure, the ground motion is separated into the SH wave and the P and SV waves as in a 1-D velocity structure[89]; however, only the SH wave ground motion is considered here for simplicity. If Fuchs' two-dimensional harmonic

[89] Section 13.4.2 of Aki and Richards [4].

function in (3.56) of Sect. 3.1.4 can be used to represent the Fourier transform of the SH wave ground motion in layer i as

$$\bar{u}_{yi}(z,x) = \int_{-\infty}^{\infty} \tilde{\upsilon}_i(z) e^{\mathrm{i}kx} dk \,, \tag{3.337}$$

the continuity condition of the ground motion at interface i is

$$\int_{-\infty}^{\infty} \tilde{\upsilon}_i(z_i) e^{\mathrm{i}kx} dk = \int_{-\infty}^{\infty} \tilde{\upsilon}_{i+1}(z_i) e^{\mathrm{i}kx} dk \,. \tag{3.338}$$

In a 1-D velocity structure, (3.338) has the obvious solution $\tilde{\upsilon}_i = \tilde{\upsilon}_{i+1}$ because z_i is constant and therefore $\tilde{\upsilon}_i$ and $\tilde{\upsilon}_{i+1}$ do not depend on x. That is, the continuity condition of the ground motion can be satisfied separately for each k (Sects. 3.1.2 and 3.1.5). However, when z_i depends on x as in (3.336), the ground motions of different k's influence each other; therefore, (3.338) has to be solved as an integral equation.

To solve this numerically using the **Aki-Larner method** [1], the solution of the ground motion in the corresponding horizontally layered structure (1-D velocity structure) based on the definition of Haskell $\tilde{\chi}_i \equiv \tilde{\upsilon}_i$ [44] (Sect. 3.1.4) and (3.46)

$$\tilde{\upsilon}_i = \tilde{\chi}_i = E_i e^{+\mathrm{i}\nu_{\beta i} z} + F_i e^{-\mathrm{i}\nu_{\beta i} z} \tag{3.339}$$

is inserted in (3.338) as a linear combination of trial functions in the method of weighted residuals. Then, by replacing the integration with a numerical integration such as (3.132), (3.338) yields the boundary condition for the boundary method:

$$B_i = \sum_{j=-N}^{N-1} \left(E_i^j e^{+\mathrm{i}\nu_{\beta i}^j z_i} + F_i^j e^{-\mathrm{i}\nu_{\beta i}^j z_i} \right) e^{\mathrm{i}k_j x} - \sum_{j=-N}^{N-1} \left(E_{i+1}^j e^{+\mathrm{i}\nu_{\beta,i+1}^j z_i} + F_{i+1}^j e^{-\mathrm{i}\nu_{\beta,i+1}^j z_i} \right) e^{\mathrm{i}k_j x}$$
$$= 0 \,, \quad \nu_{\beta i}^j = \begin{cases} \sqrt{k_{\beta i}^2 - k_j^2} \,, \; k_{\beta i} \ge k_j \\ -\mathrm{i}\sqrt{k_j^2 - k_{\beta i}^2} \,, \; k_{\beta i} < k_j \end{cases} . \tag{3.340}$$

The integration step size is constant and common on the i and $i+1$ sides, therefore this is removed from the numerical integrations in (3.340). For the weighting function $w_k(\mathbf{x})$, we do not use one of the standard functions listed in Table 3.1 but

$$w_m(\mathbf{x}) = e^{-\mathrm{i}k_m x} \,, \tag{3.341}$$

where the subscript is changed from k to m to avoid confusion with the "wavenumber". We then apply the changes of variables of $k \rightarrow m$ and $dS \rightarrow dx$ because of the two-dimensional case, and the substitutions of (3.340) and (3.341) into (3.225), so that

$$\int \sum_{j=-N}^{N-1} \left(E_i^j e^{+\mathrm{i}\nu_{\beta i}^j z_i} + F_i^j e^{-\mathrm{i}\nu_{\beta i}^j z_i} \right) e^{\mathrm{i}k_j x} e^{-\mathrm{i}k_m x} dx$$

$$= \int \sum_{j=-N}^{N-1} \left(E_{i+1}^j e^{+\mathrm{i}\nu_{\beta,i+1}^j z_i} + F_{i+1}^j e^{-\mathrm{i}\nu_{\beta,i+1}^j z_i} \right) e^{\mathrm{i}k_j x} e^{-\mathrm{i}k_m x} dx \,. \quad (3.342)$$

The integrations on both sides of (3.342) are exactly the **Fourier transform** (Sect. 4.2.2) with respect to x. That is, the weighting function of (3.341)) realizes the **collocation method** in the wavenumber domain. Substituting (3.336) into the integrand of (3.342) and separating the result into parts with and without x, we obtain

$$\left(E_i^j e^{+\mathrm{i}\nu_{\beta i}^j z_i} + F_i^j e^{-\mathrm{i}\nu_{\beta i}^j z_i} \right) e^{\mathrm{i}k_j x} = H_{i-}^j \chi_{i-}^j + H_{i+}^j \chi_{i+}^j \,.$$

$$H_{i\mp}^j = e^{\pm\mathrm{i}\nu_{\beta i}^j h(x)} e^{\mathrm{i}k_j x}, \quad \chi_{i-}^j = E_i^j e^{+\mathrm{i}\nu_{\beta i}^j z_{0i}}, \quad \chi_{i+}^j = F_i^j e^{-\mathrm{i}\nu_{\beta i}^j z_{0i}} \,. \quad (3.343)$$

We substitute this into (3.342), perform the Fourier transforms at k_m ($m = -N, -N + 1, \cdots, N - 1$), and obtain the $2N$ simultaneous equations

$$\sum_{j=-N}^{N-1} \left(H_{i-}^{mj} \chi_{i-}^j + H_{i+}^{mj} \chi_{i+}^j \right) = \sum_{j=-N}^{N-1} \left(H_{i+1,-}^{mj} \chi_{i+1,-}^j + H_{i+1,+}^{mj} \chi_{i+1,+}^j \right) ,$$

$$H_{i\mp}^{mj} = \int e^{\pm\mathrm{i}\nu_{\beta i}^j h(x)} e^{\mathrm{i}(k_j - k_m)x} dx \,, \quad (3.344)$$

where the unknowns are $\chi_{i\mp}^j$ and $\chi_{i+1,\mp}^j$, which number $4N$. We rewrite (3.344) in the form of vectors and matrices as

$$\mathbf{H}_i \mathbf{\Phi}_i = \mathbf{H}_{i+1} \mathbf{\Phi}_{i+1} \,,$$

$$\mathbf{\Phi}_i = \left(\chi_{i-}^{-N}, \chi_{i-}^{-N+1}, \cdots, \chi_{i-}^{N-1}, \chi_{i+}^{-N}, \chi_{i+}^{-N+1}, \cdots, \chi_{i+}^{N-1} \right)^{\mathrm{T}}, \quad (3.345)$$

$$\mathbf{H}_i = \begin{pmatrix} H_{i-}^{-N,-N} & H_{i-}^{-N,-N+1} & \cdots & H_{i-}^{-N,N-1} & H_{i+}^{-N,-N} & H_{i+}^{-N,-N+1} & \cdots & H_{i+}^{-N,N-1} \\ H_{i-}^{-N+1,-N} & H_{i-}^{-N+1,-N+1} & \cdots & H_{i-}^{-N+1,N-1} & H_{i+}^{-N+1,-N} & H_{i+}^{-N+1,-N+1} & \cdots & H_{i+}^{-N+1,N-1} \\ \cdot & \cdot & \cdots & \cdot & \cdot & \cdot & \cdots & \cdot \\ \cdot & \cdot & \cdots & \cdot & \cdot & \cdot & \cdots & \cdot \\ \cdot & \cdot & \cdots & \cdot & \cdot & \cdot & \cdots & \cdot \\ H_{i-}^{N-1,-N} & H_{i-}^{N-1,-N+1} & \cdots & H_{i-}^{N-1,N-1} & H_{i+}^{N-1,-N} & H_{i+}^{N-1,-N+1} & \cdots & H_{i+}^{N-1,N-1} \end{pmatrix} .$$

$\mathbf{\Phi}_i$ in (3.345) corresponds to the potential vector $\mathbf{\Phi}_i = \left(\chi_i^-, \chi_i^+ \right)^{\mathrm{T}}$ in (3.50) for the 1-D velocity structure. However, although (3.50) contains potentials only for a certain k, (3.345) includes potentials for all k's, which are $k_{-N}, k_{-N+1}, \cdots, k_{N-1}$. To distinguish this difference, we name a vector such as $\mathbf{\Phi}_i$ a supervector, and likewise, we name a matrix such as $\mathbf{H}_i$ a supermatrix [75].

The continuity condition of the stress vector at interface i is that the stress vector $\mathbf{T}_n$ in the normal direction from the interface is continuous, according to the balance of forces (1.11). If the unit vector in the normal direction is $\mathbf{n} = (n_x, 0, n_z)$ (Fig. 3.31),

$$\mathbf{T}_n = \left(\tau_{xx}n_x + \tau_{zx}n_z,\ \tau_{xy}n_x + \tau_{zy}n_z,\ \tau_{xz}n_x + \tau_{zz}n_z\right) . \tag{3.346}$$

For the two-dimensional SH wave, since $\partial/\partial y = 0$ and $u_x = u_z = 0$, $\tau_{xx} = \tau_{zx} = \tau_{xz} = \tau_{zz} = 0$ from (1.24). The x- and z-components of T_n are therefore always zero, and only the continuity of the y-component $T_{ny} = \mu\left(n_x \partial u_y/\partial x + n_z \partial u_y/\partial z\right)$ should be considered. When the Fourier transform of T_{ny} in layer i is

$$\overline{T}_{nyi}(z, x) = \int_{-\infty}^{\infty} \tilde{p}_i(z) e^{\mathrm{i}kx} dk , \tag{3.347}$$

the continuity condition at the interface i yields

$$\int_{-\infty}^{\infty} \tilde{p}_i(z_i) e^{\mathrm{i}kx} dk = \int_{-\infty}^{\infty} \tilde{p}_{i+1}(z_i) e^{\mathrm{i}kx} dk . \tag{3.348}$$

Using

$$n_x = \frac{-h_i'}{(1 + h_i'^2)^{1/2}} , \quad n_z = \frac{1}{(1 + h_i'^2)^{1/2}} , \quad h_i' = \frac{dh_i}{dx} \tag{3.349}$$

which are given for the interface represented by (3.336) [1], (3.337), and (3.339), we perform the formulation similar to the ground motion and obtain

$$\mathbf{J}_i \mathbf{\Phi}_i = \mathbf{J}_{i+1} \mathbf{\Phi}_{i+1} . \tag{3.350}$$

The supermatrix $\mathbf{J}_i$ consists of the elements

$$J_{i\mp}^{mj} = \mu_i \int \frac{-h_i'(x)\mathrm{i}k_j \pm \mathrm{i}\nu_{\beta i}^j}{(1 + h_i'^2)^{1/2}} e^{\pm \mathrm{i}\nu_{\beta i}^j h(x)} e^{\mathrm{i}(k_j - k_m)x} dx . \tag{3.351}$$

The combination of (3.345) and (3.350) results in the $4N$ simultaneous equations

$$\mathbf{K}_i \mathbf{\Phi}_i = \mathbf{K}_{i+1} \mathbf{\Phi}_{i+1} , \quad \mathbf{K}_i = \begin{pmatrix} \mathbf{H}_i \\ \mathbf{J}_i \end{pmatrix} , \tag{3.352}$$

and we can solve these for the $4N$ unknowns, which are $\chi_{i\mp}^j$ and $\chi_{i+1,\mp}^j$. However, in order to distinguish $\mathbf{K}_{i+1}$ in (3.352) for interface i from $\mathbf{K}_{i+1}$ in (3.352) for interface $i+1$, we add the interface number to $\mathbf{K}$ as a subscript and the average interface depth to $\mathbf{\Phi}_i$ as an argument. Thus, we represent (3.352) as

$$\mathbf{K}_{i,i} \mathbf{\Phi}_i(z_{0i}) = \mathbf{K}_{i,i+1} \mathbf{\Phi}_{i+1}(z_{0i}) \tag{3.353}$$

in the following.

The potential supervector around $z = z_{0,i-1}$ and the potential supervector around $z = z_{0i}$, which are both in layer i, are related as

$$\mathbf{\Phi}_i(z_{0i}) = \mathbf{E}_i \mathbf{\Phi}_i(z_{0,i-1}), \quad d_i = z_{0i} - z_{0,i-1}, \tag{3.354}$$

$$\mathbf{E}_i = \begin{pmatrix} e^{+\mathrm{i}\nu_{\beta i}^{-N} d_i} & & & & & & \\ & e^{+\mathrm{i}\nu_{\beta i}^{-N+1} d_i} & & & & & \\ & & \ddots & & & & \\ & & & e^{+\mathrm{i}\nu_{\beta i}^{N-1} d_i} & & & \\ & & & & e^{-\mathrm{i}\nu_{\beta i}^{-N} d_i} & & \\ & & & & & e^{-\mathrm{i}\nu_{\beta i}^{-N+1} d_i} & \\ & & & & & & \ddots \\ & & & & & & & e^{-\mathrm{i}\nu_{\beta i}^{N-1} d_i} \end{pmatrix}$$

using the diagonal supermatrix $\mathbf{E}_i$, the elements of which are those for the 1-D velocity structure because of the homogeneous layers. As (3.345) represents the continuity of the ground motion, the left-hand side for $i = 0$ represents the ground motion on the ground surface, which is denoted by the supervector

$$\mathbf{V} = \left(V^{-N}, V^{-N+1}, \cdots, V^{N-1}\right)^{\mathrm{T}}. \tag{3.355}$$

Similarly, as (3.350) represents the continuity of the normal stress vector, the left-hand side for $i = 0$ represents the normal stress vector on the ground surface, which is the zero supervector

$$\mathbf{0} = (0,\ 0,\ \cdots,\ 0)^{\mathrm{T}} \tag{3.356}$$

due to the stress-free condition. Substituting these and $z_{00} = 0$ (Fig. 3.31) into (3.353) for $i = 0$ gives

$$\begin{pmatrix} \mathbf{V} \\ \mathbf{0} \end{pmatrix} = \mathbf{K}_{0,1} \mathbf{\Phi}_1(0). \tag{3.357}$$

Next, substituting (3.354) for $i = 1$, and then (3.353) for $i = 1$, gives

$$\mathbf{E}_1 \mathbf{K}_{0,1}^{-1} \begin{pmatrix} \mathbf{V} \\ \mathbf{0} \end{pmatrix} = \mathbf{\Phi}_1(z_{01}),$$
$$\mathbf{K}_{1,2}^{-1} \mathbf{K}_{1,1} \mathbf{E}_1 \mathbf{K}_{0,1}^{-1} \begin{pmatrix} \mathbf{V} \\ \mathbf{0} \end{pmatrix} = \mathbf{\Phi}_2(z_{01}). \tag{3.358}$$

If this operation is repeated until $\mathbf{\Phi}_n(z_{0,n-1}) = (\boldsymbol{\chi}_{n-}, \boldsymbol{\chi}_{n+})^{\mathrm{T}}$, we have

$$\begin{pmatrix} \boldsymbol{\chi}_{n-} \\ \boldsymbol{\chi}_{n+} \end{pmatrix} = \mathbf{M} \begin{pmatrix} \mathbf{V} \\ \mathbf{0} \end{pmatrix},$$
$$\mathbf{M} = \mathbf{K}_{n-1,n}^{-1} \mathbf{K}_{n-1,n-1} \mathbf{E}_{n-1} \cdots \mathbf{K}_{2,3}^{-1} \mathbf{K}_{2,2} \mathbf{E}_2\, \mathbf{K}_{1,2}^{-1} \mathbf{K}_{1,1} \mathbf{E}_1\, \mathbf{K}_{0,1}^{-1}. \tag{3.359}$$

For example, when a plane wave is incident perpendicularly from below as shown in Fig. 3.27, among the elements of $\mathbf{\Phi}_n$, χ_{n-}^0 is the incident wave amplitude and the others are zero from the radiation boundary condition. As a result, there are $2N$ unknowns for V^j and $2N$ unknowns for χ_{n+}^j, and the total number, $4N$, agrees with the number of simultaneous equations (3.359) therefore we can solve them. If the incident wave amplitude is 1,

$$\chi_{n-} = \mathbf{1} = (0, 0, \cdots, 0, 1, 0, \cdots, 0)^{\mathrm{T}} \tag{3.360}$$

and we obtain the solution of (3.359)

$$\begin{aligned} \mathbf{V} &= \mathbf{m}_{11}^{-1} \cdot \mathbf{1} = \left(\hat{m}^{-N,0}, \hat{m}^{-N+1,0}, \cdots, \hat{m}^{N-1,0}\right)^{\mathrm{T}}, \\ \mathbf{M} &= \begin{pmatrix} \mathbf{m}_{11} & \mathbf{m}_{12} \\ \mathbf{m}_{21} & \mathbf{m}_{22} \end{pmatrix}, \quad \mathbf{m}_{11}^{-1} = \left(\hat{m}^{ij}\right). \end{aligned} \tag{3.361}$$

The solutions $V^j = \hat{m}^{j0}$ are further summed up as the wavenumber integration in (3.337) to obtain the Fourier transform of the ground motion at the ground surface. The ground motion computed by Kohketsu [70] using this method, which was termed the "2-D reflectivity method", and the ground motion computed by Bard and Bouchon [7] using the original Aki-Larner method agree well with the results of the other methods for 210 s, as shown in Fig. 3.27.

The method of matching boundary conditions at irregular boundaries, explained above, was originally used in electromagnetism. However, this method is suitable for the problem of ground motions, which have longer periods, and has been actively studied since it was introduced by Aki and Larner [1] in an appropriate form for use on computers, in which the **FFT** (Sect. 4.2.4) is used abundantly for Fourier transforms. In the Aki-Larner method, the coefficient matrix of simultaneous equations to be solved is near diagonal and numerically stable when the boundary irregularity is not strong, and the simultaneous equations can be constructed quickly using the FFT. However, when the irregularity is strong, the method of weighted residuals through the usual collocation method is more effective. Although a similar formulation is possible for the P and SV wave, the formulation that extends the **reflection/transmission matrix** (Sect. 3.1.5) to a supermatrix is desirable compared to the Haskell formulation used for the SH wave, because the subdeterminants, which are countermeasures for the **overflow** problem (Sect. 3.1.3), cannot be analytically obtained as in the 1-D velocity structure. In this case, it is also desirable to utilize a **propagation invariant** [64] to relax the computational complexity, which is significantly increased by the introduction of supermatrices. This not only makes the formulation more stable than the Haskell formulation, but also faster in some cases [75]. All these examples relate to two-dimensionally irregular interfaces, and the extension to three-dimensionally irregular interfaces has also been carried out by Horike et al. [52], etc.

The **Rayleigh ansatz** [4] is always considered a problem when considering the Aki-Larner method. Unless there is no ground motion propagating upward in the lower halfspace in Fig. 3.31, subject to the radiation boundary condition, the simultaneous equations cannot be solved. However, it is argued that the Aki-Larner method gives only an approximate solution, because there is a significant possibility of upgoing waves in scattering due to the effect of the irregular interface above the halfspace, especially when the interface is strongly inclined. As Millar [90] showed, downgoing plane waves mathematically form a complete system in the square integrable function

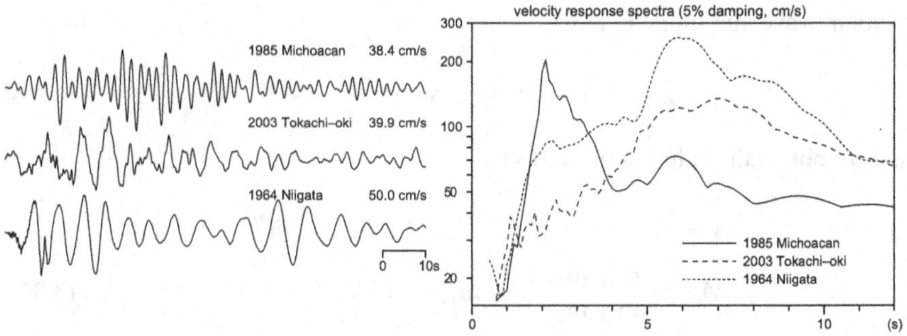

Fig. 3.32 Examples of long-period ground motion, in velocity seismograms (left) and velocity response spectra (right). Modified from Koketsu and Miyake [43] with permission of Springer Nature

space defined on the interface; therefore, it is possible in principle to approximate all wavefields with arbitrary L^2 norm accuracy using only linear combinations of the downgoing plane waves. However, even although the wavefield can be represented in principle, the computation is impossible if a nearly infinite number of terms are required. The Aki-Larner method therefore remains unsuitable for strongly irregular velocity structures.

3.3 Analysis of Propagation

3.3.1 Long-Period Ground Motion

Long-period ground motions are ground motions with periods for which the lower bound is 1 s to several seconds and the upper bound is 10 to 20 s [128]. They are also characterized by relatively long duration (Fig. 3.32). Long period ground motion is developed in the process of propagation as described below. This can be a threat to structures with long natural periods, such as high-rise buildings, long bridges, and large storage tanks. These are contemporary structures, therefore long period ground motion is a modern problem related to the propagation of ground motion [60].

In Japan, Shima [105] noted long-period ground motion for the first time in 1970 when he studied the ground motion of the 1968 **Tokachi-oki earthquake** (M 7.9) observed in Japan's first high-rise building under construction, which was located 650 km away from the epicenter. In the United States, also in 1970, Trifunac and Brune [113] noted long-period ground motion when they studied the ground motions of the **Imperial Valley earthquake** (1940, M_s 7.1) observed at stations 200 to 500 km away from the epicenter. However, the term "long-period ground motion" became established after the 2003 **Tokachi-oki earthquake** (M_w 8.3). The phenomenon was initially referred to as "slightly long period" ground motion in view of the balance

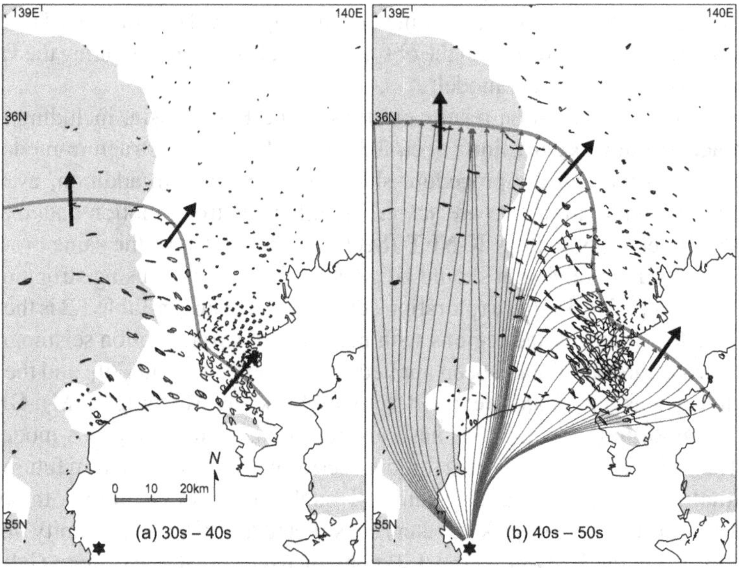

Fig. 3.33 Propagation and surface wave ray tracing for the long-period ground motions in the southwestern part of the Kanto basin (reprinted from Koketsu and Kikuchi [76] with permission of AAAS)

with "long period" in other fields of seismology [128]. Since then, research has been conducted retrospectively, and the damage from several earthquakes such as the **Michoacan earthquake** (1985, M_s 8.1) in Mexico, the **Niigata earthquake** (1964, M 7.5) in Japan, etc. is now considered to have been caused by long-period ground motion (Fig. 3.32).

Both Shima [105] and Trifunac and Brune [113] determined from their analyses that the physical entity of long-period ground motion is a surface wave (Sects. 3.1.9 and 3.1.10). As surface waves are radiated from a source, and develop during propagation, long-period ground motion is a problem related to propagation, as described in the beginning of this subsection. In a surface wave propagating over a long distance, the wavetrains at various periods are separated by **dispersion**, and the duration is necessarily lengthened. The types (Love wave or Rayleigh wave) and modes of surface waves are identified by comparing the phase velocities, group velocities, and dispersion curves computed using the methods described in Sects. 3.1.9 and 3.1.10.

This computation assumes a 1-D velocity structure, and if a difference between the results and observations remains despite sufficiently adjusting the 1-D velocity structure model, the difference shows the effect of a real 3-D velocity structure. For example, the dark gray curves in Fig. 3.33 represent the wavefronts of surface waves obtained from observed long-period ground motions (corresponding to the ellipses, as displayed in 2-D plan view), and these curves deviate from concentric circles centered at the source (marked by the star) as a result of the effect of the 3-D velocity structure in the Kanto basin. The ray tracing of surface waves was performed using the 3-D velocity structure model obtained from geophysical exploration. The results (thin long arrows) agree well with the observed wavefront, demonstrating the validity of the 3-D velocity structure model.

When various exploration results exist, as in the Kanto basin, including results from **microtremor exploration** (Sect. 3.3.2), a 3-D velocity structure model may be constructed through the procedure shown in Table 3.2. In addition, even in a region where exploration data are scarce, if sufficient strong motion data exist via observation networks such as **K-NET** (Sects. 2.3.8 and 4.1.4), the same procedure may be used but more weight should be given to Steps 6 and 7 using strong motion data than to Steps 3 to 5 using exploration data. The procedure in Table 3.2 is therefore applicable for countries and regions with networks of strong motion seismographs.

This procedure has already been applied to the whole of Japan [79], and the result was published as the Japan Integrated Velocity Structure Model (JIVSM). **Ground motion simulations** (Sect. 3.2.4) have also been conducted using this model, and hazard maps have been constructed for long-period ground motions from future **great earthquakes** (Sect. A.1.1) in the Nankai Trough region. In California, the **SCEC** (Southern California Earthquake Center) constructed the SCEC Community Velocity Model [87], and the **USGS** (Sect. 4.1.2) constructed the USGS Northern California 3D Seismic Velocity Model [14]. Although many ground motion simulations have been carried out using these velocity structure models, the current status is that the second part of the procedure detailed in Table 3.2 has not yet been applied for those regions.

If "long-period ground motion" is literally interpreted as "ground motion with long periods", **seismic pulse** (Sect. 2.3.8) near source faults due to the directivity effect should also be classified as long-period ground motions. From this standpoint, we can divide the long-period ground motion defined thus far in this subsection into two classes, which are "far-source long-period ground motion" and "near-fault long-period ground motion". After the **Kumamoto earthquake** (2016, M_w 7.0), the near-fault class is sometimes referred to as a "long-period pulse". The comparison of Fig. 3.32 with Fig. 2.41 clearly shows that the far-source and near-fault classes can be distinguished from each other by the duration of ground motion [77], even if the observation site is unknown. The short duration of the near-fault class is not only due to its proximity to a source, but also due to mainly consisting of constructively interfering body waves rather than surface waves (Sect. 2.3.8). The velocity response spectra of the far-source class are rather similar to that of the near-fault class, because, in the far-source class, the amplitudes are small due to the large distances; however the duration is long.

Table 3.2 The standard procedure for modeling a 3-D velocity structure (reprinted from Koketsu et al. [78] with permission of Elsevier). "Microtremor survey" is referred to as "microtremor exploration" in this book

Step 1: Assume an initial layered model consisting of a seismic basement and sedimentary layers from a comprehensive overview of geological information, borehole data, and exploration results.	Step 4: Compile data and information on faults and folds. Convert time sections from seismic reflection surveys and borehole logging into depth sections using the P- and S-wave velocities in Step 2.
Step 2: Assign P-wave velocities to the basement and layers based on the results of refraction and reflection surveys, and borehole logging. Assign S-wave velocities based on the results of borehole logging, microtremor surveys, spectral ratio analyses of seismograms, and empirical relationships between P- and S-wave velocities.	Step 5: Determine the shapes of interfaces between the layers and the basement by inversions of geophysical-survey data (e.g., refraction travel times and gravity anomalies). In case of insufficient data, forward modeling is carried out. The depths of faults and folds in Step 4 are introduced into the inversions as constraints, or additional data for the forward modeling.
Step 3: Obtain the velocity structure right under engineering bedrock from the results of microtremor surveys, referring to the results of borehole logging, since among 2-D and 3-D surveys only microtremor surveys are sensitive to shallow velocity distributions and the shapes of shallow interfaces.	Step 6: Calibrate the P- and S-wave velocities in Step 2 and the interface shapes in Step 5 by inversion or forward modeling of spectral features of observed seismograms, such as dominant periods of H/V (horizontal/vertical) spectral ratios.
	Step 7: Adjust the velocities and interface shapes using inversion or forward modeling of time history waveforms of observed seismograms.

3.3.2 Microtremors

The ground is continually shaken by causes other than earthquakes, and this phenomenon is called **microtremors** [117]. According to Horike [51], the efficiency of data collection for these phenomena is overwhelmingly high compared with ground motion, as the microtremors occur constantly. In addition, the microtremors themselves constitute noise, therefore the problem of the **S/N ratio** (Sect. 4.2.1) does not occur, and the microtremors can be used even in urban areas. If microtremors are used for velocity structure exploration, the cost for sources is almost zero in comparison with exploration utilizing blasting and vibrators. As described below, another advantage is that the structure of S wave velocities, which are important in ground motion seismology, can be directly obtained. For these various reasons, microtremor exploration has become the mainstream exploration technique in ground motion seismology, despite being subject to many problems such as the accuracy of results.

Although surface waves (Sects. 3.1.9 and 3.1.10) also play a central role in the analysis of microtremors, these are not uniquely surface waves as in long-period ground motions, but are mixed with body waves. The information on surface waves is therefore first extracted using Capon's [16] frequency-wavenumber analysis, etc., and the velocity structure is then modeled from this information. In terms of the information to be extracted, the **phase velocity** of the **Rayleigh wave** (Sect. 3.1.10) is the most often used, but the **H/V spectral ratio** of the Rayleigh wave and the phase velocity of the **Love wave** (Sect. 3.1.9) are also used [51]. The following formulation is based on Takeuchi and Saito [109] rather than the methods in Sects. 3.1.9 and 3.1.10, because of the need to obtain partial derivatives.

The two-dimensional plane wave problem, which involves symmetry in the y-axis direction ($\partial/\partial y = 0$), has already been dealt with in Sect. 3.1.4, and the result in (3.44) shows that the SH wave including the Love wave can be extracted from the wavefield by setting the displacement potentials to $\phi = \psi = 0$, that is $u_x = u_z = 0$. Substituting this and $\partial/\partial y = 0$ into the definitions of the stresses (1.24), we have

$$\tau_{xx} = \tau_{yy} = \tau_{zz} = 0\,, \quad \tau_{yz} = \mu\frac{\partial u_y}{\partial z}\,, \quad \tau_{zx} = 0\,, \quad \tau_{xy} = \mu\frac{\partial u_y}{\partial x}\,. \tag{3.362}$$

Assuming that the variable z in u_y remains as is because of a 1-D velocity structure depending on z, and u_y is oscillatory in the x-direction, we can set u_y to be

$$u_y = y_1(z, \omega, k)\, e^{-\mathrm{i}(kx-\omega t)}, \quad \tau_{yz} = y_2(z, \omega, k)\, e^{-\mathrm{i}(kx-\omega t)}\,, \tag{3.363}$$

and obtain

$$y_2 = \mu\frac{dy_1}{dz}\,, \quad \tau_{xy} = -\mathrm{i}k\mu y_1\, e^{-\mathrm{i}(kx-\omega t)}\,. \tag{3.364}$$

Substituting these into (1.26) gives the simultaneous ordinary differential equations[90]

$$\frac{dy_1}{dz} = \frac{1}{\mu}y_2\,, \quad \frac{dy_2}{dz} = (k^2\mu - \omega^2\rho)y_1\,. \tag{3.365}$$

Next, from the **stress-free condition** at the ground surface ($z = 0$) and the condition that both the ground motion and stress disappear at $z = \infty$ because of the surface waves, the boundary conditions of y_1 and y_2 are determined as

$$y_2(0) = 0\,, \quad y_1(\infty) = y_2(\infty) = 0\,. \tag{3.366}$$

The second equation in (3.366) indicates that in the lower halfspace, not only does the upgoing ground motion disappear due to the **radiation boundary condition**, but the downgoing ground motion also disappears. This formulation is equivalent to the formulation in Sect. 3.1.9, which makes the discontinuity vector zero, because the downgoing ground motion does not occur in the lower halfspace in the case where there is no source in the velocity structure.

Similarly, the result of Sect. 3.1.4, which is (3.44), shows that the P and SV waves including the Rayleigh waves can be extracted from the wavefield by setting the displacement potential as $\chi = 0$, that is $u_y = 0$. Substituting this and $\partial/\partial y = 0$ into the definitions of the stresses (1.24), we have

[90]These agree with Eq. (46) of Takeuchi and Saito [109] if $L = N = \mu$.

$$\tau_{xx} = (\lambda + 2\mu)\frac{\partial u_x}{\partial x} + \lambda\frac{\partial u_z}{\partial z}, \quad \tau_{yy} = \lambda\left(\frac{\partial u_x}{\partial x} + \frac{\partial u_z}{\partial z}\right),$$
$$\tau_{zz} = \lambda\frac{\partial u_x}{\partial x} + (\lambda + 2\mu)\frac{\partial u_z}{\partial z}, \quad \tau_{yz} = \tau_{xy} = 0, \quad \tau_{zx} = \mu\left(\frac{\partial u_x}{\partial z} + \frac{\partial u_z}{\partial x}\right). \tag{3.367}$$

Assuming that the variable z remains as is, and the wavefield is oscillatory in the x-direction, we can set u_x, etc. to be

$$u_x = -\mathrm{i}y_3(z, \omega, k)\, e^{-\mathrm{i}(kx-\omega t)}, \quad u_z = y_1(z, \omega, k)\, e^{-\mathrm{i}(kx-\omega t)},$$
$$\tau_{zz} = y_2(z, \omega, k)\, e^{-\mathrm{i}(kx-\omega t)}, \quad \tau_{zx} = -\mathrm{i}y_4(z, \omega, k)\, e^{-\mathrm{i}(kx-\omega t)}, \tag{3.368}$$

where

$$y_2 = (\lambda + 2\mu)\frac{dy_1}{dz} - k\lambda y_3, \quad y_4 = \mu\left(\frac{dy_3}{dz} + ky_1\right). \tag{3.369}$$

Substituting (3.368) and (3.369) into (1.26) gives the simultaneous ordinary differential equations[91]

$$\frac{dy_1}{dz} = \frac{1}{\lambda + 2\mu}(y_2 + k\lambda y_3), \quad \frac{dy_2}{dz} = -\omega^2\rho y_1 + ky_4, \quad \frac{dy_3}{dz} = -ky_1 + \frac{1}{\mu}y_4,$$
$$\frac{dy_4}{dz} = -\frac{k\lambda}{\lambda + 2\mu}y_2 + \left\{k^2\left(\lambda + 2\mu - \frac{\lambda^2}{\lambda + 2\mu}\right) - \omega^2\rho\right\}y_3. \tag{3.370}$$

The boundary conditions for y_1, y_2, y_3, and y_4 are

$$y_2(0) = y_4(0) = 0, \quad y_1(\infty) = y_2(\infty) = y_3(\infty) = y_4(\infty) = 0. \tag{3.371}$$

The boundary value problem of the simultaneous ordinary differential equations in (3.365) and (3.366), or in (3.370) and (3.371), can be solved in the same way as the boundary value problem of the ray equations for ray tracing (Sect. 3.2.3). Disper 80 [102] in *Seismological Algorithms* [30] is a group of programs that solve these problems in various coordinate systems using the **shooting method** (Sect. 3.2.3) with the phase velocity $c \equiv \omega/k$ as a parameter.

According to the analytical mechanics for elastic bodies, the **Lagrangian** L (Sect. 3.2.2) is given by

$$L = K - U, \quad K = \int_0^\infty \frac{1}{2}\rho\sum_{i=x,y,z}\left(\frac{\partial u_i}{\partial t}\right)^2 dz, \quad U = \int_0^\infty \frac{1}{2}\sum_{i=x,y,z}\sum_{j=x,y,z}\tau_{ij}e_{ij}dz, \tag{3.372}$$

which is the total kinetic energy K minus the total strain energy U. U is represented by the **strain energy function** in (1.16). According to **Hamilton's principle** (Sect. 3.2.2), the time integral of the Lagrangian must take a stationary value. The

[91] These agree with Eq. (62) of Takeuchi and Saito [109] if $L = \mu$, $F = \lambda$, and $A = C = \lambda + 2\mu$.

time integral $\int L\,dt$ should be a constant multiple of the time mean $\langle L \rangle$, so that $\int L\,dt$ in the stationary state is equivalent to $\langle L \rangle$ in the stationary state. For the Love wave, substituting (3.362), etc. into (3.372) we obtain

$$\langle L \rangle = \omega^2 I_1 - k^2 I_2 - I_3 ,$$
$$I_1 = \frac{1}{2}\int_0^\infty \rho y_1^2 dz , \quad I_2 = \frac{1}{2}\int_0^\infty \mu y_1^2 dz , \quad I_3 = \frac{1}{2}\int_0^\infty \mu \left(\frac{dy_1}{dz}\right)^2 dz , \tag{3.373}$$

where I_1, I_2, and I_3 are termed **energy integral**.[92] Using **variation** in mathematics, $\langle L \rangle$ in the stationary state is represented as [3]

$$\delta\langle L \rangle = \omega^2 \delta I_1 - k^2 \delta I_2 - \delta I_3 = 0 . \tag{3.374}$$

In addition, when the two equations in (3.365) are combined into one equation of y_1, and this equation is multiplied by y_1, and integrated by z,

$$\begin{aligned} 0 &= \int_0^\infty \left\{ \omega^2 \rho y_1^2 - k^2 \mu y_1^2 + y_1 \frac{d}{dz}\left(\mu \frac{dy_1}{dz}\right)\right\} dz \\ &= 2\omega^2 I_1 - 2k^2 I_2 - 2I_3 + \mu y_1 \frac{dy_1}{dz}\bigg|_0^\infty \end{aligned} \tag{3.375}$$

is obtained. From the boundary conditions in (3.366), the last term in the above is zero, so that

$$\omega^2 I_1 - k^2 I_2 - I_3 = 0 , \tag{3.376}$$

therefore $\langle L \rangle$ is zero [3]. If we perturb all the energy integrals and parameters in (3.376) by the variations, we have

$$(\omega + \delta\omega)^2 (I_1 + \delta I_1) - (k + \delta k)^2 (I_2 + \delta I_2) - (I_3 + \delta I_3) = 0 . \tag{3.377}$$

Substituting (3.376) itself and (3.374) into this and ignoring the second-order terms of the variations, we obtain

$$2\omega\,\delta\omega\, I_1 - 2k\,\delta k\, I_2 = 0 . \tag{3.378}$$

The **group velocity** U (Sect. 3.1.9) is defined in (3.146). Replacing the phase velocity c by ω/k, we have

$$\frac{1}{U} = \frac{1}{c}\left(1 - \frac{\omega}{c}\frac{dc}{d\omega}\right) \Rightarrow U = \frac{\omega}{k} \Big/ \left\{1 - k\left(\frac{1}{k} - \frac{\omega}{k^2}\frac{dk}{d\omega}\right)\right\} = \frac{d\omega}{dk} = \frac{\delta\omega}{\delta k} . \tag{3.379}$$

[92] This agrees with Eq. (7.66) of Aki and Richards [3] if $l_1 \rightarrow y_1$, also with Eq. (169) of Takeuchi and Saito [109] if $I_1 \rightarrow 2I_1$, $I_2 \rightarrow 2I_2 + 2I_3$, and $L = N = \mu$.

Then, substituting (3.378) into this, we obtain[93]

$$U = \frac{kI_2}{\omega I_1} = \frac{I_2}{cI_1} \,. \tag{3.380}$$

Returning to (3.376), the second to fourth equations in (3.373) are substituted for the energy integrals, and except for ω, all other parameters and y_1 are perturbed by the variations. From this result, we subtract (3.374) and (3.376), where the energy integrals are substituted with the second to fourth equations in (3.373), and obtain

$$\omega^2 \int_0^\infty y_1^2\,\delta\rho\,dz - k^2 \int_0^\infty y_1^2\,\delta\mu\,dz - 2k\,\delta k \int_0^\infty \mu\,y_1^2\,dz - \int_0^\infty \left(\frac{dy_1}{dz}\right)^2 \delta\mu\,dz = 0\,. \tag{3.381}$$

The variation of $c \equiv \omega/k$ with fixed ω is

$$\left(\frac{\delta c}{\delta k}\right)_\omega = -\frac{\omega}{k^2}\,. \tag{3.382}$$

Substituting δk obtained from (3.381) into this, we have[94]

$$\left(\frac{\delta c}{c}\right)_\omega = -\frac{\delta k}{k} = \frac{\displaystyle\int_0^\infty \left\{k^2 y_1^2 + \left(\frac{dy_1}{dz}\right)^2\right\} \delta\mu\,dz - \int_0^\infty \omega^2 y_1^2\,\delta\rho\,dz}{\displaystyle 2k^2 \int_0^\infty \mu\,y_1^2\,dz}\,. \tag{3.383}$$

Perturbing the relationship of the S wave velocity β and the rigidity μ, that is $\rho\,\beta^2 = \mu$ (the fourth equation in (1.42)), by the variations, subtracting the original relationship from the result, and ignoring the second-order terms of the variations, we obtain

$$\delta\mu = 2\rho\beta\,\delta\beta + \beta^2\delta\rho\,. \tag{3.384}$$

When this, (3.373), and (3.380) are substituted into (3.383),

$$\begin{aligned}\left(\frac{\delta c}{c}\right)_\omega = \frac{1}{4k^2 c I_1 U}\Bigg[&\int_0^\infty \left\{k^2 y_1^2 + \left(\frac{dy_1}{dz}\right)^2\right\} 2\rho\beta\,\delta\beta\,dz \\ &+ \int_0^\infty \left\{\beta^2 k^2 y_1^2 + \beta^2 \left(\frac{dy_1}{dz}\right)^2 - \omega^2 y_1^2\right\} \delta\rho\,dz\Bigg]\end{aligned} \tag{3.385}$$

is obtained.

---

[93]This agrees with Eq. (7.70) of Aki and Richards [3]. This also agrees with Eq. (184) of Takeuchi and Saito [109] because their I_3 in Eq. (183) is equivalent to I_2 here.

[94]This agrees with Eq. (7.71) of Aki and Richards [3] if $l_1 \to y_1$.

From the definition of a partial derivative, we have[95]

$$\left(\frac{\delta c}{c}\right)_\omega = \int_0^\infty \frac{\rho}{c}\left[\frac{\partial c}{\partial \rho}\right]_{\omega,\beta}\frac{\delta \rho}{\rho}dz + \int_0^\infty \frac{\beta}{c}\left[\frac{\partial c}{\partial \beta}\right]_{\omega,\rho}\frac{\delta \beta}{\beta}dz\,. \tag{3.386}$$

The comparison of (3.385) and (3.386) leads to[96]

$$\begin{aligned}\frac{\rho}{c}\left[\frac{\partial c}{\partial \rho}\right]_{\omega,\beta} &= \frac{\rho}{4k^2cI_1U}\left\{\beta^2k^2y_1^2 + \beta^2\left(\frac{dy_1}{dz}\right)^2 - \omega^2 y_1^2\right\}\\ &= \frac{1}{4\omega^2 I_1}\frac{c}{U}\left(\mu k^2 y_1^2 + \frac{1}{\mu}y_2^2 - \omega^2\rho\, y_1^2\right),\\ \frac{\beta}{c}\left[\frac{\partial c}{\partial \beta}\right]_{\omega,\rho} &= \frac{2\rho\beta^2}{4k^2cI_1U}\left\{k^2y_1^2 + \left(\frac{dy_1}{dz}\right)^2\right\} = \frac{1}{2\omega^2 I_1}\frac{c}{U}\left(\mu k^2y_1^2 + \frac{1}{\mu}y_2^2\right).\end{aligned} \tag{3.387}$$

In the notation used in (3.387), which is equivalent to the notation of Takeuchi and Saito [109], $[\partial c/\partial \rho]_{\omega,\beta}$ and $[\partial c/\partial \beta]_{\omega,\rho}$ are partial derivatives at a certain depth in the usual sense, which correspond to $\partial c/\partial \rho$ and $\partial c/\partial \beta$ [3]. In the case of a horizontally layered structure as shown in Fig. 3.1, the partial derivatives for the density ρ_i and the S wave velocity β_i of the ith layer can be obtained by integrating the right-hand side of (3.387) with the transposition of ρ/c and β/c, from the upper limit z_{i-1} to the lower limit z_i, as [50]

$$\begin{aligned}\frac{\partial c}{\partial \rho_i} &= \int_{z_{i-1}}^{z_i}\frac{1}{D}\left\{\beta_i^2k^2y_1^2 + \frac{1}{\rho_i^2\beta_i^2}y_2^2 - \omega^2y_1^2\right\}dz\\ \frac{\partial c}{\partial \beta_i} &= \int_{z_{i-1}}^{z_i}\frac{2}{D}\left\{\rho_i\beta_ik^2y_1^2 + \frac{1}{\rho_i\beta_i^3}y_2^2\right\}dz\,,\quad D = 4k^2I_1U\,.\end{aligned} \tag{3.388}$$

Equations (3.372) – (3.388) are the results of applying the **calculus of variations** to the Love wave. The calculus of variations for the Rayleigh wave is considerably more complicated because of the coupling of the P wave with the SV wave. The formulation is therefore not carried out here, but the results are shown in Equations (196) and (197) of Takeuchi and Saito [109] and Eqs. (14)–(16) of Horike [50]. Sect. 13.2 of Saito [103] also shows the explanation and results.

If the phase velocity of the microtremors is observed and the phase velocity of the surface wave in the 1-D velocity structure model and its partial derivatives are computed with the method above, a **velocity structure inversion** can be carried out based on the **nonlinear least-squares method** (Sect. 4.4) using the **Jacobian matrix** composed of the partial derivatives. This is the basis of **microtremor exploration,** which started with the work of Horike [50]. In addition, in the nonlinear least-squares

[95]This is Eq. (1) in Box 7.8 of Aki and Richards [3] with the substitution $\mu \to \beta$.

[96]These agree with Eq. (195) of Takeuchi and Saito [109] if $\beta_V \to \beta$, $\xi \equiv 1$, $L = N = \mu$, and $I_1 \to 2I_1$.

method, the **gradient method** is usually adopted. In this method, the direction in which a solution improve (the "gradient") is obtained from the linearized simultaneous equations and the solution is iteratively improved using this direction (Sect. 4.4). **Heuristic search methods** such as the **genetic algorithm** and **simulated annealing**, which search for a solution heuristically over a wide range, are also sometimes used. While the heuristic search methods can reach the true least-squares solution over a wide range, even in problems including local minima, the theoretical values corresponding to observations must be computed many times. However, the computation of the phase velocity of the surface wave is not as complicated, therefore the heuristic search methods are often used in velocity structure inversions of microtremor data [125].

3.3.3 Seismic Interferometry[97]

Nakahara [92] wrote "**seismic interferometry** is a technique to obtain the Green's function (impulse response) from the cross-correlation function of the wavefields at two stations, one of which is the source and the other an observation point." The "Green's function" refers to the **tensor Green's function** as defined in Sect. 1.3.3, and the solution $u_i = G_{in}(\mathbf{x}, t; \boldsymbol{\xi}, \tau)$ of the equations

$$\rho \frac{\partial^2 u_i}{\partial t^2} = \frac{\partial \tau_{ij}}{\partial x_j} + \rho f_i\,, \quad \tau_{ij} = C_{ijkl} \frac{\partial u_k}{\partial x_l} \tag{3.389}$$

which include (1.81) and (1.82), and

$$\rho f_i = \delta_{in} \delta(\mathbf{x} - \boldsymbol{\xi}) \delta(t - \tau)\,. \tag{3.390}$$

Also, from Sect. 1.3.3, "impulse" indicates the body force represented by (3.390), therefore $G_{in}(\mathbf{x}, t; \boldsymbol{\xi}, \tau)$ can be considered an "impulse response". As mentioned in Sect. 1.3.3, as the tensor Green's function is the effect of propagation itself, seismic interferometry is suitable for analyzing propagation. For the notation in this subsection, we follow Wapenaar and Fokkema [123], so that $G_{in}(\mathbf{x}, t; \boldsymbol{\xi}, \tau)$ for $v_i = \partial u_i / \partial t$ and $\tau = 0$ is written as $G^v_{in}(\mathbf{x}, \boldsymbol{\xi}, t)$, and its Fourier transform $\overline{G}^v_{in}(\mathbf{x}, \boldsymbol{\xi}, \omega)$ is also called a Green's function. Among these Green's functions, $G^v_{in}(\mathbf{x}, \boldsymbol{\xi}, t)$ is the solution for v_i in

$$\rho \frac{\partial v_i}{\partial t} = \frac{\partial \tau_{ij}}{\partial x_j} + \rho f_i\,, \quad \tau_{ij} = C_{ijkl} \frac{\partial}{\partial x_l} \int v_k dt\,, \quad \rho f_i = \delta_{in} \delta(\mathbf{x} - \boldsymbol{\xi}) \delta(t)\,, \tag{3.391}$$

which is obtained from (3.389) and (3.390). $\overline{G}^v_{in}(\mathbf{x}, \boldsymbol{\xi}, \omega)$ is the solution for the $\bar{v}_i$ in

[97] As in Sects. 1.3.2 and 1.3.3, Einstein's **summation convention** is adopted throughout Sect. 3.3.3.

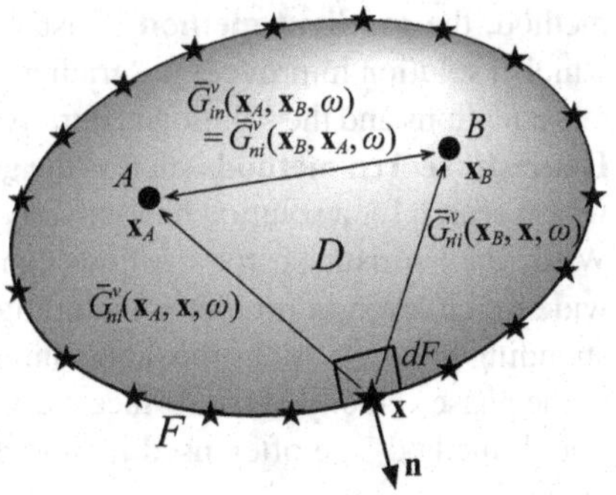

Fig. 3.34 Region D and its distant circumference F. Impulses (star marks) are distributed over F, the normal vector of which is n (based on Wapenaar and Fokkema [123] and Matsuoka and Shiraishi [89], reprinted from Koketsu [72] with permission of Kindai Kagaku)

$$\mathrm{i}\omega\rho\bar{v}_i = \frac{\partial\bar{\tau}_{ij}}{\partial x_j} + \rho\bar{f}_i\,, \quad \bar{\tau}_{ij} = \frac{1}{\mathrm{i}\omega}C_{ijkl}\frac{\partial\bar{v}_k}{\partial x_l}\,, \quad \rho\bar{f}_i = \delta_{in}\delta(\mathbf{x}-\boldsymbol{\xi})\,, \tag{3.392}$$

obtained from the Fourier transforms of both sides of (3.391).

As shown in Fig. 3.34, it is assumed that there is a region D in the velocity structure, its outer circumference F is distant, and impulses are closely distributed on F. Based on this assumption, the "wavefield" in the definition of Nakahara [92] is the total sum of ground motions from sources, which exist in great numbers at large distances. The subjects of the analysis are generally considered to be **microtremors** (Sect. 3.3.2), but if they exist in great numbers, ground motions from earthquakes can also be used.

The Fourier transform $\overline{G}^v_{ni}(\mathbf{x}_A, \mathbf{x}, \omega)$ of the ground velocity at $\mathbf{x}_A$ excited by an impulse at $\mathbf{x}$ on F is

$$\overline{G}^v_{ni}(\mathbf{x}_A, \mathbf{x}, \omega) = \overline{G}^v_{in}(\mathbf{x}, \mathbf{x}_A, \omega) = \bar{v}_i \tag{3.393}$$

from the **reciprocity relation** (1.96) of the Green's function with respect to the spatial coordinates. Similarly, for the Fourier transform $\overline{G}^v_{ni}(\mathbf{x}_B, \mathbf{x}, \omega)$ of the ground velocity at $\mathbf{x}_B$ excited by an impulse at $\mathbf{x}$ on F, we have

$$\overline{G}^v_{n'i}(\mathbf{x}_B, \mathbf{x}, \omega) = \overline{G}^v_{in'}(\mathbf{x}, \mathbf{x}_B, \omega) = \bar{w}_i\,. \tag{3.394}$$

Considering a stress impulse in which the impulse acts as a stress rather than a body force, we define the Green's function $H^v_{p,qr}(\mathbf{x}, \boldsymbol{\xi}, t)$ of the stress impulse h_{ij} to be the solution v_i in

$$\rho\frac{\partial v_i}{\partial t} = \frac{\partial\tau_{ij}}{\partial x_j}\,, \quad \tau_{ij} + h_{ij} = C_{ijkl}\frac{\partial}{\partial x_l}\int v_k dt\,, \quad h_{ij} = \delta_{iq}\delta_{jr}\delta(\mathbf{x}-\boldsymbol{\xi})\,. \tag{3.395}$$

This Green's function has the reciprocity relation

$$G^\tau_{qr,p}(\boldsymbol{\xi}_2, \boldsymbol{\xi}_1, t) = H^v_{p,qr}(\boldsymbol{\xi}_1, \boldsymbol{\xi}_2, t)\,, \tag{3.396}$$

where G^{τ} is the Green's function of the stress in response to the impulse acting as a body force [123]. The Fourier transform of the reciprocity relation (3.396) therefore yields

$$\begin{aligned}\overline{H}^{v}_{n,ij}(\mathbf{x}_A, \mathbf{x}, \omega) &= \overline{G}^{\tau}_{ij,n}(\mathbf{x}, \mathbf{x}_A, \omega) = \bar{\tau}_{ij} \,, \\ \overline{H}^{v}_{n',ij}(\mathbf{x}_B, \mathbf{x}, \omega) &= \overline{G}^{\tau}_{ij,n'}(\mathbf{x}, \mathbf{x}_B, \omega) = \bar{\sigma}_{ij}\end{aligned} \tag{3.397}$$

in the setting of Fig. 3.34.

Using the definitions of the **cross correlation** in Table 4.3 and combining two cross correlations, we define

$$\int_{-\infty}^{+\infty} \left\{G^{v}_{ni}(\mathbf{x}_A, \mathbf{x}, \tau) H^{v}_{n',ij}(\mathbf{x}_B, \mathbf{x}, t+\tau) + H^{v}_{n,ij}(\mathbf{x}_A, \mathbf{x}, \tau) G^{v}_{n'i}(\mathbf{x}_B, \mathbf{x}, t+\tau)\right\} d\tau \,. \tag{3.398}$$

We then rewrite this using the Fourier transform of cross correlation in Table 4.3 as

$$\left\{\overline{G}^{v}_{ni}(\mathbf{x}_A, \mathbf{x}, \omega)\right\}^{*} \overline{H}^{v}_{n',ij}(\mathbf{x}_B, \mathbf{x}, \omega) + \left\{\overline{H}^{v}_{n,ij}(\mathbf{x}_A, \mathbf{x}, \omega)\right\}^{*} \overline{G}^{v}_{n'i}(\mathbf{x}_B, \mathbf{x}, \omega) \,. \tag{3.399}$$

Next, by performing a surface integral on $\mathbf{x}$, (3.399) is extended to the total wavefield due to the distribution of impulses in Fig. 3.34 as

$$\iint \left[\left\{\overline{G}^{v}_{ni}(\mathbf{x}_A, \mathbf{x}, \omega)\right\}^{*} \overline{H}^{v}_{n',ij}(\mathbf{x}_B, \mathbf{x}, \omega) + \left\{\overline{H}^{v}_{n,ij}(\mathbf{x}_A, \mathbf{x}, \omega)\right\}^{*} \overline{G}^{v}_{n'i}(\mathbf{x}_B, \mathbf{x}, \omega)\right] n_j dF \,, \tag{3.400}$$

where n_j is a component of the normal vector $\mathbf{n}$ to F shown in Fig. 3.34. Substituting the reciprocity relations (3.393), (3.394), and (3.397) into (3.400), we obtain

$$\iint \left[\left\{\overline{G}^{v}_{in}(\mathbf{x}, \mathbf{x}_A, \omega)\right\}^{*} \overline{G}^{\tau}_{ij,n'}(\mathbf{x}, \mathbf{x}_B, \omega) + \left\{\overline{G}^{\tau}_{ij,n}(\mathbf{x}, \mathbf{x}_A, \omega)\right\}^{*} \overline{G}^{v}_{in'}(\mathbf{x}, \mathbf{x}_A, \omega)\right] n_j \, dF \,. \tag{3.401}$$

Using the last terms of (3.393), (3.394), and (3.397), we rewrite this in a simple form as

$$\iint \left(\bar{v}_i^{*} \bar{\sigma}_{ij} + \bar{\tau}_{ij}^{*} \bar{w}_i\right) n_j \, dF \,. \tag{3.402}$$

Here, we consider the case when the time goes in the opposite direction. This case is equivalent to the change of the time variable from t to $t' = -t$. In the equations of motion (3.389) and (3.390) with $\tau = 0$, t appears only in $\partial^2/\partial t^2$ and $\delta(t)$. Hence, even if the time variable is changed to $t' = -t$, the same solution u_i is obtained because $\partial^2/\partial t'^2 = \partial^2/\partial t^2$ and $\delta(t') = \delta(t)$. As t is not included in the second equation of (3.389), the solution remains the same for τ_{ij} with t'. In summary, neither an odd function of t nor an odd-order partial derivative of t is included in (3.389) and (3.390) with $\tau = 0$, and u_i and τ_{ij} do not change with the time reversal. If they are included, their signs are inverted by the time reversal, therefore the equation of motion changes and the same solutions cannot be obtained. If the medium has the property of **anelasticity**, the equation of motion includes a first-order partial derivative of t as

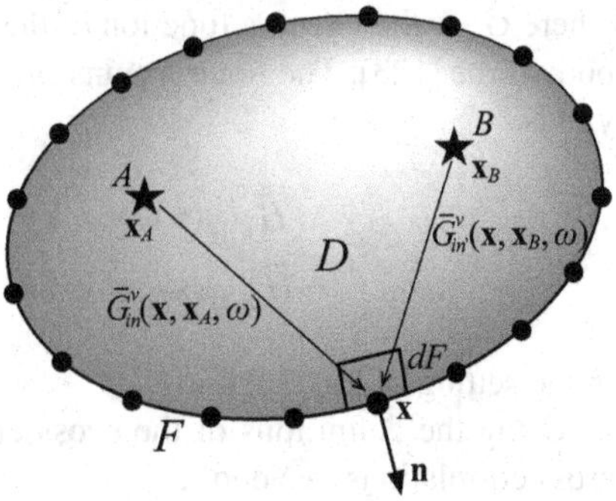

Fig. 3.35 The reciprocity relations applied to the Green's functions in Fig. 3.34. There are impulses (star marks) at points A and B, and their ground velocities are observed at points (black dots) distributed on F (reprinted from Koketsu [72] with permission of Kindai Kagaku)

in the equation of damped oscillation (1.60) because of **intrinsic attenuation** from the anelasticity, and therefore the formulation here cannot be applied, but theoretical research on alternative formulations has already begun [92].

Applying the reciprocity relations as in (3.393), (3.394), and (3.397) implies that the situation in Fig. 3.34 changes as shown in Fig. 3.35. However, the impulses return to F, because the reciprocity relations can return to the original state if the time is reversed. The boundary condition on F therefore yields an **inhomogeneous** equation, because it contains inhomogeneous terms due to the impulses. Hence, if the **reciprocity theorem** (1.96) described in Sect. 1.3.2 is applied to region D containing the body forces

$$\rho f_i = \delta_{in}\delta(\mathbf{x} - \mathbf{x}_A)\delta(t)\,, \quad \rho g_i = \delta_{in}\delta(\mathbf{x} - \mathbf{x}_B)\delta(t) \tag{3.403}$$

and the outer circumference F in Fig. 3.35, the surface integral on F remains as[98]

$$\begin{aligned} &\textstyle\iiint \left\{\rho g_i(t) * u_i(-t) - \rho f_i(-t) * v_i(t)\right\} dD \\ &= \textstyle\iint \left\{v_i(t) * \tau_{ij}(-t) - u_i(-t) * \sigma_{ij}(t)\right\} n_j\, dF\,. \end{aligned} \tag{3.404}$$

When both sides of (3.404) are partially differentiated by t and $\partial u_i(-t)/\partial t = -v(-t)$, $w_i = \partial U_i/\partial t$ is substituted, we obtain

$$\begin{aligned} &\textstyle\iiint \left\{-\rho g_i(t) * v_i(-t) - \rho f_i(-t) * w_i(t)\right\} dD \\ &= \textstyle\iint \left\{w_i(t) * \tau_{ij}(-t) + v_i(-t) * \sigma_{ij}(t)\right\} n_j\, dF\,. \end{aligned} \tag{3.405}$$

Furthermore, taking the Fourier transform of (3.405) and using the relation $f(-t) \Leftrightarrow \{F(\omega)\}^*$ in Table 4.3, we obtain

$$-\iiint \left(\rho \bar{g}_i \bar{v}_i^* + \rho \bar{f}_i^* \bar{w}_i\right) dD = \iint \left(\bar{w}_i \bar{\tau}_{ij}^* + \bar{v}_i^* \bar{\sigma}_{ij}\right) n_j\, dF\,. \tag{3.406}$$

[98] This agrees with Eq. (2.35) of Aki and Richards [3] if $V \to D$, $S \to F$, $\mathbf{f}(t) \to \rho\mathbf{f}(-t)$, $\mathbf{g}(t) \to \rho\mathbf{g}(t)$, etc. and each vector is decomposed into components.

As the right-hand side of (3.406) coincides with (3.402), this can be replaced with (3.401), which is a representation with the Green's functions. For the left-hand side multiplied by (–1), we substitute the Fourier transform of (3.403)

$$\rho \bar{f}_i = \delta_{in}\delta(\mathbf{x} - \mathbf{x}_A)\,, \quad \rho \bar{g}_i = \delta_{in}\delta(\mathbf{x} - \mathbf{x}_B)\,, \tag{3.407}$$

(3.393), and (3.394), then apply the definition of the delta function in (1.98), obtaining

$$\begin{aligned} &\iiint \left[\delta_{in'}\delta(\mathbf{x} - \mathbf{x}_B)\left\{\overline{G}^{\,v}_{in}(\mathbf{x}, \mathbf{x}_A, \omega)\right\}^* + \{\delta_{in}\delta(\mathbf{x} - \mathbf{x}_A)\}^* \overline{G}^{\,v}_{in'}(\mathbf{x}, \mathbf{x}_B, \omega)\right] dD \\ &= \left\{\overline{G}^{\,v}_{n'n}(\mathbf{x}_B, \mathbf{x}_A, \omega)\right\}^* + \overline{G}^{\,v}_{nn'}(\mathbf{x}_A, \mathbf{x}_B, \omega)\,. \end{aligned} \tag{3.408}$$

Summarizing the above, we have[99]

$$\begin{aligned} &\left\{\overline{G}^{\,v}_{n'n}(\mathbf{x}_B, \mathbf{x}_A, \omega)\right\}^* + \overline{G}^{\,v}_{nn'}(\mathbf{x}_A, \mathbf{x}_B, \omega) = \\ &-\iint \left[\left\{\overline{G}^{\,v}_{in}(\mathbf{x}, \mathbf{x}_A, \omega)\right\}^* \overline{G}^{\,\tau}_{ij,n'}(\mathbf{x}, \mathbf{x}_B, \omega) + \left\{\overline{G}^{\,\tau}_{ij,n}(\mathbf{x}, \mathbf{x}_A, \omega)\right\}^* \overline{G}^{\,v}_{in'}(\mathbf{x}, \mathbf{x}_B, \omega)\right] n_j\, dF\,. \end{aligned} \tag{3.409}$$

To simplify the right-hand side of (3.409), the **scalar potential** ϕ and the **vector potential** $\boldsymbol{\psi} = (\psi_i)$ defined in Sect. 1.2.4 are introduced, and ϕ and ψ_1, ψ_2, ψ_3 are collectively called φ_0, φ_1, φ_2, and φ_3. According to Wapenaar and Haimé [122],[100] if the medium is homogeneous and isotropic in the small region around point A in Fig. 3.35, then the Fourier transform of the body force is given as

$$\rho \bar{f}_i^{\varphi} = \rho\alpha^2 \frac{\partial}{\partial x_i}\delta(\mathbf{x} - \mathbf{x}_A) \tag{3.410}$$

when the impulse of the P wave represented by $\phi = \varphi_0$ acts on point A. This can be related to the Fourier transform of the body force $\rho \bar{f}_i$. According to the **linearity** of the equation of motion (Sect. 1.3.1), this relation holds between the Fourier transform of the ground velocity due to $\rho \bar{f}_i^{\varphi}$ and the Fourier transform of the ground velocity due to $\rho \bar{f}_i$. As mentioned earlier, the solution $\bar{v}_i$ of (3.392), which is the Fourier transform of the ground velocity due to $\rho \bar{f}_i$ with $\mathbf{x}_A = \boldsymbol{\xi}$, is the Green's function $\overline{G}^{\,v}_{in}(\mathbf{x}, \boldsymbol{\xi}, \omega)$. The Green's function for the body force, the Fourier transform of which is represented by (3.410), is termed the P wave Green's function and is denoted by $\overline{G}^{\,\varphi}_{0n}$. Using the above, Wapenaar and Haimé [122] obtained

[99]This agrees with Eq. (62) of Wapenaar and Fokkema [123] if $\partial D \to F$, $p \to n$, and $q \to n'$.

[100]The first author of this paper is "C. P. A. Wapenaar", while the first author of Wapenaar and Fokkema [123] is "Kees Wapenaar". They are the same person, because Kees Wapenaar's home page includes the both papers.

$$\overline{G}^{\varphi}_{0n}(\mathbf{x}, \mathbf{x}_A, \omega) = -\frac{\rho\alpha^2}{\mathrm{i}\omega}\frac{\partial}{\partial x_i}\overline{G}^{v}_{in}(\mathbf{x}, \mathbf{x}_A, \omega) . \tag{3.411}$$

We substitute the reciprocity relation in (1.106) for the right-hand side of (3.411) and obtain the reciprocity relation of the P wave Green's function

$$\overline{G}^{\varphi}_{0n}(\mathbf{x}, \mathbf{x}_A, \omega) = \overline{G}^{\varphi}_{n0}(\mathbf{x}_A, \mathbf{x}, \omega) . \tag{3.412}$$

For the P wave Green's function at point B, we also have

$$\overline{G}^{\varphi}_{0n'}(\mathbf{x}, \mathbf{x}_B, \omega) = \overline{G}^{\varphi}_{n'0}(\mathbf{x}_B, \mathbf{x}, \omega) . \tag{3.413}$$

Similarly, according to Wapenaar and Haimé [122], when the impulse of the S wave represented by $\psi_k = \varphi_k$ $(k = 1, 2, 3)$ acts on point A, the Fourier transform of the body force is given as

$$\rho \bar{f}_i = -\rho\beta^2 \varepsilon_{kij}\frac{\partial}{\partial x_j}\delta(\mathbf{x} - \mathbf{x}_A) . \tag{3.414}$$

ε_{kij} are elements of the **alternating tensor**, which has the antisymmetric property $\varepsilon_{kij} = -\varepsilon_{kji}$. Wapenaar and Haimé [122] again obtained the S wave Green's functions

$$\overline{G}^{\varphi}_{kn}(\mathbf{x}, \mathbf{x}_A, \omega) = \frac{\rho\beta^2}{\mathrm{i}\omega}\varepsilon_{kji}\frac{\partial}{\partial x_j}\overline{G}^{v}_{in}(\mathbf{x}, \mathbf{x}_A, \omega) , \tag{3.415}$$

the reciprocity relations at point A

$$\overline{G}^{\varphi}_{kn}(\mathbf{x}, \mathbf{x}_A, \omega) = \overline{G}^{\varphi}_{nk}(\mathbf{x}_A, \mathbf{x}, \omega) , \tag{3.416}$$

and those at point B

$$\overline{G}^{\varphi}_{kn'}(\mathbf{x}, \mathbf{x}_B, \omega) = \overline{G}^{\varphi}_{n'k}(\mathbf{x}_B, \mathbf{x}, \omega) \tag{3.417}$$

for $k = 1, 2, 3$.

Returning now to (3.402), Wapenaar and Haimé [122] proved that if F is a horizontal plane where $\mathbf{n} = (0, 0, 1)$ and the ground motion consists of body waves, -1 times (3.402) is

$$-\iint \left(\bar{v}_i^* \bar{\sigma}_{ij} + \bar{\tau}_{ij}^* \bar{w}_i\right) n_j \, dF = \frac{2}{\mathrm{i}\omega\rho}\iint \left[\frac{\partial}{\partial x_i}\left\{\overline{G}^{\varphi}_{0n}(\mathbf{x}, \mathbf{x}_A, \omega)\right\}^* \overline{G}^{\varphi}_{0n'}(\mathbf{x}, \mathbf{x}_B, \omega) + \frac{\partial}{\partial x_i}\left\{\overline{G}^{\varphi}_{kn}(\mathbf{x}, \mathbf{x}_A, \omega)\right\}^* \overline{G}^{\varphi}_{kn'}(\mathbf{x}, \mathbf{x}_B, \omega)\right] n_i \, dF. \tag{3.418}$$

In the notation of Wapenaar and Fokkema [123], the right-hand side of (3.418) is further simplified to

$$\frac{2}{\mathrm{i}\omega\rho}\iint\frac{\partial}{\partial x_i}\left\{\overline{G}^{\varphi}_{Kn}(\mathbf{x},\mathbf{x}_A,\omega)\right\}^*\overline{G}^{\varphi}_{Kn'}(\mathbf{x},\mathbf{x}_B,\omega)\,n_i\,dF\,,\quad K=0,1,2,3\,. \tag{3.419}$$

Substituting (3.412), (3.413), (3.416), and (3.417) into this, (3.418) yields

$$-\iint\left(\bar{v}_i^*\bar{\sigma}_{ij}+\bar{\tau}_{ij}^*\bar{w}_i\right)n_j\,dF= \tag{3.420}$$
$$\frac{2}{\mathrm{i}\omega\rho}\iint\frac{\partial}{\partial x_i}\left\{\overline{G}^{\varphi}_{nK}(\mathbf{x}_A,\mathbf{x},\omega)\right\}^*\overline{G}^{\varphi}_{n'K}(\mathbf{x}_B,\mathbf{x},\omega)\,n_i\,dF.$$

In the case of an approximately stationary integrand in (3.420), Wapenaar and Fokkema [123] showed that (3.420) is valid for body waves if F is an arbitrary surface, and that (3.420) is valid for surface waves if F is a closed surface as shown in Figs. 3.34 and 3.35.

The left-hand side of (3.420) is –1 times (3.402), and the right-hand side of (3.409) is –1 times (3.401). Since it has already been shown that (3.402) and (3.401) are equivalent, it is possible to substitute (3.420) for the right-hand side in (3.409), so that

$$\left\{\overline{G}^{v}_{n'n}(\mathbf{x}_B,\mathbf{x}_A,\omega)\right\}^*+\overline{G}^{v}_{nn'}(\mathbf{x}_A,\mathbf{x}_B,\omega)= \tag{3.421}$$
$$\frac{2}{\mathrm{i}\omega\rho}\iint\frac{\partial}{\partial x_i}\left\{\overline{G}^{\varphi}_{nK}(\mathbf{x}_A,\mathbf{x},\omega)\right\}^*\overline{G}^{\varphi}_{n'K}(\mathbf{x}_B,\mathbf{x},\omega)\,n_i\,dF$$

is obtained and the problem setting changes from Fig. 3.35 back to Fig. 3.34. As $\overline{G}^{\varphi}_{nK}(\mathbf{x}_A,\mathbf{x},\omega)$ on the right-hand side of (3.421) are the Fourier transforms of the P wave and S wave Green's functions, they should be solutions of the **Helmholtz equation** (Sect. 3.1.1), which is the Fourier transform of the wave equation. A special solution of the Helmholtz equation is the harmonic function, which is an extension of (3.45) to three dimensions

$$\mathcal{H}=e^{-\mathrm{i}k_1x_i}e^{-\mathrm{i}k_2x_2}e^{-\mathrm{i}\nu_\upsilon x_3},\quad k_\upsilon^2=k_1^2+k_2^2+k_3^2,\quad k_\upsilon=\frac{\omega}{\upsilon},\quad \upsilon=\alpha\text{ or }\beta\,, \tag{3.422}$$

where only outward propagation is considered. If F is sufficiently distant to give $k_i/k_\upsilon\sim n_i$,

$$\frac{\partial\mathcal{H}}{\partial x_i}n_i=-\mathrm{i}k_1\frac{k_1}{k_\upsilon}-\mathrm{i}k_2\frac{k_2}{k_\upsilon}-\mathrm{i}k_3\frac{k_3}{k_\upsilon}\mathcal{H}=-ik_\upsilon\mathcal{H} \tag{3.423}$$

and therefore

$$\frac{\partial}{\partial x_i}\overline{G}^{\varphi}_{nK}(\mathbf{x}_A,\mathbf{x},\omega)\,n_i=-ik_\upsilon\overline{G}^{\varphi}_{nK}(\mathbf{x}_A,\mathbf{x},\omega) \tag{3.424}$$

holds also for $\overline{G}^{\varphi}_{nK}(\mathbf{x}_A,\mathbf{x},\omega)$, which is the superposition of the special solutions $\mathcal{H}$. υ is α when $K=0$ or otherwise β when $K=1,2,3$. Substituting this into the right-hand side of (3.421), swapping the terms on the left-hand side, and applying

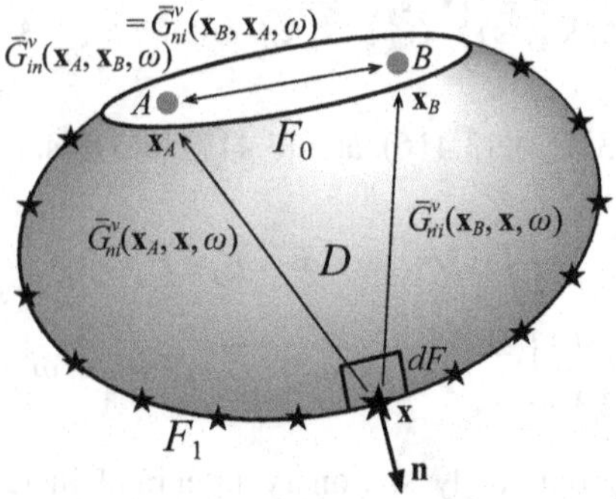

Fig. 3.36 The outer circumference F includes the ground surface F_0 (white ellipse) in a realistic situation. The impulses (star marks) are distributed in the region but excluding the ground surface (F_1), and points A and B (gray circles) are located in D immediately below F_0 (based on Matsuoka and Shiraishi [89], reprinted from Koketsu [72] with permission of Kindai Kagaku)

the reciprocity relation (1.96), we obtain[101]

$$\begin{aligned}\overline{G}^{v}_{nn'}(\mathbf{x}_A,\mathbf{x}_B,\omega)+\left\{\overline{G}^{v}_{nn'}(\mathbf{x}_A,\mathbf{x}_B,\omega)\right\}^*\\=\frac{2}{\rho\upsilon}\iint\left\{\overline{G}^{\varphi}_{nK}(\mathbf{x}_A,\mathbf{x},\omega)\right\}^*\overline{G}^{\varphi}_{n'K}(\mathbf{x}_B,\mathbf{x},\omega)\,dF\,.\end{aligned}\tag{3.425}$$

In a realistic situation as shown in Fig. 3.36, the outer circumference F contains the ground surface F_0 (white ellipse), and the observation point is often placed on the ground surface. Even in such a case, supposing that points A and B are located in region D immediately below the ground surface, we can keep (3.425) by simply replacing the surface integral $\int dF$ with $\int dF_1$ [89]. When F and F_1 are collectively referred to as F and the Kth type impulses are distributed on F with the density function $N_K(\mathbf{x},t)$ for point A, and with $N_L(\mathbf{x}',t)$ for point B, the ground velocities observed at points A and B are given as

$$\begin{aligned}\bar{v}^{\mathrm{obs}}_{n}(\mathbf{x}_A,\omega)&=\int\overline{G}^{\varphi}_{nK}(\mathbf{x}_A,\mathbf{x},\omega)\overline{N}_K(\mathbf{x},\omega)\,dF\\\bar{v}^{\mathrm{obs}}_{n'}(\mathbf{x}_B,\omega)&=\int\overline{G}^{\varphi}_{n'L}(\mathbf{x}_B,\mathbf{x}',\omega)\overline{N}_L(\mathbf{x}',\omega)\,dF'\,.\end{aligned}\tag{3.426}$$

Assuming that $\overline{N}_K$ and $\overline{N}_L$ are independently distributed and uncorrelated, and that both power spectra are the same except for the normalization coefficients $\rho\alpha/\rho\upsilon$, we have

$$\left\langle\left\{\overline{N}_K(\mathbf{x},\omega)\right\}^*\overline{N}_L(\mathbf{x}',\omega)\right\rangle=\frac{\rho\alpha}{\rho\upsilon}\delta_{KL}\delta(\mathbf{x}-\mathbf{x}')\,\overline{S}(\omega)\,.\tag{3.427}$$

[101] This agrees with Eq. (76) of Wapenaar and Fokkema [123] if $\partial D\to F$, $p\to n$, and $q\to n'$.

$\langle\ \rangle$ stands for the **ensemble average** with respect to the spatial coordinate $\mathbf{x}$, and $\overline{S}(\omega)$ is the power spectrum of the time function of the impulse. Equations (3.426) and (3.427) yield

$$\left\langle \left\{ \bar{v}_n^{\mathrm{obs}}(\mathbf{x}_A,\omega) \right\}^* \bar{v}_{n'}^{\mathrm{obs}}(\mathbf{x}_B,\omega) \right\rangle = \frac{\rho\alpha}{\rho\upsilon} \int \left\{ \overline{G}_{nK}^{\varphi}(\mathbf{x}_A,\mathbf{x},\omega) \right\}^* \overline{G}_{n'K}^{\varphi}(\mathbf{x}_B,\mathbf{x},\omega)\, \overline{S}(\omega)\, dF \,, \tag{3.428}$$

and by substituting this into (3.425) we obtain[102]

$$\left[\overline{G}_{nn'}^{v}(\mathbf{x}_A,\mathbf{x}_B,\omega) + \left\{ \overline{G}_{nn'}^{v}(\mathbf{x}_A,\mathbf{x}_B,\omega) \right\}^* \right] \overline{S}(\omega) = \frac{2}{\rho\alpha} \left\langle \left\{ \bar{v}_n^{\mathrm{obs}}(\mathbf{x}_A,\omega) \right\}^* \bar{v}_{n'}^{\mathrm{obs}}(\mathbf{x}_B,\omega) \right\rangle . \tag{3.429}$$

Unknown constants such as $\overline{S}(\omega)$ remaining in (3.429) can be canceled by normalization as follows [126]. As $n' \to n$ and $B \to A$ in (3.429), we obtain[103]

$$C^{-1}\overline{S}(\omega) = \frac{2}{\rho\alpha} \left\langle \left| \bar{v}_n^{\mathrm{obs}}(\mathbf{x}_A,\omega) \right|^2 \right\rangle , \quad C^{-1} = \overline{G}_{nn}^{v}(\mathbf{x}_A,\mathbf{x}_A,\omega) + \left\{ \overline{G}_{nn}^{v}(\mathbf{x}_A,\mathbf{x}_A,\omega) \right\}^* . \tag{3.430}$$

This is substituted back into (3.429), and then we have

$$C \left[\overline{G}_{nn'}^{v}(\mathbf{x}_A,\mathbf{x}_B,\omega) + \left\{ \overline{G}_{nn'}^{v}(\mathbf{x}_A,\mathbf{x}_B,\omega) \right\}^* \right] = \frac{\left\langle \left\{ \bar{v}_n^{\mathrm{obs}}(\mathbf{x}_A,\omega) \right\}^* \bar{v}_{n'}^{\mathrm{obs}}(\mathbf{x}_B,\omega) \right\rangle}{\left\langle \left| \bar{v}_n^{\mathrm{obs}}(\mathbf{x}_A,\omega) \right|^2 \right\rangle} . \tag{3.431}$$

The denominator on the right-hand side of (3.431) corresponds to the **power spectrum** (Sect. 4.2.2) of the ground velocities observed at point A. If the wavefield is stationary and the power spectrum is almost constant, the right-hand side of (3.431) can be approximated to

$$\frac{\left\langle \left\{ \bar{v}_n^{\mathrm{obs}}(\mathbf{x}_A,\omega) \right\}^* \bar{v}_{n'}^{\mathrm{obs}}(\mathbf{x}_B,\omega) \right\rangle}{\left\langle \left| \bar{v}_n^{\mathrm{obs}}(\mathbf{x}_A,\omega) \right|^2 \right\rangle} \sim \left\langle \frac{\left\{ \bar{v}_n^{\mathrm{obs}}(\mathbf{x}_A,\omega) \right\}^* \bar{v}_{n'}^{\mathrm{obs}}(\mathbf{x}_B,\omega)}{\left| \bar{v}_n^{\mathrm{obs}}(\mathbf{x}_A,\omega) \right|^2} \right\rangle . \tag{3.432}$$

We substitute (3.432) for (3.431), take the inverse Fourier transforms of both sides of the result, and use the formula in Table 4.3 to obtain[104]

$$C \left[G_{nn'}^{v}(\mathbf{x}_A,\mathbf{x}_B,t) + G_{nn'}^{v}(\mathbf{x}_A,\mathbf{x}_B,-t) \right] = \left\langle \mathcal{F}^{-1} \frac{\left\{ \bar{v}_n^{\mathrm{obs}}(\mathbf{x}_A,\omega) \right\}^* \bar{v}_{n'}^{\mathrm{obs}}(\mathbf{x}_B,\omega)}{\left| \bar{v}_n^{\mathrm{obs}}(\mathbf{x}_A,\omega) \right|^2} \right\rangle , \tag{3.433}$$

where $\mathcal{F}^{-1}$ is the operator of the inverse Fourier transform.

---

[102] This agrees with Eq. (86) of Wapenaar and Fokkema [123] if $\partial D \to F$, $p \to n$, and $q \to n'$.

[103] This agrees with Eq. (6) of Yokoi and Margaryan [126] if $z \to n, n'$.

[104] After $G_{nn'}^{v}(\mathbf{x}_A,\mathbf{x}_B,-t)$ is included in $G_{nn'}^{v}(\mathbf{x}_A,\mathbf{x}_B,t)$ and $C = 1$, this agrees with Eq. (1) of Viens et al. [118] in the case of $i \to n$, $j \to n'$, $S \to A$, $R \to B$, and $\xi = 0$.

Furthermore, assuming that the power spectrum is perfectly constant, we can include it in the constant as

$$C' = C\left\langle\left|\bar{v}_n^{\mathrm{obs}}(\mathbf{x}_A, \omega)\right|^2\right\rangle. \tag{3.434}$$

Taking the inverse Fourier transforms of both sides of (3.431), we obtain

$$C'\left[G^v_{nn'}(\mathbf{x}_A, \mathbf{x}_B, t) + G^v_{nn'}(\mathbf{x}_A, \mathbf{x}_B, -t)\right] = \left\langle\int v_n^{\mathrm{obs}}(\mathbf{x}_A, \tau)\, v_{n'}^{\mathrm{obs}}(\mathbf{x}_B, t+\tau)\, d\tau\right\rangle. \tag{3.435}$$

On the right-hand side of (3.435), the terms in [ ] are the Green's functions between points A and B in Fig. 3.36, and the integral in <> on the right-hand side is the cross correlation of the ground velocities observed at points A and B. In addition, taking an ensemble mean <> is equivalent to **stacking** observation records. In other words, (3.435) shows that stacking of the cross correlation of observation records such as microtremors at two points yields a form of Green's function between the two points, and (3.433) has the same physical meaning. Hence, (3.435) or (3.433) is the basic principle of **seismic interferometry**. However, it should be noted that the Green's function also contains the time-reversed $G^v_{nn'}(\mathbf{x}_A, \mathbf{x}_B, -t)$. In addition, because the constant C' or C cannot be calculated from the records of microtremors, etc., it must be determined from the comparison of several earthquake records [118].

When the ground motion (ground displacement) and its Green's function are denoted by u_n and $G_{nn'}$, $\bar{v}_n = \mathrm{i}\omega\bar{u}_n$ and $\overline{G}^v_{nn'} = \mathrm{i}\omega\overline{G}_{nn'}$. Hence, by substituting these and (3.434) into (3.431),

$$\begin{aligned} C'\mathrm{i}\omega&\left[\overline{G}_{nn'}(\mathbf{x}_A, \mathbf{x}_B, \omega) - \left\{\overline{G}_{nn'}(\mathbf{x}_A, \mathbf{x}_B, \omega)\right\}^*\right] \\ &= -(\mathrm{i}\omega)^2\left\langle\left\{\bar{u}_n^{\mathrm{obs}}(\mathbf{x}_A, \omega)\right\}^* \bar{u}_{n'}^{\mathrm{obs}}(\mathbf{x}_B, \omega)\right\rangle \end{aligned} \tag{3.436}$$

is obtained. If we divide both sides of (3.436) by $\mathrm{i}\omega$ and consider the inverse Fourier transform, we obtain[105]

$$-C'\left[G_{nn'}(\mathbf{x}_A, \mathbf{x}_B, t) - G_{nn'}(\mathbf{x}_A, \mathbf{x}_B, -t)\right] = \frac{\partial}{\partial t}\left\langle\int u_n^{\mathrm{obs}}(\mathbf{x}_A, \tau)\, u_{n'}^{\mathrm{obs}}(\mathbf{x}_B, t+\tau)\, d\tau\right\rangle. \tag{3.437}$$

In many cases, (3.437) is used as a basic principle, instead of (3.435) or (3.433).

3.3.4 *Seismic Tomography*

A 3-D velocity structure is analyzed using **seismic tomography**, which is an extension of the ray theory (Sect. 3.2.2) and ray tracing (Sect. 3.2.3). The term "tomogra-

[105] This agrees with Eq. (1) of Stehly et al. [106] if the cross correlation is written down, the constant term → C', $\mathbf{r} \to \mathbf{x}$, and $\tau \leftrightarrow t$.

Fig. 3.37 Cross-sectional view of a 3-D velocity structure divided into blocks along the Cartesian coordinate system. The rays of seismic waves pass through the gray blocks

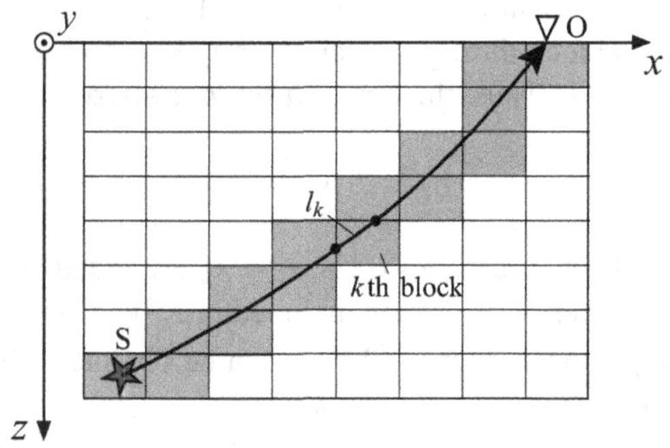

phy" is coined by combining "tomo", which is derived from Greek, meaning cross section or cutting, and "graphy", also derived from Greek, meaning a technique for producing images. According to Herman [47], "tomography" is "image reconstruction from projections", and is a technology developed independently in various fields. The imaging of a cross section of the human body by X-ray projection, developed in the field of diagnostic medicine, has had a great impact on the world. This imaging technique is called "computerized tomography", or **CT** for short. G. N. Hounsfield, who developed CT equipment, and A. M. Cormack, who proposed the theory of CT analysis, won the Nobel prize for physiology or medicine in 1979. In CT, the X-ray sources are arranged on the circumference of a circle orthogonal to the human body, and the cross section of the human body intersected by the circle is the object to be analyzed. Although seismic waves are used in seismic tomography and natural earthquakes are their sources, it is impossible to find a sufficient number of natural earthquakes distributed in a similar manner as in the X-ray sources in CT. Hence, in seismic tomography, velocity structures, which are three-dimensional parts of the Earth, are targeted, and earthquakes sparsely distributed within them are used. In addition, the method of A. M. Cormack is not used; instead, the least-squares method is applied after the discretization of a model, as in other analyses in Chaps. 2 and 3.

The **3-D velocity structure** in Fig. 3.37 is represented by a set of hexahedron blocks along the Cartesian coordinate system. A block is numbered as the kth block, in which the constant velocity is v_k and the slowness is $s_k = 1/v_k$. Assume that the **hypocenter** S of an earthquake and the observation point O exist in this model. The **hypocenter determination** (Sect. 2.3.1) is performed using arrival time t at O for the earthquake, and the position (x_S, y_S, z_S) and the **origin time** t_S for S are obtained. The ray tracing (Sect. 3.2.3) is performed for a seismic wave propagating from S to O, and the position of its ray and theoretical travel time T are obtained as shown in Fig. 3.37. The ray is divided into segments at the boundaries of the blocks, and when a segment is in the kth block, the length of the segment is assumed to be l_k.

The above configuration is expanded to a case with many observation points $i = 1, 2, \cdots, N$ and many earthquakes $j = 1, 2, \cdots, M$. Let t_{ij} be the arrival time observed for the combination of the ith observation point and the jth earthquake, and the ray (i, j) connecting this pair has a theoretical travel time T_{ij}. The length of the segment in the kth block of the ray (i, j) is calculated to be l_{ijk}, and the length in a block through which the ray does not pass (white blocks in Fig. 3.37) is set to $l_{ijk} = 0$. Thus, the **observation equation** (Sect. 4.4) can be expressed as

$$t_{ij} \simeq T_{ij}(x_{Sj}, y_{Sj}, z_{Sj}, s_k) + t_{Sj}\,, \quad k = 1, 2, \cdots, K \tag{3.438}$$

so that the variables $s_1, s_2, \cdots, s_K$ are added to (2.136) for the hypocenter determination and x_S, y_S, z_S are included in T_{ij} in a nonlinear form. Therefore, an observation equation similar to (2.137) can be obtained using the estimates $x_S^{(0)}$, $y_S^{(0)}$, $z_S^{(0)}$, $t_S^{(0)}$, and $s_k^{(0)}$ $(k = 1, 2, \cdots, K)$ as[106]

$$\begin{aligned}
&\Delta t_{ij} \simeq \frac{\partial T_{ij}^{(0)}}{\partial x_S}\Delta x_{Sj} + \frac{\partial T_{ij}^{(0)}}{\partial y_S}\Delta y_{Sj} + \frac{\partial T_{ij}^{(0)}}{\partial z_S}\Delta z_{Sj} + \Delta t_{Sj} + \sum_k l_{ijk}\Delta s_k\,,\\
&\Delta t_{ij} = t_{ij} - T_{ij}^{(0)} - t_{Sj}^{(0)},\ \Delta x_{Sj} = x_{Sj} - x_{Sj}^{(0)},\ \Delta y_{Sj} = y_{Sj} - y_{Sj}^{(0)},\ \Delta z_{Sj} = z_{Sj} - z_{Sj}^{(0)},\\
&\Delta t_{Sj} = t_{Sj} - t_{Sj}^{(0)},\ \Delta s_k = s_k - s_k^{(0)}\ (k = 1, 2, \cdots, K)\,.
\end{aligned} \tag{3.439}$$

The **least-squares method** in the seismic tomography is then formulated as

$$\begin{aligned}
&S = \sum_{i=1}^{N}\sum_{j=1}^{M}\frac{1}{\sigma_{ij}^2}\left(\Delta t_{ij} - \frac{\partial T_{ij}^{(0)}}{\partial x_{Sj}}\Delta x_{Sj} - \frac{\partial T_{ij}^{(0)}}{\partial y_{Sj}}\Delta y_{Sj} - \frac{\partial T_{ij}^{(0)}}{\partial z_{Sj}}\Delta z_{Sj} - \Delta t_{Sj} - \sum_k l_{ijk}\Delta s_k\right)^2,\\
&\frac{\partial S}{\partial x_{Sj}} = 0,\ \frac{\partial S}{\partial y_{Sj}} = 0,\ \frac{\partial S}{\partial z_{Sj}} = 0,\ \frac{\partial S}{\partial t_{Sj}} = 0\ (j = 1, 2, \cdots, M),\\
&\frac{\partial S}{\partial s_k} = 0\ (k = 1, 2, \cdots, K)\,.
\end{aligned} \tag{3.440}$$

Using the definitions

$$\begin{aligned}
\Delta\mathbf{x} &= (\Delta x_{S1}, \Delta y_{S1}, \Delta z_{S1}, \Delta t_{S1}, \cdots, \Delta s_1, \Delta s_2, \cdots, \Delta s_K)^T,\\
\Delta\mathbf{y} &= (\Delta t_{11}, \Delta t_{12}, \cdots\cdots\cdots, \Delta t_{MN})^T,\\
\mathbf{A} &= \begin{pmatrix}
\frac{\partial T_{11}}{\partial x_S} & \frac{\partial T_{11}}{\partial y_S} & \frac{\partial T_{11}}{\partial z_S} & 1 \cdots & l_{111} & l_{112} & \cdots & l_{11K}\\
0 & 0 & 0 & 0 \cdots & l_{121} & l_{122} & \cdots & l_{12K}\\
\vdots & \vdots & \vdots & \vdots & \vdots & \vdots & & \vdots\\
0 & 0 & 0 & 0 \cdots & l_{MN1} & l_{MN2} & \cdots & l_{MNK}
\end{pmatrix},
\end{aligned} \tag{3.441}$$

the observation equation in (3.439) yields

$$\Delta\mathbf{y} \simeq \mathbf{A}\,\Delta\mathbf{x}\,. \tag{3.442}$$

As this agrees with (4.80), the seismic tomography is accomplished by solving the simultaneous equations along the way explained in Sect. 4.4.1. Since x_S, y_S, z_S are included in T_{ij} in the nonlinear form, **iterative refinement** (Sect. 4.4.1) should be performed including the ray tracing, but this is not often done.

106This agrees with Eq. (1) of Aki and Lee [2] if $T_{ij}^{(k)} \to l_{ijk}s_k$ and $F_k \to \Delta s_k/s_k$.

For $\partial T_{ij}/\partial x_S$, $\partial T_{ij}/\partial y_S$, and $\partial T_{ij}/\partial z_S$ in **A**, (2.156) in the hypocenter determination can be used if the initial model is a 1-D velocity structure. It is also possible to remove t_S from the unknowns by applying the **centering** in the hypocenter determination. The scale of simultaneous equations in (3.441) is notably larger than that in (2.139) for the hypocenter determination. As the number of earthquakes increases to M, the number of equations increases to M times N, and the number of unknowns about the sources also increases to M times 4. However, it is the number of unknowns in the velocity structure model, Δs_k, that has the largest effect on the size of the system of simultaneous equations. For a realistic velocity structure model, a very large K is required and $K > MN$ may occur. For such a large **underdetermined system** of simultaneous equations, it is necessary to introduce constraints into the formulation of the least-squares method (Sect. 4.4.2), as well as more contrivances into the computation of the least-squares method (Sect. 4.4.1) to efficiently obtain a stable solution.

The computational approaches described in Sect. 4.4.1 fall into the class of **direct methods**, because they directly obtain solutions through a series of operations. In addition, there is another class of approaches called **iterative methods**, where a series of operations for an approximate solution is repeated until the best solution is obtained. The iterative methods are effective for large-scale simultaneous equations, especially **sparse** problems with many zero elements in the coefficient matrix, such as seismic tomography [82]. In the field of CT, an iterative method called **ART** (Algebraic Reconstruction Techniques) is often used, and Hirahara [48] extends this to the seismic tomography. ART is an iterative method for the observation equation, but there also exist iterative methods for the normal equation, as the direct methods. Among them, the **conjugate gradient method** is widely used (e.g., Zhao et al. [129]), because the LSQR algorithm for the least-squares method was available from early on in various forms, such as a Fortran subroutine.

We here explain the ART for the constrained least-squares method as follows. We have the simultaneous linear equations $\mathbf{Bx} = \mathbf{d}$, where the coefficient matrix $\mathbf{B}$ is an I by J matrix and the transpose of the ith row of $\mathbf{B}$ is represented by the vector $\mathbf{b}_i$. According to Herman et al. [46], an advanced algorithm was proposed by S. Kaczmarz in 1937 as

$$\mathbf{x}^{(n+1)} = \mathbf{x}^{(n)} + \frac{d_i - \mathbf{b}_i^{\mathrm{T}}\mathbf{x}^{(n)}}{\|\mathbf{b}_i\|^2}\mathbf{b}_i \ , \quad i = (n \bmod I) + 1 \ , \tag{3.443}$$

where $n = 0, 1, 2, \cdots$. From an initial estimate $\mathbf{x}^{(0)}$, the vectors $\mathbf{x}^{(0)}, \mathbf{x}^{(1)}, \mathbf{x}^{(2)}, \cdots$ converge to solutions of $\mathbf{Bx} = \mathbf{d}$. $\| \ \|$ represents the L2 norm of a vector, so that $\| \mathbf{z} \| = \left(z_1^2 + z_2^2 + \cdots\right)^{1/2}$. $\left\| \mathbf{x} - \mathbf{x}^{(n)} \right\| < \mu^n \left\| \mathbf{x} - \mathbf{x}^{(0)} \right\|$ and $0 \le \mu < 1$ are also proven. Therefore, the solutions are considered to be the least-squares solutions.

We add a constraint $\mathbf{x} = \mathbf{x}_0$ such that the solutions are close to the estimates $\mathbf{x}_0$. Herman et al. [46] also showed a method to obtain the least-squares solutions satisfying both the constraints and the simultaneous linear equations using the Kaczmarz algorithm. The algorithm for the constrained least-squares method is given as

$$\mathbf{x}^{(0)} = \mathbf{x}_0\,, \quad \mathbf{u}^{(0)} = \mathbf{0}\,,$$
$$\mathbf{x}^{(n+1)} = \mathbf{x}^{(n)} + r^2 \mathbf{c}^{(n)} \mathbf{b}_i\,, \quad \mathbf{u}^{(n+1)} = \mathbf{u}^{(n)} + \mathbf{c}^{(n)} \mathbf{e}_i\,,$$
$$\mathbf{c}^{(n)} = \frac{d_i - \mathbf{b}_i^{\mathrm{T}} \mathbf{x}^{(n)} - \mathbf{u}_i^{(n)}}{1 + r^2 \|\mathbf{b}_i\|^2}\,, \quad i = (n \bmod I) + 1\,, \tag{3.444}$$

where r is the relative weight of the simultaneous linear equations to the constraints and $\mathbf{u}$ is an auxiliary variable. Similar to $\mathbf{b}_i$ in $\mathbf{B}$, the vector $\mathbf{e}_i$ represents the transpose of the ith row of the I by I unit matrix $\mathbf{I}$ (due to the unit matrix, $\mathbf{e}_i$ is the same as the ith column of $\mathbf{I}$).

In seismic tomography, the observation equations are given in (3.442), and the constraints are similar to the above, that the solutions $\Delta\mathbf{x}$ are close to the estimates $\Delta\mathbf{x}_0$. The constrained least-squares method is represented by S_c in (4.97), where the first term S_{c1} is related to the observation equations and the second term S_{c2} is related to the constraints. S_{c1} indicates that the reciprocals of the observational errors σ_i $(i = 1, 2, \cdots, MN)$ are given as weights to the observation equations. When all σ_i are equal to σ_d, the **variance matrix** (Sect. 4.4.1), the diagonal elements of which are σ_i^2, yields $\mathbf{\Sigma}_d = \sigma_d^2 \mathbf{I}$. Then, S_{c2} indicates that the reciprocals of the standard deviations σ_{mj} $(j = 1, 2, \cdots, J)$ of the unknowns are given as weights to the constraints. In (4.97), an application to source inversion is assumed, and the difference of the standard deviation between the unknowns in the source inversion is small so that all σ_{mj} are equal to ρ. However, in seismic tomography, the difference of the standard deviation is large, so that σ_{mj} $(j = 1, 2, \cdots, J)$ remain as they are, and the **variance matrix** is

$$\mathbf{\Sigma}_m = \begin{pmatrix} \sigma_{m1}^2 & & & \\ & \sigma_{m2}^2 & & \\ & & \ddots & \\ & & & \sigma_{mJ}^2 \end{pmatrix}. \tag{3.445}$$

If we use these, set

$$\mathbf{B} = \mathbf{\Sigma}_d^{-1/2} \mathbf{A}\,, \quad \mathbf{d} = \mathbf{\Sigma}_d^{-1/2} \Delta\mathbf{y} \tag{3.446}$$

as in (4.83), and replace $\Delta\mathbf{x}$ with $\mathbf{x}$ and $\Delta\mathbf{x}_0$ with $\mathbf{x}_0$, the observation equations and their constraints are exactly the same as the simultaneous linear equations $\mathbf{Bx} = \mathbf{d}$ and their constraints. Therefore, the least square solutions can be obtained by the algorithm in (3.444). Assuming that the transpose of the ith row of $\mathbf{A}$ is a vector $\mathbf{a}_i$, $\mathbf{b}_i = \mathbf{a}_i / \sigma_d$ is obtained from (3.446). The relative weight r of the simultaneous linear equations to the constraints should be a matrix $\mathbf{r}$, where $\mathbf{r} = \mathbf{\Sigma}_d^{-1/2} / \mathbf{\Sigma}_m^{-1/2}$. However, $\mathbf{r} = \mathbf{\Sigma}_m^{1/2}$ because $\mathbf{\Sigma}_d$ is already incorporated into the observation equations according to (3.446). Substituting these for (3.444), we obtain[107]

[107] These agree with Eq. (8) of Hirahara [48] if $k \to n$ and $G_{ij} \to A_{ij}$.

$$
\begin{aligned}
&\Delta\mathbf{x}^{(0)} = \Delta\mathbf{x}_0 \,, \quad \mathbf{u}^{(0)} = \mathbf{0} \,, \\
&\Delta\mathbf{x}^{(n+1)} = \Delta\mathbf{x}^{(n)} + \mathbf{c}^{(n)}\boldsymbol{\Sigma}_m \cdot \mathbf{a}_i/\sigma_d \,, \quad \mathbf{u}^{(n+1)} = \mathbf{u}^{(n)} + \mathbf{c}^{(n)}\mathbf{e}_i \,, \\
&\mathbf{c}^{(n)} = \frac{(\Delta y_i - \mathbf{a}_i^{\mathrm{T}}\Delta\mathbf{x}^{(n)})/\sigma_d - \mathbf{u}_i^{(n)}}{1 + \left\|\boldsymbol{\Sigma}_m^{1/2} \cdot \mathbf{a}_i/\sigma_d\right\|^2} \,, \quad i = (n \bmod I) + 1 \,, \\
&\frac{\boldsymbol{\Sigma}_m \cdot \mathbf{a}_i}{\sigma_d} = \frac{1}{\sigma_d}\sum_{j=1}^{J} \sigma_{mj}^2 A_{ij} \,, \quad \frac{\boldsymbol{\Sigma}_m^{1/2} \cdot \mathbf{a}_i}{\sigma_d} = \frac{1}{\sigma_d}\sum_{j=1}^{J} \sigma_{mj} A_{ij} \,.
\end{aligned}
\tag{3.447}
$$

In this book, as in typical seismic tomography, the case of a block model and travel time data has been introduced. However, seismic tomography that models interfaces in an **irregularly layered structure** has also been performed. Zhao et al. [129] also show such results for the Moho discontinuity[108] and the **Conrad discontinuity**.[109] The upper surface of the **seismic basement**, which is attributed to the uppermost part of the crust, under a sedimentary basin is analyzed by e.g., Koketsu and Higashi [74].

The travel times of seismic waves are most often used as the data for seismic tomography, but waveforms have also been used since they have different sensitivity to velocity structure. Seismic tomography with waveforms is termed **full waveform inversion** [120]. This approach exploits a comparison of a **synthetic seismogram** $u_k(\mathbf{x}_j, t)$ at the jth observation point in the kth component computed assuming a velocity structure model with the corresponding observation. The **forward problem** (Sect. 2.3.4) requires extensive numerical calculations, using either the finite difference method (Sect. 3.2.4) or the finite element method (Sect. 3.2.5). In the **inverse problem** (Sect. 2.3.4), the parameters of the velocity structure model such as density ρ and elastic constants C_{ijkl} are determined from **observed seismogram** $u_k^o(\mathbf{x}_j, t)$ using sampled times t_i, (2.175), and the nonlinear least-squares method (Sect. 4.4.1), so that $S = \sum_i \sum_j \sum_k \left\{u_k^o(\mathbf{x}_j, t_i) - u_k(\mathbf{x}_j, t_i)\right\}^2$ is minimized (here for simplicity, all weights are set to 1). However, unlike travel time tomography the Jacobian matrix (Sect. 4.4.1) cannot be obtained analytically from results of the forward problem. This means that in the worst case the numerical computation needs to be repeated with a variation of each parameter.

The **adjoint method**, which has been used in meteorology etc., is utilized in order to avoid this difficulty. Since the summation $\sum_i$ in S multiplied by the sampling interval Δt approximates an integral with respect to t, we can use

[108]The Moho discontinuity is the boundary between the **crust** and the **mantle**, and is named after A. Mohorovičić who discovered it in 1909.

[109]The Conrad discontinuity is the boundary between the upper and lower **continental crust**, which was discovered by V. Conrad in the 1920s.

$$\chi = \frac{1}{2}\sum_j \int_0^T \left\|\mathbf{u}^o(\mathbf{x}_j, t) - \mathbf{u}(\mathbf{x}_j, t)\right\|^2 dt = \frac{1}{2}\sum_j \sum_k \int_0^T \left\|u_k^o(\mathbf{x}_j, t) - u_k(\mathbf{x}_j, t)\right\|^2 dt$$
$$\simeq \frac{1}{2}\sum_i \sum_j \sum_k \left\{u_k^o(\mathbf{x}_j, t_i) - u_k(\mathbf{x}_j, t_i)\right\}^2 \Delta t = \frac{1}{2} S \Delta t \quad (3.448)$$

rather than S in the inverse problem. The synthetic seismogram $\mathbf{u} = (u_k)$ must satisfy the equation of motion (1.101) and the initial condition $\mathbf{u} = \dot{\mathbf{u}} = 0$. Since the method of Lagrangian multipliers works for obtaining the extremum of a function under strong constraints, we adopt this method to obtain the minimum of χ under the constraint of satisfying the equation of motion and the initial condition. We use a vector $\boldsymbol{\lambda}$ of Lagrangian multipliers. Then, (1.101) is rewritten with a fixed source term $\mathbf{F} = (F_i)$ as

$$\rho\ddot{\mathbf{u}} - \boldsymbol{M}(\mathbf{u}) = 0\,, \quad \boldsymbol{M}(\mathbf{u}) = (M_i)\,, \; M_i = \frac{\partial}{\partial x_j}\left(C_{ijkl}\frac{\partial u_k}{\partial x_l}\right) + F_i\,, \quad (3.449)$$

and integrated over time as for χ. We use the integrated motion term as a constraint and define

$$\mathcal{L} = \chi + \int_0^T \langle \boldsymbol{\lambda}, \rho\ddot{\mathbf{u}} - \boldsymbol{M}(\mathbf{u})\rangle\, dt\,, \quad (3.450)$$

where $\langle \mathbf{a}, \mathbf{b}\rangle = \iiint \mathbf{a}^* \cdot \mathbf{b}\, dV$ indicates the inner product of $\mathbf{a}$ and $\mathbf{b}$ over the whole velocity structure V. Using the method of Lagrangian multipliers, the variation of $\mathcal{L}$, $\delta\mathcal{L}$ will be zero around $\hat{\mathbf{u}}$, where χ is minimized under the constraint.

We first consider the case where the ground motion $\mathbf{u}$ is perturbed but the velocity structure is stationary, as in common applications of the adjoint method [85]. The variation of the first term on the right-hand side of (3.450) yields

$$\delta\chi = \delta\int_0^T X\, dt = \int_0^T \langle \nabla_{\mathbf{u}} X, \delta\mathbf{u}\rangle\, dt\,, \quad X = \frac{1}{2}\sum_j \left\|\mathbf{u}^o(\mathbf{x}_j, t) - \mathbf{u}(\mathbf{x}_j, t)\right\|^2, \quad (3.451)$$

and the variation of the second term is obtained as

$$\delta\int_0^T \langle \boldsymbol{\lambda}, \rho\ddot{\mathbf{u}} - \boldsymbol{M}(\mathbf{u})\rangle\, dt = \int_0^T \langle \delta\boldsymbol{\lambda}, \rho\ddot{\mathbf{u}} - \boldsymbol{M}(\mathbf{u})\rangle\, dt$$
$$+ \int_0^T \langle \boldsymbol{\lambda}, \rho\frac{\partial^2 \delta\mathbf{u}}{\partial t^2}\rangle\, dt - \int_0^T \langle \boldsymbol{\lambda}, \mathbf{M}\delta\mathbf{u}\rangle\, dt \quad (3.452)$$

using $\delta\boldsymbol{M}(\mathbf{u}) \sim \mathbf{M}\delta\mathbf{u}$, where $\mathbf{M}$ is the linearization of $\boldsymbol{M}$ around $\hat{\mathbf{u}}$. We then perform integration by parts twice on the second term on the right-hand side of (3.452) obtaining

$$\int_0^T \langle \boldsymbol{\lambda}, \rho \frac{\partial^2 \delta \mathbf{u}}{\partial t^2} \rangle \, dt = \langle \boldsymbol{\lambda}, \rho \frac{\partial \delta \mathbf{u}}{\partial t} \rangle \Big|_0^T - \int_0^T \langle \frac{\partial \boldsymbol{\lambda}}{\partial t}, \rho \frac{\partial \delta \mathbf{u}}{\partial t} \rangle \, dt$$
$$= \langle \boldsymbol{\lambda}, \rho \frac{\partial \delta \mathbf{u}}{\partial t} \rangle \Big|_0^T - \langle \frac{\partial \boldsymbol{\lambda}}{\partial t}, \rho \delta \mathbf{u} \rangle \Big|_0^T + \int_0^T \langle \frac{\partial^2 \boldsymbol{\lambda}}{\partial t^2}, \rho \delta \mathbf{u} \rangle dt \,. \tag{3.453}$$

The third term on on the right-hand side of (3.452) yields

$$-\int_0^T \langle \boldsymbol{\lambda}, \mathbf{M} \delta \mathbf{u} \rangle \, dt = -\int_0^T \langle \mathbf{M}^* \boldsymbol{\lambda}, \delta \mathbf{u} \rangle \, dt = -\int_0^T \langle \mathbf{M} \boldsymbol{\lambda}, \delta \mathbf{u} \rangle \, dt \tag{3.454}$$

using the definition of the inner product and $M^* = M$. Summarizing (3.450) – (3.454), we have

$$\delta \mathcal{L} = \int_0^T \langle \rho \ddot{\boldsymbol{\lambda}} - \mathbf{M} \boldsymbol{\lambda} + \nabla_{\mathbf{u}} X, \delta \mathbf{u} \rangle \, dt + \int_0^T \langle \delta \boldsymbol{\lambda}, \rho \ddot{\mathbf{u}} - \boldsymbol{M}(\mathbf{u}) \rangle \, dt \tag{3.455}$$
$$+ \rho \langle \lambda(T), \delta \dot{\mathbf{u}}(T) \rangle - \rho \langle \lambda(0), \delta \dot{\mathbf{u}}(0) \rangle - \rho \langle \dot{\lambda}(T), \delta \mathbf{u}(T) \rangle + \rho \langle \dot{\lambda}(0), \delta \mathbf{u}(0) \rangle \,.$$

As $\delta \mathcal{L} = 0$ for any $\delta \mathbf{u}$ and $\delta \boldsymbol{\lambda}$ around $\hat{\mathbf{u}}$,

$$\rho \ddot{\mathbf{u}} = \boldsymbol{M}(\mathbf{u}) = \mathbf{M}\mathbf{u} + \mathbf{F} \,, \tag{3.456}$$
$$\rho \ddot{\boldsymbol{\lambda}} = \mathbf{M} \boldsymbol{\lambda} - \nabla_{\mathbf{u}} X \,, \; \boldsymbol{\lambda}(T) = 0 \,, \; \dot{\boldsymbol{\lambda}}(T) = 0 \,. \tag{3.457}$$

Equation (3.456) is the equation of motion (3.449) itself. Equation (3.457), which is termed the **adjoint equation**, is also in the form of the equation of motion for $\boldsymbol{\lambda}$, but the source term is replaced by $-\nabla_{\mathbf{u}} X$ from $\mathbf{F}$. This adjoint source term is[110]

$$-\nabla_{\mathbf{u}} X = -\nabla_{\mathbf{u}} \frac{1}{2} \sum_j \left\| \mathbf{u}^o(\mathbf{x}_j, t) - \mathbf{u}(\mathbf{x}, t) \right\|^2 = \sum_j \left\{ \mathbf{u}^o(\mathbf{x}_j, t) - \mathbf{u}(\mathbf{x}_j, t) \right\} \delta(\mathbf{x} - \mathbf{x}_j) \tag{3.458}$$

from (3.451) and $\nabla_{\mathbf{u}} \mathbf{u}(\mathbf{x}_j, t) = \delta(\mathbf{x} - \mathbf{x}_j)$. From (3.458) we see that the difference between synthetic and observed seismograms plays a role of a source at $\mathbf{x}_j$. $\boldsymbol{\lambda}(T) = 0$ and $\dot{\boldsymbol{\lambda}}(T) = 0$ in (3.457) are the initial conditions for the adjoint equation. $\mathbf{u}(\mathbf{x}, t)$ is obtained from forward simulation based on the equation of motion (3.456), and $\boldsymbol{\lambda}(\mathbf{x}, T - t)$ is obtained by reverse simulation using the adjoint equation (3.457).

We next consider the case where the ground motion $\mathbf{u}$ is stationary but the velocity structure is perturbed, as the second step of the full waveform inversion [85]. We again take the variation of $\mathcal{L}$ perturbing only ρ as

$$\delta \mathcal{L} = \delta \chi + \delta \rho \int_0^T \langle \boldsymbol{\lambda}, \ddot{\mathbf{u}} \rangle \, dt = 0 \,. \tag{3.459}$$

[110]This agrees with Eq. (9) of Tromp et al. [114] if $(s_i) \to \mathbf{u}$ and $(d_i) \to \mathbf{u}^o$. However, $\mathbf{u}$ and $\mathbf{u}^o$ are swapped as the Lagrangian multiplier is changed from $-\lambda$ to λ.

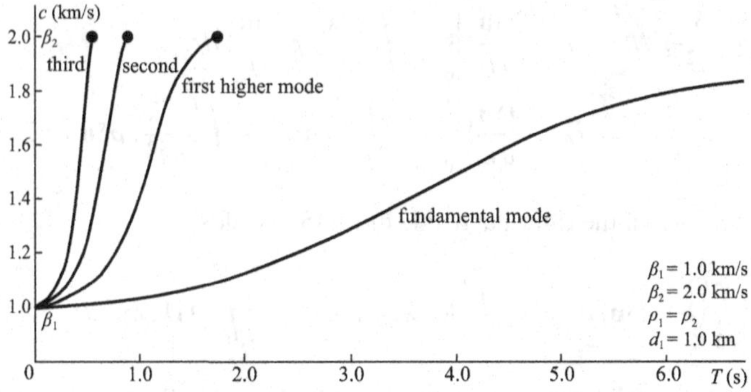

Fig. 3.38 Dispersion curves with respect to the period T in the two-layer structure (same as Fig. 3.14, reprinted from Koketsu [72] with permission of Kindai Kagaku)

We accordingly obtain the partial derivative of χ with respect to ρ[111]

$$\frac{\partial \chi}{\partial \rho} \sim \frac{\delta \chi}{\delta \rho} = -\int_0^T \langle \boldsymbol{\lambda}, \ddot{\mathbf{u}} \rangle \, dt = -\int_0^T \langle \boldsymbol{\lambda}, \frac{\partial^2 \mathbf{u}}{\partial t^2} \rangle \, dt \, . \tag{3.460}$$

The partial derivatives of χ with respect to C_{ijkl} can be obtained in a similar way. Since χ represents S as shown in (3.448), we can construct a matrix related to the Jacobian matrix for the nonlinear least-squares method, and formulate the full waveform inversion using only two numerical simulations. These are the forward simulation based on the equation of motion and the reverse simulation based the adjoint equation using the residual between observed and synthetic seismograms as a source.

Problems

3.1 Write a computer program in the C or FORTRAN language, to find the phase velocities of the Love wave in the two-layer structure shown in the lower right of Fig. 3.38 using the equation in (3.138), and the bracketing and the bisection method in Sect. 3.2.3. The program has to include the parts for input of periods and output of phase velocities. Compile and execute the program to obtain the phase velocities at periods of 1, 2, and 3 s.

3.2 Search for the genetic algorithm and the simulated annealing (Sect. 3.3.2) in the literature, and explain what they are and how to use them in a least-squares problem.

---

[111]This agrees with $-K_\rho$ in Eq. (21) of Liu and Tromp [85] if $\mathbf{s} \rightarrow \mathbf{u}$ and $\mathbf{s}^\dagger \rightarrow \boldsymbol{\lambda}$. However, the sign is switched as the Lagrangian multiplier is changed from $-\lambda$ to λ.

Based on your explanation, write a computer program in the C or FORTRAN language, to determine the S wave velocity in the first layer of the two-layer structure in Problem 3.1 using the three phase velocities obtained in Problem 3.1.

References

1. Aki, K., & Larner, K. L. (1970). Surface motion of a layered medium having an irregular interface due to incident plane SH waves. *Journal of Geophysical Research, 75*, 933–954.
2. Aki, K., & Lee, W. H. K. (1976). Determination of three-dimensional velocity anomalies under a seismic array using first P arrival times from local earthquakes. 1. A homogeneous initial model. *Journal of Geophysical Research, 81*, 4381–4399.
3. Aki, K., & Richards, P. G. (2002). *Quantitative seismology* (2nd ed., p. 700). Sausalito: University Science Books.
4. Aki, K., & Richards, P. G. (1980). *Quantitative seismology* (Vol. II, pp. 559–932). San Francisco: W. H. Freeman and Company.
5. Arfken, G. B., & Weber, H. J. (1995). *Mathematical methods for physicists* (4th ed., p. 1029). San Diego: Academic Press.
6. Auld, B. A. (1973). *Acoustic fields and waves in solids I* (p. 423). New Jersey: Wiley.
7. Bard, P. Y., & Bouchon, M. (1980). The seismic response of sediment-filled valleys. Part 1. The case of incident SH waves. *Bulletin of the Seismological Society of America, 70*, 1263–1286.
8. Bathe, K.-J. (1996). *Finite element procedures* (p. 1037). Upper Saddle River: Prentice-Hall.
9. Baumgardt, D. R. (1980). Errors in matrix element expressions for the reflectivity method. *Journal of Geophysics, 48*, 124–125.
10. Ben-Menahem, A., & Singh, S. J. (1981). Appendices A, D, and G. *Seismic waves and sources* (pp. 967–983). Berlin: Springer.
11. Boore, D. M. (1972). Finite difference methods for seismic wave propagation in heterogeneous materials. *Seismology: Surface waves and earth oscillations* (pp. 1–37). New York: Academic Press.
12. Bouchon, M., & Aki, K. (1977). Discrete wave-number representation of seismic-source wave fields. *Bulletin of the Seismological Society of America, 67*, 259–277.
13. Bouchon, M. (1981). A simple method to calculate Green's functions for elastic layered media. *Bulletin of the Seismological Society of America, 71*, 959–971.
14. Brocker, T., Aaggard, B., Simpson, R., & Jachens, R. (2006). The new USGS 3D seismic velocity model for Northern California (abstract). *Seismological Research Letters, 77*, 271.
15. Cagniard, L. (1939). *Réflexion et réfraction des ondes seismiques progressives* (p. 255). Paris: Gauthier-Villars.
16. Capon, J. (1969). High-resolution frequency wavenumber spectrum analysis. *Proceedings of the IEEE, 57*, 1408–1418.
17. Červený, V. (1974). Reflection and transmission coefficients for transition layers, Studia Geophys. *Geodaet, 18*, 59–68.
18. Červený, V. (1983). Synthetic body wave seismograms for laterally varying layered structures by the Gaussian beam method. *Geophysical Journal of the Royal Astronomical Society, 73*, 389–426.
19. Červený, V. (2001). *Seismic ray theory* (p. 713). Cambridge: Cambridge University Press.
20. Červený, V., Molotkov, I. A., & Pšeník, I. (1977). *Ray method in seismology* (p. 214). Prague: Univerzita Karlova.
21. Červený, V., & Ravindra, R. (1971). *Theory of seismic head waves* (p. 312). Tronto: University of Tronto Press.
22. Chapman, C. H. (1978). A new method for computing synthetic seismograms. *Geophysical Journal of the Royal Astronomical Society, 58*, 481–518.

23. Chapman, C. H. (1982). Body-wave seismograms in inhomogeneous media using Maslov asymptotic theory. *Bulletin of the Seismological Society of America, 72*, S277–S317.
24. Choy, G. L., Cormier, V. F., Kind, R., Müller, G., & Richards, P. G. (1980). A comparison of synthetic seismograms of core phases generated by the full wave theory and by the reflectivity method. *Geophysical Journal of the Royal Astronomical Society, 61*, 21–39.
25. Cormier, V. F., & Richards, P. G. (1977). Full wave theory applied to a discontinuous velocity increase: The inner core boundary. *Journal of Geophysics, 43*, 3–31.
26. Courant, R., Friedrichs, K., & Lewy, H. (1928). Über die partiellen Differenzengleichungen der mathematischen Physik. *Mathematische Annalen, 100*, 32–74.
27. Dahlen, F. A., & Tromp, J. (1998). *Theoretical global seismology* (p. 1025). Princeton: Princeton University Press.
28. de Hoop, A. T. (1960). A modification of Cagniard's method for solving seismic pulse problems. *Applied Scientific Research Sect. B, 8*, 349–356.
29. Diao, H., Kobayashi, H., & Koketsu, K. (2018). Rupture process of the 2016 Meinong, Taiwan, earthquake and its effects on strong ground motions. *Bulletin of the Seismological Society of America, 108*, 163–174.
30. Doornbos, D. J. (ed.). (1988). *Seismological algorithms* (p. 469). New York: Academic Press.
31. Dunkin, J. W. (1965). Computation of modal solutions in layered, elastic media at high frequencies. *Bulletin of the Seismological Society of America, 55*, 335–358.
32. Finlayson, B. A. (1972). *The method of weighted residuals and variational principles* (p. 412). New York: Academic Press.
33. Fletcher, C. A. J. (1984). *Computational Galerkin Methods* (p. 310). Berlin: Springer.
34. Fuchs, K. (1968). Das Reflexions-und Transmissionsvermögen eines geschichteten Mediums mit beliebiger Tiefenverteilung der elastischen Moduln und der Dichte für schrägen Einfall ebener Wellen. *Zeitschrift für Geoplzysik, 34*, 389–413.
35. Fuchs, K. (1968). The reflection of spherical waves from transition zones with arbitrary depth-dependent elastic moduli and density. *Journal of Physics of the Earth, 16*, 27–41.
36. Fuchs, K., & Müller, G. (1971). Computation of synthetic seismograms with the reflectivity method and comparison with observations. *Geophysical Journal of the Royal Astronomical Society, 23*, 417–433.
37. Furumura, T., Kennett, B. L. N., & Takenaka, H. (1998). Parallel 3-D pseudospectral simulation of seismic wave propagation. *Geophysics, 63*, 279–288.
38. Furumura, T., Koketsu, K., & Takenaka, H. (2000). A hybrid PSM/FDM parallel simulation for large-scale 3-D seismic (acoustic) wavefield. *Geophysics Exploration, 53*, 294–308. [J]
39. Gilbert, F., & Backus, G. E. (1966). Propagator matrices in elastic wave and vibration problems. *Geophysics, 31*, 326–332.
40. Graves, R. W. (1996). Simulating seismic wave propagation in 3D elastic media using staggered-grid finite differences. *Bulletin of the Seismological Society of America, 86*, 1091–1106.
41. Haddon, R. A. W., & Buchen, P. W. (1981). Use of Kirchhoff's formula for body wave calculations in the Earth. *Geophysical Journal of the Royal Astronomical Society, 67*, 587–598.
42. Harkrider, D. G. (1964). Surface waves in multilayered elastic media. I. Rayleigh and Love waves from buried sources in a multilayered elastic half-space. *Bulletin of the Seismological Society of America, 54*, 627–679.
43. Harvey, D. J. (1981). Seismogram synthesis using normal mode superposition: the locked mode approximation. *Geophysical Journal of the Royal Astronomical Society, 66*, 37–69.
44. Haskell, N. A. (1953). The dispersion of surface waves in multilayered media. *Bulletin of the Seismological Society of America, 43*, 17–34.
45. Helmberger, D. V. (1968). The crust-mantle transition in the Bering Sea. *Bulletin of the Seismological Society of America, 58*, 179–214.
46. Herman, G. T., Hurwitz, H., Lent, A., & Lung, H.-P. (1979). On the Bayesian approach to image reconstruction. *Information and Control, 42*, 60–71.

47. Herman, G. T. (1980). *Image reconstruction from projections* (p. 316). New York: Academic Press.
48. Hirahara, K. (1988). Detection of three-dimensional velocity anisotropy. *Physics of the Earth and Planetary Interiors, 51*, 71–85.
49. Hisada, Y. (1995). An efficient method for computing Green's functions for a layered half-space with sources and receivers at close depths (Part 2). *Bulletin of the Seismological Society of America, 85*, 1080–1093.
50. Horike, M. (1985). Inversion of phase velocity of long-period microtremors to the S-wave-velocity structure down to the basement in urbanized areas. *Journal of Physics of the Earth, 33*, 59–96.
51. Horike, M. (1993). Studies on microtremors. *Zisin (Journal of Seismological Society of Japan), 46*, 343–350. [J]
52. Horike, M., Uebayashi, H., & Takeuchi, Y. (1990). Seismic response in three-dimensional sedimentary basin due to plane S wave incidence. *Journal of Physics of the Earth, 38*, 261–284.
53. Howard, J. N. (1964). John William Strutt, third Baron Rayleigh. *Applied Optics, 3*, 1091–1101.
54. Ikegami, Y. (2009). *Ground motion simulation with voxel FEM including the effects of broadband attenuation, topography, and oceans*, Ph.D. thesis, University of Tokyo, 130pp. [J]
55. Inui, T. (1957). *Partial differential equations and their applications* (p. 366). Tokyo: Corona Publishing. [J]
56. Jacob, K. H. (1970). Three-dimensional seismic ray tracing in a laterally heterogeneous spherical earth. *Journal of Geophysical Research, 75*, 6675–6689.
57. Jeffreys, H. (1925). On certain approximate solutions of linear differential equations of the second order. *Proceedings of the London Mathematical Society, s2-23*, 428–436.
58. Jeffreys, H., & Bullen, K. E. (1948). *Seismological tables* (p. 50). London: Office of the British Association.
59. Kanai, K. (1969). *Engineering seismology* (p. 176). Tokyo: Kyoritsu Shuppan. [J]
60. Kanamori, H. (1979). A semi-empirical approach to prediction of long-period ground motions from great earthquakes. *Bulletin of the Seismological Society of America, 69*, 1645–1670.
61. Kawasaki, I., Suzuki, Y., & Sato, R. (1973). Seismic waves due to a shear fault in a semi-infinite medium. Part I: Point source. *Journal of Physics of the Earth, 21*, 251–284.
62. Kennett, B. L. N. (1975). The effects of attenuation on seismograms. *Bulletin of the Seismological Society of America, 65*, 1643–1651.
63. Kennett, B. L. N. (1980). Seismic waves in a stratified half space -II. Theoretical seismograms. *Geophysical Journal of the Royal Astronomical Society, 61*, 1–10.
64. Kennett, B. L. N. (1983). *Seismic wave propagation in stratified media* (p. 342). Cambridge: Cambridge University Press.
65. Kennett, B. L. N. (2001). *The seismic wavefield* (Vol. 1, p. 370). Cambridge: Cambridge University Press.
66. Kikuchi, M. (1995). *Earthquake source process* (p. 99). JICA International Center: Training Course on Seismology and Earthquake Engineering II.
67. Kikuchi, M., & Kanamori, H. (2006). *Note on teleseismic body-wave inversion program.* http://wwweic.eri.u-tokyo.ac.jp/ETAL/KIKUCHI/ .
68. Kind, R. (1976). Computation of reflection coefficients for layered media. *Journal of Geophysics, 42*, 191–200.
69. Kind, R. (1978). The reflectivity method for a buried source. *Journal of Geophysics, 44*, 603–612.
70. Kohketsu, K. (1985). The extended reflectivity method for synthetic near-field seismograms. *Journal of Physics of the Earth, 33*, 121–131. https://doi.org/10.4294/jpe1952.33.121.
71. Kohketsu, K. (1987). 2-D reflectivity method and synthetic seismograms in irregularly layered structures - I. SH-wave generation. *Geophysical Journal of the Royal Astronomical Society, 89*, 821–838.
72. Koketsu, K. (2018). *Physics of seismic ground motion* (p. 353). Tokyo: Kindai Kagaku. [J]

73. Koketsu, K., Fujiwara, H., & Ikegami, Y. (2004). Finite-element simulation of seismic ground motion with a voxelmesh. *Pure and Applied Geophysics, 161*, 2183–2198.
74. Koketsu, K., & Higashi, S. (1992). Three-dimensional topography of the sediment/basement interface in the Tokyo metropolitan area, central Japan. *Bulletin of the Seismological Society of America, 82*, 2328–2349.
75. Koketsu, K., Kennett, B. L. N., & Takenaka, H. (1991). 2-D reflectivity method and synthetic seismograms in irregularly layered structures - II. Invariant embedding approach. *Geophysical Journal International, 105*, 119–130.
76. Koketsu, K., & Kikuchi, M. (2000). Propagation of seismic ground motion in the Kanto basin. *Japan, Science, 288*, 1237–1239.
77. Koketsu, K., & Miyake, H. (2008). A seismological overview of long-period ground motion. *Journal of Seismology, 12*, 133–143.
78. Koketsu, K., Miyake, H., Afnimar, & Tanaka, Y. (2009). A proposal for a standard procedure of modeling 3-D velocity structures and its application to the Tokyo metropolitan area, Japan. *Tectonophysics, 472*, 290–300.
79. Koketsu, K., Miyake, H., & Suzuki, H. (2012). Japan integrated velocity structure model version 1. *Proceedings of the 15th World Conference on Earthquake Engineering*, Paper No. 1773.
80. Koketsu, K., & Sekine, S. (1998). Pseudo-bending method for three-dimensional seismic ray tracing in a spherical earth with discontinuities. *Geophysical Journal International, 132*, 339–346.
81. Komatitsch, D., & Tromp, J. (1999). Introduction to the spectral element method for three-dimensional seismic wave propagation. *Geophysical Journal International, 139*, 806–822.
82. Kreyszig, E. (1999). *Advanced Engineering Mathematics* (8th ed., p. 1156). New York: Wiley.
83. Lamb, H. (1904). On the propagation of tremors over the surface of an elastic solid. *Philosophical Transactions of the Royal Society of London Series A, 203*, 1–42.
84. Landau, L. D., & Lifshitz, E. M. (1973). *Mechanics* (3rd ed., p. 224). Oxford: Butterworth-Heinemann.
85. Liu, Q., & Tromp, J. (2006). Finite-frequency kernels based on adjoint methods. *Bulletin of the Seismological Society of America, 96*, 2383–2397.
86. Love, A. E. H. (1911). *Some problems in geodynamics* (p. 180). Cambridge: Cambridge University Press.
87. Magistrale, H., Day, S., Clayton, R. W., & Graves, R. (2000). The SCEC southern California reference three-dimensional seismic velocity model version 2. *Bulletin of the Seismological Society of America, 90*, S65–S76.
88. Mathematical Society of Japan (ed.) (1968). *Dictionary of mathematics* (2nd ed., p. 1140). Tokyo: Iwanami Shoten. [J]
89. Matsuoka, T., & Shiraishi, K. (2008). Synthesis of Green's function by seismic interferometry and subsurface imaging. *Geophysics Exploration, 61*, 133–144. [J]
90. Millar, R. F. (1973). Rayleigh hypothesis and a related least-squares solution to scattering problems for periodic surfaces and other scatterers. *Radio Science, 8*, 785–796.
91. Moriguchi, S., Udagawa, K., & Hitotsumatsu, S. (1957). *Mathematical formulae II* (p. 328). Iwanami Shoten, Tokyo. [J]
92. Nakahara, H. (2015). Seismic interferometry, (1) Historical development and principles. *Zisin (Journal of Seismological Society of Japan), 68*, 75–82. [J]
93. Paige, C. C., & Saunders, M. A. (1982). LSQR: An algorithm for sparse linear equations and sparse least squares. *ACM Transactions on Mathematical Software, 8*, 43–71.
94. Papoulis, A. (1962). *The Fourier integral and its applications* (p. 318). New York: McGraw-Hill.
95. Pereyra, V., Lee, W. H. K., & Keller, H. B. (1980). Solving two-point seismic-ray tracing problems in a heterogeneous medium, Part 1. *Bulletin of the Seismological Society of America, 70*, 79–99.
96. Phinney, R. A. (1965). Theoretical calculation of the spectrum of first arrivals in layered elastic mediums. *Journal of Geophysical Research, 70*, 5107–5123.

97. Press, W. H., Flannery, B. P., Teukolsky, S. A., & Vetterling, W. T. (1988). *Numerical recipes in C: The art of scientific computing* (p. 735). Cambridge: Cambridge University Press.
98. Physics Dictionary Editorial Committee (ed.) (1992). *Physics dictionary* (rev. ed., p. 2465). Tokyo: Baifukan.
99. Rayleigh, L. (1885). On waves propagated along the plane surface of an elastic solid. *Proceedings of the London Mathematical Society, 17*, 4–11.
100. Richards, P. G. (1974). Weakly coupled potentials for high-frequency elastic waves in continuously stratified media. *Bulletin of the Seismological Society of America, 64*, 1575–1588.
101. Saito, M. (1966). *Introduction to linear algebra* (p. 279). Tokyo: University of Tokyo Press.
102. Saito, M. (1988). Disper80: A subroutine package for the calculation of seismic normal-mode solutions. *Seismological algorithms* (pp. 293–319). New York: Academic Press.
103. Saito, M. (2016). *The theory of seismic wave propagation* (p. 473). Tokyo: TERRAPUB.
104. Satô, Y. (1978). *Elastic wave theory* (p. 454). Tokyo: Iwanami Shoten.
105. Shima, E. (1970). Seismic surface waves detected by the strong motion acceleration seismograph. *3rd Japan Earthquake Engineering Symposium* (pp. 277–284). [J]
106. Stehly, L., Campillo, M., Froment, B., & Weaver, R. L. (2008). Reconstructing Green's function by correlation of the coda of the correlation (C^3) of ambient seismic noise. *Journal of Geophysical Research, 113*, B11306.
107. Sun, Y. (1993). Ray tracing in 3-D media by parameterized shooting. *Geophysical Journal International, 114*, 145–155.
108. Takahashi, R., & Tanamachi, Y. (1991). *Finite difference method* (p. 323). Tokyo: Baifukan.
109. Takeuchi, H., Saito, M. (1972). Seismic surface waves. In *Seismology: Surface waves and earth oscillations* (pp. 217–295). New York: Academic Press.
110. Terasawa, K. (1954). *Introduction to mathematics for natural scientists* (rev. ed., p. 722). Tokyo: Iwanami Shoten. [J]
111. Thomson, C. J., & Chapman, C. H. (1985). An introduction to Maslov's asymptotic method. *Geophysical Journal of the Royal Astronomical Society, 83*, 143–168.
112. Thomson, W. T. (1950). Transmission of elastic waves through a stratified solid. *Journal of Applied Physics, 21*, 89–93.
113. Trifunac, M. D., & Brune, J. N. (1970). Complexity of energy release during the Imperial Valley, California, earthquake of 1940. *Bulletin of the Seismological Society of America, 60*, 137–160.
114. Tromp, J., Tape, C., & Liu, Q. (2005). Seismic tomography, adjoint methods, time reversal and banana-doughnut kernels. *Geophysical Journal International, 160*, 195–216.
115. Um, J., & Thurber, C. (1987). A fast algorithm for two-point seismic ray tracing. *Bulletin of the Seismological Society of America, 77*, 972–986.
116. USGS (United States Geological Survey) (Cited in 2017). Earthquake glossary. https://earthquake.usgs.gov/learn/glossary/ .
117. Utsu, T. (2001). *Seismology* (3rd ed., p. 376). Tokyo: Kyoritsu Shuppan. [J]
118. Viens, L., Miyake, H., & Koketsu, K. (2016). Simulations of long-period ground motions from a large earthquake using finite rupture modeling and the ambient seismic field. *Journal of Geophysical Research, 121*, 8774–8791.
119. Virieux, J. (1984). SH-wave propagation in heterogeneous media: Velocity-stress finite-difference method. *Geophysics, 49*, 1933–1957.
120. Virieux, J., & Operto, S. (2009). An overview of full-waveform inversion in exploration geophysics. *Geophysics, 74*(6), WCC1–WCC26.
121. Wang, R. (1999). A simple orthonormalization method for stable and efficient computation of Green's functions. *Bulletin of the Seismological Society of America, 89*, 733–741.
122. Wapenaar, C. P. A., & Haimé, G. C. (1990). Elastic extrapolation of primary seismic P- and S-waves. *Geophysical Prospecting, 38*, 23–60.
123. Wapenaar, K., & Fokkema, J. (2006). Green's function representations for seismic interferometry, *Geophysics, 71*, SI33–SI46.
124. Washizu, K., Miyamoto, H., Yamada, Y., Yamamoto, Y., & Kawai, D. (eds.) (1981). *Handbook of finite element method, Part I: Foundation* (p. 443). Tokyo: Baifukan. [J]

125. Yamanaka, H. (2005). Comparison of performance of heuristic search methods for phase velocity inversion in shallow surface wave method. *Journal of Environmental and Engineering Geophysics*, *10*, 163–173.
126. Yokoi, T., & Margaryan, S. (2008). Consistency of the spatial autocorrelation method with seismic interferometry and its consequence. *Geophysical Prospecting*, *56*, 435–451.
127. Yoshida, S., Koketsu, K., Shibazaki, B., Sagiya, T., Kato, T., & Yoshida, Y. (1996). Joint Inversion of near- and far-field waveforms and geodetic data for the rupture process of the 1995 Kobe earthquake. *Journal of Physics of the Earth*, *44*, 437–454.
128. Zama, S. (1993). Long-period strong ground motion. *Zisin (Journal of Seismological Society of Japan)*, *46*, 329–342. [J]
129. Zhao, D., Hasegawa, A., & Horiuchi, S. (1992). Tomographic imaging of P and S wave velocity structure beneath northeastern Japan. *Journal of Geophysical Research*, *97*, 19909–19928.
130. Zhu, L., & Rivera, L. A. (2002). A note on the dynamic and static displacements from a point source in multilayered media. *Geophysical Journal International*, *148*, 619–627.

Chapter 4
Observation and Processing

Abstract In this chapter explains the observation and processing techniques required for the study of ground motion. Section 4.1 first describes the tools used for ground motion observation, which are referred to as "seismographs". The "instrumentation of seismographs" is explained and various types of seismographs are introduced, such as "strong motion seismographs", "electromagnetic seismographs", and "servo mechanisms". Section 4.2 then explains the "spectral processing" of ground motion observed by seismographs, which are termed seismograms. Seismograms are processed using the "A/D conversion", "Fourier transform", "discrete Fourier transform", and "FFT" methods. Section 4.3 next explains another technique of seismogram processing referred to as "filtering". "Filters and windows" are defined, and "low-pass filters", "high-pass and band-pass filters" are introduced. Finally, Sect. 4.4 describes the "least-squares method", as the most commonly adopted statistical method for analyzing earthquake sources and velocity structures using ground motion data. The computation and constraints involved in this method are explained in detail.

Keywords Seismographs · Spectral processing · Fourier transform · Least-squares method · Constraints

4.1 Seismographs [1]

4.1.1 Instrumentation of Seismographs

A **seismograph** or **seismometer**[2] observes ground motions according to the following principle. The **pendulum**, which has a large moment of inertia and a long period, does not move immediately when its fulcrum moves, but behaves almost like a fixed point. If the seismograph is composed of a recorder connected to the ground and a

[1]General references relating to Sect. 4.1 include Hamada [14] and Utsu [37].

[2]The word "seismometer" was originally referred to the sensing part of an electromagnetic seismograph (Sect. 4.1.3), but in recent years this has often been used to refer to a "seismograph" as a whole.

K. Koketsu, *Ground Motion Seismology*, Advances in Geological Science,
https://doi.org/10.1007/978-981-15-8570-8_4

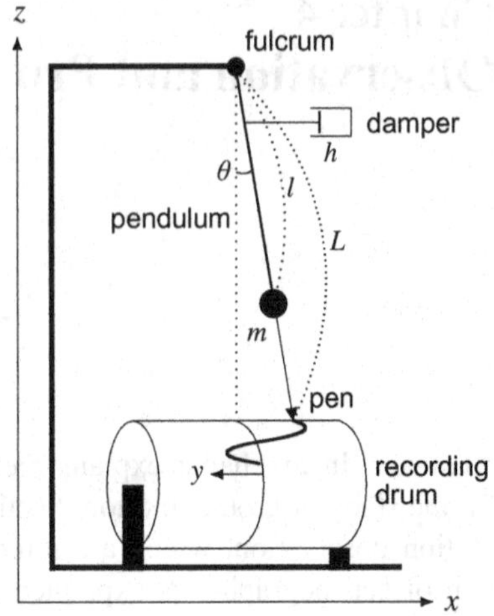

Fig. 4.1 The principle of operation of a seismograph with a simple pendulum and a recording drum (modified from Koketsu [21] with permission of Kindai Kagaku)

pendulum made of a spring such as a leaf spring, the relative movement of the pendulum with respect to the recorder is recorded in the reverse direction, the **ground motion** (Sect. 1.1), which is the movement of the ground, can then be obtained.

In practice, even if the pendulum is not a fixed point, it is possible to restore the ground motion in the case that the relationship between the movement of the pendulum and the ground motion is known. It is, therefore, not necessary to strictly require that the pendulum maintains a fixed point. For example, for a **horizontal component seismograph** that observes ground motions in the horizontal direction, consider a seismograph in which a simple pendulum of length l and weight m is attached with a pen giving the total length L, and the movement of the pen is recorded on a recording drum, as shown in Fig. 4.1. In quiet conditions during which no external force other than gravity operates, the simple pendulum should swing in **simple harmonic oscillation** with the period

$$T_0 = \frac{2\pi}{n}, \quad n^2 = \frac{g}{l}, \tag{4.1}$$

which is the **natural period** of the seismograph and n is the corresponding angular frequency. However, in the case of a pendulum close to the fixed point, because of its large moment of inertia, the pendulum repeats the simple harmonic oscillation even after the ground motion ends. A damper is, therefore, added to the seismograph as shown in Fig. 4.1, to suppress the simple harmonic oscillation. When the damper gives a resistance force in proportion to the change $d\theta/dt$ of the angle θ for the pendulum and the proportional constant is κ, we obtain the equation of motion of the pendulum in the tangential direction:

$$m \frac{d^2(l\theta)}{dt^2} + \kappa \frac{d\theta}{dt} + mg \sin\theta = 0. \tag{4.2}$$

If θ is small, then $\sin\theta \sim \theta$, and (4.2) yields

$$\frac{d^2\theta}{dt^2} + 2hn\frac{d\theta}{dt} + n^2\theta = 0\,, \quad h = \frac{\kappa}{2nml}\,. \tag{4.3}$$

As this is a kind of equation of motion for **damped oscillation** (Sect. 1.2.6), the solution can be given as

$$\theta = Ae^{-hnt}\sin\left(\sqrt{1-h^2}\,nt + \gamma\right)\,. \tag{4.4}$$

h is a constant called the **damping constant**, and the amplitude of the oscillation decays at the rate e^{-hnt} with time t compared with simple harmonic oscillation. The period of the damped oscillation $T_0' = T_0/\sqrt{1-h^2}$ is longer than the natural period T_0.

During an earthquake, the fulcrum of the pendulum and the recording drum move together with the ground following the ground motion. If x stands for this movement, the equation of motion of the pendulum in the tangential direction in the static coordinate system is given as

$$m\,\frac{d^2}{dt^2}(x\,\cos\theta + l\theta) + \kappa\,\frac{d\theta}{dt} + mg\,\sin\theta = 0\,. \tag{4.5}$$

Furthermore, as for (4.3), if θ is small, $\cos\theta \sim 1$ and $\sin\theta \sim \theta$, so that (4.5) yields

$$\frac{d^2\theta}{dt^2} + 2hn\frac{d\theta}{dt} + n^2\theta = -\frac{1}{l}\frac{d^2x}{dt^2}\,. \tag{4.6}$$

As described above, when the pendulum is close to the fixed point, the relative motion of the pendulum is opposite to the ground motion or the motion of the recording drum. Hence, if the y coordinate on the paper of the recording drum is taken in the direction opposite to x, because of $y = -L\theta$, we obtain

$$\frac{d^2y}{dt^2} + 2hn\frac{dy}{dt} + n^2y = V\frac{d^2x}{dt^2}\,, \qquad V = \frac{L}{l}\,. \tag{4.7}$$

This is the basic equation of the seismograph for the input x and output y, and V is referred to as the **static magnification**. As it can be noted from (4.7), the characteristics of the seismograph are determined by the three constants $T_0 = 2\pi/n$, h, and V.

In a **vertical component seismograph**, the observation of the ground motion in the vertical direction is realized by suspending a pendulum in a horizontal direction supported by a spring. In addition, the ground motion in three-dimensional space can be completely observed by two horizontal and one vertical component seismographs, the directions of which are orthogonal to each other. Thus, to obtain a full observation of the ground motion, we can use a **three-component seismograph**, in which three such seismographs are packaged together.

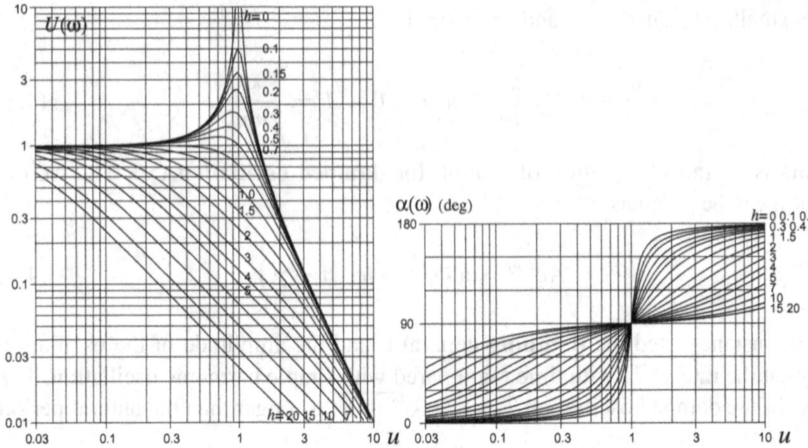

Fig. 4.2 Normalized amplitude characteristics $U(\omega)$ (left) and phase characteristics $\alpha(\omega)$ (right) of a seismograph for the dimensionless period u (based on Utsu [37], reprinted from Koketsu [21] with permission of Kindai Kagaku)

In such a system, y/x, where the output y varies with the input x, is called a response. In particular, the response obtained when a **sine wave** $e^{\mathrm{i}\omega t}$ is applied to the input is called the **frequency response** or **frequency characteristics**. When $x = e^{\mathrm{i}\omega t}$ and $y = \Lambda(\omega)\, x$ are substituted into (4.7), the frequency response

$$\Lambda(\omega) = \frac{V\omega^2}{\omega^2 - n^2 - 2hn\omega\mathrm{i}} \tag{4.8}$$

is obtained.

In addition, using the dimensionless quantity $u = T/T_0 = n/\omega$, where $T = 2\pi/\omega$ is the period of the input ground motion and T_0 is the natural period, we can decompose the frequency behavior into the **amplitude characteristics** and the **phase characteristics** as $\Lambda(\omega) = A(\omega)\, e^{\mathrm{i}\alpha(\omega)}$, and

$$A(\omega) = VU(\omega)\,, \quad U(\omega) = \frac{1}{\sqrt{(1-u^2)^2 + 4h^2n^2}}$$
$$\alpha(\omega) = \tan^{-1}\frac{2hu}{1-u^2}\,, \quad 0 < \alpha(\omega) < \pi\,. \tag{4.9}$$

Figure 4.2 shows a graph of the normalized amplitude characteristics $U(\omega)$ with respect to the dimensionless period u on a logarithmic scale and a graph of the phase characteristics on a semi-logarithmic scale, for various h.

As can be noted from this figure, at $T \gg T_0$ and $u \gg 1$, if h is not too large, $U(\omega)$ asymptotically approaches a straight line of slope -2 through the point $(1, 1)$ and $\alpha(\omega)$ approaches $\pi\,(180°)$. This implies that

Table 4.1 Natural periods and damping constants of the three types of seismographs

Type	Natural period	Damping constant
Displacement seismograph	$T \ll T_0$	$h \sim 1$
Velocity seismograph	$T \sim T_0$	$h \gg 1$
Acceleration seismograph	$T \gg T_0$	$h \sim 1$

$$\Lambda(\omega) \sim \frac{V e^{\mathrm{i}\pi}}{u^2} = \frac{-Vn^2}{\omega^2}, \quad T \gg T_0 \ (\omega \ll n), \quad h \sim 1. \tag{4.10}$$

In this case, if the ground motion is $x = e^{\mathrm{i}\omega t}$, the output of a seismograph is given as

$$y = \Lambda(\omega)\, x \sim -\frac{Vn^2 e^{\mathrm{i}\omega t}}{\omega^2} \Rightarrow Vn^2 \frac{d^2x}{dt^2}, \tag{4.11}$$

so that the output is proportional to the acceleration of the ground motion d^2x/dt^2 and an **acceleration seismograph**, which is also called an **accelerograph** or **accelerometer**, is obtained.

In addition, for $u \ll 1$ ($T \ll T_0$), if h is not too large, the curves in the left panel of Fig. 4.2 approach the flat straight line $U(\omega) = 1$, and the curves in the right panel approach $\alpha(\omega) = 0$. As the output of the seismograph is $y \sim Vx$, the output is proportional to the ground motion, that is the ground displacement, and a **displacement seismograph** is obtained. Moreover, around $u \sim 1$ ($T \sim T_0$), if h is large enough, $U(\omega)$ is a straight line with a slope of -1, therefore, $U(\omega) \propto u^{-1} \propto \omega$ and the phase characteristics are close to $\alpha(\omega) = \pi/2$ (90°). The output is, therefore, given as $y \propto \mathrm{i}\omega\, e^{\mathrm{i}\omega t} \Rightarrow dx/dt$, so that the output is proportional to the ground velocity and a **velocity seismograph** is obtained. These seismographs and their characteristics are summarized in Table 4.1.

A seismograph with a natural period of 1 s or shorter is commonly called a **short-period seismograph**, whereas a seismograph with a natural period of 10 s or longer is called a **long-period seismograph**, but these are not strict definitions. A seismograph with a natural period between these two categories is sometimes called an **intermediate-period seismograph**.

4.1.2 Strong Motion Seismographs[3]

A seismograph that can record **strong motions** (strong ground motions) is termed a **strong motion seismograph**. The most basic definition of a strong motion seismograph is, therefore, a seismograph of low magnification ($V = 1$ to around 20) that is not saturated even during strong ground motions (as opposed to a seismograph

[3]For more information on Sect. 4.1.2, see Kudo *et al.* [23].

of high magnification, which is referred to as a **high-sensitivity seismograph**). In engineering, the dominant idea was that building damage was caused by the force of ground motion, and the acceleration directly connected to the force was an appropriate index of strong motion. Thus, even in seismology, an acceleration seismograph of low magnification that can directly measure the acceleration of ground motion has long been a synonym for a strong motion seismograph.

In the history of Japanese seismology, the **Omori-type strong motion seismograph** and its successor, the **Ichibai strong motion seismograph**, were used by the **JMA** (Appendix A.2.1) for a long time following World War II, and represent displacement seismographs with a small static magnification and long natural periods. For example, the Ichibai strong motion seismograph has the characteristics $V = 1$ ("ichibai" means static magnification of 1 in Japanese) and $T_0 = 5 \sim 6$ s. Acceleration seismographs were also developed since the earliest times in the world. In Japan, the **Ishimoto-type accelerograph** was developed in 1930 [16]. Using records produced by this type of seismograph, Ishimoto established for the first time the relationship between the **peak ground acceleration** (PGA, the maximum amplitude of the acceleration recording) and seismic intensity (Appendix A.2).

However, a certain degree of robustness is required to observe strong motions larger than 100 gal. As such strong motions rarely occur, it is also necessary to include a **trigger mechanism** to begin recording when strong motions greater than a certain level are sensed, in order to facilitate maintenance. The United States of America manufactured a strong motion seismograph that could withstand such practical use and strong motion was observed for the first time using this seismograph. In 1931, K. Suyehiro, then director of the **Earthquake Research Institute**, Tokyo Imperial University (currently, University of Tokyo) was invited by the American Society of Civil Engineers, USA, and gave a lecture on **engineering seismology**, a field that encompasses research areas relating to strong motion seismology and earthquake engineering, or that lie at the boundary between these topics. In this lecture, Suyehiro introduced the Ishimoto-type acceleration seismograph and explained the importance of acceleration seismographs.

In response to Suyehiro's lecture, the United States Coast and Geodesy Survey (USCGS), the predecessor of the United States Geological Survey (**USGS**), developed a strong motion seismograph in 1932 called the **USCGS standard type** [8]. Strong ground motions from the **Long Beach earthquake** (1933, M_w 6.4) and the **Imperial Valley earthquake** (1940, M_w 6.9) were observed using this seismograph. In particular, the records from the latter event (Fig. 4.3) reached a peak ground acceleration of 341.7 cm/s^2, and are referred to as the **El Centro seismograms** after the name of the observation site. These records are still used at present as input ground motions for the structural design of buildings.

In Japan, the momentum for the development of a genuine strong motion seismograph grew following the **Fukui earthquake** (1948, M 7.1, M_s 7.3), and the **Strong Motion Accelerometer Committee** was set up to carry out this development. A prototype was completed in 1953 and the final strong motion seismograph with $V = 16$ and $T_0 = 0.1$ s was called the **SMAC type** seismograph, from the initial letters of the committee name. The SMAC type seismograph has maintained its status in Japan

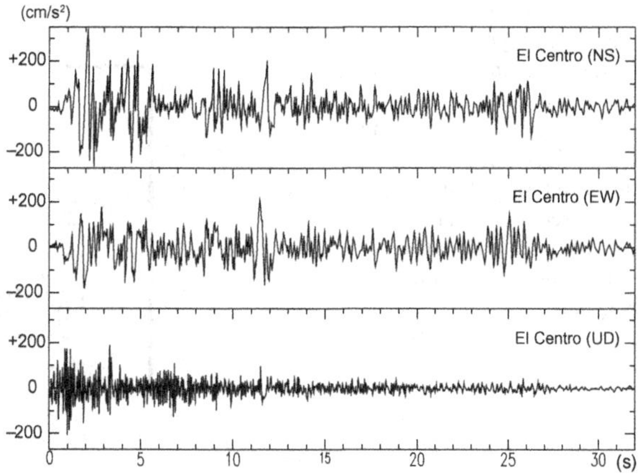

Fig. 4.3 The "El Centro seismograms", which represent the ground accelerations of the Imperial Valley earthquake recorded by the USCGS standard type strong motion seismograph at El Centro (reprinted from Koketsu [21] with permission of Kindai Kagaku)

since then as a representative strong motion seismograph. This seismograph has been gradually improved over time, and observation records from this instrument have been widely used in engineering. However, the SMAC type seismograph has some fatal drawbacks for seismology, such as:

(1) Since there is no **chronograph** and time calibration unit (which are collectively called a **clock**), the absolute time is not known;
(2) In spite of the trigger mechanism, there is no **delay unit** to preserve the part of the record just before the trigger, therefore the P wave, etc. are easily lost;
(3) Since its use was typically aimed at engineering applications, it was often installed in buildings and structures in urban areas, and the effects of this appear in the records.

The SMAC type seismograph has, therefore, rarely been used in seismology, compared with the Ichibai strong motion seismograph, which has a low-accuracy clock and a continuous recording system.

4.1.3 Electromagnetic Seismographs

An older type of seismograph, which can be represented as shown in Fig. 4.1, is called a **mechanical seismograph**. This type of seismograph records the movement of a pendulum by drawing on paper with a pen or by scratching soot on a drum with a needle. A seismograph in which light is reflected by a mirror attached to the pendulum and is recorded on a photographic film is called an **optical seismograph**. The **Wood–Anderson seismograph** (Appendix A.1.1), which was used to define M for the first time, was an optical seismograph. The USCGS standard type and its commercial successor, **SMA-1** by Kinemetrics, are also optical seismographs [8].

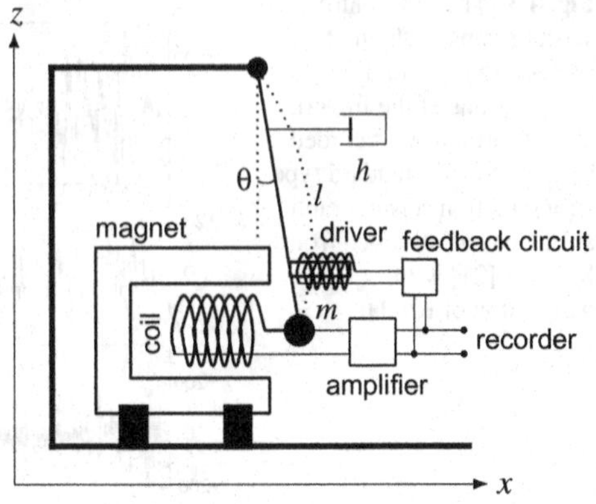

Fig. 4.4 The principle of the moving coil type seismometer in an electromagnetic seismograph. The driver and feedback circuit are also displayed for instruments with a servo mechanism (modified from Koketsu [21] with permission of Kindai Kagaku)

In addition, a seismograph in which the movement of the pendulum is converted into electric signals that are recorded is called an **electromagnetic seismograph**. In the modern age, since the development of electronics, the electric signals can be electromagnetically recorded on a storage medium and the subsequent processing can be done based on electronics. The adjustment of the seismograph itself can also be done based on electronics, and miniaturization and weight reduction are possible. Consequently, almost all seismographs are now electromagnetic, and strong motion seismographs are no exception.

In an electromagnetic seismograph, the part that performs the conversion is called a **transducer**, and the pendulum and the transducer are collectively referred to as a **seismometer**. Two main types of seismometer are used. The first is called a **moving coil type** seismometer, which involves a combination of a pendulum with a coil and a magnet, using the electromotive force generated by electromagnetic induction when the coil moves with respect to the permanent magnet. In the second type of seismometer, the transducer consists of a capacitor, and the change of capacitance due to the movement of the pendulum is measured [20].

The pen in Fig. 4.1 is replaced with a coil of radius a and number of turns N, surrounded by a permanent magnet of magnetic flux density B (Fig. 4.4). The coil moves following the motion of the pendulum, and the electromotive force

$$E = 2\pi a N B \frac{d(l\theta)}{dt} = G \frac{d\theta}{dt}, \qquad G = 2\pi a N B l \tag{4.12}$$

is generated by electromagnetic induction in proportion to the velocity $d(l\theta)/dt$ for the motion of the coil. If the coil and the other part of the seismometer have resistances of R and r, respectively, this electromotive force has a current $i = E/(R + r)$, which generates a force $2\pi a N B i = G i$ to suppress the motion of the coil. Thus, by adding this resistance force to (4.5), we obtain the equation of motion of the pendulum

$$m\,\frac{d^2}{dt^2}(x\,\cos\theta + l\theta) + \kappa\,\frac{d\theta}{dt} + \frac{G^2}{(R+r)}\frac{d\theta}{dt} + mg\,\sin\theta = 0\,. \tag{4.13}$$

As for (4.5), assuming $\theta \sim 0$ we further obtain

$$\frac{d^2\theta}{dt^2} + 2(h+h_e)n\frac{d\theta}{dt} + n^2\theta = -\frac{1}{l}\frac{d^2x}{dt^2}\,, \qquad h_e = \frac{G^2}{2nml(R+r)}\,. \tag{4.14}$$

Comparing this with (4.6), we find that the damping constant increases by h_e due to the electromagnetic induction.

When the coil is connected to the recorder via an amplifier, a current proportional to $i = E/(R+r)$ flows according to the amplification factor of the amplifier. Since E is proportional to $d\theta/dt$ from (4.12), the output electric signal corresponds to the first-order derivative of the angle θ. Hence, if the pendulum and damper are of the type corresponding to a displacement seismograph (Table 4.1), the output corresponding to the velocity is recorded, and therefore, the transducer in a moving coil type seismometer is sometimes referred to as a **velocity detection type**. Static displacements cannot be detected because the output is proportional to $d\theta/dt$. Conversely, for the case of a condenser transducer in the second type of seismometer, if the pendulum and damper are of the type corresponding to a displacement seismograph, the output corresponds to the displacement and the transducer is referred to as a **displacement detection type** [20].

4.1.4 Servo Mechanisms[4]

As noted from Fig. 4.2 and Table 4.1, in the case of acceleration seismographs, the output is proportional to the acceleration from around $T = T_0$ ($u = 1$) to $T = \infty$ ($u = \infty$). Thus, in designing an acceleration seismograph, if the natural period T_0 is made to be short, the output proportional to the acceleration is obtained in a broader period band, and the design result becomes **broadband**. Although this broadband response can be achieved by changing the mechanical structure of the pendulum and damper, in the electromagnetic seismograph, the broadband can also be realized by a **servo mechanism** used in control engineering. In practice, since broadband systems using a servo mechanism can be produced easily and inexpensively, not only strong motion seismographs but also most recent broadband seismographs use servo mechanisms. In addition, as described later, the servo mechanism in a seismograph works to suppress the amplitude of the pendulum swing, so that it is especially convenient in the case of a strong motion seismograph, where emphasis is put on avoiding saturation for very strong ground motion.

There exist only a few examples of broadbands produced without using a servo mechanism, such as the **Muramatsu-type seismograph** in which a pendulum is

[4]For more information on Sect. 4.1.4, see Kinoshita [23].

immersed in silicon oil to increase the damping constant. In addition, some strong motion seismographs do not have a servo mechanism if they do not require a broadband, or if they sacrifice the broadband to some extent due to requirements of portability and low operation power (*e.g.*, Kudo *et al.* [24]). In this case, a large h is produced by a strong magnet (**over-damping**), and a velocity type pendulum with flat characteristics up to a period of around 10 s is used.

As shown in Fig. 4.4, the servo mechanism in a seismograph is generally composed of a **feedback circuit** and a **driver**. The feedback circuit deforms the output of a seismometer by using amplification and differentiation circuits. The driver, which is composed of a magnet and a coil, transmits the deformed output to the pendulum by the principle opposite to that of the moving coil transducer. The servo mechanism is called a **displacement feedback**, **velocity feedback**, or **acceleration feedback**, depending on whether the force transmitted to the pendulum is proportional to θ, $d\theta/dt$, or $d^2\theta/dt^2$ of the pendulum. A range of variations could, therefore, be considered within the electromagnetic seismograph, through the combination of different types of transducers and servo mechanisms, however, the following three combinations are mainly used.

For strong motion seismographs, the most widely used combination involves the displacement detection and displacement feedback types in an acceleration seismograph. In this case, the force transmitted to the pendulum is proportional to the pendulum displacement θ, and this force is applied in the opposite direction of θ. If the proportionality constant is G_f, (4.6) yields

$$\frac{d^2\theta}{dt^2} + 2hn\frac{d\theta}{dt} + (n^2 + G_f)\theta = -\frac{1}{l}\frac{d^2x}{dt^2}, \tag{4.15}$$

where the factor of the term related to θ increases from n^2 to $n^2 + G_f$. According to (4.1), the natural period T_0 changes from $2\pi/n$ to $2\pi/(n^2 + G_f)^{1/2}$, so that a short natural period is obtained. This type of acceleration seismograph is also called a **force balance type** seismograph, because the servo mechanism acts as a force to pull the pendulum back to the balance point. The strong motion seismographs by Kinemetrics [18], such as K2 and ETNA, are of the displacement detection and displacement feedback types. The **JMA-87 type** strong motion seismograph, which has been used by the **JMA** (Appendix A.2.1) since 1987 in place of the Ichibai strong motion seismograph [17], and the **K-NET95** [19], which was used in the nationwide network **K-NET** by the **NIED** (Sect. 2.3.8), are also of displacement detection and displacement feedback types. The appearance of the latter is shown in Fig. 4.5.

Next, we review the case of **broadband seismograph**, such as STS-1, 2 and CMG-3T, 40T. When an instrument is simply called a "broadband seismograph", this typically indicates a broadband **high-sensitivity seismograph** for **teleseismic earthquake** (Sect. 3.1.11), with displacement detection and velocity feedback types. Since the pendulum has an inverse force proportional to the pendulum velocity $d\theta/dt$, if the constant of this proportionality is represented again by G_f, (4.6) yields

Fig. 4.5 Appearance of the K-NET 95. The main unit with a communication modem is shown on the left, and the battery is shown on the right (reprinted from NIED [29])

$$\frac{d^2\theta}{dt^2} + 2\left(h + \frac{G_f}{2n}\right)n\frac{d\theta}{dt} + n^2\theta = -\frac{1}{l}\frac{d^2x}{dt^2}, \tag{4.16}$$

where the factor of the term related to $d\theta/dt$ increases from $2hn$ in (4.6) to $2(h + G_f/2n)n$. This increase is equivalent to the increase of h by $G_f/2n$. As noted from Fig. 4.2 and Table 4.1, when h becomes large, the seismograph works as a velocity seismograph in a broadband around T_0. Moreover, if the output of the feedback circuit is taken out, the acceleration, which is the differential of the velocity, can be obtained. The velocity types of strong motion seismographs such as VSE-11 and 12 are produced using similar servo mechanisms.

Some acceleration seismographs of velocity detection and velocity feedback types also exist. When feedback is applied to velocity, the pendulum behaves as a broadband velocity seismometer as above, but the output to the recorder is the acceleration because of the differentiation by the velocity detection type transducer, therefore, the seismograph is an acceleration seismograph. In this type of instrument, it is possible to design the feedback circuit, which is sensitive to lightning, to be separate from the transducer and the driver. This type of instrument is, therefore, often used for borehole observations, because the feedback circuit can be placed on the ground and easily maintained [20].

In either case, the driver must act in the direction opposite to the motion of the pendulum, therefore, the servo mechanism of a seismograph is also called a **negative feedback mechanism**. As the movement of the pendulum is suppressed, even if a large ground motion arrives, the pendulum seldom exceeds the mechanical movable range, and if a servo mechanism is provided, the measuring range of a seismograph, the so-called **dynamic range**, can be expanded.

Modern strong motion seismographs are now equipped with clocks, and not only acceleration types but also velocity types are now used. In addition, seismographs that are used in other fields of seismology, such as microearthquake observation and global seismology, have been equipped with trigger mechanisms. Thus, the difference between **strong motion observation** and other observations is now small and can be said to consist in what is considered as an observation target.

4.2 Spectral Processing

4.2.1 A/D Conversion

If computer analysis is essential, the data must be provided in **digital** form. As the electric signals of an electromagnetic seismograph are **analog**, a format which can take continuous values, the conversion from analog to digital is carried out as a part of the analysis. However, as a matter of course, if the output of a seismograph itself is digital, this conversion is unnecessary and analysis is simple, therefore, the recorders in recent seismographs have **A/D converters** for converting from analog to digital signals, that is, performing **A/D conversion**. The A/D conversion corresponds to **sampling** digital values from the analog quantity, and the inverse operation of the A/D conversion is called a **D/A conversion**.

A/D converters used to be assembled by circuits with discrete components, however in recent years, one-chip LSIs have become commercially available, and various types exist as shown in Table 4.2 [12, 27] with performances listed as of the time of the publication. The rate of A/D conversion in earthquake observation is smaller than that for consumer use such as audio, and around a maximum of 1 kHz is sufficient. Thus, according to Table 4.2, when a resolution of about 16 bit is sufficient, the successive approximation type of A/D converter can be used, but the delta-sigma modulation type must be used for a recorder with a higher resolution of 24 bit or thereabouts.

In the **successive approximation type**, the input voltage is compared with the reference voltage corresponding to each bit in order to determine whether the bit is on or off. This type has a built-in D/A converter. At first, only the most significant bit is turned on, and the output through the D/A converter is compared with the input voltage. If the input voltage is larger, the most significant bit is left as it is, and if the input voltage is smaller, the bit is cleared off. Next, the second bit is turned on with the result of the most significant bit, and the output is compared with the input voltage again. The second bit is left as it is if the input voltage is larger, and is cleared

Table 4.2 Types of A/D converter [12, 27]

Conversion method	Resolution	Conversion rate
Time comparison method (high resolution & accuracy)		
Integration type	~22 bit	~several kHz
Delta-sigma modulation type	~24 bit	~several tens of kHz
Voltage comparison method (high or medium rate)		
Successive approximation type	~16 bit	~several hundreds of kHz
Series-parallel type	~ 14 bit	~ several tens of MHz
All-parallel type	~ 10 bit	~ several hundreds of MHz

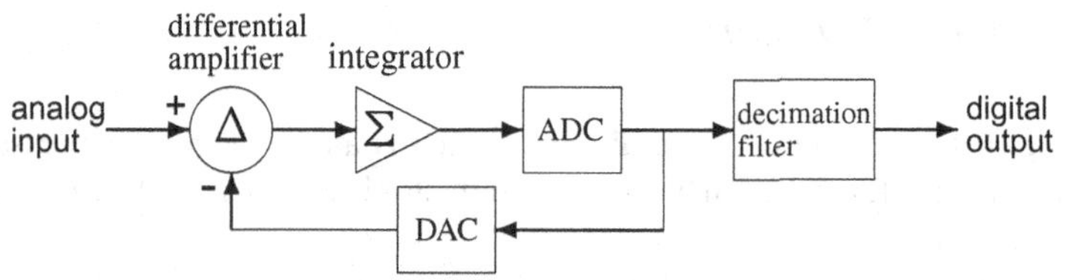

Fig. 4.6 Configuration of the delta-sigma modulation type A/D converter. ADC and DAC are low resolution A/D and D/A converters of about 1–4 bit

off if the input voltage is smaller. In this successive approximation type, a digital value is obtained by repeating this operation successively for all bits.

In the **delta-sigma modulation type**, A/D conversion of low resolution (1 to 4 bit) is performed by high-rate sampling (**oversampling**), and the result is averaged by a **low-pass filter** and decimated to obtain a digital value of high resolution. The internal A/D converter is combined with the D/A converter, and the low resolution digital value is fed back through the D/A converter. The difference with the next analog input is taken and integrated, and then the A/D conversion is performed (Fig. 4.6). For the low resolution digital value, the **decimation filter** performs averaging through a low pass filter and decimation, and the high resolution digital value is obtained at the prescribed sampling rate. If negative feedback is not performed, resolution improvement cannot be obtained even in the case of oversampling and averaging by a low-pass filter. Since the delta-sigma modulation type does not include analog processing except for the differential amplifier and integrator, a stable performance can be obtained. As the digital processing of the internal A/D converter and the decimation filter regulates the whole performance, it is not difficult to design an appropriate algorithm for achieving a high resolution.

Because the degradation of a signal with recording is virtually eliminated by digital recording, the "signal-to-noise ratio" (**S/N ratio**, the ratio of the maximum signal amplitude to the noise level) of the recorder is dramatically improved compared with analog recording in which the electrical noise of the recorder passes directly to the output. For example, the S/N ratio of a 16 bit or 24 bit digital recorder is theoretically 2^{15} or 2^{23}, which can reach as high as 90 dB or 140 dB in decibel (**dB**, 20 times the common logarithm of the ratio). However, in practice, when the accuracy of the A/D conversion itself and the electric noise mixed in the conversion are considered, the 24 bit recorder used to have only about 120 dB, which is an S/N ratio of about 20 bit. The minimum measurement range of the seismograph can be defined by the S/N ratio of the recorder. Based on this definition, the **dynamic range** of the seismograph, which is the ratio of the maximum measurement range to the minimum measurement range, is improved by the digitization of the recorder, even though the practical resolution of a 24 bit recorder is 20 bit.

4.2.2 Fourier Transform

Assuming that a continuous **time history**, which is analog, can be expressed by a function $f(t)$, this function can be decomposed into **sine wave** $e^{\mathrm{i}\omega t}$ of various angular frequencies ω as

$$f(t) = \frac{1}{2\pi}\int_{-\infty}^{+\infty} F(\omega)e^{\mathrm{i}\omega t}d\omega\,, \tag{4.17}$$

$F(\omega)$, which is the amplitude of each sine wave, is given by

$$F(\omega) = \int_{-\infty}^{+\infty} f(t)e^{-\mathrm{i}\omega t}dt \tag{4.18}$$

and is called the **Fourier transform** of $f(t)$, and conversely, the relationship in (4.17) is called the **inverse Fourier transform** (*e.g.*, Papoulis [30]). When $F(\omega)$, which is generally a complex function, is divided into amplitude and phase as $F(\omega) = A(\omega)e^{\mathrm{i}\phi(\omega)}$, $A(\omega)$ is called the **Fourier spectrum** [30] or **amplitude spectrum**, and $\phi(\omega)$ is called the **phase angle** or **phase spectrum**. The entire $F(\omega)$ is called the **frequency spectrum**, or simply **spectrum**.

Here, a time series with time as an independent variable is used as an example, but the Fourier transform and inverse Fourier transform can be defined using sine waves e^{ikx} for a continuous function with space coordinate x as an independent variable. The spectrum in this case is called a **wavenumber spectrum**, where the **wavenumber** is explained in Sect. 2.1.5. In addition, in the case of the Hankel transform

$$F(k) = \int_{0}^{\infty} f(r)J_l(kr)rdr\,, \quad f(r) = \int_{0}^{\infty} F(k)J_l(kr)kdk\,, \tag{4.19}$$

where cylindrical waves represented by $J_l(kr)$ are used as in (3.5) instead of sine waves e^{ikx}, k is also called the "wavenumber" (Sect. 2.2.1).

The correspondence between a function and its Fourier transform is expressed as $f(t) \Leftrightarrow F(\omega)$, and has the properties shown in Table 4.3. First, **convolution**

$$f_1(t) * f_2(t) = \int_{-\infty}^{+\infty} f_1(\tau)f_2(t-\tau)d\tau = \int_{-\infty}^{+\infty} f_1(t-\tau)f_2(\tau)d\tau\,, \tag{4.20}$$

which is an operation that convolves two functions into one function, and its Fourier transform plays an important role in seismology (*e.g.*, Sect. 1.3.2) as well as in various other fields. Secondly, ground motion is a physical phenomenon and the time function of a physical phenomenon should be a real function, therefore, $f(t)$ for the ground motion and its Fourier transform $F(\omega)$ satisfy the conditions of a "real function" as listed in Table 4.3. The condition for $F(\omega)$ indicates that the real and imaginary parts of $F(\omega)$ should be even and odd functions, respectively. Thirdly, if the ground motion starts at $t = 0$, $f(t) = 0$ should be satisfied for $t < 0$. This condition is called

Table 4.3 Properties of the Fourier transform [9, 30]. F^* is the complex conjugate of F

linearity	$\sum a_i f_i(t) \Leftrightarrow \sum a_i F_i(\omega)$	symmetry	$F(t) \Leftrightarrow 2\pi f(-\omega)$
time scaling	$f(at) \Leftrightarrow \frac{1}{\lvert a\rvert} F\left(\frac{\omega}{a}\right)$	frequency scaling	$\frac{1}{\lvert a\rvert} f\left(\frac{t}{a}\right) \Leftrightarrow F(a\omega)$
time shift	$f(t-t_0) \Leftrightarrow F(\omega)e^{-i\omega t_0}$	frequency shift	$f(t)e^{i\omega_0 t} \Leftrightarrow F(\omega-\omega_0)$
time derivative	$\frac{d^n f}{dt^n} \Leftrightarrow (i\omega)^n F(\omega)$	frequency derivative	$(-it)^n f(t) \Leftrightarrow \frac{d^n F(\omega)}{d\omega^n}$
time integral	$\int f(t)(dt)^n \Leftrightarrow \frac{F(\omega)}{(i\omega)^n}$	frequency integral	$\frac{f(t)}{(-it)^n} \Leftrightarrow \int F(\omega)(d\omega)^n$
complex conjugate	$\{f(t)\}^* \Leftrightarrow \{F(-\omega)\}^*$	time reversal	$f(-t) \Leftrightarrow \{F(\omega)\}^*$

convolution	$\int_{-\infty}^{+\infty} f_1(\tau)f_2(t-\tau)d\tau = f_1(t) * f_2(t) \Leftrightarrow F_1(\omega)F_2(\omega)$
frequency convolution	$f_1(t)f_2(t) \Leftrightarrow \frac{1}{2\pi} F_1(\omega) * F_2(\omega)$
cross correlation	$\int_{-\infty}^{+\infty} f_1(\tau)f_2(t+\tau)d\tau \Leftrightarrow \{F_1(\omega)\}^* F_2(\omega)$
auto correlation, power spectrum	$\int_{-\infty}^{+\infty} f(\tau)f(t+\tau)d\tau \Leftrightarrow \lvert\{F(\omega)\rvert^2$
time moment	$\int_{-\infty}^{+\infty} f(t)\, t^n dt \Leftrightarrow \frac{1}{(-i)^n} \frac{d^n F(0)}{d\omega^n}$
real function	$\mathrm{Im}\, f(t) = 0 \Leftrightarrow F(-\omega) = \{F(\omega)\}^*$
causality	$f(t) = 0,\ t < 0 \Leftrightarrow \mathrm{Re}\, F(\omega) = \frac{1}{\pi} \int_{-\infty}^{+\infty} \frac{\mathrm{Im}\, F(y)}{\omega - y} dy$

causality, and under this condition the real and imaginary parts are the **Hilbert transform** of each other. The last item listed in Table 4.3 indicates this, and the proofs were provided by Papoulis [30] and in Box 5.8 of Aki and Richards [3].

As useful examples, the Fourier transforms of the representative hyperfunctions, which are the **delta function** $\delta(t)$, **step function** $H(t)$, **ramp function** $U(t)$, and **triangular function** $V(t)$ in Fig. 4.7, are shown in Table 4.4.[5] The ramp and triangle functions are commonly used for such functions as the moment time function and moment rate function (Sect. 2.1.2).

4.2.3 *Discrete Fourier Transform*

When an analog quantity $f(t)$ is A/D-converted into a digital quantity by **sampling** with a constant time interval Δt, the digital quantity can be represented by the function of t:

$$\tilde{f}(t) = f(t)\, \Delta t \sum_{n=-\infty}^{+\infty} \delta(t - n\Delta t) \tag{4.21}$$

[5] The items from the top of Table 4.4 agree with (3–3), (3–6), (3–27), and (2–58) of Papoulis [30] if $T \to T/2$.

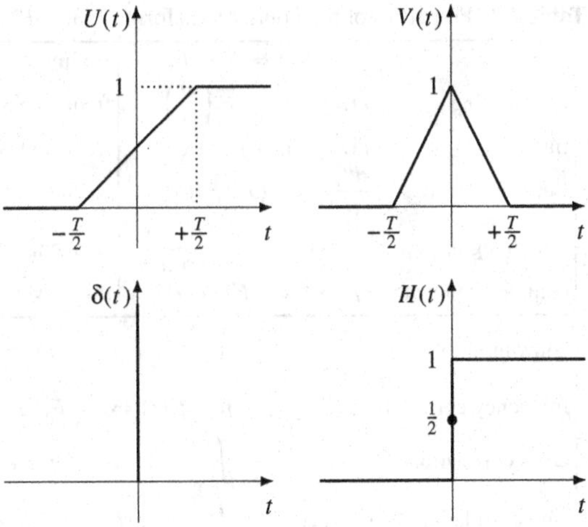

Fig. 4.7 The delta function $\delta(t)$, step function $H(t)$, ramp function $U(t)$, and triangular function $V(t)$ (reprinted from Koketsu [21] with permission of Kindai Kagaku)

Table 4.4 Fourier transforms of hyperfunctions [30]

delta function	$\delta(t) \Leftrightarrow 1$
frequency delta function	$1 \Leftrightarrow 2\pi\delta(\omega)$
step function	$H(t) \Leftrightarrow \pi\delta(\omega) + \dfrac{1}{\mathrm{i}\omega}$
ramp function	$U(t) \Leftrightarrow \pi\delta(\omega) + \dfrac{2\sin(\omega T/2)}{\mathrm{i}\omega^2 T}$
triangular function	$V(t) \Leftrightarrow \dfrac{8\sin^2(\omega T/4)}{\omega^2 T}$

where $\delta(t)$ is the delta function. The factor Δt is given so that (4.21) agrees with the equation obtained from (4.18) using numerical integration. From the Fourier transform $F(\omega)$ of $f(t)$ and the definition of the delta function in (1.98), which is $\int_{-\infty}^{+\infty} \phi(x)\delta(x - x_0)dx = \phi(x_0)$, we have

$$F(\omega) * \delta(\omega - \omega_0) = \int_{-\infty}^{+\infty} F(\omega - w)\delta(w - \omega_0)dw = F(\omega - \omega_0)\,. \tag{4.22}$$

For the Fourier transform of the **delta function sequence** and the inverse Fourier transform of the frequency delta function sequence, we also have

$$\Delta t \sum_{n=-\infty}^{+\infty} \delta(t - n\Delta t) \Leftrightarrow 2\pi \sum_{n=-\infty}^{+\infty} \delta\left(\omega - \frac{2\pi n}{\Delta t}\right) \tag{4.23}$$

the proof of which was given by Papoulis [30]. Using the property of frequency convolution in Table 4.3, (4.21), (4.22), and (4.23), we obtain the Fourier transform

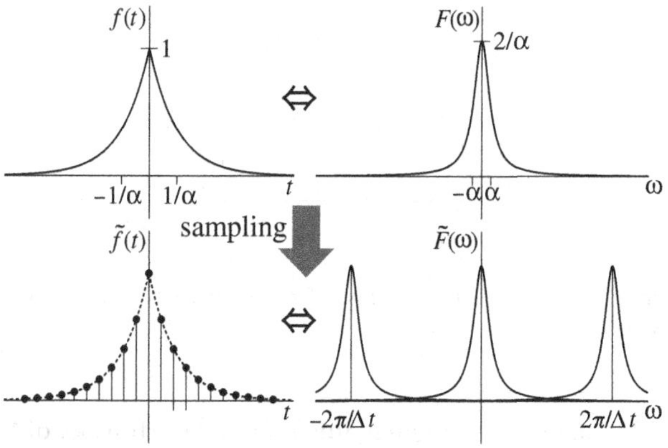

Fig. 4.8 Schematic representation of "aliasing" assuming $f(t) = e^{-\alpha|t|}$ (based on Brigham [9], reprinted from Koketsu [21] with permission of Kindai Kagaku)

$\breve{F}(\omega)$ of $\tilde{f}(t)$ as

$$\breve{F}(\omega) = \frac{1}{2\pi} F(\omega) * 2\pi \sum_{n=-\infty}^{+\infty} \delta\left(\omega - \frac{2\pi n}{\Delta t}\right) = \sum_{n=-\infty}^{+\infty} F\left(\omega - \frac{2\pi n}{\Delta t}\right). \quad (4.24)$$

Equation (4.24) does not simply indicate that $\breve{F}(\omega)$ is equal to $F(\omega)$, but that $\breve{F}(\omega)$ is the sum of $F(\omega)$ at angular frequencies distributed at an interval of $2\omega_c = 2\pi/\Delta t$ (or at frequencies distributed at an interval of $2f_c = 1/\Delta t$), as shown in Fig. 4.8. This repeated distribution is called **aliasing**. If $F(\omega) \neq 0$ for $|\omega| > \omega_c$, adjacent $F(\omega)$ and $F(\omega \pm \omega_c)$ overlap. Conversely, when $f(t)$ has frequency components up to f_c, sampling at a time interval longer than $1/2f_c$ leads to a spectrum distorted by aliasing. The sampling frequency $2f_c$ corresponding to this time interval is the **Nyquist sampling rate**. If the sampling rate cannot be changed, an **anti-aliasing filter**, which excludes frequency components above f_c, must be applied to records before the A/D conversion.

It is impossible to perform infinite sampling in (4.21), and in reality we have to stop at some finite time limits such as $[-T, T]$. This truncation is equivalent to

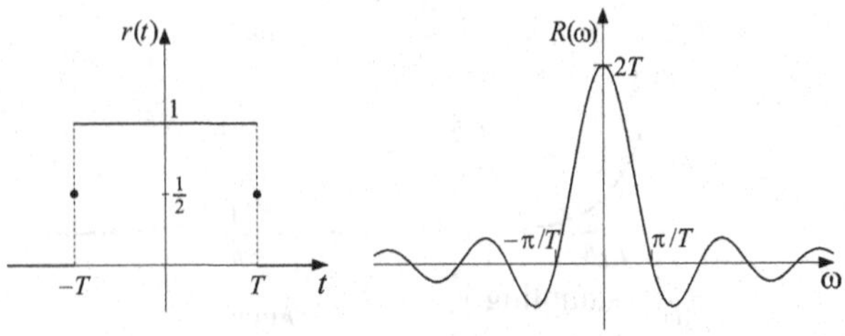

Fig. 4.9 Rectangular function and its Fourier transform (reprinted from Koketsu [21] with permission of Kindai Kagaku)

convolving (4.21) and the **rectangular function** in the left panel of Fig. 4.9. Since the spectrum of the rectangular function $r(t)$:

$$R(\omega) = \frac{2\sin\omega T}{\omega} \tag{4.25}$$

is an oscillatory function gradually damped toward $\omega = \pm\infty$ as shown in the right panel of Fig. 4.9, this oscillation produces a **ripple** on $\breve{F}(\omega)$. We next consider the rectangular function $r(\omega)$ in the frequency domain. Using (4.25) and the property of symmetry in Table 4.3, the inverse Fourier transform of $r(\omega)$ is given as

$$\frac{R(t)}{2\pi} = \frac{\sin\Omega t}{\pi t}, \tag{4.26}$$

where T is replaced with Ω. Thus, if the spectrum distributed over a wide frequency band is truncated by a **filter** (Sect. 4.3.1), a ripple also appears in the time history obtained by taking the inverse Fourier transform of the spectrum. These ripples are called **Gibbs' phenomenon**. However, if the truncation range is sufficiently wide, the spectrum of the rectangular function approaches that of the delta function and the ripples can be suppressed.

To further make use of computers, it is also desirable to digitize $\breve{F}(\omega)$, because $F(\omega)$ is analog, and therefore with $n?k$, $\breve{F}(\omega)$ is analog. Similarly to $\tilde{f}(t)$ but using the right-hand side of (4.23), with $n \to k$, $F(\omega)$ is digitized as

$$\tilde{F}(\omega) = F(\omega)\sum_{k=-\infty}^{+\infty}\delta(\omega - k\Delta\omega'), \quad \Delta\omega' = \frac{2\pi}{\Omega}. \tag{4.27}$$

Since the rectangular function must be convolved as previously mentioned, the sampling rate of the angular frequency, $\Delta\omega'$ in (4.27), is determined from the length of the rectangular function, Ω. We replace $F(\omega)$ in (4.24) with $\tilde{F}(\omega)$ in (4.27), take the inverse Fourier transform using the property of convolution in Table 4.3 and (4.23) with $\Delta t' = 2\pi/\Delta\omega' = \Omega$, and obtain

$$\breve{f}(t) = f(t)\,\Delta t \sum_{n=-\infty}^{+\infty} \delta(t - n\Delta t) * \frac{\Delta t'}{2\pi} \sum_{k=-\infty}^{+\infty} \delta(t - k\Delta t') \,. \tag{4.28}$$

By applying the rectangular function $R(t - T)$ with $\Omega = 2T = N\Delta t$, $\sum_{n=-\infty}^{+\infty} \delta(t - n\Delta t)$ in (4.28) yields $\sum_{n=0}^{N-1} \delta(t - n\Delta t)$, because $f(t - k\Delta t)$ for k outside $[0, N-1]$ is not contributed due to $R(t - T)$.

We then apply the definition of the Fourier transform in (4.18) to $\breve{f}(t)$ in (4.28), use (4.27) and the definition of the delta function in (1.98), and obtain digital $\breve{F}(\omega)$ as

$$\begin{aligned}
\breve{F}(\omega) &= \Delta t \sum_{n=0}^{N-1} \int_{-\infty}^{+\infty} f(t)\delta(t - n\Delta t)e^{-\mathrm{i}\omega t} dt \cdot \sum_{k=-\infty}^{+\infty} \delta\left(\omega - \frac{2\pi k}{\Omega}\right) \\
&= \Delta t \sum_{n=0}^{N-1} f(n\Delta t)e^{-\mathrm{i}\omega n\Delta t} \cdot \sum_{k=-\infty}^{+\infty} \delta\left(\omega - \frac{2\pi k}{\Omega}\right) \\
&= \sum_{k=-\infty}^{+\infty} F_k\,\delta\left(\omega - \frac{2\pi k}{N\Delta t}\right), \quad F_k = \Delta t \sum_{n=0}^{N-1} f(n\Delta t)e^{-2\pi \mathrm{i}kn/N}.
\end{aligned} \tag{4.29}$$

This is termed the **discrete Fourier transform**, and F_k constitutes the **Fourier series** of $f(t)$. In addition, the **discrete inverse Fourier transform**

$$\tilde{f}(t) = \sum_{k=-\infty}^{+\infty} f_k\,\delta(t - k\Delta t)\,, \quad f_k = \frac{1}{N\Delta t} \sum_{n=0}^{N-1} F\left(\frac{2\pi n}{N\Delta t}\right) e^{+2\pi \mathrm{i}kn/N} \tag{4.30}$$

is obtained by almost the same derivation. However, in many publications such as Brigham [9] and Press et al. [31], the factor Δt is removed from (4.29) and the factor $1/N\Delta t$ in (4.30) is replaced with $1/N$, so that

$$F_k = \sum_{n=0}^{N-1} f(n\Delta t)e^{-2\pi \mathrm{i}kn/N}, \quad f_k = \frac{1}{N} \sum_{n=0}^{N-1} F\left(\frac{2\pi n}{T}\right) e^{+2\pi \mathrm{i}kn/N}. \tag{4.31}$$

If the ground motion $f(t)$ is a real function, its digital quantity $\breve{f}(t)$ is also a real function. Thus, using the following property of real functions listed in Table 4.3, that is $F_{-N/2+k} = F^*_{N/2-k}$ (* stands for complex conjugate), and the fact that $\breve{f}(t)$ and $e^{-2\pi ikn/N}$ are both periodic functions, we have the property

$$F_{N/2+k} = F^*_{N/2-k} \,. \tag{4.32}$$

In other words, the discrete Fourier transform is folded at $N/2$ and its two halves are complex conjugates to each other and the spectrum for which the discrete inverse Fourier transform is to be performed must satisfy this property. The discrete Fourier transform represented by (4.31) must also satisfy this complex conjugate condition.

4.2.4 FFT

Although the computation of the discrete Fourier transform in (4.29) or the discrete inverse Fourier transform (4.30) is easy for a modern computer, faster algorithms were always required in the age in which high-speed computers were expensive, and the "fast Fourier transform" (**FFT**) is one such algorithm. As the representation in (4.31) is used in the following section, the result obtained by the method explained hereafter becomes the Fourier transform after being multiplied by Δt, or the inverse Fourier transform after being multiplied by $1/\Delta t$.

For simplicity, we consider the discrete Fourier transform of $N = 2^2 = 4$. Its simple computation through (4.31) can be represented in a matrix form with $W = e^{-2\pi/4}$ as

$$\begin{pmatrix} F_0 \\ F_1 \\ F_2 \\ F_3 \end{pmatrix} = \begin{pmatrix} W^0 & W^0 & W^0 & W^0 \\ W^0 & W^1 & W^2 & W^3 \\ W^0 & W^2 & W^4 & W^6 \\ W^0 & W^3 & W^6 & W^9 \end{pmatrix} \begin{pmatrix} f_0 \\ f_1 \\ f_2 \\ f_3 \end{pmatrix} \,. \tag{4.33}$$

This simple computation requires 4×4 complex multiplications such as $W^1 \times f_1$, although the number is reduced to 3×3 if $W^0 = 1$ is used.

In the FFT formulation, k and n of $e^{-2\pi ikn/N}$ in (4.31) are expressed using the binary representations $k = 2k_1 + k_0, n = 2n_1 + n_0$ where k_0, k_1, n_0, and n_1 take only 0 or 1. As $W^{4n_1k_1} \equiv 1$, we have

$$e^{-2\pi ikn/4} = W^{kn} = W^{(2k_1+k_0)(2n_1+n_0)} = W^{2n_0k_1} W^{(2n_1+n_0)k_0} \,. \tag{4.34}$$

$F_k = F_{2k_1+k_0}$ can, therefore, be computed through the two-stage matrix operation:

$$F_{2k_1+k_0} = \sum_{n_0=0}^{1} \left(\sum_{n_1=0}^{1} f_{2n_1+n_0} W^{2k_0n_1} \right) W^{(2k_1+k_0)n_0} \,. \tag{4.35}$$

If the matrices are written explicitly, we obtain

$$\begin{pmatrix} F_0 \\ F_1 \\ F_2 \\ F_3 \end{pmatrix} = \begin{pmatrix} 1 & W^0 & 0 & 0 \\ 0 & 0 & 1 & W^1 \\ 1 & W^2 & 0 & 0 \\ 0 & 0 & 1 & W^3 \end{pmatrix} \begin{pmatrix} 1 & 0 & W^0 & 0 \\ 0 & 1 & 0 & W^0 \\ 1 & 0 & W^2 & 0 \\ 0 & 1 & 0 & W^2 \end{pmatrix} \begin{pmatrix} f_0 \\ f_1 \\ f_2 \\ f_3 \end{pmatrix}, \tag{4.36}$$

where half of the non-zero elements are 1, and therefore, the number of required complex multiplications such as $W^2 \times f_2$ is only 4×2.

In the case of $N = 2^2$, the difference between 4×4 and 4×2 is small, but in the case of $N = 2^3$, the computation is decomposed into the three-stage matrix operation, and the number of complex multiplications becomes 8×3. Compared to 8×8 (or 7×7) for the simple computation, the effect of using (4.35) is greater. In the general case of $N = 2^\gamma$, the computation is decomposed into a matrix operation of γ stages, and the number of complex multiplications is $N \times \gamma$, which is $N \log_2 N$, while the number in the simple computation of (4.31) is N^2 or $(N-1)^2$. Thus, as the complex multiplication is the most time-consuming operation in algebra, the larger N is, the greater the increase in speed gained by the FFT formulation [9].

This is the basic principle of FFT. In addition to this,

$$\begin{pmatrix} F_0 \\ F_2 \\ F_1 \\ F_3 \end{pmatrix} = \begin{pmatrix} 1 & W^0 & 0 & 0 \\ 1 & W^2 & 0 & 0 \\ 0 & 0 & 1 & W^1 \\ 0 & 0 & 1 & W^3 \end{pmatrix} \begin{pmatrix} 1 & 0 & W^0 & 0 \\ 0 & 1 & 0 & W^0 \\ 1 & 0 & W^2 & 0 \\ 0 & 1 & 0 & W^2 \end{pmatrix} \begin{pmatrix} f_0 \\ f_1 \\ f_2 \\ f_3 \end{pmatrix} \tag{4.37}$$

is constructed by exchanging the second and third rows of the second-stage matrix in (4.36). In order to examine the number of complex multiplications required for the computation by (4.37), we let

$$\begin{pmatrix} f'_0 \\ f'_1 \\ f'_2 \\ f'_3 \end{pmatrix} = \begin{pmatrix} 1 & 0 & W^0 & 0 \\ 0 & 1 & 0 & W^0 \\ 1 & 0 & W^2 & 0 \\ 0 & 1 & 0 & W^2 \end{pmatrix} \begin{pmatrix} f_0 \\ f_1 \\ f_2 \\ f_3 \end{pmatrix} \tag{4.38}$$

at the first stage. $f'_0 = f_0 + W^0 f_2$ is computed by one complex multiplication (W^0 is not reduced to unity in order to develop a generalized result). $f'_1 = f_1 + W^0 f_3$ is also determined by one complex multiplication. No complex multiplication is required to compute f'_2, because $W^0 = W^2$, hence $f'_2 = f_0 + W^2 f_2 = f_0 - W^0 f_2$ where the complex multiplication $W^0 f_2$ has already been computed in the determination of f'_0. By the same reasoning, f'_3 is computed with no multiplication. Two complex multiplications are then required for the first stage. We next move on to the second stage, letting

$$\begin{pmatrix} F_0 \\ F_2 \\ F_1 \\ F_3 \end{pmatrix} = \begin{pmatrix} 1 & W^0 & 0 & 0 \\ 1 & W^2 & 0 & 0 \\ 0 & 0 & 1 & W^1 \\ 0 & 0 & 1 & W^3 \end{pmatrix} \begin{pmatrix} f'_0 \\ f'_1 \\ f'_2 \\ f'_3 \end{pmatrix}. \tag{4.39}$$

```
      subroutine nlogn(n,x,sign)
      complex   x,wk,hold,q
      dimension m(25),x(2)
      lx=2**n
      do 1 i=1,n
    1 m(i)=2**(n-i)
      do 4 l=1,n
      nblock=2**(l-1)
      lblock=lx/nblock
      lbhalf=lblock/2
      k=0
      do 4 iblock=1,nblock
      fk=k
      flx=lx
      v=sign*6.2831853e0*fk/flx
      wk=cmplx(cos(v),sin(v))
      istart=lblock*(iblock-1)
      do 2 i=1,lbhalf
      j=istart+i
      jh=j+lbhalf
      q=x(jh)*wk
      x(jh)=x(j)-q
      x(j)=x(j)+q
    2 continue
      do 3 i=2,n
      ii=i
      if(k.lt.m(i)) go to 4
    3 k=k-m(i)
    4 k=k+m(ii)
      k=0
      do 7 j=1,lx
      if(k.lt.j) go to 5
      hold=x(j)
      x(j)=x(k+1)
      x(k+1)=hold
    5 do 6 i=1,n
      ii=i
      if(k.lt.m(i)) go to 7
    6 k=k-m(i)
    7 k=k+m(ii)
      if(sign.lt.0.0) return
      do 8 i=1,lx
    8 x(i)=x(i)/flx
      return
      end
```

Fig. 4.10 Example of Fortran subroutine performing the FFT (modified from NLOGN of Robinson [33] with permission of Springer Netherlands)

$F_0 = f'_0 + W^0 f'_1$ is determined by one complex multiplication, and $F_2 = f'_0 + W^2 f'_1$ is computed with no complex multiplications because $W^0 = -W^2$. By a similar reasoning, F_1 is determined by one complex multiplication, and F_3 requires no complex multiplications. Two complex multiplications are again required for the second stage. In total, the number of complex multiplications in (4.37) is four, which is a half of the eight for (4.36) [9].

In the general case of $N = 2^\gamma$, the number of complex multiplications is further reduced from $N \log_2 N$ to $N \log_2 N/2$ because of row exchanges at γ stages. These row exchanges result in the scrambled order of F_n, but the order can be recovered by performing a bit-reversal permutation on n, such as $2^{\gamma-1}n_{\gamma-1} + \cdots + 2^1 n_1 + 2^0 n_0 \rightarrow 2^{\gamma-1}n_0 + 2^{\gamma-2}n_1 + \cdots 2^0 n_{\gamma-1}$. A well-known example in which the above algorithms are implemented in the Fortran language is shown in Fig. 4.10. This algorithm is called "nlogn", which came from the number of complex multiplications $N \log_2 N$.

4.3 Filtering

4.3.1 Filters and Windows

The operation that takes out the component of specific frequencies from an observed or synthetic waveform of ground motion is termed **filtering**, and the equipment or software that actually performs the operation is termed **filter**. Filters are also divided into analog filters for analog electrical signals and digital filters for digital quantities after A/D conversion. However, in the present situation where most seismographs have A/D converters, the filter is typically a digital filter, except for analog filters that perform a pretreatment of the A/D conversion, such as an anti-aliasing filter.

When filters are classified by their functions, a filter that extracts only the frequency components above a certain frequency is called a **high-pass filter** or **low-cut filter**, and a filter that extracts only the frequency components below a certain frequency is called a **low-pass filter** or **high-cut filter**. A filter that extracts the frequency components within a band from a certain frequency to another frequency is called a **band-pass filter**.

Intuitively, the most obvious way to perform the filtering is to apply a **window** to the Fourier transform and perform the inverse Fourier transform. However, if the **rectangular function** is simply used as a window to apply a band-pass filter, **Gibbs' phenomenon** appears after the inverse Fourier transform, disturbing the waveform as described in Sect. 4.2.3. The shape of the window must, therefore, be considered, avoiding blocking the spectrum suddenly at the edges of the window as in the rectangular function. In many cases, the window is chosen to decay the spectrum gradually within a certain width around its edges.

Among this type of window, the **Hanning window**:

$$w(\omega) = \begin{cases} \dfrac{1}{2}\left(1 + \cos\dfrac{\pi\omega}{\Omega}\right), & |\omega| < \Omega \\ 0, & |\omega| \geq \Omega \end{cases} \tag{4.40}$$

is the best known. Using the inverse Fourier transform $R(t)$ in (4.26) for the rectangular function $r(\omega)$ in the frequency domain, and $\delta(t - t_0) \Leftrightarrow e^{-\mathrm{i}\omega t_0}$ from Tables 4.3 and 4.4, we obtain the inverse Fourier transform of the Hanning window $W(t)/2\pi$, where[6]

$$W(t) = \frac{R(t)}{2} + \frac{R(t + \pi/\Omega)}{4} + \frac{R(t - \pi/\Omega)}{4}\,. \tag{4.41}$$

As shown in Fig. 4.11, **ripples** in $W(t)$ are much smaller than those in $R(t)$. If it is problematic that the Hanning window distorts the whole range of $[-\Omega,\ \Omega]$, the **cosine-tapered window**

[6] $W(t)$ agrees with Equation (11.57) of Hino [15] if $Q_0 \to R$, $f \to t$, and $\pi/\Omega \to 1/2\tau_m$.

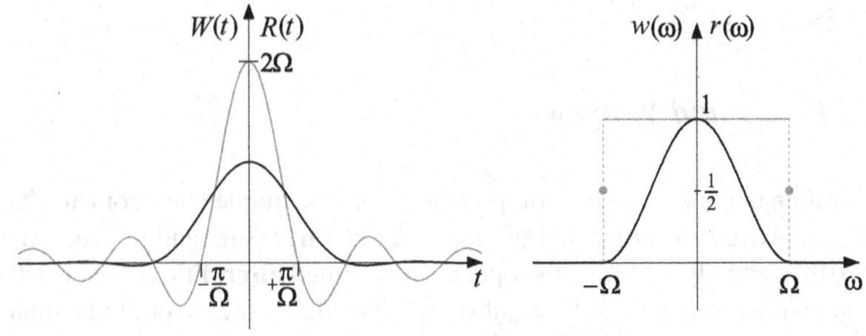

Fig. 4.11 Hanning window $w(\omega)$ in the frequency domain (right) and $W(t)$ in its inverse Fourier transform (left). For comparison, the rectangular function $r(\omega)$ and its inverse Fourier transform $R(t)$ are also shown in gray. Modified from Koketsu [21] with permission of Kindai Kagaku

$$w(\omega) = \begin{cases} 1\,, & |\omega| \le \Omega - \Delta\Omega \\ \dfrac{1}{2}\left(1 + \cos\dfrac{\pi(|\omega| - \Omega + \Delta\Omega)}{\Delta\Omega}\right), & \Omega - \Delta\Omega < |\omega| < \Omega \\ 0\,, & |\omega| \ge -\Omega \end{cases} \tag{4.42}$$

is used, which limits the area to be distorted to the width $\Delta\Omega$ at both ends.

4.3.2 *Low-Pass Filters*

The Fourier transform and inverse Fourier transform operations are necessary for filtering with a window in the frequency domain, however, various problems with the discrete transforms can also occur (Sect. 4.2.3). It is, therefore, desirable, both in terms of the filtering performance and computation time, for a filter to be applied via algebraic operations in the time domain, where possible. These time domain filters are called **recursive filter** because they are generally in the form of **recursive equation**. This subsection explains the theory and design of the recursive filter following the treatment of Saito [34].

When the filter $H(\omega)$ given as the ratio of the quadratic functions of $e^{-\mathrm{i}\omega\Delta t}$ (hereinafter this is referred to as in "quadratic form"):

$$\frac{a_0 + a_1 e^{-\mathrm{i}\omega\Delta t} + a_2 e^{-2\mathrm{i}\omega\Delta t}}{1 + b_1 e^{-\mathrm{i}\omega\Delta t} + b_2 e^{-2\mathrm{i}\omega\Delta t}} \tag{4.43}$$

is applied to the spectrum $F(\omega)$ of the time series $f(t)$,

$$(1 + b_1 e^{-\mathrm{i}\omega\Delta t} + b_2 e^{-2\mathrm{i}\omega\Delta t})G(\omega) = (a_0 + a_1 e^{-\mathrm{i}\omega\Delta t} + a_2 e^{-2\mathrm{i}\omega\Delta t})F(\omega) \tag{4.44}$$

holds between $F(\omega)$ and $G(\omega)$, which is the spectrum of the output time series $g(t)$. Here, we use the property of the time shift for the Fourier transform as listed in

Table 4.3, which is $f(t-t_0) \Leftrightarrow F(\omega)e^{-\mathrm{i}\omega t_0}$, and obtain the inverse Fourier transform of (4.44) as

$$g(t) + b_1 g(t-\Delta t) + b_2 g(t-2\Delta t) = a_0 f(t) + a_1 f(t-\Delta t) + a_2 f(t-2\Delta t)\,. \tag{4.45}$$

Next, we digitize $f(t)$ and $g(t)$ using the sampling rate Δt as $f_i = f(i\Delta t)$ and $g_i = g(i\Delta t)$, then (4.45) yields the recursive equation

$$g_i = a_0 f_i + a_1 f_{i-1} + a_2 f_{i-2} - b_1 g_{i-1} - b_2 g_{i-2}\,. \tag{4.46}$$

The design of the recursive filter is now the determination of coefficients a_0, a_1, a_2, b_1, and b_2 of this recursive equation ensuring filtering performance and computation efficiency. Undefinable f_{-1}, f_{-2}, g_{-1}, and g_{-2} are set to 0.

When quadratic-form filters, which are filters in the form of (4.43), are continuously applied, the number of filters is n, and the coefficients of the jth filter are a_{0j}, a_{1j}, a_{2j}, b_{1j}, and b_{2j}, we have

$$H(\omega) = \prod_{j=1}^{n} H_j^0(\omega)\,, \quad H_j^0(\omega) = \frac{a_{0j} + a_{1j}e^{-\mathrm{i}\omega\Delta t} + a_{2j}e^{-2\mathrm{i}\omega\Delta t}}{1 + b_{1j}e^{-\mathrm{i}\omega\Delta t} + b_{2j}e^{-2\mathrm{i}\omega\Delta t}}\,. \tag{4.47}$$

Following Saito [34], we now normalize the coefficients so that $a'_{0j} = 1$, $a'_{1j} = a_{1j}/a_{0j}$, and $a'_{2j} = a_{2j}/a_{0j}$, and rewrite (4.47) as

$$H(\omega) = G_0 \prod_{j=1}^{n} H_j(\omega)\,,$$
$$G_0 = \prod_{j=1}^{n} a_{0j}\,, \quad H_j(\omega) = \frac{1 + a'_{1j}e^{-\mathrm{i}\omega\Delta t} + a'_{2j}e^{-2\mathrm{i}\omega\Delta t}}{1 + b_{1j}e^{-\mathrm{i}\omega\Delta t} + b_{2j}e^{-2\mathrm{i}\omega\Delta t}}\,. \tag{4.48}$$

When the recursive equations equivalent to these filters:

$$g_{i,0} = f_i\,, \quad g_{i,j} = g_{i,j-1} + a'_{1j}g_{i-1,j-1} + a'_{2j}g_{i-2,j-1} - b_{1j}g_{i-1,j} - b_{2j}g_{i-2,j} \tag{4.49}$$

are applied to f_i $(i+1 = 1, 2, \ldots, N)$, $g_{i,j}$ $(j+1 = 1, 2, \ldots, n)$ are sequentially obtained, and finally $g_{i,n}$ multiplied by G_0 yields the result of this filtering. In comparison with the case where (4.49) is not normalized, the number of multiplications in this case is reduced from five to four, and computation time can be saved accordingly.

The **Butterworth filter** [10] is defined as

$$|H(\omega)|^2 = \frac{1}{1 + \sigma^{2n}(\omega)}\,, \tag{4.50}$$

where $\sigma(\omega)$ is a real function that varies monotonically with ω [34]. Using the solutions σ_j of $1+\sigma^{2n}=0$, (4.50) is rewritten as

$$|H(\omega)|^2 = \prod_{j=1}^{2n} \frac{1}{\sigma - \sigma_j}, \quad \sigma_j = \exp\left(\mathrm{i}\frac{2j-1}{2n}\pi\right) = s_j + \mathrm{i}t_j . \tag{4.51}$$

Dividing the total product $\prod$ on the right-hand side into the parts for $j = 1, 2, \ldots, n$ and $j = n+1, n+2, \ldots, 2n$ and using $\mathrm{i} \cdot (-\mathrm{i}) = 1$ and $\sigma_j^* = \sigma_{2n-j+1}$, we then have

$$\begin{aligned} \prod_{j=1}^{2n} \frac{1}{\sigma - \sigma_j} &= \prod_{j=1}^{n} \frac{1}{\mathrm{i}(\sigma - \sigma_j)} \cdot \prod_{j=n+1}^{2n} \frac{1}{-\mathrm{i}(\sigma - \sigma_j)} \\ &= \prod_{j=1}^{n} \frac{1}{\mathrm{i}(\sigma - \sigma_j)} \cdot \prod_{j=1}^{n} \frac{1}{-\mathrm{i}(\sigma - \sigma_j^*)} . \end{aligned} \tag{4.52}$$

Comparing $|H(\omega)|^2 = H(\omega) \cdot H^*(\omega)$ with (4.52), we obtain

$$H(\omega) = \prod_{j=1}^{n} \frac{1}{\mathrm{i}(\sigma - \sigma_j)} . \tag{4.53}$$

As $\sigma_1, \sigma_2, \ldots, \sigma_n$ are all in the upper half of the complex plane, $H(\omega)$ is analytic in the lower half-plane. This indicates that $H(\omega)$ is the Fourier transform of a causal function.[7]

For $\sigma(\omega)$, we choose the real function that increases monotonically with ω:

$$\sigma(\omega) = c \tan \frac{\omega \Delta t}{2} = \frac{c}{\mathrm{i}} \cdot \frac{1 - e^{-\mathrm{i}\omega\Delta t}}{1 + e^{-\mathrm{i}\omega\Delta t}} . \tag{4.54}$$

$|H(\omega)|^2 = 1/(1+\sigma^{2n}(\omega))$ in (4.50) with $\sigma(\omega)$ in (4.54) is 1 for $\omega = 0$, and approaches 0 for $\omega \to \omega_c = \pi/\Delta t$ (Sect. 4.2.3), therefore, a **low-pass filter** is realized. As $\sigma_{n-j+1} = -\sigma_j^*$ from (4.51), combining every two terms of $\sigma - \sigma_j$ gives

$$H(\omega) = \prod_{j=1}^{n/2} \frac{(1+e^{-\mathrm{i}\omega\Delta t})^2}{(c+t_j)^2 + s_j^2 - 2(c^2 - |\sigma_j|^2)e^{-\mathrm{i}\omega\Delta t} + ((c-t_j)^2 + s_j^2)e^{-2\mathrm{i}\omega\Delta t}} \tag{4.55}$$

in the quadratic form. However, if n is an odd number, the $(n-1)/2+1$-st term remains alone, and a linear-form filter is, therefore, added.

[7] Page 214 of Papoulis [30] states that the condition for this is that $H(p/\mathrm{i})$ is analytic with $\mathrm{Re}\, p \geq 0$, and $\mathrm{Re}\, p \geq 0$ equals $\mathrm{Im}\, \omega \leq 0$ because $\omega = p/\mathrm{i}$.

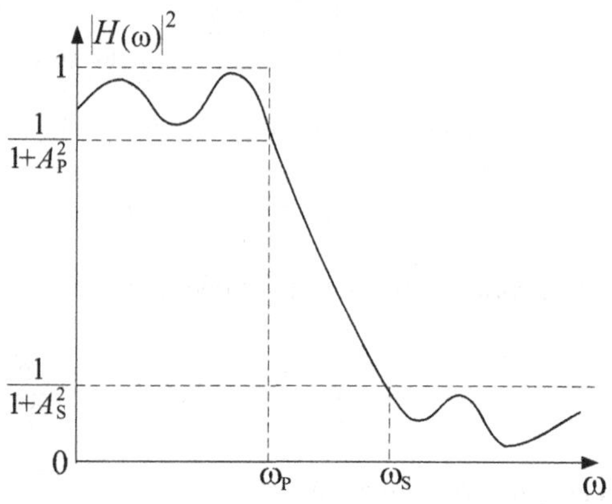

Fig. 4.12 Design of a low-pass filter (modified from Saito [34] with permission of the author)

The parameters n and c in (4.55) are determined from the required filtering performance. In order to design an ideal low-pass filter close to the step function, many quadratic-form filters must be combined with a large n. However, in practical applications, such a realistic function as shown in Fig. 4.12 is sufficient, and this is expressed, using the parameters A_P and A_S, by the allowable range in the passband ($\omega \leq \omega_P$):

$$|H(\omega)|^2 \geq \frac{1}{1 + A_P^2} \tag{4.56}$$

and the allowable range in the stopband of $\omega \geq \omega_S$:

$$|H(\omega)|^2 \leq \frac{1}{1 + A_S^2} . \tag{4.57}$$

Since $|H(\omega)|^2 = 1/(1 + \sigma^{2n}(\omega))$ in (4.50) generally decreases with ω as described above, the minimum value of $|H(\omega)|^2$ in the passband is given by $\omega = \omega_P$. Thus, the condition that (4.56) always holds in the passband is $1/(1 + A_P^2) \leq 1/(1 + \sigma^{2n}(\omega_P))$, that is

$$A_P \geq \sigma^n(\omega_P) = \sigma_P^n . \tag{4.58}$$

Similarly, the condition under which (4.57) always holds in the stopband is

$$A_S \leq \sigma^n(\omega_S) = \sigma_S^n . \tag{4.59}$$

The condition obtained by combining (4.58) and (4.59) is, therefore,

$$\left(\frac{\sigma_S}{\sigma_P}\right)^n \geq \frac{A_S}{A_P} . \tag{4.60}$$

Using (4.60) and (4.54), we obtain the condition for n:

$$n \geq \frac{\ln(A_S/A_P)}{\ln(\sigma_S/\sigma_P)} = \frac{\ln(A_S/A_P)}{\ln(\tan(\omega_S\Delta t/2)\cdot\cot(\omega_P\Delta t/2))} . \tag{4.61}$$

In addition, using (4.58), (4.59), and (4.54), we also obtain the condition for c

$$A_S^{1/n}\cot(\omega_S\Delta t/2) \leq c \leq A_P^{1/n}\cot(\omega_P\Delta t/2) . \tag{4.62}$$

The geometric mean of the upper and lower limits of this inequality:

$$c = \left[(A_P A_S)^{1/n}\cot\frac{\omega_P\Delta t}{2}\cdot\cot\frac{\omega_S\Delta t}{2}\right]^{1/2} \tag{4.63}$$

can, therefore, be adopted.

4.3.3 High-Pass and Band-Pass Filters

For $\sigma(\omega)$, we next select

$$\sigma(\omega) = -c\cot\frac{\omega\Delta t}{2} = \frac{c}{\mathrm{i}}\cdot\frac{1+e^{-\mathrm{i}\omega\Delta t}}{1-e^{-\mathrm{i}\omega\Delta t}} \tag{4.64}$$

instead of (4.54). As $\sigma(\omega)$ approaches $-\infty$ for $\omega \to 0$, and 0 for $\omega \to \omega_c = \pi/\Delta t$ (Sect. 4.2.3), $|H(\omega)|^2 = 1/(1+\sigma^{2n}(\omega))$ in (4.50) approaches 0 for $\omega \to 0$, and 1 for $\omega \to \omega_c$. A **high-pass filter** is, therefore, realized by the Butterworth filter. The filter can be designed in the same way as the low-pass filter. In order to gain the filtering performance $|H(\omega)|^2 \geq 1/(1+A_P^2)$ and $|H(\omega)|^2 \leq 1/(1+A_S^2)$ in the passband of $\omega > \omega_P$ and the stopband of $\omega < \omega_S$, respectively, we adopt

$$n \geq \frac{\ln(A_S/A_P)}{\ln(\cot\omega_S\Delta t/2\cdot\tan\omega_P\Delta t/2)} , \quad c = \left[(A_P A_S)^{1/n}\tan\frac{\omega_P\Delta t}{2}\cdot\tan\frac{\omega_S\Delta t}{2}\right]^{1/2} . \tag{4.65}$$

In comparison to the low-pass and high-pass filters, the realization of a **band-pass filter** by the Butterworth filter is complicated. We first perform the change of variables from σ to λ, which is defined by the quadratic equation

$$\sigma(\omega) = \frac{\lambda^2(\omega)-\lambda_0^2}{\lambda(\omega)} , \quad \lambda(\omega) = c\tan\frac{\omega\Delta t}{2} = \frac{c}{\mathrm{i}}\cdot\frac{1-e^{-\mathrm{i}\omega\Delta t}}{1+e^{-\mathrm{i}\omega\Delta t}} . \tag{4.66}$$

Using the solutions λ_H and $-\lambda_L$ of (4.66) with $\sigma(\omega_P) = \sigma_P$, the passband of $|\sigma| < \sigma_P$ is mapped to the λ band of $\lambda_L < |\lambda| < \lambda_H$. We also have

$$\sigma_P = \lambda_H - \lambda_L , \quad \lambda_0 = \lambda_H\lambda_L \tag{4.67}$$

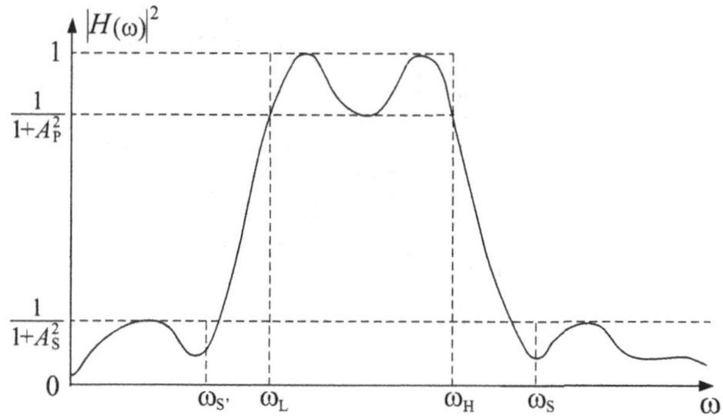

Fig. 4.13 Design of a band-pass filter (modified from Saito [34] with permission of the author)

from (4.66) with $\sigma(\omega_P) = \sigma_P$. Similarly, using the solutions λ_S and $-\lambda_{S'}$ of (4.66) with $\sigma(\omega_S) = \sigma_S$, the stopband of $|\sigma| > \sigma_S$ is mapped to the λ bands of $|\lambda| > \lambda_S$ and $|\lambda| < \lambda_{S'}$. We also have

$$\sigma_S = \lambda_S - \lambda_{S'} \;, \quad \lambda_0 = \lambda_S \lambda_{S'} \tag{4.68}$$

from (4.66) with $\sigma(\omega_S) = \sigma_S$. If λ_H, λ_L, λ_S, and $\lambda_{S'}$ correspond to the angular frequencies ω_H, ω_L, ω_S, and $\omega_{S'}$, respectively (Fig. 4.13), the second equation in (4.66), (4.67), and (4.68) result in

$$\begin{aligned}&\sigma_P = c(\tan(\omega_H \Delta t/2) - \tan(\omega_L \Delta t/2)) \;, \quad \sigma_S = c(\tan(\omega_S \Delta t/2) - \tan(\omega_{S'} \Delta t/2)) \;, \\ &\lambda_0^2 = c^2 \tan(\omega_H \Delta t/2) \tan(\omega_L \Delta t/2) = c^2 \tan(\omega_S \Delta t/2) \tan(\omega_{S'} \Delta t/2) \;. \end{aligned} \tag{4.69}$$

However, since the number of angular frequencies is one redundant compared with the number of equations, for example, $\omega_{S'}$ is automatically determined if ω_H, ω_L, and ω_S are determined. Substituting (4.66) for $\sigma - \sigma_j$ in (4.51), the numerator of the result yields $\lambda^2 - \sigma_j \lambda - \lambda_0^2 = 0$, and we let $\lambda_{jk} = \mu_{jk} + \mathrm{i}\nu_{jk}$ $(k = 1, 2)$ be the solutions of this quadratic equation. The quadratic equation similarly obtained from $\sigma - \sigma_{n-j+1}$ in (4.51) has the solutions $-\lambda_{jk}^* = -\mu_{jk} + \mathrm{i}\nu_{jk}$ $(k = 1, 2)$. We substitute the second equation in (4.66), $(\sigma - \sigma_j) = \prod_{k=1}^{2} (\lambda - \lambda_{jk})$, etc. into $H(\omega)$ and combine every two terms of $\sigma - \sigma_j$, so that we obtain

$$\begin{aligned}&H(\omega) = \\ &\prod_{j=1}^{n/2} \prod_{k=1}^{2} \frac{c(1 - e^{-2\mathrm{i}\omega \Delta t})}{(c + \nu_{jk})^2 + \mu_{jk}^2 - 2(c^2 - |\lambda_{jk}|^2) e^{-\mathrm{i}\omega \Delta t} + ((c - \nu_{jk})^2 + \mu_{jk}^2) e^{-2\mathrm{i}\omega \Delta t}}\end{aligned} \tag{4.70}$$

in the quadratic form. If n is an odd number, the $(n-1)/2+1$-st term remains alone, and a linear-form filter is, therefore, added.

```
      SUBROUTINE BNDPAS(AP, AS, FFL, FFH, FFS, DT, NPTS, Y, Z)
      DIMENSION  Z(NPTS),Y(NPTS),H(50)
      FL = FFL*DT
      FH = FFH*DT
      FS = FFS*DT
      CALL BUTPAS(H,M,GN,N1,FL,FH,FS,AP,AS)
      CALL TANDEM(Z,Y,NPTS,H,M, 0)
      CALL TANDEM(Y,Z,NPTS,H,M,-1)
      DO 10 IJK=1,NPTS
        Z(IJK) = Z(IJK)*GN*GN
   10 CONTINUE
      RETURN
      END
```

Fig. 4.14 Fortran program that executes the band-pass Butterworth filter, where the subroutines BUTPAS and TANDEM by Saito [34] are called

Based on the above formulation, a Fortran program for performing the band-pass Butterworth filter using the subroutines TANDEM and BUTPAS written by Saito [34] in the Fortran language is shown in Fig. 4.14. Recursive filters generally cause a **phase shift**, which means distorting the **phase spectrum**. However, the phase shift is canceled by executing the recursive equations (4.49) in the direction $i + 1 = 1 \rightarrow N$ and then again in the opposite direction $i + 1 = N \rightarrow 1$, although the causality becomes unsatisfied. The program given in Fig. 4.14 shows this case. Note that depending on how AP (A_{P}), AS (A_{S}), FFH ($\omega_{\mathrm{H}}/2\pi$), FFL ($\omega_{\mathrm{L}}/2\pi$), and FFS ($\omega_{\mathrm{S}}/2\pi$) are given, n calculated by (4.65) becomes large, and the dimension of the array H in which a_i and b_i are stored may fall short of 50, although 50 is sufficient for the few examples given by Saito [34].

There exist various recursive filters other than the Butterworth filter, and the **Chebyshev filter**:

$$|H(\omega)|^2 = \frac{1}{1 + \epsilon^2 T_n^2(\sigma(\omega))},$$

$$H(\omega) = \frac{1}{2^{n-1}\epsilon} \prod_{j=1}^{n} \frac{1}{\mathrm{i}(\sigma - \sigma_j)}, \quad \phi = \frac{1}{n} \ln \frac{1 + \sqrt{1 + \epsilon^2}}{\epsilon},$$

$$\sigma_j = \cos\left(\mathrm{i}\frac{2j - 1}{2n}\pi\right) \cosh\phi + \mathrm{i}\sin\left(\mathrm{i}\frac{2j - 1}{2n}\pi\right) \sinh\phi \qquad (4.71)$$

is also often used. The name of this filter originates from the first equation, which includes the Chebyshev polynomial $T_n(\sigma) = \cos(n \cos^{-1}\sigma)$ of order n, but the form of the total product $\prod$ in the second equation is similar to that of the Butterworth filter. In this filter, the parameters n and ϵ are determined from the requirements for the filtering performance.

4.4 Least-Squares Method

4.4.1 Computation in Least-Squares Method[8]

The **least-squares method** is the statistical method commonly used to estimate parameters of an earthquake source and a velocity structure from observed ground motions, as mentioned in Sects. 2.3.1, 2.3.4, 2.3.6, 3.3.2, and 3.3.4. We first explain the least-squares method by taking the position-fixed **CMT inversion** (Sect. 2.3.4) as an example, because its formulation is not very complex. The observed seismogram $F_k^o(t_i, \mathbf{x}_j)$ of the kth component of the ground motion at the ith time t_i and the jth observation point at $\mathbf{x}_j$ is rearranged from the three-dimensional subscript (i, j, k) into the one-dimensional subscript n as y_n $(n = 1, 2, \ldots, N)$. The corresponding synthetic seismogram $F_k(t_i, \mathbf{x}_j)$ is also rearranged to be $f_n(x_m)$ $(m = 1, 2, \ldots, M)$, where x_m are unknown parameters to be determined by the least squares method. In the case of the CMT inversion, they are elements of the moment tensor M_m $(m = 1, 2, \ldots, 6)$ in (2.173). From (2.174), $f_n(x_m)$ is

$$f_n(x_1, x_2, \ldots, x_M) = \sum_{m=1}^{M} A_{nm} x_m \,, \quad A_{nm} = f_{mk}(t_i, \mathbf{x}_j) \tag{4.72}$$

where $f_{mk}(t_i, \mathbf{x}_j)$ are the Green's functions (Sect. 2.3.4), which are rearranged in a similar way as the synthetic seismogram. $M_m = x_m$ are determined so that y_n and $f_n(x_m)$ are statistically equal as much as possible, and this is written formally as

$$\mathbf{y} \simeq \mathbf{f}(\mathbf{x}) \,, \quad \mathbf{y} = (y_n) \,, \; \mathbf{x} = (x_m) \,, \; \mathbf{f} = (f_n) \,. \tag{4.73}$$

Equation (4.73) agrees with Equation (3.34) of Nakagawa and Oyanagi [28], which was termed the **observation equation**. Furthermore, substituting (4.72) for (4.73) gives

$$\mathbf{y} \simeq \mathbf{A}\mathbf{x} \,, \quad \mathbf{A} = (A_{nm}) \,. \tag{4.74}$$

Equation (4.74) is consistent with Equation (3.39) of Nakagawa and Oyanagi [28], therefore, the explanation here will follow that in their book, as follows. Observation is always accompanied by **observational error**, which are represented as $\boldsymbol{\sigma} = (\sigma_n)$. An observation of the ground motion y_n is distributed around its true value y_n^0 with the **normal distribution**

$$L_n = \frac{1}{(2\pi)^{1/2}\sigma_n} \exp\left\{ \frac{(y_n - y_n^0)^2}{2\sigma_n^2} \right\} , \tag{4.75}$$

[8] Among many references relating to Sect. 4.4.1, Nakagawa And Oyanagi [28] provide a full explanation of the practical aspects.

and the Green's function correctly represents the true value as $y_n^0 = f_n(\mathbf{x})$. The overall probability of the problem is called the **likelihood** and given as

$$L = \prod_{n=1}^{N} L_n = \frac{1}{(2\pi)^{N/2}} \prod_{n=1}^{N} \frac{1}{\sigma_n} \exp\left\{ -\sum_{n=1}^{N} \frac{(y_n - f_n(\mathbf{x}))^2}{2\sigma_n^2} \right\} . \tag{4.76}$$

According to the **maximum likelihood estimation**, $\hat{\mathbf{x}}$, which maximizes the likelihood L in (4.76), gives the most probable estimate of $\mathbf{x}$. In other words, $\hat{\mathbf{x}}$ can also be called the **maximum likelihood estimate**. As noted from the form of (4.76), the maximization of L is equivalent to

$$S(\hat{\mathbf{x}}) = \sum_{n=1}^{N} \frac{(y_n - f_n(\hat{\mathbf{x}}))^2}{\sigma_n^2} = \min . \tag{4.77}$$

In (4.77), $y_n - f_n(\hat{\mathbf{x}})$ is a **residual**, and the sum of the squares of residuals is minimized, i.e. this is made "the least", therefore, the estimation by (4.77) is termed the least-squares method. As

$$\frac{\partial S}{\partial x_m} = 0 , \quad m = 1, 2, \ldots, M \tag{4.78}$$

are necessary conditions for (4.77), substituting (4.72) for (4.77) and taking the partial derivatives in (4.78) gives the equations to be solved in the least-squares method. In (4.72), $f_n(x_m)$ is a linear polynomial $\sum_{m=1}^{M} A_{nm} x_m$ (or $\mathbf{f}(\mathbf{x}) = \mathbf{A}\mathbf{x}$ in the format of (4.74)), and the least-squares method in this case is termed the **linear least-squares method**. As well as the position-fixed CMT inversion, the **source inversion** with the **multi-time window** formulation (Sect. 2.3.6) is solved by the linear least-squares method, although the problem is much more complex.

However, in the cases of the variable-position CMT inversion (Sect. 2.3.4), the hypocenter determination (Sect. 2.3.1), or the velocity structure inversion (Sect. 3.3.2), f_n is not a linear polynomial as in (4.72), and the **nonlinear least-squares method** must be adopted. If the approximation $\mathbf{x}^{(0)}$ of $\mathbf{x}$ is known in some way, and the first-order Taylor expansion at $\mathbf{x}^{(0)}$ can approximate $f_n(\mathbf{x})$,

$$f_n(\mathbf{x}) = f_n(\mathbf{x}^{(0)}) + \frac{\partial f_n(\mathbf{x}^{(0)})}{\partial \mathbf{x}} \left(\mathbf{x} - \mathbf{x}^{(0)} \right) . \tag{4.79}$$

Substituting this into the original observation equation (4.73), we obtain

$$\Delta\mathbf{y} \simeq \mathbf{A}\,\Delta\mathbf{x} , \quad \Delta\mathbf{y} = \mathbf{y} - \mathbf{f}(\mathbf{x}^{(0)}) , \quad \Delta\mathbf{x} = \mathbf{x} - \mathbf{x}^{(0)} , \quad \mathbf{A} = \left(\frac{\partial f_n(\mathbf{x}^{(0)})}{\partial \mathbf{x}} \right) . \tag{4.80}$$

The matrix $\mathbf{A}$ consisting of $\partial f_n/\partial x_m$ is in the same form as a transformation matrix between two coordinate systems. As the determinant of a transformation matrix is termed the **Jacobian** (Sect. 3.2.2), $\mathbf{A}$ is sometimes called a **Jacobian matrix** [25]. (4.80) yields (4.74) if $\Delta\mathbf{y} \to \mathbf{y}$ and $\Delta\mathbf{x} \to \mathbf{x}$, and the nonlinear least-squares method, therefore, yields the linear least-squares method. However, since (4.80) is approximate, $\mathbf{x}^{(1)} = \mathbf{x}^{(0)} + \Delta\mathbf{x}$ where $\Delta\mathbf{x}$ is the solution of (4.80) is rarely a correct solution for the problem, and it is necessary to repeat **iterative refinement** to obtain a new $\Delta\mathbf{x}$ by using $\mathbf{x}^{(1)}$ as a new approximation. This iterative procedure is termed the **Gauss–Newton method**.

Hence, all the least-squares methods including the nonlinear least-squares method use the formulation of the linear least-squares method, therefore, we now return to the linear least-squares method. Substituting (4.72) for (4.77) and performing partial derivatives in (4.78), we obtain the simultaneous linear equations

$$\begin{cases} B_{11}x_1 + B_{12}x_2 + \cdots + B_{1M}x_M &= b_1 \\ B_{21}x_1 + B_{22}x_2 + \cdots + B_{2M}x_M &= b_2 \\ \qquad \vdots \\ B_{M1}x_1 + B_{M2}x_2 + \cdots + B_{MM}x_M &= b_M \end{cases},$$

$$B_{mm'} = \sum_{n=1}^{N} \frac{A_{nm}A_{nm'}}{\sigma_n^2}, \quad b_m = \sum_{n=1}^{N} \frac{A_{nm}}{\sigma_n^2} y_n . \tag{4.81}$$

We rewrite (4.81) using the vectors and the matrix in (4.73) and (4.74) as

$$\mathbf{Bx} = \mathbf{b}, \quad \mathbf{B} = \mathbf{A}^{\mathrm{T}}\boldsymbol{\Sigma}^{-1}\mathbf{A}, \quad \mathbf{b} = \mathbf{A}^{\mathrm{T}}\boldsymbol{\Sigma}^{-1}\mathbf{y}, \quad \boldsymbol{\Sigma} = \begin{pmatrix} \sigma_1^2 & & & \\ & \sigma_2^2 & & \\ & & \ddots & \\ & & & \sigma_N^2 \end{pmatrix} . \tag{4.82}$$

($\boldsymbol{\Sigma}$) is a variance matrix in Sect. 3.3.4.
($\boldsymbol{\Sigma}$) is the bold symbol on the left-hand side of the fourth equation of (4.82).
We further rewrite (4.82) using $\mathbf{A}' = \boldsymbol{\Sigma}^{-1/2}\mathbf{A}$ and $\mathbf{y}' = \boldsymbol{\Sigma}^{-1/2}\mathbf{y}$ as

$$\mathbf{A}'^{\mathrm{T}}\mathbf{A}'\mathbf{x} = \mathbf{A}'^{\mathrm{T}}\mathbf{y}' . \tag{4.83}$$

Equations (4.81), (4.82), and (4.83) are termed **normal equation**. As the normal equations are usual simultaneous linear equations, a general solution, for example, the **Gaussian elimination**,[9] which is one of the most basic solutions, can be applied. Returning to (4.81), we subtract B_{m1}/B_{11} times the first equation from every mth equation ($m = 2, 3, \ldots, M$). Through this subtraction, in the mth equation ($m = 2, 3, \ldots, M$), the first term is deleted and the second and third terms yield $B^1_{m2} = B_{m2} - B_{m1}/B_{11} \cdot B_{12}$ and $B^1_{m3} = B_{m3} - B_{m1}/B_{11} \cdot B_{13}$, respectively.

[9] Kreyszig [22] termed this "Gauss elimination", although "Gaussian elimination" is more popular.

We then subtract B^1_{m2}/B^1_{22} times the second equation from every mth equation ($m = 3, 4, \ldots, N$). Through this subtraction, in the mth equation ($m = 2, 3, \ldots, M$), the second term is deleted, and the third term yields $B^2_{m3} = B^1_{m3} - B^1_{m2}/B^1_{22} \cdot B^1_{23}$. If this operation is repeated ($M - 1$) times, (4.81) yields

$$\begin{cases} B_{11}x_1 + B_{12}x_2 + \cdots + B_{1M}x_M = & b_1 \\ \qquad B^1_{22}x_2 + \cdots + B^1_{2M}x_M = & b^1_2 \\ \qquad\qquad \vdots & \\ \qquad\qquad B^{M-1}_{MM}x_M = b^{M-1}_M & \end{cases} . \tag{4.84}$$

From the Mth equation, $x_M = b^{M-1}_M / B^{M-1}_{MM}$ is obtained. From this and the $(M-1)$-th equation $x_{M-1} = (b^{M-2}_{M-1} - B^{M-2}_{M-1,M}x_M)/B^{M-2}_{M-1,M-1}$ is then obtained. In the reverse direction such as these, x_m ($m = M, M-1, \ldots, 1$) are sequentially obtained.

We can rewrite (4.84) in the form of (4.82) as

$$\mathbf{B}'\mathbf{x} = \mathbf{b}', \quad \mathbf{B}' = \begin{pmatrix} B_{11} & B_{12} & \cdots & B_{1M} \\ 0 & B^1_{22} & \cdots & B^1_{2M} \\ & & \ddots & \vdots \\ 0 & 0 & 0 \cdots 0 & B^{M-1}_{MM} \end{pmatrix} \tag{4.85}$$

so that the Gaussian elimination is to transform $\mathbf{B}$ into an upper triangular matrix $\mathbf{B}'$ by repeated "addition of a constant multiple of one row to another row", among the **elementary operations** for matrices [22]. If B_{11}, B^1_{22}, etc. are close to zero, constants in the operations such as B_{m1}/B_{11} and B^1_{m2}/B^1_{22} become very large, and **rounding error** [22] can occur during computation. In the above, when performing the elementary operation of the mth row, the denominator of the constant is fixed to B^{m-1}_{mm}. However, in order to avoid rounding errors, the element that gives the largest absolute value in the mth row is found, moved to the position of B^{m-1}_{mm} by another elementary operation consisting of the "interchange of two columns", and used as the denominator of the constant. This procedure is termed **pivoting** [22, 28]. Furthermore, in order to absorb the difference between rows as much as possible, it is desirable to normalize all rows by elementary operations consisting of "multiplication of a row by a non-zero constant". This procedure is termed **scaling** [22].

Taking advantage of the fact that the coefficient matrix is in the form of $\mathbf{A}'^{\mathrm{T}}\mathbf{A}'$ in the normal equation of (4.83), several approaches to reduce the rounding errors of the Gaussian elimination have been proposed. Among these, the modified Gram–Schmidt algorithm [28] is shown here. When $\mathbf{A}'$ is a real matrix of N rows and M columns, $\mathbf{A}'$ can be factorized as

$$\mathbf{A}' = \mathbf{Q}\mathbf{R} \quad \mathbf{Q}^{\mathrm{T}}\mathbf{Q} = \mathbf{I} \quad \mathbf{R} = \begin{pmatrix} R_{11} & R_{12} & \cdots & R_{1M} \\ 0 & R_{22} & \cdots & R_{2M} \\ & & \ddots & \vdots \\ 0 & 0 & 0 \cdots 0 & R_{MM} \end{pmatrix} \tag{4.86}$$

using an orthogonal matrix $\mathbf{Q}$ of N rows and M columns and an upper triangular matrix $\mathbf{R}$ of M rows and M columns [36], which is termed the **QR-factorization** [22]. Since $\mathbf{A}'$ is generally a real matrix in problems discussed in Sects. 2.3.1, 2.3.4, 2.3.6, 3.3.2, and 3.3.4, (4.86) holds so that this can be substituted into (4.83) and

$$\mathbf{R}^{\mathrm{T}}\mathbf{Q}^{\mathrm{T}}\mathbf{Q}\mathbf{R}\,\mathbf{x} = \mathbf{R}^{\mathrm{T}}\mathbf{Q}^{\mathrm{T}}\mathbf{y} \;\Rightarrow\; \mathbf{R}^{\mathrm{T}}\mathbf{R}\,\mathbf{x} = \mathbf{R}^{\mathrm{T}}\mathbf{Q}^{\mathrm{T}}\mathbf{y} \;\Rightarrow\; \mathbf{R}\,\mathbf{x} = \mathbf{Q}^{\mathrm{T}}\mathbf{y} \tag{4.87}$$

is obtained. Equation (4.87) has the same form as (4.85) for the Gaussian elimination, therefore, if $\mathbf{Q}$ and $\mathbf{R}$ are obtained, x_m are sequentially obtained in the reverse order of $m = M, M-1, \ldots, 1$.

When $\mathbf{A}'$ and $\mathbf{Q}$ are represented as

$$\mathbf{A}' = (\mathbf{a}_1, \mathbf{a}_2, \ldots, \mathbf{a}_M), \quad \mathbf{Q} = (\mathbf{q}_1, \mathbf{q}_2, \ldots, \mathbf{q}_M) \tag{4.88}$$

using N-elements column vectors, the following procedure can be considered for the QR factorization from the form of (4.86) [28]:

1. $\mathbf{q}_1$ is given the normalized $\mathbf{a}_1$, so that $R_{11} = |\mathbf{a}_1|$ and $\mathbf{q}_1 = \mathbf{a}_1/R_{11}$.
2. Subtracting the component parallel to $\mathbf{a}_1$ from $\mathbf{a}_2$ gives the vector orthogonal to $\mathbf{a}_1$ and the normalized vector orthogonal to $\mathbf{q}_1$, so that $R_{12} = \mathbf{q}_1^{\mathrm{T}}\mathbf{a}_2$, $R_{22} = |\mathbf{a}_2 - \mathbf{q}_1 R_{12}|$, and $\mathbf{q}_2 = \mathbf{a}_2 - \mathbf{q}_1 R_{12}/R_{22}$.
3. Subtracting the components parallel to $\mathbf{a}_1$ and $\mathbf{a}_2$ from $\mathbf{a}_3$ gives the vector orthogonal to $\mathbf{a}_1$ and $\mathbf{a}_2$, and the normalized vector orthogonal to $\mathbf{q}_1$ and $\mathbf{q}_2$, so that $R_{13} = \mathbf{q}_1^{\mathrm{T}}\mathbf{a}_3$, $R_{23} = \mathbf{q}_2^{\mathrm{T}}\mathbf{a}_3$, and $R_{33} = |\mathbf{a}_3 - \mathbf{q}_1 R_{13} - \mathbf{q}_2 R_{23}|$, $\mathbf{q}_3 = (\mathbf{a}_3 - \mathbf{q}_1 R_{13} - \mathbf{q}_2 R_{23})/R_{33}$.
4. Similarly, subtracting the components parallel to $\mathbf{a}_1, \mathbf{a}_2, \ldots, \mathbf{a}_{k-1}$ gives the vector orthogonal to $\mathbf{a}_1, \mathbf{a}_2, \ldots, \mathbf{a}_{k-1}$ and the normalized vector orthogonal to $\mathbf{q}_1, \mathbf{q}_2, \ldots, \mathbf{q}_{k-1}$, so that $R_{jk} = \mathbf{q}_j^{\mathrm{T}}\mathbf{a}_k$ $(j = 1, 2, \ldots, k-1)$, $R_{kk} = \left|\mathbf{a}_k - \sum_{j=1}^{k-1} \mathbf{q}_j R_{jk}\right|$, and $\mathbf{q}_k = \left(\mathbf{a}_k - \sum_{j=1}^{k-1} \mathbf{q}_j R_{jk}\right)/R_{kk}$.

This is the original version of the Gram–Schmidt algorithm (the "classical Gram–Schmidt algorithm"[10]). Although the algorithm is clear, it has the following drawbacks with respect to numerical errors [28]. Even after $\mathbf{q}_2$ is made orthogonal to $\mathbf{q}_1$ in step 2 of the above procedure, a small component parallel to $\mathbf{q}_1$ still remains in $\mathbf{q}_2$ due to numerical errors such as rounding errors, and $\mathbf{q}_2$ is assumed to be $\mathbf{q}_2 + \epsilon\mathbf{q}_1$. We then perform the computation of R_{23} in step 3 and obtain

$$R_{23} = (\mathbf{q}_2 + \epsilon\mathbf{q}_1)^{\mathrm{T}}\mathbf{a}_3 = \mathbf{q}_2^{\mathrm{T}}\mathbf{a}_3 + \epsilon\mathbf{q}_1^{\mathrm{T}}\mathbf{a}_3\,. \tag{4.89}$$

[10] Although there is an extensive literature mentioning the classical Gram–Schmidt algorithm, few sources show the original papers. Axelsson [4] is one of the few exceptions, citing Schmidt [35] on page 71. However, the formulation of the classical Gram–Schmidt algorithm is not shown there, but is given in the paper by the Danish J. P. Gram [13], which is further cited therein.

This equation indicates that the error ϵ propagates into R_{23}.

To avoid such **error propagation**, Björk [7] modified the above procedure as follows and called the new procedure the **modified Gram–Schmidt algorithm** [28]:

1. $\mathbf{q}_1$ is given the normalized $\mathbf{a}_1$, so that $R_{11} = |\mathbf{a}_1|$ and $\mathbf{q}_1 = \mathbf{a}_1/R_{11}$. In addition, the components parallel to $\mathbf{a}_1$ are subtracted from $\mathbf{a}_j$ $(j = 2, 3, \ldots, M)$ so that $R_{1j} = \mathbf{q}_1^{\mathrm{T}}\mathbf{a}_j$ and $\mathbf{a}_j^{(1)} = \mathbf{a}_j - \mathbf{q}_1 R_{1j}$ $(j = 2, 3, \ldots, M)$.
2. $\mathbf{q}_2$ is given the normalized $\mathbf{a}_2^{(1)}$, so that $R_{22} = \left|\mathbf{a}_2^{(1)}\right|$ and $\mathbf{q}_2 = \mathbf{a}_2^{(1)}/R_{22}$. In addition, the components parallel to $\mathbf{q}_2$ are subtracted from $\mathbf{a}_j^{(1)}$ $(j = 3, 4, \ldots, M)$ so that $R_{2j} = \mathbf{q}_2^{\mathrm{T}}\mathbf{a}_j^{(1)}$ and $\mathbf{a}_j^{(2)} = \mathbf{a}_j^{(1)} - \mathbf{q}_2 R_{2j}$ $(j = 3, 4, \ldots, M)$.
3. Similarly, $\mathbf{q}_k$ $(k = 3, 4, \ldots, M-1)$ are given the normalized $\mathbf{a}_k^{(k-1)}$ so that $R_{kk} = \left|\mathbf{a}_k^{(k-1)}\right|$ and $\mathbf{q}_k = \mathbf{a}_k^{(k-1)}/R_{kk}$. In addition, the components parallel to $\mathbf{q}_k$ are subtracted from $\mathbf{a}_j^{(k-1)}$ $(j = k+1, k+2, \ldots, M)$ so that $R_{kj} = \mathbf{q}_k^{\mathrm{T}}\mathbf{a}_j^{(k-1)}$ and $\mathbf{a}_j^{(k)} = \mathbf{a}_j^{k-1} - \mathbf{q}_k R_{kj}$ $(j = k+1, k+2, \ldots, M)$.
4. Finally, $\mathbf{q}_M$ is given the normalized $\mathbf{a}_M^{(M-1)}$, so that $R_{MM} = \left|\mathbf{a}_M^{(M-1)}\right|$ and $\mathbf{q}_M = \mathbf{a}_M^{(M-1)}/R_{MM}$.

In the modified Gram–Schmidt algorithm also, a small component parallel to $\mathbf{q}_1$ still remains in $\mathbf{q}_2$ in step 2 due to the numerical errors, and $\mathbf{q}_2 + \epsilon\mathbf{q}_1$ is assumed (although $\mathbf{q}_2$ is orthogonal to $\mathbf{q}_1$). Using this, the computation of R_{23} in step 2 is performed. We then substitute $\mathbf{q}_1^{\mathrm{T}}\mathbf{q}_1 = 1$, $\mathbf{q}_2^{\mathrm{T}}\mathbf{q}_1 = 0$, $R_{13} = \mathbf{q}_1^{\mathrm{T}}\mathbf{a}_3$, and $\mathbf{a}_3^{(1)} = \mathbf{a}_3 - \mathbf{q}_1 R_{13}$ from step 1, obtaining

$$\begin{aligned} R_{23} &= (\mathbf{q}_2 + \epsilon\mathbf{q}_1)^{\mathrm{T}}\mathbf{a}_3^{(1)} = (\mathbf{q}_2^{\mathrm{T}} + \epsilon\mathbf{q}_1^{\mathrm{T}})(\mathbf{a}_3 - \mathbf{q}_1 R_{13}) \\ &= \mathbf{q}_2^{\mathrm{T}}\mathbf{a}_3 - \mathbf{q}_2^{\mathrm{T}}\mathbf{q}_1 R_{13} + \epsilon\mathbf{q}_1^{\mathrm{T}}\mathbf{a}_3 - \epsilon\mathbf{q}_1^{\mathrm{T}}\mathbf{q}_1 R_{13} = \mathbf{q}_2^{\mathrm{T}}\mathbf{a}_3 \,. \end{aligned} \tag{4.90}$$

In this equation, $+\epsilon\mathbf{q}_1^{\mathrm{T}}\mathbf{a}_3 = +\epsilon R_{13}$ and $-\epsilon\mathbf{q}_1^{\mathrm{T}}\mathbf{q}_1 R_{13} = -\epsilon R_{13}$ are canceled by each other, so that the QR factorization may be performed with high accuracy, which further leads to a least-squares solution with high accuracy. In order to further increase the accuracy, **pivoting** and **scaling** are effective, as in the Gaussian elimination [28].

The situation in which a solution is only approximate due to numerical errors in the linear least-squares method is similar to the situation in which $\mathbf{x}^{(1)} = \mathbf{x}^{(0)} + \Delta\mathbf{x}$ using the solution of (4.80) is only approximate in the nonlinear least-squares method. Thus, assuming the first solution of the linear least-squares method for the observation equation $\mathbf{y} \simeq \mathbf{A}\mathbf{x}$ in (4.74) is $\mathbf{x}^{(0)}$, the following observation equation is newly derived:

$$\Delta\mathbf{y} \simeq \mathbf{A}\,\Delta\mathbf{x}\,, \quad \Delta\mathbf{y} = \mathbf{y} - \mathbf{A}\mathbf{x}^{(0)}\,, \quad \Delta\mathbf{x} = \mathbf{x} - \mathbf{x}^{(0)}\,, \tag{4.91}$$

and a solution including numerical errors can be improved by **iterative refinement** using this.

4.4.2 Constraints in Least-Squares Method[11]

As shown in Sect. 4.4.1, problems due to numerical errors in the application of least-squares methods can be avoided by computational procedures such as pivoting, scaling, the modified Gram–Schmidt algorithm, and iterative refinement. However, issues arising from the characteristics of the problems to be solved cannot be avoided by employing these computational procedures. For example, in a CMT inversion or source inversion, instability may occur when observed seismograms contain large observational errors, or when synthetic seismograms contain large errors due to the limitations of the velocity structures used for seismogram synthesis. Other examples include the instability which can occur in a source inversion or velocity structure inversion when a large number of unknown parameters are assigned (e.g., Sect. 2.3.6), or in a nonlinear least-squares method when nonlinearity is strong or when good initial values of unknown parameters are not obtained.

In order to avoid such instabilities due to the characteristics of the problems to be solved, **constraints** are introduced. For example, $\mathbf{x} \simeq \mathbf{x}_0$ is one of the simplest constraints when the approximate estimates $\mathbf{x}_0$ of unknown parameters $\mathbf{x}$ are known. We include this as well as various other constraints and collectively represent them as

$$\mathbf{h} \simeq \mathbf{g}(\mathbf{x}), \quad \mathbf{h} = (h_{n'}), \ \mathbf{g} = (g_{n'}), \ n' = 1, 2, \ldots, N' \tag{4.92}$$

where $h_{n'} - g_{n'}(\mathbf{x})$ has a normal distribution with a common standard deviation ρ. The overall probability of all the constraints is thus given as

$$L' = \prod_{n'=1}^{N'} L'_{n'} = \frac{1}{(2\pi)^{N'/2}\rho^{N'}} \exp\left\{-\sum_{n'=1}^{N'} \frac{(h_{n'} - g_{n'}(\mathbf{x}))^2}{2\rho^2}\right\}. \tag{4.93}$$

If event A occurs with event H as its cause, and if H_i ($i = 1, 2, \ldots, I$) are all possible causes of H, H_i are independent of each other. If each probability is then $P(H_i)$, $P(H_i)$ is termed a "prior probability". The conditional probability in which the result A occurs from each cause H_i is denoted by $P(A|H_i)$, and the conditional probability in which the cause is H_i when A occurs is denoted by $P(H_i|A)$, and $P(H_i|A)$ is termed a "posterior probability". **Bayes' theorem** states that the posterior probability is given as

$$P(H_i|A) = \frac{P(A|H_i)\,P(H_i)}{\sum_{i=1}^{I} P(A|H_i)\,P(H_i)}. \tag{4.94}$$

[11] Among the extensive literature relating to Sect. 4.4.2, Bishop [6] explains the theoretical aspects in the most detail.

Bayes [5] proved (4.94) in several cases. If event H is represented as a continuum H rather than a discrete quantity H_i ($i = 1, 2, \ldots, I$), then the posterior probability $P(H|A)$ is represented as

$$P(H|A) = \frac{P(A|H)\,P(H)}{\int P(A|H)\,P(H)\,dH} \tag{4.95}$$

using the prior probability $P(H)$ [26]. The denominators on the right-hand sides of (4.94) and (4.95) are constants for the normalization, therefore, the posterior probability on the left-hand side is a probability density function.

For the **inversions** in this book, because an earthquake source or a velocity structure is a cause H and ground motions are results A (Sect. 2.3.4), the probabilities of the constraints on the model of an earthquake source or a velocity structure are prior probabilities, and the probability that both the constraint and the observation equation in (4.74) are satisfied is a posterior probability. In the problems in this book, the prior probability $P(H)$ is, therefore, the normal distribution L' in (4.93), and the posterior probability $P(A|H)$ is the probability distribution L_c of a whole constrained inversion. Since $P(A|H)$ is the probability of the observation equation given the constraints, this is the normal distribution L in (4.76), and it is also called the **likelihood** as for L. Substituting the above into (4.95), we obtain

$$L_c = \frac{L\,L'}{D}\,, \quad D = \int L\,L' d\mathbf{x}\,,$$
$$L\,L' = \frac{1}{(2\pi)^{(N+N')/2}\rho^{N'}} \prod_{n=1}^{N} \frac{1}{\sigma_n} \exp\left\{ -\sum_{n=1}^{N} \frac{(y_n - f_n(\mathbf{x}))^2}{2\sigma_n^2} - \sum_{n'=1}^{N'} \frac{(h_{n'} - g_{n'}(\mathbf{x}))^2}{2\rho^2} \right\}, \tag{4.96}$$

where D is the normalization constant in the denominator of (4.95). An integral with respect to a stochastic variable is termed a **marginal probability** [6], therefore, D is a marginal probability with respect to the model parameters $\mathbf{x}$.

As with the **maximum likelihood estimation** for the likelihood L, we assume that $\hat{\mathbf{x}}$, which maximizes the posterior probability L_c, is the most probable $\mathbf{x}$. This is termed the "maximum posterior estimation", which is abbreviated as the **MAP estimation** ("A" is derived from the Latin "a posteriori", from which "posterior" comes), and $\hat{\mathbf{x}}$ is the **maximum posterior estimate**. The MAP estimation is applied to (4.96), and as noted from the form of (4.96), the maximization of L_c is equivalent to the least-squares method represented by

$$S_c(\hat{\mathbf{x}}) = \sum_{n=1}^{N} \frac{(y_n - f_n(\hat{\mathbf{x}}))^2}{\sigma_n^2} + \sum_{n'=1}^{N'} \frac{(h_{n'} - g_{n'}(\hat{\mathbf{x}}))^2}{\rho^2} = \min\,. \tag{4.97}$$

Furthermore, $\mathbf{f}(\mathbf{x}) = \mathbf{A}\mathbf{x}$ in (4.74) is substituted for (4.97), and the linear relation $\mathbf{g}(\mathbf{x}) = \mathbf{G}\mathbf{x}$ is assumed for the constraints, and this is also substituted for (4.97). We then construct the simultaneous linear equations from

$$\frac{\partial S_c}{\partial x_m} = 0\,, \quad m = 1, 2, \ldots, M \tag{4.98}$$

and obtain the normal equations for the constrained linear least-squares method as

$$\mathbf{B}_c\,\mathbf{x} = \mathbf{b}_c\,, \quad \mathbf{B}_c = \mathbf{A}^{\mathrm{T}}\boldsymbol{\Sigma}^{-1}\mathbf{A} + \frac{1}{\rho^2}\mathbf{G}^{\mathrm{T}}\mathbf{G}\,, \ \mathbf{b}_c = \mathbf{A}^{\mathrm{T}}\boldsymbol{\Sigma}^{-1}\mathbf{y} + \frac{1}{\rho^2}\mathbf{G}^{\mathrm{T}}\mathbf{h}\,. \tag{4.99}$$

$\boldsymbol{\Sigma}$ is the variance matrix given in (4.82).

ρ is a type of parameter that controls, in $\mathbf{B}_c$ and $\mathbf{b}_c$, the relative weight of the second terms of the constraints to the first terms of the data (observations). This type of parameter is referred to as a **hyperparameter**, because it has different properties from the parameters of a model. For example, if we give a constraint that the slip distribution is smooth, as in the source inversion of Sect. 2.3.6, and a hyperparameter makes the weight of the constraints too large, we obtain a solution that is unrealistically smooth and has no significant feature. Conversely, if the weight is too small, numerical instability cannot be removed.

To avoid such a **tradeoff** and obtain the most probable model, the following approach is taken. The marginal probability of the likelihood and the prior probability is called the **marginal likelihood**, and the marginal likelihood can be regarded as the probability that the data are generated when the model parameters are randomly sampled from the prior probability [6]. The parameters with the maximum marginal likelihood, therefore, become the most probable model. If we take the position that the hyperparameter is not a stochastic variable, the prior probability, likelihood, and model parameters are L', L, and $\mathbf{x}$, respectively, so that the marginal likelihood is

$$\mathcal{L} = \int L\,L' d\mathbf{x}\,, \tag{4.100}$$

which agrees with D in (4.96). Conversely, if we take the position that the hyperparameter is also a stochastic variable according to the Bayesian statistics, we must give a prior probability $P(\rho)$ to ρ and integrate not only $\mathbf{x}$ but also ρ as

$$\mathcal{L} = \iint L(\mathbf{x})L'(\mathbf{x}, \rho)P(\rho)\,d\mathbf{x}\,d\rho\,. \tag{4.101}$$

However, if $P(\rho)$ is sharply peaked around a certain value, (4.101) can be approximated by (4.100). This approximation is termed **evidence approximation**. In the **empirical Bayes estimation**, the model parameters and hyperparameter are determined to maximize the marginal likelihood with the evidence approximation [6].

According to Yabuki and Matsu'ura [38], the maximization of $\mathcal{L}$ is formulated based on the empirical Bayes estimation, as follows. The observational errors are represented by σ, $\boldsymbol{\Sigma}$ in (4.82) is replaced with $\sigma^2\mathbf{I}$ ($\mathbf{I}$ is a unit matrix), and σ is also assumed to be a hyperparameter. Performing a change of variables from ρ to $\alpha = \sigma/\rho$ and multiplying the normal equation in (4.99) by σ^2, the equation yields

$$\mathbf{B}_c \mathbf{x} = \mathbf{b}_c\,, \quad \mathbf{B}_c = \mathbf{A}^{\mathrm{T}}\mathbf{I}^{-1}\mathbf{A} + \alpha^2\mathbf{G}^{\mathrm{T}}\mathbf{G}\,,\ \mathbf{b}_c = \mathbf{A}^{\mathrm{T}}\mathbf{I}^{-1}\mathbf{y} + \alpha^2\mathbf{G}^{\mathrm{T}}\mathbf{h}\,. \tag{4.102}$$

Solving this with the procedures in Sect. 4.4.1, we obtain the least-squares solution $\hat{\mathbf{x}}$ and $S_c(\hat{\mathbf{x}}) = s(\hat{\mathbf{x}})/\sigma^2$. In addition, performing the integration $\mathcal{L} = \int L'L\,d\mathbf{x}$ in (4.96) analytically, we obtain

$$\mathcal{L} = \frac{\alpha^{N'}}{(2\pi\sigma^2)^{(N+N'-M)/2}} ||\,\mathbf{I}\,||^{-1/2} ||\mathbf{A}^{\mathrm{T}}\mathbf{I}^{-1}\mathbf{A} + \alpha^2\mathbf{G}^{\mathrm{T}}\mathbf{G}||^{-1/2} \exp\left\{-\frac{1}{2\sigma^2}s(\mathbf{x})\right\}, \tag{4.103}$$

where $||\ ||$ indicates the absolute value of a determinant. Among the necessary conditions for maximizing the marginal likelihood $\mathcal{L}(\hat{\mathbf{x}})$ for the least-squares solution, that related to σ can be solved analytically as

$$\frac{\partial \mathcal{L}(\hat{\mathbf{x}})}{\partial \sigma^2} = 0 \ \Rightarrow\ \hat{\sigma}^2 = \frac{s(\hat{\mathbf{x}})}{N + N' - M}\,, \tag{4.104}$$

and this solution is substituted for $\mathcal{L}(\hat{\mathbf{x}})$ to obtain $\mathcal{L}(\hat{\mathbf{x}};\hat{\sigma})$. Since there is no analytical solution for α, $\check{\alpha}$ which maximizes $\mathcal{L}(\hat{\mathbf{x}};\hat{\sigma})$ or $\log\mathcal{L}(\hat{\mathbf{x}};\hat{\sigma})$ is searched for numerically. If α changes upon searching, the normal equation (4.102) changes, therefore, $\hat{\mathbf{x}}$ and $s(\hat{\mathbf{x}})$ must be found again each time. The least-squares solution $\check{\mathbf{x}}$ corresponding to $\check{\alpha}$ is the solution of the constrained linear inversion. The formulation of the constrained nonlinear inversion is presented by Koketsu and Higashi [21].

Independent of the studies on the empirical Bayes estimation, Akaike [2] proposed a method to select hyperparameters by minimizing the "Akaike Bayesian Information Criterion" (**ABIC**):

$$\mathrm{ABIC} = -2\log\mathcal{L} + 2N_h\,, \quad N_h = \text{number of hyperparameters}\,. \tag{4.105}$$

In 1974, the same author [1] proposed the "Akaike Information Criterion" (**AIC**):

$$\mathrm{AIC} = -2\log\mathcal{L} + 2N_p\,, \quad N_p = \text{number of model parameters} \tag{4.106}$$

as a criterion for selecting a model in an inversion. The ABIC is an extension of the AIC to a constrained inversion. Since $\mathcal{L}$ in (4.105) is the same as the marginal likelihood $\mathcal{L}$ in (4.100), substituting (4.103) results in[12]

$$\mathrm{ABIC} = (N + N' - M)\log s(\hat{\mathbf{x}}) - N'\log\alpha^2 + \log||\mathbf{A}^{\mathrm{T}}\mathbf{I}^{-1}\mathbf{A} + \alpha^2\mathbf{G}^{\mathrm{T}}\mathbf{G}|| + C \tag{4.107}$$

---

[12] This agrees with Equation (49) of Yabuki and Matsu'ura [38] and Equation (20) of Fukahata [11] if $g \to N'$ and $\mathbf{G} \to \mathbf{G}^{\mathrm{T}}\mathbf{G}$.

for the constrained linear inversion herein (C is a constant independent of α). However, in many cases, the number of hyperparameters is not changed, therefore, minimizing the ABIC maximizes only the marginal likelihood $\mathcal{L}$, and the method of Akaike [2] is equivalent to the empirical Bayes estimation. In the work of Akaike [2], the ABIC is related to information entropy, which provides a theoretical basis for the empirical Bayes estimation. The ABIC has the feature that it can be used when designing a constrained linear inversion by varying the number of hyperparameters, and this is an original way to use the ABIC, as in the AIC [28].

Problems

4.1 Consider an electromagnetic seismograph consisting of a pendulum, condenser transducer, amplifier, differentiator, and driver, as shown in Fig. 4.15, which is used for a velocity strong motion seismograph with a servo mechanism (Sect. 4.1.4).

The condenser transducer and amplifier convert $l\theta$ to the input voltage V_{in} with amplification factor A. The differentiator is a circuit that is designed such that the output voltage is proportional to the rate of change of V_{in} with a negative sign, using a capacitor C and a resistor R, as $V_{\mathrm{out}} = -CR\, dV_{\mathrm{in}}/dt$. The driver with a resistance r generates a force $G\,i$ where $i = V_{\mathrm{out}}/r$.

Obtain an equation of motion similar to (4.6) for the pendulum using the above relations and $U(\omega)$ in (4.9) for the equation. Show that $VU(\omega)$ approaches $\dfrac{mr}{GCR}$ if the amplifier is set so that $V = A$ and A becomes large.

4.2 Prove the property of causality for the Fourier transform in Table 4.3:

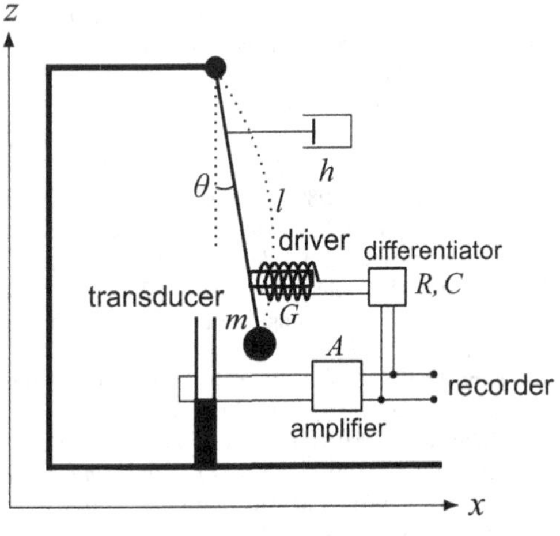

Fig. 4.15 An electromagnetic seismograph consisting of a pendulum, condenser transducer, amplifier, differentiator, and driver

$$f(t) = 0\,,\; t < 0 \Leftrightarrow \mathrm{Re}\,F(\omega) = \frac{1}{\pi}\int_{-\infty}^{+\infty}\frac{\mathrm{Im}\,F(y)}{\omega - y}dy$$

after Papoulis [30] and Aki and Richards [3].

References

1. Akaike, H. (1974). A new look at the statistical model identification. *IEEE Transactions on Automatic Control, 19*, 716–723.
2. Akaike, H. (1980). Likelihood and Bayes procedure. In J. M. Bernards, et al. (Eds.), *Bayesian statistics* (pp. 143–166). Valencia: University Press.
3. Aki, K., & Richards, P. G. (2002). *Quantitative seismology* (2nd ed., p. 700). Sausalito: University Science Books.
4. Axelsson, O. (1996). *Iterative solution methods* (Paperback ed., p. 654). Cambridge: Cambridge University Press.
5. Bayes, T. (1763). An essay towards solving a problem in the doctrine of chance. *Philosophical Transactions of the Royal Society, 53*, 370–418.
6. Bishop, C. M. (2006). *Pattern recognition and machine learning* (p. 738). Berlin: Springer.
7. Björk, Å. (1967). Solving linear least squares problems by Gram-Schmidt orthogonalization. *BIT, 7*, 1–21.
8. Brady, A. G. (2009). Strong-motion accelerographs: Early history. *Earthquake Engineering & Structural Dynamics, 38*, 1121–1134.
9. Brigham, E. O. (1974). *The fast Fourier transform and its applications* (p. 252). Englewood Cliffs: Prentice Hall.
10. Butterworth, S. (1930). On the theory of filter amplifiers. *Experimental wireless & the radio engineer, 7*, 536–541.
11. Fukahata, Y. (2009). Development of study of inversion analyses Using ABIC in seismology. *Zisin (Journal of the Seismological Society of Japan), 61*, S103–S113. [J]
12. Goebuchi, T. (1997). Delta-sigma modulation A/D converters. *Geophysical exploration, 50*, 332–334. [J]
13. Gram, J. P. (1883). Ueber die Entwickelung reeller Functionen in Reihen mittelst der Methode der kleinsten Quadrate. *Journal für die reine und angewandte Mathematik, 94*, 41–73.
14. Hamada, N. (1987). Seismographs. *Encyclopedia of Earthquakes* (2nd ed., pp. 18–31). Tokyo: Asakura Shoten. [J]
15. Hino, M. (1977). *Spectral analysis* (p. 300). Tokyo: Asakura Shoten. [J]
16. Ishimoto, M. (1931). Un sismographe accélérométrique et ses enregistrements. *Bulletin of the Earthquake Research Institute, 9*, 316–332.
17. Katsumata, A. (1989). Comparison of the JMA-87 type and JMA mechanical strong motion seismograph. *1989 Seismological Society of Japan Fall Meeting*, 245. [J]
18. Kinemetrics. (2019). *A legacy of success*, https://kinemetrics.com/about-us/our-history/.
19. Kinoshita, S., Uehara, M., Tozawa, T., Wada, Y., & Ogue, Y. (1997). Recording characteristics of the K-NET95 strong-motion seismograph. *Zisin (Journal of the Seismological Society of Japan), 49*, 467–481. [J]
20. Kinoshita, S. (1998). Negative feedback seismometers. *Zisin (Journal of the Seismological Society of Japan), 50*, 471–483. [J]
21. Koketsu, K. (2018). *Physics of seismic ground motion* (p. 353). Tokyo: Kindai Kagaku. [J]
22. Kreyszig, E. (1999). *Advanced engineering mathematics* (8th ed., p. 1156). New York: Wiley.
23. Kudo, K. (2001). Intensity of ground motion. *Encyclopedia of earthquakes* (2nd ed., pp. 358–386). Tokyo: Asakura Shoten. [J]

24. Kudo, K., Kakuma, H., Tsuboi, D., Sasatani, T., Takahash, M., & Kanno, T. (1997). A new highly overdamped moving coil accelerometer for mobile strong motion observation and its performance tests. *1997 Seismological Society of Japan Fall Meeting*, P12. [J]
25. Mathematical Society of Japan (ed.). (1968). *Dictionary of mathematics* (2nd ed. [J], p. 1140). Tokyo: Iwanami Shoten. [J]
26. Matsubara, N. (1992). Bayesian decision. *Natural science statistics* (pp. 251–276). Tokyo: University of Tokyo Press. [J]
27. Miyazaki, H., Uemae, T., Fujimori, H., & Onuki, A. (1996). A-D conversion methods and design techniques. *Transistor Technology, 33*(2), 223–289. [J]
28. Nakagawa, T., & Oyanagi, Y. (1982). *Experimental data analysis by the least-squares method* (p. 206). Tokyo: University of Tokyo Press. [J]
29. NIED (National Research Institute for Earth Science and Disaster Resilience) (2018). *Strong motion seismograph characteristics diagram*, https://www.kyoshin.bosai.go.jp/kyoshin/seismo/knet95/knet95.shtml.
30. Papoulis, A. (1962). *The Fourier integral and its applications* (p. 318). New York: McGraw-Hill.
31. Press, W. H., Flannery, B. P., Teukolsky, S. A., & Vetterling, W. T. (1988). *Numerical recipes in C: The art of scientific computing* (p. 735). Cambridge: Cambridge University Press
32. Physics Dictionary Editorial Committee (ed.). (1992). *Physics dictionary* (rev ed., p. 2465). Tokyo: Baifukan.
33. Robinson, E. A. (1978). *Multichannel time series analysis with digital computer programs* (rev. ed., p. 298). San Francisco: Holden-Day.
34. Saito, M. (1978). An automatic design algorithm for band selective recursive digital filter.*Geophysical Exploration, 31*, 112–135. [J]
35. Schmidt, E. (1907). Zur Theorie der linearen und nichtlinearen Integralgleichungen. I. Teil: Entwicklung willkürlicher Funktionen nach Systemen vorgeschriebener. *Mathematische Annalen, 63*, 433–476.
36. Togawa, H. (1971). *Numerical computation of matrices* (p. 323). Tokyo: Ohm Publishing.
37. Utsu, T. (2001). *Seismology* (3rd ed., p. 376). Tokyo: Kyoritsu Shuppan. [J]
38. Yabuki, T., & Matsu'ura, M. (1992). Geodetic data inversion using a Bayesian information criterion for spatial distribution of fault slip. *Geophysical Journal International, 109*, 363–375.

Appendix

A.1 Magnitude

A.1.1 Definition of Magnitude

The scale of an earthquake was empirically evaluated using the intensity of ground motion at a certain distance from the earthquake before the related theories were established in seismology. Then, it was theoretically established that the origin of an earthquake is **faulting** on a plane in the ground, and the seismic moment M_0 generated from this faulting represents the scale of the earthquake (Sect. 2.1.1). However, before the CMT inversion was developed, it would take a long time to calculate M_0 because a variety of information and complicated processing were required. Empirical methods are therefore still used even today, and the scale of an earthquake so obtained is simply called **magnitude** (M for short).

The first M scale was conceived by Richter [27] for an earthquake in southern California, approximately 30 years before the theory of M_0 was established [2]. Therefore, media in USA and Europe often call M the **Richter scale**. Richter's M was defined as the common logarithm of the maximum amplitude A in μm, in a horizontal component of the **Wood-Anderson seismograph** (natural period 0.8 s, damping constant 0.8, static magnification 2,800) located at an **epicentral distance** Δ (Sect. 1.1) of 100 km. This is the local magnitude (M_L for short). In reality, there is not necessarily a seismograph at $\Delta = 100$ km. Therefore, a correction term C_L is given for Δ as

$$M_L = \log A + C_L \ . \tag{A.1}$$

Subsequent definitions of M follow this way. However, in Richter [27], A is measured in units of mm, and the sign of the correction term is inverted. Table I in this paper is rewritten for C_L as in Table A.1. The paper also shows that the correction term from 200 km $< \Delta <$ 600 km can be approximated by a linear function of $\log \Delta$. When this is applied to Table A.1, we obtain the approximate formula

K. Koketsu, *Ground Motion Seismology*, Advances in Geological Science,
https://doi.org/10.1007/978-981-15-8570-8

Table A.1 Correction term given to M_L for Δ in km (based on Utsu [37])

Δ (km)	30	50	100	150	200	250	300	400	500	600
C_L	−0.90	−0.37	0.00	0.29	0.53	0.79	1.02	1.46	1.74	1.94

$$C_L = 3 \log \Delta - 6.37 . \tag{A.2}$$

The approximate formula for the correction term is called a **calibration function** [37].

As can be seen from Table A.1, the M_L considers only relatively short-range earthquakes with Δ less than 600 km. For **teleseismic earthquakes** (Sect. 3.1.11), Gutenberg [7] proposed another magnitude. He noticed that the **surface waves** (Sect. 1.2.4) with periods of around 20 s prevailed when a teleseismic earthquake was observed. He then defined the **surface wave magnitude**

$$M_s = \log A + C_s , \tag{A.3}$$

where A is the maximum amplitude of the surface wave horizontal ground motion. The ground motion was observed by the two horizontal components of a **long-period seismograph** (Sect. 4.1.1) at that time, and the amplitudes measured on the two components in μm were just divided by the static magnification of the seismograph. A is obtained by taking the square root of the sum of squares, i.e., the vector sum, of the maximum amplitudes of the two components [37]. The correction term C_s is given in Table A.2 for Δ measured in degrees (°), and this can be approximated by the calibration function

$$C_s = 1.656 \log \Delta + 1.818 \tag{A.4}$$

in the range $15° \leq \Delta \leq 130°$ [7].

However, the surface wave does not develop even for a teleseismic earthquake if the depth h of the earthquake exceeds several ten kilometers; therefore, M must be decided from **body wave** (Sect. 1.2.4) such as P and S waves. In the recording of an **intermediate-period seismograph** (Sect. 4.1.1) with a natural period of several to 10 s, the maximum amplitude A in μm and the period T in s of a body wave are obtained, both of which are measured on the recording paper. The maximum amplitude of an S wave is the vector sum of the two horizontal components, as for M_s by surface waves. Using this resultant, Gutenberg [8] and Gutenberg and Richter

Table A.2 Correction term given to M_s for Δ in degrees (°) (based on Gutenberg [7] and Utsu [37])

Δ (°)	20	30	40	50	60	70	80	90	100	120	140	160
C_s	3.97	4.26	4.47	4.63	4.76	4.87	4.97	5.05	5.13	5.26	5.33	5.35

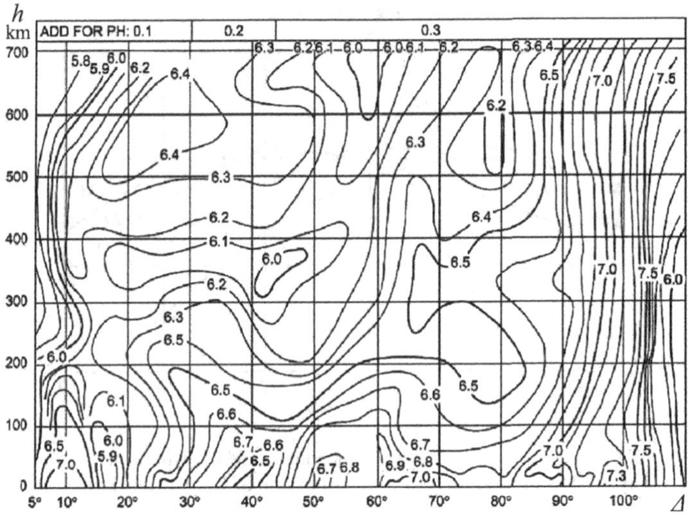

Fig. A.1 The correction term $q(\Delta, h)$ of m_B (based on Gutenberg and Richter [10], reprinted from Koketsu [20] with permission of Kindai Kagaku)

[10] defined the **body wave magnitude**

$$m_B = \log\left(\frac{A}{T}\right) + q(\Delta, h) \tag{A.5}$$

in addition to M_s. The correction term $q(\Delta, h)$ in (A.5) is a very complex function of the epicentral distance Δ and the depth h, and is displayed in Fig. A.1 for the P wave part of a vertical ground motion. When horizontal ground motion by the S wave is used, PH written at the top of Fig. A.1 is added.

C_s in M_s and $q(\Delta, h)$ in m_B should have been given so that their values would be consistent with M_L ($h = 18$ km is assumed for m_B) when the calculations were performed for a same earthquake. It is known, however, that they deviate systematically due to the difference in the period band of ground motions employed (see Fig. A.2 in the next section).

In their book "*Seismicity of the Earth*" [9], Gutenberg and Richter give the M's of major earthquakes around the world from 1904 to 1952. This is the de facto standard of M, and various subsequent formulas of magnitude are often adjusted to give values similar to those in "*Seismicity of the Earth*" for the same earthquakes. It is not necessarily clear how M values in "*Seismicity of the Earth*" were determined. According to Abe [1], it is assumed that M_s was used when the depth of an earthquake was less than 40 km and m_B was used when the depth was equal to or more than 40 km. As described above, in the process of proposing various M's, for example, by combining (A.1) and (A.2) for M_L, we obtain

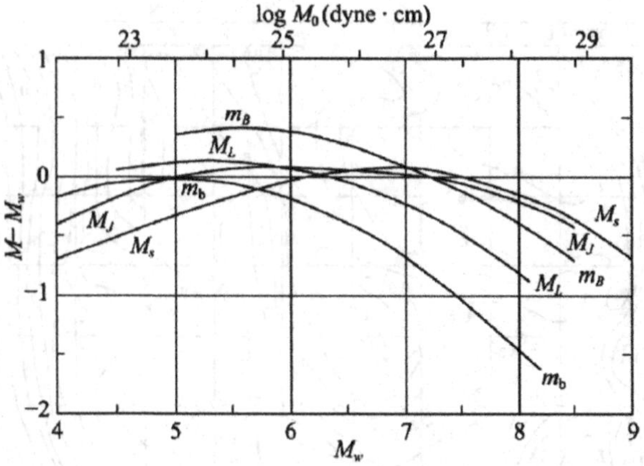

Fig. A.2 Relationships of various M's with seismic moment M_0 and moment magnitude M_w (based on Utsu [37], reprinted from Koketsu [20] with permission of Kindai Kagaku)

$$\log A = a\,M - b\,\log X + c\,, \quad M = M_L\,,\; X = \Delta\,,\; a = 1\,,\; b = 3\,,\; c = 6.37\,. \tag{A.6}$$

Such equations as (A.6) are termed **attenuation relation**. Since then, many studies have been performed (e.g., Si and Midorikawa [30]) using statistical approaches and sophisticated methodologies have been developed for various ground motion amplitudes A, magnitudes M, and distances X [24].

To classify earthquakes by M, the terms, which are **large earthquake** ($M \geq 7$), **moderate earthquake** ($5 \leq M < 7$), **small earthquake** ($3 \leq M < 5$), and **microearthquake** ($M < 3$), as well as the recent term, which is **great earthquake** ($M \geq$ around 8[1]), are used in Japan [36]. Utsu [36, 37] states that "they are not accepted internationally"; however, generally similar terms and classifications are used internationally.

A.1.2 Recent Magnitudes

When "*Seismicity of the Earth*" was published, in Japan, the **JMA** (Sect. A.2.1) used the **Wiechert seismograph** (natural period 5 s, damping constant 0.55, static magnification 80) and the **Ichibai strong motion seismograph** (Sect. 4.1.2) as displacement seismographs. Because these were different in characteristics from the seismographs used by Gutenberg and Richter [9], in 1954, Tsuboi [33] derived **Tsuboi's formula**

$$M = \log A + 1.73 \log \Delta - 0.83 \tag{A.7}$$

[1] The reason why this range has "around" is to include the Kanto earthquake (1923, M 7.9).

Table A.3 Correction term in (A.8) for Δ and h in km (based on Katsumata [18])

$h\backslash\Delta$	100	200	300	400	500	600	700	800	900	1000	1200	1400
50	2.58	3.14	3.40	3.69	3.90	4.08	4.23	4.29	4.41	4.54	4.68	4.83
100	2.65	3.19	3.38	3.73	3.99	4.18	4.38	4.41	4.55	4.74	4.83	5.04
150	2.85	3.31	3.43	3.77	4.01	4.18	4.40	4.45	4.58	4.76	4.85	5.07
200	3.11	3.47	3.54	3.83	4.01	4.15	4.35	4.43	4.53	4.78	4.79	4.98
250	3.39	3.64	3.68	3.89	4.01	4.10	4.27	4.38	4.44	4.56	4.70	4.85
300	3.67	3.80	3.85	3.97	4.03	4.08	4.21	4.33	4.36	4.44	4.61	4.71
350	3.90	3.95	4.02	4.07	4.07	4.10	4.18	4.29	4.31	4.36	4.55	4.60
400	4.09	4.08	4.17	4.19	4.16	4.18	4.21	4.36	4.30	4.33	4.53	4.55
450	4.22	4.20	4.30	4.32	4.29	4.30	4.29	4.27	4.35	4.37	4.56	4.57
500	4.30	4.34	4.39	4.48	4.46	4.45	4.41	4.31	4.44	4.47	4.64	4.65
550	4.35	4.51	4.44	4.65	4.66	4.61	4.54	4.38	4.57	4.61	4.74	4.78
600	4.41	4.77	4.42	4.84	4.87	4.74	4.64	4.46	4.72	4.77	4.83	4.93

to obtain similar M's to theirs for the earthquakes around Japan in "*Seismicity of the Earth*". For M_s, the amplitudes in μm are measured on the two horizontal components of a seismograph above and divided by the static magnification of the seismograph. A in (A.7) is the vector sum of the maximum amplitudes of the two components.

The Wiechert seismographs were then abandoned, so that the Ichibai strong motion seismographs and intermediate-period seismographs with similar frequency characteristics were used. Since 1994, in JMA, M has been determined by (A.7) using the **D93-type seismograph**, which are acceleration seismographs with displacement output by digital integration. Their natural periods are 10 s, and they are equipped with a filter that blocks signals with frequencies higher than 10 Hz. This magnitude is called **JMA magnitude** and sometimes abbreviated as M_J. As the natural period of the Wiechert seismograph is 5 s while that of the D93-type seismograph is 10 s, maximum amplitudes should be read after removing components with periods longer than 5 s from records [36]; however, this was not considered until 2000.

Furthermore, if the depth h of an earthquake exceeds 60 km, the formula by Katsumata [18]:

$$M = \log A + K(\Delta, h) \tag{A.8}$$

is used. The correction term $K(\Delta, h)$ is given in Table A.3. (A.7) and (A.8) should be determined to match their M's with M's in "Seismicity of the Earth", which were calculated by (A.3) and (A.5). However, the correction terms in (A.7) and (A.8) are smaller by about two than those in (A.3) and (A.5). The reason for this is that Δ's in the former are given in km whereas those in the latter are given in degrees (°).

In addition, short-period components are predominant in ground motions when the scale of an earthquake decreases (Sect. 2.3.5). Therefore, if the above M using intermediate-period seismographs such as the Wiechert seismograph is smaller than 5.5, the maximum amplitude A_V in 10^{-5} m/s of vertical ground motion by short-period seismographs is used and M is calculated by the formula of Kanbayashi and Ichikawa [15]:

$$M = \log A_V + 1.64 \log \Delta + C \tag{A.9}$$

where C is 0.22 or 0.44 depending on the type of short-period velocity seismograph. However, if M using the intermediate-period seismographs is larger than M using the short-period seismograph by 0.5 or more, the former is adopted as it is [25].

The problem of the direct use of records of the D93-type seismograph mentioned above became apparent during the **Tottori earthquake** (2000, M_w 6.7). The M_J of this earthquake calculated from the records was announced to be 7.3; however, the scale of the damage and the spread of aftershocks were less than those of the **Kobe earthquake** (1995, M_w 6.9, M_J 7.2 at that time). Thus, the validity of the value of M_J was questioned, and a review committee was established by the JMA. By the detailed examination in the committee, it was proven that M_J was larger by 0.06 on average due to the difference in the natural period of a used seismograph. In addition, what was more important than this difference was the difference in the installation placement of a seismograph. While the previous seismographs were installed at JMA offices in urban areas, the D93-type seismographs are installed in suburban areas. An urban area is often located in the center of a sedimentary basin, and ground motion amplification frequently occurs. However, this is suppressed in a suburban area so that M_J is smaller by 0.22 on average. Based on these results, the filter in the D93-type seismograph was modified and 0.22 has been added to M_J since then. For significant earthquakes from 1994 to then, the M_J's were recalculated using records observed by **seismic intensity meters** (Sect. A.2.3) installed at JMA offices. According to this recalculation, the M_J of the Kobe earthquake was revised from 7.2 to 7.3, while the M_J of the Tottori earthquake remained unchanged [2] [14].

ISC and **USGS** determine M_s and m_B of earthquakes in the world by using records such as observed by the **WWSSN** (World-Wide Standard Seismograph Network), which was developed for the purpose of detecting nuclear tests. In accordance with the **IASPEI** (International Association of Seismology and Physics of the Earth's Interior) recommendation in 1967, the determination is performed using the formula of Vaněk et al. [38]:

$$M_s = \log\left(\frac{A}{T}\right) + 1.66 \log \Delta + 3.30\,, \quad 20^\circ \le \Delta \le 160^\circ \tag{A.10}$$

where A and T are the vector-sum maximum amplitude in μm and the period in s of a surface wave record observed by the **Press-Ewing seismograph** (natural period 15 ~ 30 s) etc. The above A is the vector sum of the two horizontal components. However, the vertical component of the record can be used instead, because A becomes smaller while T also becomes smaller on the vertical component. Comparing M_s in (A.10) with M_s in (A.3) by Gutenberg [7], it should be noted that Vaněk's M_s

[2] There should remain something different because the M_w 6.7 of the Tottori earthquake is smaller by 0.2 than the M_w 6.9 of the Kobe earthquake.

is approximately 0.2 larger than Gutenberg's M_s in the case of $T = 20$ s, which is the definition of the surface wave for Gutenberg's M_s [25]. In addition, Utsu [36] mentioned that the maximum A/T should be searched in Vaněk et al. [38] while in reality the above measurement is performed.

For body wave magnitudes, ISC and USGS also apply the formula in (A.5) for Gutenberg's m_B to body wave records observed by **short-period seismograph** (Sect. 4.1.1) such as **Benioff seismograph** (natural period 1 s). However, as A is measured only within 5 s from the initial motion of the P wave, the result is often quite different from m_B due to this and the difference in the natural period of the seismograph used. It is therefore abbreviated as m_b to distinguish it from m_B. In particular, for large earthquakes, the short-period components of ground motions are relatively small (Sect. 2.3.5) so that m_b is much smaller than m_B [37].

M is basically determined from the amplitudes of ground motion records, although the frequency characteristics of ground motions differ depending on the scale of an earthquake. M can therefore be deviated if the frequency characteristics of a seismograph used or ground motions covered by M are different from the frequency characteristics associated with the scale of an earthquake. In general, as the scale of an earthquake increases, the long-period component of ground motion increases, but the short-period component does not increase as much. For example, when an M 8 earthquake is compared with an M 6.2 earthquake, according to the **scaling law** of the ground motion spectrum (Sect. 2.3.5), the long-period component is 500 times larger, while the short-period component is only 10 times larger. Thus, m_B and m_b, which are determined from body waves mainly composed of short-period components and observed by a short-period seismograph, are not appropriate for representing the magnitude of an M 8 class earthquake, as they saturate around M 6. Conversely, M_s, which is determined from surface waves mainly composed of long-period components and observed by a long-period seismograph, gives a smaller magnitude for an M 5 class earthquake. Saturation also occurs even in M_s for a great earthquake exceeding M 8.3 $\sim$ 8.5. The JMA magnitude M_J, which uses either an intermediate-period seismograph or a short-period seismograph depending on the magnitude of an earthquake, has a wider application range than M_s, m_B, and m_b (Fig. A.2).

In any case, all M's calculated from the amplitudes of seismograms have limitations. Therefore, Kanamori [16] returned to the **seismic moment** M_0 (2.1.2), which directly represents the magnitude of an earthquake, and proposed the **moment magnitude** M_w [3]

$$\log M_0 = 1.5\ M_w + 16.1 \tag{A.11}$$

so that this matches the conventional magnitudes, especially M_s of a plate-boundary earthquake (Sect. 2.1.1). Here, M_0 takes a value in dyne·cm, but if it is given in N·m, 16.1 is replaced with 9.1. By the advance of analysis methods such as the CMT

[3] Although (A.11) and the symbol M_w first appeared in Kanamori [16], the phrase "moment magnitude" first appeared in Kanamori [17]. Kanamori [16] used the subscript in italic as in this book; however, Kanamori [17] used that in roman.

inversion, M_0 would be determined daily and M_w would be mainly used. However, as even for the determination of M_0 seismograms are mostly used, saturation may occur unless records of very long period seismographs such as **gravimeter** and crustal deformation data are used additionally. Correct M_w cannot be obtained for small earthquakes unless seismographs sensitive to short periods are used.

In the case of earthquakes in Japan, the M with the most complete dataset is of course the **JMA magnitude** M_J. Therefore, a relationship between M_J and M_0 is often required. Takemura [32] obtained $\log M_0 = 1.17\ M_J + 17.72$ for **crustal earthquake** (Sect. 2.1.1) in Japan. This was rounded up to the first decimal place as

$$\log M_0 = 1.2\ M_J + 17.7\ , \tag{A.12}$$

which is more often used. For earthquakes along the Japan Trench and the eastern margin of the Sea of Japan, Sato [29] obtained

$$\log M_0 = 1.5\ M_J + 16.2\ . \tag{A.13}$$

(A.13) is almost identical to (A.11), so that $M_J \sim M_w$ holds in the sea areas around Japan. However, a different formula is necessary in the land areas because **surface wave** developed with intermediate periods result in larger displacement amplitudes.

A.2 Seismic Intensity

A.2.1 Characteristics of Seismic Intensity

Seismic intensity is "the intensity of ground motion at a certain place is represented by several grades categorized according to the human sensation, the size of the effect on surrounding objects, structures or nature, etc." [37] As can be understood from this definition, the seismic intensity can be determined instantaneously with no special equipment, because it is determined from what observers feel and what they see. In addition, because there is little possibility for the seismic intensity to be misreported or mismeasured and the seismic intensity holds **robustness** as data, this is published earliest among earthquake information, and many disaster prevention organizations use this as a standard when taking emergency measures such as personnel allocation and facility inspection [12]. In general, the higher the level of information, the more complicated the procedures needed. Therefore, it is more likely that unexpected values or null results occur when information is needed urgently, and so it is difficult to use complex procedures in disaster prevention organizations. For example, we cannot say that automated systems for determining information such as hypocenter location and fault mechanism are currently robust and accurate.

However, seismic intensity forcibly expresses the ground motion, combining various information such as amplitude, frequency, and duration, to give a simple integer

Table A.4 Rossi-Forel intensity scale [5, 6] (rewritten by Richter [28])

I	Microseismic shock	Recorded by a single seismograph or by seismographs of the same model, but not by several seismographs of different kinds; the shock felt by an experienced observer
II	Extremely feeble shock	Recorded by several seismographs of different kinds; felt by a small number of persons at rest
III	Very feeble shock	Felt by several persons at rest; strong enough for the direction or duration to be appreciable
IV	Feeble shock	Felt by persons in motion; disturbance of movable objects, doors, windows, cracking of ceilings
V	Shock of moderate intensity	Felt generally by everyone; disturbance of furniture, beds, etc., ringing of some bells
VI	Fairly strong shock	General awakening of those asleep; general ringing of bells; oscillation of chandeliers; stopping of clocks; visible agitation of trees and shrubs; some startled persons leaving their dwellings
VII	Strong shock	Overthrow of movable objects; fall of plaster; ringing of church bells; general panic, without damage to buildings
VIII	Very strong shock	Fall of chimneys; cracks in the walls of buildings
IX	Extremely strong shock	Partial or total destruction of some buildings
X	Shock of extreme intensity	Great disaster; ruins; disturbance of the strata, fissures in the ground, rock falls from mountains

scale using ambiguous filters such as human sensation, objects, and structures. Therefore, the seismic intensity is not suitable for modern seismology based on quantitative evaluation and theoretical reproduction. However, the history of seismic observation that can support modern seismology is only about 100 years, and the accumulation of data is insufficient, as this time period is remarkably short compared with the occurrence rate of strong ground motion at any particular place. This means that there is still a role for seismic intensity as data. The seismic intensity also has another importance as an interface with the society, including its use in disaster prevention organizations.

According to the simple measurement of the seismic intensity, the origin of this can be traced back to the 19th century, and the **seismic intensity scale**, which are tables referred to for determining the seismic intensity, were born simultaneously worldwide. In Europe, de Rossi [5] in Italy and F. A. Forel in Switzerland started independently, and in 1884, the **Rossi-Forel intensity scale** [6] (Table A.4) was made by the cooperation of both. However, this scale has defects such as that it was outdated by the later advance of technology, that the area covered by X of the highest grade was too wide, and that the descriptions depended too much on the situation of Europe [28]. Mercalli [23] proposed to improve the Rossi-Forel intensity scale by resolving these defects. This was further revised by Cancani [3] and Sieberg [31] and resulted in the **Mercalli intensity scale**.

Table A.5 History of the seismic intensity scale in Japan (based on JMA [12]). "(0) and "(−)" indicate "(unfelt)" and "(weaker)", respectively. Arabic numerals in italic *0, 1, 2,* ⋯, *6* represent Chinese numerals. "+" and "-" in the last column indicate "Upper" and "Lower", respectively

1884	1898	1908	1936	1949	1996
	Feeble (0)	*0*	Unfelt	O	0
Feeble	Feeble	*1*	I	I	1
Weak	Weak (−)	2	II	II	2
	Weak	*3*	III	III	3
Strong	Strong (−)	*4*	IV	IV	4
	Strong	*5*	V	V	5−
					5+
Severe	Severe	*6*	VI	VI	6−
					6+
				VII	7

In Japan, K. Sekiya, who was one of the founders of the first seismological society in the world and later the first professor of seismology in the world, defined the seismic intensity scale in Column 1 of Table A.5 which consists of four grades of "feeble", "weak", "strong", and "severe" in 1884 [12]. The annual reports of the Geographic Bureau of the Ministry of Home Affairs and later the **Central Meteorological Observatory**, which was independent from the Bureau, included seismic intensities based on this. Since 1898, in annual reports of the Central Meteorological Observatory, "feeble (unfelt)" was added under "feeble". "weak" was divided into "weak (weaker)" and just "weak". "strong" was replaced with "strong (weaker)", and "severe" was divided into "strong" and "severe". As a result, the seismic intensity scale was expanded to seven grades (Column 2 of Table A.5). After 1908, the seven grades were written using Chinese numerals corresponding to Arabic numerals 0, 1, 2, ⋯, 6 (Column 3 of Table A.5).

Seismic intensity observation in Japan is determined by law to be the business of the Central Meteorological Observatory, and after the reorganization in 1956 to be the responsibility of the Japan Meteorological Agency (**JMA**). Seven-grade intensity scales have been in use from 1898 to the present in Japan, though, since 1936, Roman numerals have been used instead of Chinese numerals (Column 4 of Table A.5), and Arabic numerals are used afterwards. There were also revisions in 1949 and 1996 (Columns 5 and 6 of Table A.5), which are explained in Sect. A.2.3.

A.2.2 Sensory Seismic Intensity

The Mercalli intensity scale, whose history is explained in the previous section, has spread to the United States, and the Modified Mercalli intensity scale ("MM intensity scale" for short, 12 grades), which Wood and Neumann [40] improved further, has been widely used worldwide. The MM intensity scale simplified by Richter [28] is shown in Table A.6. Conventional seismic intensities such as represented by the MM intensity scale are here called **sensory seismic intensity**.

However, in Europe, where the Mercalli intensity scale originated, there was a movement to internationally standardize seismic intensity scale on a basis other than the MM intensity scale. In 1964, UNESCO recommended the use of the MSK intensity scale (12 grades; same as the MM intensity scale) by Medvedev, Sponheuer, and Kárník [22]. The MSK intensity scale has been adopted in Europe and has developed into the **European Macroseismic Scale** (EMS). However, this approach was not taken up much outside of Europe, because of differences in architectural style and lifestyle of each country and the use of compilations of prior seismic intensity data. Although Japan investigated this scale, the MSK intensity scale was not adopted because detailed damage investigation is often necessary and hence the approach is not suitable for quick reports right after earthquakes [12, 21].

The **JMA intensity scale** is next explained as a sensory seismic intensity scale used in Japan. The columns of Table A.5 show its history: Column 5 indicates the addition of intensity VII as the 7th grade in 1949. This addition was triggered by the **Fukui earthquake** (1948, M 7.1, M_s 7.3) in June of the preceding year [37]. This earthquake caused 3,769 fatalities and 36,184 collapsed houses [35], and it was the most damaging earthquake after the Kanto earthquake (1923) until the Kobe earthquake. In particular, in the Fukui basin just above the source region of the 1948 event, a collapse rate of houses over 90% was observed everywhere. At that time, the officers of the Central Meteorological Observatory shared the feeling that the description of intensity VI: "Severe shaking. Shaking such that houses are collapsed, landslides and cracks in the ground are caused" was insufficient for these observations. Then, in January 1949, the operational rules for earthquakes and tsunamis were revised, and intensity VII was added: "Extreme shaking. The collapse rate of houses reaches 30% or over. Landslides, cracks and faults in the ground are caused." (Table A.7). After the addition, intensity VII was not announced for a long time, but was announced with an assigned area for the first time for the **Kobe earthquake**.

For **historical earthquakes**, in the period during which there were no seismographs, the only data available are sensory seismic intensities and arrival times, which are roughly estimated from descriptions in historical documents. However, buildings in the historic periods are likely to be more weakly constructed than modern buildings. Therefore, seismic intensities are determined by comparing the descriptions with Table A.5 or A.6 and can be adjusted by subtracting certain values. For example, Usami [34] used $0.5 \sim 1$ as the certain values for historical earthquakes in Japan and seismic intensities in the JMA intensity scale. If a distribution of concentric circles appears by plotting the adjusted intensities on a map and drawing **isoseismals**, which show the range of each intensity, it can be assumed that the vicinity of the circle center is the epicenter, and the magnitude of the historical earthquake can also be obtained from relationships between epicentral distances and seismic intensities.[4] Usami [35] is a compilation of such works in Japan.

[4]Many relationships are listed in Sect. 3.5 of Utsu [36].

Table A.6 Modified Mercalli intensity scale [40] with the simplified descriptions [28]

I	Not felt. Marginal and long-period effects of large earthquakes
II	Felt by persons at rest, on upper floors, or favorably placed
III	Felt indoors. Hanging objects swing. Vibration like passing of light trucks. Duration estimated. May not be recognized as an earthquake
IV	Hanging objects swing. Vibration like passing of heavy trucks; or sensation of a jolt like a heavy ball striking the walls. Standing motor cars rock. Windows, dishes, doors rattle. Glasses clink. Crockery clashes. In the upper range of IV wooden walls and frame creak
V	Felt outdoors; direction estimated. Sleepers wakened. Liquids disturbed, some spilled. Small unstable objects displaced or upset. Doors swing, close, open. Shutters, pictures move. Pendulum clocks stop, start, change rate
VI	Felt by all. Many frightened and run outdoors. Persons walk unsteadily. Windows, dishes, glassware broken. Knickknack, books, etc., off shelves. Pictures off walls. Furniture moved or overturned. Weak plaster and masonry D cracked. Small bells ring (church, school). Trees, bushes shaken (visibly, or heard to rustle)
VII	Difficult to stand. Noticed by drivers of motor cars. Hanging objects quiver. Furniture broken. Damage to masonry D, including cracks. Weak chimneys broken at roof line. Fall of plaster, loose bricks, stones, tiles, cornices (also unbraced parapets and architectural ornaments). Some cracks in masonry C. Waves on ponds; water turbid with mud. Small slides and caving in along sand or gravel banks. Large bells ring. Concrete irrigation ditches damaged
VIII	Steering of motor cars affected. Damage to masonry C; partial collapse. Some damage to masonry B; none to masonry A. Fall of stucco and some masonry walls. Twisting, fall of chimneys, factory stacks, monuments, towers, elevated tanks. Frame houses moved on foundations if not bolted down; loose panel walls thrown out. Decayed piling broken off. Branches broken from trees. Changes in flow or temperature of springs and wells. Cracks in wet ground and on steep slopes
IX	General panic. Masonry D destroyed; masonry C heavily damaged, sometimes with complete collapse; masonry B seriously damaged. (General damage to foundations) Frame structures, if not bolted, shifted off foundations, Frames racked. Serious damage to reservoirs. Underground pipes broken. Conspicuous cracks in ground. In alleviated areas sand and mud ejected, earthquake fountains, sand craters
X	Most masonry and frame structures destroyed with their foundations. Some well-built wooden structures and bridges destroyed. Serious damage to dams, dikes, embankments. Large landslides. Water thrown on banks of canals, rivers, lakes, etc. Sand and mud shifted horizontally on beaches and flat land. Rails bent slightly
XI	Rails bent greatly. Underground pipelines completely out of service
XII	Damage nearly total. Large rock masses displaced. Lines of sight and level distorted. Objects thrown into the air
Masonry A, B, C, D:	
A	Good workmanship, mortar, and design; reinforced, especially laterally, and bound together by using steel, concrete, etc.; designed to resist lateral forces
B	Good workmanship and mortar; reinforced, but not designed in detail to resist lateral forces
C	Ordinary workmanship and mortar; no extreme weaknesses like failing to tie in at corners, but neither reinforced nor designed against horizontal forces
D	Weak materials, such as adobe; poor mortar; low standards of workmanship; weak horizontally

Table A.7 The JMA intensity scale revised in 1949 [4] (at the time of revision, JMA was called the Central Meteorological Observatory). The supplements were added in 1978 [11]

Grade	Description	Supplement
O	Unfelt. The degree to which something is recorded on a seismograph without being felt by a human body	Even if the slight shaking of a hanging object is seen or a rattling sound is heard, it is unfelt that a human body does not feel the shaking
I	Feeble shaking. Shaking felt only by people at rest and especially people with keen sense of shaking	People at rest slightly feel it but do not feel it for a long time. Most standing people do not feel it
II	Light shaking. Shaking felt by many people, and the slight movement of a sliding door can be seen	People can sense the movement of a hanging object, and people can slightly feel the shaking when standing, but hardly feel it when moving. Some wake up even when sleeping
III	Weak shaking. Shaking such that houses shake, sliding doors rattle, hanging objects like a light shake considerably, and the movement of the water surface in a vessel is recognized	So startling that people asleep are waken up but do not rush out of houses, and feel no fear. Many people sense it outdoors, but some walking do not
IV	Moderate shaking. Shaking such that houses shake severely, unstable vases etc. fell down, and the water in vessels overflows. People walking also sense it, and many people rush out of houses	People asleep jump out of bed and feel fear. People can see utility poles and trees shaking. In ordinary houses, even if some roof tiles are shifted, it is not likely to be damaged yet. People feel dizzy
V	Strong shaking. Shaking such that cracks are made in walls, gravestones and stone lanterns fall down, and chimneys, stone walls, etc. are damaged	It is quite difficult to stand. Minor damage to ordinary houses begins to occur. Soft ground can crack or collapse. Unstable furniture falls
VI	Severe shaking. The collapse rate of houses is less than 30%. Shaking such that landslides and cracks in the ground are caused, and many people cannot stand	Walking is difficult and crawling is required to move
VII	Extreme shaking. The collapse rate of houses reaches 30% or over. Landslides, cracks and faults in the ground are caused	

A.2.3 *Instrumental Seismic Intensity*

In Japan and Taiwan, seismic intensity observations are performed by governments nationwide, and intensity data have been positively utilized for measures following earthquakes. For the official observations in Japan, observers at JMA offices determine seismic intensities by comparing their own sensations and observations of indoor and outdoor with the descriptions and supplements of the JMA intensity scale shown in Table A.7 (Sect. A.2.2). This determination leads to the characteristics of sensory seismic intensity such as simplicity, immediacy, and robustness, but it works in reverse causing the following problems [12]:

(a) Because the intensity of ground motion is strongly affected by the ground, there is no guarantee that the seismic intensity of a JMA office is a representative value of the surrounding region.
(b) It is inevitable that differences arise depending on the type of building in which the observer is located.
(c) There are individual differences among observers.
(d) Sensory seismic intensity cannot be observed in an unmanned place.

In addition, after the Kobe earthquake, in which intensity VII was announced for the first time since its establishment in 1949, the JMA pointed out the following problems [12]:

(e) Because the definition of seismic intensity VII is strict as 30% or more of the collapse rate of houses, several days or more are required for detailed investigation (in the Kobe earthquake, it took three days, and several more days for its affected area).
(f) For seismic intensities V and VI, the ranges of the corresponding damage are too large and it is difficult to perform appropriate measures.

To solve these problems of the sensory seismic intensities, the JMA developed **seismic intensity meters**, which measures the seismic intensity, and distributed many of them nationwide. This solves the problems of individual difference, unmanned observation, and immediacy of intensity VII (problems (c), (d), and (e)). If the seismic intensity meters are densely distributed, the seismic intensity is observed at points with various ground conditions, so that the problem (a) is solved. The problem (b) is also solved, if the seismic intensity meters are located away from buildings.

The seismic intensity meter is a device to calculate seismic intensity from strong motion records, and its specifications are determined so that calculated seismic intensity can reproduce sensory seismic intensity. Therefore, a new processing procedure is established by collecting strong motion records at sites where sensory seismic intensities are obtained. Using this procedure, a seismic intensity is calculated so that the result is close to the sensory seismic intensity. To increase the accuracy by increasing the data, strong motion records at locations other than the JMA offices are used as far as possible when the **questionnaire seismic intensity** [26] are available [12].

The **instrumental seismic intensity** thus measured is a continuous quantity, and is rounded up to the 1st decimal place. The integer value is then obtained by rounding the instrumental seismic intensity off (6.5 or more and less than 0.5 are regarded as 7 and 0, respectively). This has been the official seismic intensity of JMA in Arabic numeral form since 1996. In addition, to cope with the problem (f), intensities 5 and 6 were divided into "Upper" and "Lower", and finally the JMA intensity scale reached 10 grades (Column 6 of Table A.5). The above is the major revision of the JMA intensity scale after the Kobe earthquake. While the descriptions in an intensity scale play a deciding role in the determination of sensory seismic intensity,

Fig. A.3 95-type seismic intensity meter consisting of the measurement unit (left) and processing unit (right) (reprinted from the JMA website)

the processing procedure instead plays this role in the determination of instrumental seismic intensity. However, the descriptions of the JMA intensity scale survive in the explanatory table after being revised to be suitable for Japan in 1996 [13].

The seismic intensity meter, which is the key component of the JMA intensity scale revised in 1996, consists of measurement and processing units, as shown in Fig. A.3. Because the seismic intensity meter observes strong motions and calculates seismic intensities from the observed records, the measurement unit is actually a **strong motion seismograph** (Sect. 4.1.2). The seismic intensity meters deployed nationwide by the JMA are called **95-type seismic intensity meter** and are designed according to the following specifications [12]. Seismic intensity meters used by local governments are also designed according to these specifications. For the measurement unit,

(1) Three-component **acceleration seismograph** with **servo mechanism**, which can observe up to ±2048 gal of each component and has an effective resolution of 8 mgal or less, and flat frequency characteristics for acceleration at 0 ~ 50 Hz.
(2) Trigger judgment is performed every 10 s, and transfer to the processing unit and data storage are performed every 60 s (therefore, the delay time is 10 s at the longest).
(3) Transfer to the processing unit and data storage are performed digitally (Sect. 4.2), and the records for about 50 minutes can be stored in an IC memory card.
(4) The communication function is by ground line or satellite line.
(5) By receiving 40 kHz standard radio wave JG2AS, continuous time calibration with accuracy within 0.01 s is possible.
(6) With an external battery, everything including the processing unit can be operated for at least 3 hours even when the power fails.
(7) Everything including the processing unit can withstand intensity 7.

The calculation of the instrumental seismic intensity by the processing unit is summarized as follows [12].

(1) The **frequency spectrum** (Sect. 4.2) is taken for each component of **ground acceleration**.
(2) The filter (Sect. 4.3.1) in Fig. A.4 is applied to the spectrum of (1).
(3) The **inverse Fourier transform** (Sect. 4.2.2) is applied to the filtered spectrum of (2) for the filtered acceleration waveforms of the three components.

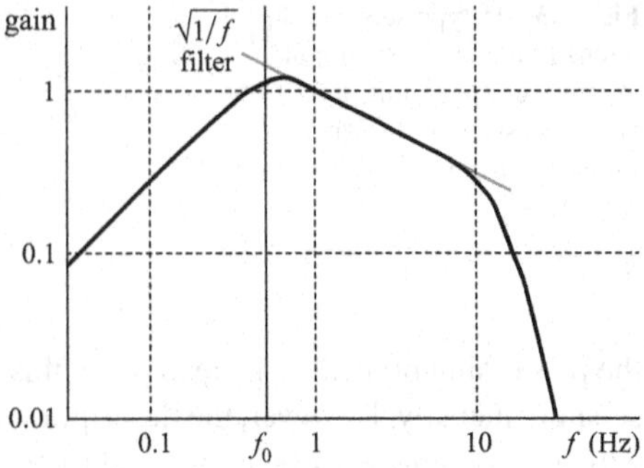

Fig. A.4 The filter applied to ground accelerations for calculating instrumental seismic intensities (based on JMA [12], modified from Koketsu [20] with permission of Kindai Kagaku)

(4) The vector sum of filtered accelerations in the three components is calculated in gal at every t (time) and the maximum a_0 is obtained under the condition that the total time when the vector sum is equal to or greater than a_0 is 0.3 s or more.
(5) $I = 2\log a_0 + 0.94$ is rounded to the second decimal place and the result is then rounded down to the second decimal place to obtain the instrumental seismic intensity. Intensities 5 and 6 are designated as "Upper" if they are 5.0 and 6.0 or above, or "Lower" for other cases.

The filter of (2) (Fig. A.4) is the combination of the three filters:

· The filter of $(1/f)^{1/2}$
· **high-cut filter** $(1 + 0.694X^2 + 0.241X^4 + 0.0557X^2 + 0.009664X^8 + 0.00134X^{10} + 0.000155X^{12})^{-1/2}$, $X = f/f_c$, $fc = 10$ Hz
· **low-cut filter** $(1 - \exp(f/f_0)^3))^{1/2}$, $f_0 = 0.5$ Hz

where f is frequency.

In this way, the instrumental seismic intensity is not simply calculated from the maximum acceleration, but is calculated by adding the effects of the frequency and duration of strong motion, which affect human sensation and damage, using the filter of (2) and the total time of a_0 of (4). In particular, the fact that the center frequency f_0 of the filter was taken to be 0.5 Hz, which is a considerably low frequency, seems to be a result of emphasizing the correlation with building damage for large seismic intensity.

The controversy over whether acceleration, which represents the force of strong motion, or velocity, which represents energy, most affects damage and human sensation, has not been settled. It can be understood that the use of the $(1/f)^{1/2}$ filter, which is not the use of no filter for acceleration nor the use of the $1/f$ filter for velocity, is at an intermediate position between the two.

Kawasumi [19] obtained $I' = 2\log a + 1.2$ where I' is the boundary between the intensities I and $I - 1$ and a is the maximum acceleration observed by **Ishimoto-type accelerograph** (Sect. 4.1.2). Utsu [37] converted this into $I = 2\log a + 0.7$ by subtracting 0.5. $I = 2\log a_0 + 0.94$ in (5) is similar to this equation; however, the constant changes from 0.7 to 0.94. This difference should be due to the difference in seismographs and the detailed investigation for the instrumental seismic intensity (Sect. A.2.3) [12].

Outside of Japan, although no special instrument such as the seismic intensity meter is prepared, many strong motion seismographs have been deployed, and the **instrumental seismic intensity** is calculated using their observational records. However, as ordinary equipment without processing units is used, complicated processing such as the JMA requests, is not possible; thus, simple indexes such as the **peak ground acceleration** (PGA, Sect. 4.1.2) and **peak ground velocity** (PGV, maximum amplitude of a velocity record) are used. For example, in California, the intensity I in the MM intensity scale can be calculated from

$$\begin{aligned} I &= 3.66 \log \mathrm{PGA} - 1.66\,, \quad \mathrm{V} \le I \le \mathrm{VIII}\,, \\ I &= 3.47 \log \mathrm{PGV} + 2.35\,, \quad \mathrm{V} \le I \le \mathrm{IX} \end{aligned} \tag{A.14}$$

as shown by Wald et al. [39], and (A.14) is widely used not only in California but also in other regions and countries.

References

1. Abe, K. (1981). Magnitude of large earthquakes from 1904 to 1980. *Physics of the Earth and Planetary Interiors*, *27*, 72–92.
2. Aki, K. (1966). Generation and propagation of G waves from the Niigata earthquake of June 16, 1964. Part 2. Estimation of earthquake moment, released energy, and stress-strain drop from the G wave spectrum. *Bulletin of the Earthquake Research Institute*, *44*, 73–88.
3. Cancani, A. (1904). Sur l'emploi d'une double échelle sismique des intensités, G. Beitr. *Ergänzungsband*, *2*, 281–283.
4. Central Meteorological Observatory. (1952). *Seismic observation method* (4th ed., p. 216). Tokyo: Central Meteorological Observatory.
5. de Rossi, M. S. (1883). Programma dell'osservatorio ed archivo centrale geodinamico. *Boll. del vulcanismo italiano*, *10*, 3–124.
6. Forel, F. A. (1884). Les tremblements de terre étudiés par la commission sismologique suisse pendant l'année 1881; 2me rapport. *Archives des Sciences Physiques et Naturelles*, *11*, 147–182.
7. Gutenberg, B. (1945). Amplitudes of surface waves and magnitudes of shallow earthquakes. *Bulletin of the Seismological Society of America*, *35*, 3–12.
8. Gutenberg, B. (1945). Amplitudes of P, PP, and S and magnitudes of shallow earthquakes. *Bulletin of the Seismological Society of America*, *35*, 57–69.
9. Gutenberg, B., Richter, C. F. (1949). *Seismicity of the Earth* (p. 273). Princeton: Princeton University Press.
10. Gutenberg, B., & Richter, C. F. (1956). Magnitude and energy of earthquakes. *Annals of Geofis*, *9*, 1–15.
11. JMA. (1978). *Seismic observation guidelines (Observation edition)* (6th ed., p. 166). Tokyo: JMA. [J]
12. JMA. (1996). *Knowing seismic intensity* (p. 238). Tokyo: Gyosei. [J]
13. JMA. (1996). *Summary of tables explaining the JMA seismic intensity scale*. https://www.jma.go.jp/jma/en/Activities/intsummary.pdf. [J]
14. JMA. (2001). *Results of the JMA magnitude review committee* (p. 8). https://www.jma.go.jp/jma/press/0104/23a/mate00.pdf. [J]

15. Kanbayashi, Y., & Ichikawa, M. (1977). A method for determining magnitude of shallow earthquakes occurring in and near Japan. *Quarterly Journal of Seismology, 41*, 57–61. [J]
16. Kanamori, H. (1977). The energy release in great earthquakes. *Journal of Geophysical Research, 82*, 2981–2987.
17. Kanamori, H. (1983). Magnitude scale quantification of earthquakes. *Tectonophysics, 93*, 185–199.
18. Katsumata, M. (1964). A method to determine the magnitude of deep-focus earthquakes in and near Japan. *Zisin (Journal of Seismological Society of Japan), 17*, 158–165. [J]
19. Kawasumi, H. (1943). Seismic intensity and seismic intensity scale. *Zisin (Journal of Seismological Society of Japan), 15*, 6–12.
20. Koketsu, K. (2018). *Physics of Seismic Ground Motion* (p. 353). Tokyo: Kindai Kagaku. [J]
21. Kudo, K. (2001). Intensity of ground motion. *Encyclopedia of earthquakes* (2nd ed., pp. 358–386). Tokyo: Asakura Shoten. [J]
22. Medvedev, S. V., & Sponheuer, W. (1969). Scale of seismic intensity. *Proceedings of the 4th World Conference On Earthquake Engineering, 1*, A-2, 143–153.
23. Mercalli, G. (1902). Sulle modificazioni proposte alla scala sismica De Rossi-Forel. *Bollettino della Societa Entomologica Italiana, 8*, 184–191.
24. Midorikawa, S. (2009). Ground motion attenuation relations. *Zisin (Journal of Seismological Society of Japan), 61*, S471–S477. [J]
25. Nishide, N. (2001). Determination of magnitude. *Encyclopedia of earthquakes* (2nd ed., pp. 53–63). Tokyo: Asakura Shoten. [J]
26. Ohta, Y., Koyama, M., & Nakagawa, K. (1998). Revision of algorithm for seismic intensity determination by questionnaire survey - in high intensity range -. *Japan Society for Journal of Natural Disaster Science, 16*, 307–323. [J]
27. Richter, C. F. (1935). An instrumental magnitude scale. *Bulletin of the Seismological Society of America, 25*, 1–32.
28. Richter, C. F. (1958). *Elementary seismology* (p. 768). San Francisco: Freeman.
29. Sato, R. (1989). *Handbook of earthquake fault parameters in Japan* (p. 390). Tokyo: Kajima Institute Publishing.
30. Si, H., & Midorikawa, S. (1999). New attenuation relationships for peak ground acceleration and velocity considering effects of fault type and site condition. *Journal of Structural and Construction Engineering AIJ, 523*, 63–70. [J]
31. Sieberg, A. (1923). *Geologische, physikalische und angewandte Erdbebenkunde* (p. 572). Jena: G. Fischer.
32. Takemura, M., Ikeura, T., & Sato, R. (1990). Scaling relation for source parameters and magnitude of earthquakes in the Izu Peninsula region. *Japan, Tohoku Geophysical Journal, 32*, 77–89.
33. Tsuboi, C. (1954). Determination of the Gutenberg-Richter's magnitude of earthquakes occurring in and near Japan. *Zisin (Journal of Seismological Society of Japan), 7*, 185–193. [J]
34. Usami, T. (2001). Survey of historical earthquakes. *Encyclopedia of earthquakes* (2nd ed., pp. 70–76). Tokyo: Asakura Shoten. [J]
35. Usami, T. (2003). *Materials for comprehensive list of destructive earthquakes in Japan* (latest ed., p. 605). Tokyo: University of Tokyo Press. [J]
36. Utsu, T. (1999). *Seismicity studies: A comprehensive review* (p. 876). Tokyo: University of Tokyo Press. [J]
37. Utsu, T. (2001). *Seismology* (3rd ed., p. 376). Tokyo: Kyoritsu Shuppan. [J]
38. Vaněk, J., Zátopek, A., Kárník, V., Kondorskaya, N. V., Riznichenko, Y. V., Savarensky, E. F., et al. (1962). Standardization of magnitude scales. *Izvestiya Academic Science USSR, 2*, 153–157.
39. Wald, D. J., Quitoriano, V., Heaton, T. H., Kanamori, H., Scrivner, C. W., & Worden, C. B. (1999). TriNet "shakeMaps": Rapid generation of peak ground motion and intensity maps for earthquakes in southern California. *Earthquake Spectra, 15*, 537–555.
40. Wood, H. O., & Neumann, F. (1931). Modified Mercalli intensity scale of 1931. *Bulletin of the Seismological Society of America, 21*, 277–283.

Index

K. Koketsu, *Ground Motion Seismology*, Advances in Geological Science,
https://doi.org/10.1007/978-981-15-8570-8

Printed in the United States
by Baker & Taylor Publisher Services